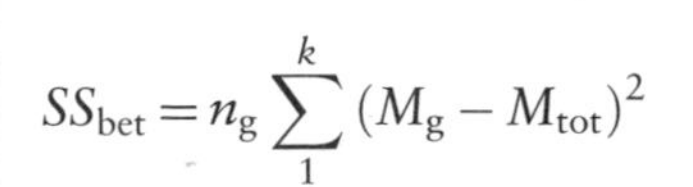

$$SS_{\text{bet}} = n_g \sum_{1}^{k} (M_g - M_{\text{tot}})^2$$

ANOVA sums of squares between, deviation formula

$$SS_{\text{bet}} = \sum_{1}^{k} \frac{\left(\sum X_g\right)^2}{n_g} - \frac{\left(\sum X_{\text{tot}}\right)^2}{N}$$

ANOVA sums of squares between, computational (raw score) formula

$$MS = \frac{SS}{df}$$

mean squa

$$F = \frac{MS_{\text{bet obs}} - MS_{\text{bet exp}}}{MS_{\text{with}}}$$

F test

$$\text{HSD} = q\sqrt{\frac{MS_{\text{with}}}{n_g}}$$

Tukey post

$$\chi^2 = \sum \left[\frac{f_o - f_e}{f_e}\right]^2$$

chi-square

$$E = \frac{RT \times CT}{N}$$

expected frequencies for chi-square test of independence

$$d = \frac{M_1 - M_2}{s_{DV}}$$

Cohen's *d* effect size, for *t* test

$$\text{Effect size } r = \sqrt{\frac{t^2}{t^2 + df}}$$

effect size *r*, for *t* test

$$\text{Effect size } \eta = \sqrt{\frac{SS_{\text{bet}}}{SS_{\text{tot}}}}$$

effect size eta, for ANOVA

$$\text{Effect size } r = \sqrt{\frac{\chi^2}{(N)(C-1)}}$$

effect size *r*, for chi-square goodness of fit

$$\phi = \sqrt{\frac{\chi^2}{N}}$$

effect size phi, for 2 × 2 chi-square test of independence

$$V = \sqrt{\frac{\phi^2}{(\text{the smaller of } R \text{ or } C) - 1}}$$

Cramer's *V* effect size, for chi-square test of independence of any size

$$r_{XY} = \frac{\sum z_X z_Y}{N}$$

definitional formula for Pearson *r*

$$r_{XY} = \frac{\sum (X - M_X)(Y - M_Y)}{N s_X s_Y}$$

deviation formula for Pearson *r*

$$r_{XY} = \frac{N\sum XY - \left(\sum X\right)\left(\sum Y\right)}{\sqrt{\left[N\left(\sum X^2\right) - \left(\sum X\right)^2\right]\left[N\left(\sum Y^2\right) - \left(\sum Y\right)^2\right]}}$$

computational (raw score) formula for Pearson *r*

$$Y = bX + a$$

equation for any straight line

$$Y' = r_{XY}\left(\frac{s_Y}{s_X}\right)X - r_{XY}\left(\frac{s_Y}{s_X}\right)M_X + M_Y$$

equation for predicting *Y* from *X*

$$s_{YX} = s_Y\sqrt{1 - r_{XY}^2}$$

standard error of prediction

$$Y' = b_1X_1 + b_2X_2 + b_3X_3 + \ldots + a$$

equation for multiple regression

Statistics *Alive!*

2 EDITION

Seldom do most of us meet anyone in our lifetimes as kind, helpful, and guileless as my husband.
I had the good fortune not only to meet such a person but also to marry him.

To you, Bob Sanford, this textbook is lovingly dedicated.

Statistics *Alive!*

Wendy J. Steinberg

Affiliate Faculty, Division of Educational Psychology and Methodology, University at Albany, State University of New York

Los Angeles | London | New Delhi
Singapore | Washington DC

For information:

SAGE Publications, Inc.
2455 Teller Road
Thousand Oaks, California 91320
E-mail: order@sagepub.com

SAGE Publications Ltd.
1 Oliver's Yard
55 City Road
London EC1Y 1SP
United Kingdom

SAGE Publications India Pvt. Ltd.
B 1/I 1 Mohan Cooperative Industrial Area
Mathura Road, New Delhi 110 044
India

SAGE Publications Asia-Pacific Pte. Ltd.
33 Pekin Street #02-01
Far East Square
Singapore 048763

Printed in the United States of America

Library of Congress Cataloging-in-Publication Data

Steinberg, Wendy J.
Statistics alive!/Wendy J. Steinberg.—2nd ed.
p. cm.
Includes bibliographical references and index.
ISBN 978-1-4129-7950-4 (pbk.)
1. Statistics. 2. Social sciences—Statistical methods. I. Title.

HA29.S7974 2011
519.5—dc22 2010005922

This book is printed on acid-free paper.

11 12 13 14 10 9 8 7 6 5 4 3

Acquisitions Editor: Vicki Knight
Associate Editor: Lauren Habib
Editorial Assistant: Sarita Sarak
Production Editor: Astrid Virding
Copy Editor: QuADS Prepress (P) Ltd.
Typesetter: C&M Digitals (P) Ltd.
Proofreader: Christina West
Accuracy Checker: Tom DeHardt
Indexer: Diggs Publication Services
Cover Designer: Bryan Fishman
Marketing Manager: Stephanie Adams

Brief Contents

Detailed Contents

List of Figures

List of Tables

Preface

I did poorly many years ago when I took my first undergraduate statistics course. I never understood what we were doing nor why we were doing it. How ironic it is that I then went on to earn a PhD in measurement and that my favorite course to teach is statistics. I believe that memory of that first dismal undergraduate course, together with graduate coursework in educational and cognitive psychology, has helped me understand what works and what doesn't work for effective statistical instruction.

Given my eye toward effective instruction, I was never able to find a statistics textbook that suited me. Many textbooks are mathematically dense, spending too much time deriving formulas and not enough time on the formulas' logic and practical use. Other textbooks go to the opposite extreme, being mere consumer-oriented overviews. As a result of my dissatisfaction with the available textbooks, over the years I developed dozens of pages of handouts, exercises, and mini-lessons, which I provided to students as a supplement to whatever textbook I was using. To my surprise, students soon told me that my lectures and supplemental materials were all that they needed to understand the material and that they had given up reading the textbook altogether! Finally, it dawned on me: Why not incorporate my teaching lectures and supplemental materials into a textbook of my own? Hence, this textbook.

I believe that even students with modest mathematical backgrounds (not beyond basic algebra) can master both the mechanics and the underlying logic of hypothesis testing, which is the heart of statistics. This is the type of student I have been unusually successful in reaching, without compromising course content or rigor. Despite the light tone throughout, this textbook is definitely *not* "Stats Lite." Coverage is comprehensive, and students are well prepared for advanced courses.

Content Distinctives

The audience for this textbook is undergraduates taking a first course in statistics. Frankly, content doesn't vary much from one introductory textbook to another. Thus, this textbook includes frequency distributions and their shapes; measurement scales; measures of central tendency and dispersion; the normal curve and z scores; hypothesis formation; inferential statistics such as t, F, and χ^2; correlation; and bivariate regression. However, in this textbook, the concepts of sampling error, significant differences, and Type 1 and Type 2 errors are stressed throughout. Also, effect size and power, often shortchanged in other textbooks, each get substantive treatment. New to this edition is a conceptual overview of multiple regression as a bridge to higher-level courses.

A first course in statistics should focus on univariate inferential hypothesis testing. Without a solid understanding of inferential logic, the student is not prepared to know when to use which statistic or understand the meaning of any statistic he or she does compute. Thus, the primary emphasis in this textbook is on the underlying logic of hypothesis testing. Every topic in the textbook meets this criterion: Is understanding this necessary and helpful to mastering inferential hypothesis testing? Toward that end, the following steps have been taken:

- *Probability theory is minimized.* Many statistics textbooks spend up to one fourth of their content covering probability topics. Although probability theory and integral calculus underlie the curves by which statistical inferences are interpreted, deriving the probabilities is daunting and unnecessary in a social science statistics course. Social science students need only enough probability theory to understand that the values in the inferential charts they consult are theory based. If you teach probability as part of your statistics course, I have included enough theory and practice to make the connection from the bar chart for hit-or-miss binary events to the continuous normal curve formed by connecting the midpoints of those discrete probabilities. If you are nonplussed by probability theory, you may skip the topic and its modules without loss to the textbook's flow or inferential logic.
- *Mathematical proofs are minimized.* Few social science students are prepared to understand that the mean is the vertex of the parabola of squared errors. Few seek to assure themselves that the raw score, deviation score, and *z*-score correlation formulas are mathematically equivalent. So why intimidate otherwise capable students? My experience is that such instruction is counterproductive. Therefore, you will find few proofs in this textbook.
- The proper sample size in the denominator for the standard deviation when using a statistic inferentially is $n - 1$. On that, we all agree. Debate arises when the statistic is being used descriptively: Shall we use N? Or shall we use $n - 1$, being as we will have to use that for later inferential instruction, anyhow? A survey of textbooks currently on the market shows about a 50/50 split in presentation in this matter. A similar preference split likely exists among instructors. Handheld calculators give the user a choice: One button says $/N$ and the other button says $/n - 1$. SPSS, on the other hand, does not give the user a choice. The SPSS algorithm uses $n - 1$ as the sample size in the denominator for the standard deviation (and the subsequent z score), whether the statistic is being used descriptively or inferentially. In this edition of the textbook, I have chosen not to take sides in the debate. Thus, while inferential statistics of course are calculated using $n - 1$, I present *descriptive standard deviation solutions using N and also using n – 1*. Instructors can take their pick. Instructors using SPSS will need to teach the $n - 1$ algorithm if hand-calculated answers are to agree with the software answers.
- After learning about measures of central tendency and dispersion, students are shown the *effects of score transformations* (adding/subtracting and multiplying/dividing) on the mean and standard deviation. This leads to the understanding that area in the normal curve is constant regardless of actual mean and standard deviation values. This understanding is helpful for most statistics students, and it is essential for those who will later study or interpret standardized educational or psychological tests that are scored in *T*, CEEB, or similar rescaled units. Depending on your goals for the course, you may include or omit this topic and its modules.
- The z score is the bridge to inferential statistics, in that the tabled percentage for any given score is based on infinite sampling. Thus, after learning how to compute and interpret individual z scores, students *reconceptualize the z score as a test of a null hypothesis:* "There is no significant difference between this person's score and the average of the population of scores from which this score came." Tabled percentages then take on a new meaning—from percentages actually scoring above or below a given score to the probability that a given score actually came from a given population. In this way, the z formula is reconceptualized using inferential logic.

- This textbook emphasizes the *logical similarity* of all experimental test statistics. That is, from z to one-sample t, to two-sample t, to F, to χ^2, the *numerator always* is the difference between what you got and what you expected to get, and the *denominator always* scales that difference in terms of random dispersion units. Once that logic is grasped, students should be able to interpret just about any inferential statistic they later run into, even one they have not previously learned. This logic is the missing element in so many statistics textbooks. Thus, each time the student learns a new inferential test statistic and computes an example, I ask the same two questions: (1) "What did you expect to get, and did you get it?" (found in the numerator, and the answer is no, probably not), followed by (2) "Is the observed difference from expectation a lot, or is it only a little?" (scaled against the sampling error term in the denominator, then checked in a table). I use the same systematic approach in teaching the sampling distributions themselves, building from a raw score distribution to a sampling distribution of the mean, to a sampling distribution of the difference between the means, to a within-groups mean square. For inferential statistics, the logic is the message.
- Correlation and prediction, even when used inferentially, follow a different logic (scaled over total variance) than does traditional hypothesis testing (scaled over error variance). In order not to break the logic that I repeat throughout the modules, *correlation and linear prediction are given their own section* at the completion of traditional hypothesis testing. However, the correlation and prediction modules are written in such a way that the instructor who prefers to teach this topic immediately after descriptive statistics can do so without an awkward break in the flow.
- New to this edition is an *introduction to multiple regression and the General Linear Model (GLM)*. This module follows naturally from bivariate regression at the conclusion of the textbook. The topic is presented and discussed conceptually only; examples are shown in SPSS but not calculated. The GLM follows the proportional logic discussed above, which is very different from the logic of hypothesis testing using t and F. Because the GLM is nearly always the focus of a second course in statistics, many students struggle in their second course without this conceptual bridge. Instructors in programs requiring that students take a second course in statistics are advised to assign this module for preparatory purposes. Instructors in terminal programs requiring only a single course in statistics may omit this module with no deficit to instruction.
- Throughout, I show the *silliness of selecting a dichotomous all-or-none decision criterion*. Computers have access to all possible Type 1 probabilities, not just the two or three probability columns provided in textbook tables. Thus, computers give the exact probability of having made a Type 1 error and then leave it to the user to determine whether the hypothesis was met, using whatever criterion the user deems appropriate. Journals, too, do not report dichotomous decisions. Rather, they report incurred error levels and let the readers determine if the study meets their own statistical and practical significance standards. Thus, after calculating the test statistic, I guide students in interpolating the incurred Type 1 error, which inevitably falls between the tabled levels. This extra instruction, rare in introductory textbooks, makes clear the meaning of the p values that students will inevitably see in computer output and in journal articles. It also leads naturally to a discussion of confidence levels, confidence intervals, and design implications.
- Many statistics textbooks still do not address *effect size*. This is an unfortunate omission because, after spending so much of the course learning how to calculate statistical significance, students tend to go away thinking that statistical significance means

that the treatment "worked." This, of course, is not necessarily so. Thus, once students understand the logic, calculations, and interpretations of hypothesis testing, I discuss effect size. This includes the difference between statistical significance and practical significance, various measures of effect size, and how much of an effect size is "enough." Moreover, textbooks that do discuss effect size tend to discuss only one measure: Cohen's d. This is surprising, because Cohen's d can be used only with t tests, not with analysis of variance or with chi-square. This textbook presents several measures of effect size—at least one for each major test statistic.

- I take the same comprehensive approach in discussing *power*. This includes the factors that affect power, where those factors "sit" in the test statistic formulas (numerator vs. denominator), how much power is enough, and sampling and design flaws that increase and decrease power. I then interrelate alpha, power, and effect size—the push-and-pull effect they have on each other. Students practice estimating power under different conditions of alpha, effect size, and N.
- I also discuss typical journal *publication guidelines* for statistical significance, power, and effect size, so that students can more effectively distinguish a publishable study from one that is not publishable. And I let them know that these are the considerations that the researcher needs to take into account when designing the study rather than be disappointed at the data analysis stage.
- A significant addition to this edition is *SPSS instruction and output*. Examples are not only solved manually within the textbook narrative but also shown as software output in the new "SPSS Connection" sections. These sections not only show output as the student would see it, but also give detailed point-and-click instructions for obtaining the output. Indeed, the instructions are sufficiently detailed that instructors may find that no separate SPSS instruction manual is needed. Placement of the output and its instructions at the end of the modules allows instructors not teaching SPSS to easily skip over these pages.
- In the first edition of this textbook, answers to all exercises were provided to the instructor, who chose whether or not to provide them to the students. In this second edition, answers to the odd-numbered exercises are provided at the rear of the textbook for the student's convenience, with answers to the even-numbered exercises reserved for the instructor. The number of exercises also has been doubled, giving students additional practice.
- PowerPoint slides for instructional purposes have been redone to more fully capture key instructional points and to allow for presentation of important figures and diagrams.

Style Distinctives

This textbook is written in short *modules* rather than long chapters. Studies of today's students have shown that they would rather read many short modules than fewer long chapters. This allows you to assign modules that better match the intended instruction. It also permits students to master a distinct block of information before progressing to the next block.

This is not to say that you can select modules at random. Statistics "builds." Hence, most modules must be taken in a prescribed order. Still, a few modules can be omitted or their order changed. For example, you may choose to omit the modules on score transformations, probability, one-sample t tests, and variations on two-sample t tests, and you may even stop short of ANOVA—all with minimal loss to the overall logic. Finally, I chose to place the correlation modules after completing the t, F, and χ^2 instruction in order not to interrupt the repetitive inferential logic. However, because some instructors prefer to teach correlation right after descriptive statistics, I intentionally worded the correlation modules so that they can be taught in either location without awkward transitions.

The textbook contains the following additional features:

- Each module begins with a set of *learning objectives* and a list of *terms and symbols* unique to that module. These assist the student by providing both a scaffold for what to expect of that day's reading and a reference for where to find key information in the future. These features also assist you in knowing where key concepts first appear.
- Frequent *Check Yourself!* boxes throughout the text *reinforce learning* soon after key concepts have been taught. Thus, students quickly test their understanding as they progress.
- *Practice exercises are dispersed* throughout the modules as subtopics are covered rather than appended to the end of each module. Also, rather than overload the textbook with exercises, a modest number of exercises appear in the textbook, with additional exercises appearing in the Instructor Resource Guide and many more in the Student Study Guide. Answers to all textbook exercises are located in the Instructor's Resources, thereby allowing you to choose between using textbook exercises for homework and testing, or providing students with the exercise answer file for self-study purposes.
- Tone is in first person and *conversational.* I ask questions: "If it were up to you, which of the three formulas would you want to use?" And I confirm students' competence: "By now you are so proficient at this task that you are probably already looking for a table in which to look it up." The intent is for me to appear to be right beside the student, in the manner of a classroom teacher. Indeed, students have said that I write like I talk.
- I make extensive use of *humor.* Cartoons are dispersed throughout, and quips are placed in the margin sidebars as their associated topics are covered. This is admittedly a bit of form over substance, but it serves as a stress buster (a person cannot be amused and anxious at the same time) and also appeals to the quick-transitions learning style of today's student. In trial runs, even graduate students enjoyed the humor and reported that they skimmed each module for its jokes before reading the module's content.

Final Note

It is my sincere hope that you and your students find this book to be, as the title states, *Statistics Alive!* If, in using the materials, you find any errors, please let me know. I also welcome any other feedback (compliments or criticism) regarding your experience using this textbook. It would be my pleasure to receive your comments and to incorporate your suggestions into the next edition. You may use the feedback card at the end of this textbook to contact me.

Sincerely,
Wendy J. Steinberg

Supplemental Material for Use With *Statistics Alive!*

Student Study Guide

This affordable Student Study Guide to accompany *Statistics Alive* (Second Edition) will help students get the added review and practice they need to improve their skills and master the material for the course. The newly revised study guide is broken down by module and includes the following: summaries, learning objectives, and practice exercises (which consist of computation, true/false, short-answer, and multiple-choice questions).

Student Study Site

An open access, free Student Study Site provides additional support to students to help them prepare for class and exams. Flashcards are included on the site to review terms and symbols introduced at the beginning of each module. Access the student study site at: www.sagepub.com/steinberg2e.

Instructor's Resource Site

Helpful teaching aids are provided for professors, particularly those new to teaching statistics and to using *Statistics Alive!* (Second Edition). Included on the password protected portion of the companion website are:

- Teaching tips with suggested class activities
- Sample midterm (covering half the text)
- Sample final (covering the entire text)
- Test bank (including computation, true/false, and multiple choice questions)
- Answers to all test bank questions, textbook exercises, and even-numbered questions
- Web links to additional resources

Visit the study site at www.sagepub.com/steinberg2e.

Acknowledgments

Only an intravenous administration of Valium will get me through this course!

—Comment written by
Shelly W. on an index card that I asked
students to complete on the first day of class

This book was inspired by all the Shelly Ws who have suffered through statistics courses over the years. May this textbook forever end your suffering.

—Wendy J. Steinberg

I would like to thank specific people who contributed to the completion of this second edition.

To my past students: You endured corny jokes and error-riddled drafts. Still, you learned. Bless you!

To Astrid Virding, production editor for the textbook; to Lauren Habib associate editor for the Student Study Guide and for the Instructor's Resources; to Sarita Sarak, editorial assistant and cartoon copyright acquisitions sleuth; and to Shankaran Srinivasan and his copyediting team at QuADS: Oh, my, what a job you each had! Without you, no way.

To Vicki Knight, senior acquisitions editor at Sage Publications: Thank you for hanging in there as we endured lost files, corrupted files, and myriad other production crises. Her knowledge and experience was invaluable.

And special thanks to Matthew Price, author of the Student Study Guide (SSG). Prior to the first edition of this textbook, I selected Matthew from among other potential authors because his sample materials seemed to best capture my textbook's "voice." I was not disappointed. I am indebted to him for a comprehensive and high quality SSG.

Wendy J. Steinberg
Spring 2010

The author and Sage Publications gratefully acknowledge the contributions of the following reviewers for the first edition:

Katherine Appleton, *Queens University*

Timothy A. Brown, *Boston University*

Matthew Chin, *University of Central Florida*

Michael A. A. Cox, *University of Newcastle Upon Tyne*

John M. Dossett, *Tennessee State University*

Renee V. Galliher, *Utah State University*

Elizabeth Mason, *California University of Pennsylvania*

Matthew Price, *Georgia State University*

Suzanne Rathbun, *University of Texas Permian Basin*

The author and Sage Publications gratefully acknowledge the contributions of the following reviewers for the second edition:

Daniel J. Dickman, *Ivy Tech Community College of Indiana Region 12*

Christopher J. Ferguson, *Texas A&M International University*

Donald A. Hantula, *Temple University*

Joseph H. Porter, *Virginia Commonwealth University*

Mrinal Sinha, *California State University, Monterey Bay*

About the Author

Wendy J. Steinberg entered academia midcareer, having spent the first part of her career in high-stakes test development. She holds a PhD in educational psychology with dual concentrations, one in measurement and the other in development and cognition. Teaching is her passion. She views education as a sacred task that teachers and students alike should treat with reverence. She wants this textbook in the hands of every statistics student so that tears will be banished forever from the classroom. A portion of the sale of each textbook goes to charity.

PART I

Preliminary Information

"First Things First"

1

Math Review, Vocabulary, and Symbols

Terms: case, subject, sample, population, statistic, parameter, variable, constant, summation

Symbols: X, X_1, X^2, $\bar{X}$, M, p, q, N, n, $\approx$, $<$, $>$, $||$, Σ, $\sqrt{\ }$

Learning Objectives:

- Become familiar with common statistical terms and symbols
- Review common arithmetic rules and functions

Getting Started

Statistics is a language. As with any language, before we can communicate effectively, we must master that language's vocabulary and symbol system. This first module, then, introduces essential statistical vocabulary and symbols. Additional vocabulary and symbols will be introduced as you progress through the textbook.

> There still remain three studies suitable for man. Arithmetic is one of them.
>
> —Plato

Following the vocabulary and symbols, the remainder of this module is a math refresher. A surprising proportion of students need a math review or, at the least, benefit from it. Note that the review is basic. It is intended merely to make sure that everyone using this textbook—including the math phobic and those who have been out of school for some time—remembers the fundamentals.

Nevertheless, do not be misled by this math review into thinking that this is a "remedial math" type of textbook. It is not. This textbook is a comprehensive, college-level treatment of statistical concepts and calculations. While the first few modules may seem elementary, the material quickly builds in difficulty. Still, the mathematics never becomes overwhelming. The difficulty of statistics seldom lies with the mathematics. Rather, the difficulty lies with the inferential logic. Even in its most difficult modules, this textbook requires only elementary algebra. In most units, basic arithmetic is sufficient. You are fully capable of mastering the material.

Your classroom teacher is, of course, the primary instructor for your course. However, I have written this textbook in a conversational tone, as if I were standing alongside you, guiding you in your understanding of statistics. In that sense, I, too, am your teacher. So, as your coteacher, let me offer a few words of advice.

- First, carefully read all the assigned text material. Because statistics is logic driven, missing a piece of the argument can be deadly to your later understanding. Think of this textbook as an integral part of your class instruction.
- Second, take notes. Have a highlighter and a pen nearby when reading the textbook.
- Third, when you come up against a *Check Yourself!* box, don't just skip over it. Rather, check yourself! The boxes are strategically placed throughout the text to ensure mastery of important material before progressing to the next topic.
- Finally, take time to laugh. Despite rumors to the contrary, the study of statistics doesn't have to be boring. I have peppered the text with cartoons and the margins with quips and quotes. Let these be tension breakers, reminding you that even statistics has a lighter side. For example, did you know that one of the statistics used to test experimental hypotheses was originally created to test the quality of beer? Or that the only certain conclusion with any inferential statistic is that we're not certain? Or that from knowledge of the amount of ice cream eaten by members of a community, we can predict the number of drownings in that community quite accurately?

How can these things be? You will have to read on to find out. I hope that you find your study to be . . . *Statistics Alive!*

SOURCE: Peanuts: © United Feature Syndicate, Inc.

Vocabulary and Symbols

To work with statistics, we need to use the notation commonly accepted by those in the field. Most statistical formulas or problems contain one or more of these terms or symbols.

Case. This is an individual unit under study. It could be a person, an animal, a car, a soft drink, a lightbulb, and so on. For example, a study of car-stopping distances might include 300 cases (cars).

Subject. When the cases are human beings, they are often referred to as subjects. For example, a study of running speed in college sophomores might include 80 subjects (80 sophomore students).

Participant. This is the American Psychological Association's (APA) preferred word for "subjects," defined above.

Sample. This is the group of participants in a study. Usually, they are a subset of a larger group. For example, we might measure the height of a sample of 100 elderly women.

Population. This is the larger group of participants about which we want to draw a conclusion. For example, although we may have sampled only 100 elderly women, we might want to draw a conclusion about the height for the whole population of elderly women of which the 100 women were a part.

Statistic. This is a summary number (e.g., an average) for a sample. Our statistical average might be 63 in.

Parameter. This is a summary number (e.g., an average) for a population. Our parametric average might be 63.5 in.

Variable. When the value of the trait being measured varies from case to case, that trait is referred to as a variable. For example, when measuring the running speed of college sophomores, running speed is a variable because each student is expected to run at a different speed.

Constant. When the value of the trait being measured is the same for all cases, that trait is referred to as a constant. In a mechanized (machine-driven) study of car-stopping distance, for example, researchers might figure how long it takes the average person to raise a foot to the brake pedal and then add that value as a constant to the measured stopping distances.

Uppercase Letters. These usually represent variables, scores that vary from case to case. An example would be X, which might stand for each subject's running score. Every subject would have a different X score.

Lowercase Letters. These usually stand for constants, values that are the same for each case. An example would be c, which, in the study of car-stopping distance, stands for the average time it takes to raise a foot to the brake pedal.

Bar Over a Letter. This represents the average of a variable. An example would be "$\bar{X}$" (pronounced "X-bar"), which is the average score on the variable X.

M. This letter is reserved for the mean, known in lay language as the average. Wait. Didn't I say above that $\bar{X}$ is the symbol for the average? Yes, either symbol, $\bar{X}$ or M, is used to represent an average. However, M is the more recent symbol and is the only symbol now accepted by American Psychological Association (APA) journals. Nevertheless, you should learn to recognize the $\bar{X}$ symbol, as it will appear in older textbooks, in textbooks aimed at students outside the social sciences, and in journal articles published prior to the change.

p. This letter is reserved for the probability of an event occurring. An example would be the probability of rolling a 3 on a standard die. Because a die has six sides, numbered one through six, the probability of rolling any given number is 1/6. In decimal form, the probability is .167.

q. This letter is reserved for the probability of an event *not* occurring, or in other words, $1 - p$. An example would be the probability of not rolling a 3 on a standard die. The probability is $1 - p$, which is $1 - 1/6$, which is 5/6. In decimal form, the probability is .833.

N, n. This letter is reserved for the number of cases. An example would be the number of students in your statistics class. If there are 50 people in your class, then $N = 50$. Typically, n is the number of cases in a sample, and N is the number of cases in a population.

Subscripts. Subscripts refer to particular subjects or cases. Continuing with the variable "X," "X_1" would be the score of the first subject, "X_2" the score of the second subject, and so on.

Wavy Parallel Lines (≈). This symbol means *about* or *approximately*. For example, we might say that the coat cost ≈\$200 or that you expect to experience a snack attack in ≈15 min.

Less Than and Greater Than (< and >). These symbols mean *less than* and *greater than*, respectively. For example, we might say that the coat cost <\$200 or that you expect >15 min to pass before you experience a snack attack.

Summation (Σ). This means to add the scores of all cases. For example, ΣX means to add up all scores on the variable X.

Multiplication Indicators. There are several ways to indicate that two numbers are to be multiplied. As a child, you learned to write "3 × 4." However, when substituting letters for numerals in algebra, the use of "×" to indicate multiplication becomes confusing. Thus, a second method to indicate multiplication is to put parentheses around each number individually. For example, "(3)(4)" means to multiply 3 and 4. Another way is to put a midlevel dot or asterisk between the numbers. For example, "3 * 4" means to multiply 3 and 4. A fourth way, which can be used only when the numbers are symbolized by letters, is to write the numbers next to each other without a space. For example, "*ab*" means to multiply *a* and *b*.

Q: When was math first mentioned in the Bible?

A: In Genesis, God told Adam and Eve to go forth and multiply.

Reciprocal. A reciprocal is "one divided by the number." Thus, the reciprocal of 3 is 1/3, and the reciprocal of 6 is 1/6.

Superscripted Number or Exponent. This indicates the number of times a number should be multiplied by itself. For example, 3^2 (pronounced "three squared") means to multiply 3 twice, which is 3 * 3, which is 9.

Radical Sign ($\sqrt{\ }$). This says to find the square root of the number under the radical sign. The square root is the number that when multiplied by itself yields the number under the radical sign. For example, "$\sqrt{4}$" is 2, because 2 times itself is 4. And "$\sqrt{25}$" is 5, because 5 times itself is 25.

CHECK YOURSELF!

Which terms in this section were new to you? Reread the definition for any term that you did not already know.

Some Rules and Procedures

A few rules and procedures are fundamental to the study of statistics. Again, this review is intended for those whose math skills are rusty.

- *Multiplying by a reciprocal*: Dividing by a number is the same as multiplying by its reciprocal. For example, 6 divided by 3 is equal to 6 times 1/3. Both are equal to 2. Some statistical formulas substitute one function for the other. If you are expecting division and don't see it, check for multiplication by a reciprocal. For example, some formulas multiply by $1/N$ rather than divide by N.
- If the signs of two numbers being multiplied differ, the result will be negative. However, if the signs of the two numbers are the same (whether positive or negative), the result will be positive. You may have learned this rule as "a negative times a negative is a positive." For example, (+2)(+4) = +8, and (–2)(–4) = +8. But (+2)(–4) = –8, because the signs of the numbers being multiplied differ.
- *Converting between fractions, decimals, and percentages:* To convert a fraction to a decimal, divide the numerator by the denominator. For example, 1/4 is 1 divided by 4, which, when you carry out the division, is 0.25. To convert a decimal to a fraction, remove the decimal point and place the number over 1 followed by as many zeros as there were

Q: Where do mathematicians shop?

A: At the decimall.

original digits. For example, 0.25 is 25/100, which further reduces to 1/4. To convert a decimal to a percentage, multiply by 100 and add a percent sign. (To multiply by 100, move the decimal place two places to the right.) For example, 0.25 is 25%.

- *Deciding on the number of decimal places*: In math, "precision" refers to the exactness of a calculation. In the social sciences, two decimal places are usually considered adequate for a final answer. Answers carried to additional decimal places imply a degree of precision that we rarely have in the social sciences. However, most of the formulas in this textbook require that you perform a series of operations before arriving at a final answer. Rarely will either the final answer or the intervening steps be in whole numbers. If final answers are to be reported to two decimal places, you should carry all the preliminary steps to three decimal places and then round the final answer to two decimal places. Never round to whole numbers during the preliminary steps, as a series of such roundings may significantly alter the final answer.

> A small error in the beginning is a great one in the end.
>
> —St. Thomas Aquinas

- *Rounding numbers*: If the last digit is greater than 5, round up by adding one to the preceding digit. If the last digit is less than 5, leave the preceding digit unchanged. For example, 46.268 rounds up to 46.27, and 46.263 rounds down to 46.26. If the last digit is exactly 5, whether or not you round depends on the preceding digit. If the preceding digit is odd, round up by adding one to it; if the preceding digit is even, leave it unchanged. For example, 32.635 rounds to 32.64, but 32.645 also rounds to 32.64.

> Round numbers are always false.
>
> —Samuel Johnson

- *Ordering operations*: When several arithmetic operations are called for in a formula, perform them in the following order: (1) squares or square roots (exponents), (2) multiplication or division, and (3) addition or subtraction. However, parentheses override that order. Operations within parentheses are always done first, regardless of what those operations are. If parentheses are stacked, do operations within the innermost parentheses first and work your way outward. You may have learned the order of operations as "Please Excuse My Dear Aunt Sally." The first letters of that mnemonic stand for parentheses, exponents, multiplication, division, addition, and subtraction. For example, $4 + 2 * 3 = 4 + 6 = 10$. But $(4 + 2) * 3 = 6 * 3 = 18$. For another example, $2 + 3^2/2 = 2 + (3^2/2) = 2 + (9/2) = 2 + 4.5 = 6.5$. But $(2 + 3^2)/2 = (2 + 9)/2 = 11/2 = 5.5$.
- *Reordering terms to solve for an unknown*: If the quantity for which we want to solve is not alone on one side of an equation, we must first isolate it. For example, if we have the equation $24 = 16 + a$, we must isolate a on one side or the other of the equal sign. We do this by subtracting 16 from both sides. This gives us $24 - 16 = a$. This, of course, is 8.
    - Sometimes, a formula is given to solve for one term, but we instead want to solve for some other term within the formula. Again, we must reorder the terms to isolate the desired term. Take the formula for a z score, which you will come across in Module 8:

$$z = \frac{X - M}{s}$$

The formula solves for z. But what if we know z and want to instead solve for X? We have to reorder the terms to isolate X. First, we get rid of the s in the denominator by multiplying both sides of the equation by s. This gives us $zs = X - M$. Then, we add M to both sides of the equation. This gives us $M + zs = X$. If we like, we can flip the terms on both sides of the equation so that X is to the left of the equal sign: $X = M + zs$. Either version is correct.

CHECK YOURSELF!

Which rules and procedures in this section were new to you? Reread the explanation for any rule or procedure that you did not already know.

CHECK YOURSELF!

Here are two formulas for a variance (a statistic that you will study later in the course):

$$s^2 = \sum(X - M)^2(1/n) \quad \text{and} \quad s^2 = \frac{\sum(X - M)^2}{n}$$

Explain why both formulas lead to the same answer.

SOURCE: Reprinted with permission of the Carolina Biological Supply Co.

PRACTICE

1. Complete the following chart:

Fraction	*Percentage*	*Decimal*
≈33 1/3		
	12.5%	
		0.667

2. Complete the following chart:

Fraction	*Percentage*	*Decimal*
1/6		
	10%	
		0.143

3. Complete the following chart:

Fraction	*Percentage*	*Decimal*
1/20		
	95%	
		0.25

4. Complete the following chart:

Fraction	*Percentage*	*Decimal*
1/50		
	.99%	
		0.01

5. Round the following numbers to two decimal places:
   a. 26.412 ___________
   b. 62.745 ___________
   c. 36.846 ___________

6. Round the following numbers to two decimal places:
   a. 77.935 ___________
   b. 1086.267 ___________
   c. 39.633 ___________

7. Round the following numbers to two decimal places:
   a. 95.555 ___________
   b. 00.023 ___________
   c. 48.950 ___________

8. Round the following numbers to two decimal places:
   a. 110.001 ___________
   b. 12.635 ___________
   c. 276.772 ___________

9. Solve the following equations applying the rules of order of operations:
   a. $4 * 5 + 3 * 2 =$ ___________
   b. $4(5 + 3) * 2 =$ ___________
   c. $((4 * 5) + 3)(2) =$ ___________
   d. $\sqrt{(4(5+3))(2)} =$ ___________
   e. $4^2(5 + 3)(2) =$ ___________

10. Solve the following equations applying the rules of order of operations:
    a. $4\sqrt{5^2}(3 * 2) =$ ___________
    b. $(4)(5 + 3 * 2) =$ ___________
    c. $4\sqrt{5} + (3 * 2) =$ ___________
    d. $4^2(5 + 3 * 2) =$ ___________
    e. $(4)(5 + 3)2^2 =$ ___________

11. Solve the following equations applying the rules of order of operations:
    a. $4\sqrt{36} * (2 * 3) + 2 =$
    b. $6^2/12 - 4/2 =$
    c. $\sqrt{8 - 4} * 6/2 - 1 =$
    d. $[(30)\ (.5) - 6] * 2 + 4 =$
    e. $5 + 3 * 7 - 4 =$

12. Solve the following equations applying the rules of order of operations:
    a. $\sqrt{5 * 6 + 6 - 3^2} - 2 =$
    b. $(4 + 7^2 - 17)/6 =$
    c. $6 - 2 * 2 + 9/3 =$
    d. $100 - 10 * 5 - 5 =$
    e. $\sqrt{(12 - 8) * 16} - 2 =$

13. Reexpress the following equations, substituting reciprocals for division. Then, solve in decimal form.

Equation	*Reexpressed in Reciprocals*		*Solution*
a. 16/5 − 0.246 =	______________	=	______
b. 68 + 68/3 =	______________	=	______
c. 2/3 − 1/5 =	______________	=	______

14. Reexpress the following equations, substituting reciprocals for division. Then, solve in decimal form.

Equation	*Reexpressed in Reciprocals*		*Solution*
a. 98 − 75/3 =	______________	=	______
b. 672 + 2/6 =	______________	=	______
c. 8/3 + 5/4 =	______________	=	______

15. Reexpress the following equations, substituting reciprocals for division. Then, solve in decimal form.

Equation	*Reexpressed in Reciprocals*		*Solution*
a. 6/3 − 0.95 =	______________	=	______
b. 12.50 − 8/2 =	______________	=	______
c. 25/5 + 6 =	______________	=	______

16. Reexpress the following equations, substituting reciprocals for division. Then, solve in decimal form.

Equation	*Reexpressed in Reciprocals*		*Solution*
a. 155 + 5/2 =	______________	=	______
b. 3.28 − 3/2 =	______________	=	______
c. 720/9 − 45 =	______________	=	______

17. Rearrange the equation to solve for the indicated unknown.
    a. $64/4 + b = 30$ Solve for b: ____________
    b. $0.30c = 20$ Solve for c: ____________
    c. $Y = bX + a$ Solve for a: ____________

18. Rearrange the equation to solve for the indicated unknown.
    a. $b + 20 = 0.5$ Solve for b: ____________
    b. $F = 1.8c + 32$ Solve for c: ____________
    c. $50 = 60/a$ Solve for a: ____________

19. Rearrange the equation to solve for the indicated unknown.
    a. $b + 24/8 = 3$ Solve for b: ____________
    b. $\mathrm{T} = 12c - 4$ Solve for c: ____________
    c. $12 = 3a$ Solve for a: ____________

20. Rearrange the equation to solve for the indicated unknown.
    a. $7 = .5b + 2.5$ Solve for b: ____________
    b. $64 = 4c$ Solve for c: ____________
    c. $3a = .25\mathrm{V}$ Solve for a: ____________

More Rules and Procedures

In this textbook, derivations and proofs are kept to a minimum. However, a fuller understanding of the formulas and of the relationship between various statistics requires mathematical manipulations beyond the basic rules just presented. If your instructor intends to derive one formula from another (say, a raw score formula from a deviation score formula) or prove theorems (say, the binomial theorem), some additional mathematical rules are necessary. Knowing these rules is especially appropriate for students intending to take additional courses in statistics. Let your instructor be your guide on the need for this section.

- Rules for summation across and within parentheses
  1. The summation of a variable is the sum of that variable.

     $\Sigma(X) = \Sigma X$

     For example, if $X_1 = 2$, $X_2 = 4$, and $X_3 = 6$, then

     $\Sigma(X) = \Sigma X = 2 + 4 + 6 = 12$

  2. The summation of a constant is N times the constant.

     $\Sigma(a) = Na$

     For example, if $a = 6.24$ and there are 3 subjects, then

     $\Sigma(a) = Na = 3 * 6.24 = 18.72$

  3. The summation of a constant times a variable is the constant times the sum of the variable.

     $\Sigma(aX) = a\Sigma X$

     For example, if $a = 6.24$ and if $X_1 = 2$, $X_2 = 4$, and $X_3 = 6$, then

     $\Sigma(aX) = a\Sigma X = 6.24 * (2 + 4 + 6) = 6.24 * 12 = 74.88$

Do not worry about your difficulties in mathematics. I assure you that mine are greater.

—Albert Einstein

4. The summation of two or more terms within parentheses is the same as their independent summation.

   $\Sigma(X + Y) = \Sigma X + \Sigma Y$

   $\Sigma(X + a) = \Sigma X + Na$

   $\Sigma(X + aY) = \Sigma X + a\Sigma Y$

   For example, if $X_1 = 2$, $X_2 = 4$, and $X_3 = 6$ and if $Y_1 = 1$, $Y_2 = 3$, and $Y_3 = 5$ and if $a = 6.24$, then

   $\Sigma(X + Y) = \Sigma X + \Sigma Y = (2 + 4 + 6) + (1 + 3 + 5) = 12 + 9 = 21$

   $\Sigma(X + a) = \Sigma X + Na = (2 + 4 + 6) + (3 * 6.24) = 12 + 18.72 = 30.72$

   $\Sigma(X + aY) = \Sigma X + a\Sigma Y = (2 + 4 + 6) + (6.24)\ (1 + 3 + 5) = 12 + (6.24)(9)$
   $= 12 + 56.16 = 68.16$

- Rules for expanding binomials

  1. $(ab)^2 = (ab)(ab) = aa * bb = a^2 * b^2$
     Thus, $(ab)^2 = a^2 * b^2$
     For example, $(4 * 3)^2 = 4^2 * 3^2 = 16 * 9 = 144$.

  2. $(a + b)^2 = aa + ab + ab + bb = a^2 + 2ab + b^2 = a^2 + b^2 + 2ab$
     Thus, $(a + b)^2 = a^2 + b^2 + 2ab$
     For example, $(4 + 3)^2 = 4^2 + 3^2 + 2(4 * 3) = 16 + 9 + 2(12) = 25 + 24 = 49$

  3. $(a - b)^2 = aa - ab - ab + bb = a^2 - 2ab + b^2 = a^2 + b^2 - 2ab$
     Thus, $(a - b)^2 = a^2 + b^2 - 2ab$
     For example, $(4 - 3)^2 = 4^2 + 3^2 - 2(4 * 3) = 16 + 9 - 2(12) = 25 - 24 = 1$

PRACTICE

21. Expand the following expressions using the rules for summation within parentheses and for binomial expansion (*Note*: Numerals are always constants; thus, *N* is always a constant):

    a. $\Sigma(X + Y)^2$ ________________

    b. $\Sigma(bXY + 1)^2$ ________________

    c. $\Sigma(XY + 1)$ ________________

22. Expand the following expressions, using the rules for summation within parentheses and for binomial expansion (*Note*: Numerals are always constants; thus, *N* is always a constant):

    a. $\Sigma(1 + N)$ ________________

    b. $\Sigma(2X + Y)$ ________________

    c. $\Sigma(aX)^2$ ________________

That should be all the math you need in this textbook. Of course, higher math is necessary for the fullest understanding of statistical formulas and for further study in statistics. However, a first course in statistics rarely goes beyond what is presented here.

Visit the study site at www.sagepub.com/steinberg2e for practice quizzes and other study resources.

2

Measurement Scales

Terms: nominal, ordinal, interval, ratio, continuous, discrete, real limits

Symbols: LL, UL

Learning Objectives:

- Classify data according to their level of measurement
- Distinguish between discrete and continuous scores
- Establish real limits for continuously scored data

What Is Measurement?

Measurement is the process of assigning numbers to the observations on some variable. However, not all numbers have the same numeric properties. For example, one football player may wear a jersey with No. 7 on it, and a second player may wear a jersey with No. 18 on it. However, you would not say that the second player has 11 points more of any trait than the first player. The jersey numbers simply don't mean that. Or you may earn the top score on a statistics test, your friend Lauren the second highest score, and your friend Sidney the third highest score. However, you cannot assume that the score difference between you and Lauren is the same as the score difference between Lauren and Sidney. Again, the rankings simply don't mean that. Or one person may score 75 on an intelligence test, and a second person may score 150 on the same intelligence test. While we would agree that the second person is probably considerably more intelligent than the first one, you cannot say that the second person is twice as intelligent as the first person. Again, the scores simply don't mean that.

Scales of Measurement

What, then, can we say about each of these situations? What we can say depends on the scale on which the data were measured. Stevens (1946) suggested that variables are measured on one of four scales: nominal, ordinal, interval, or ratio. Each of these scales allows us to draw different conclusions about the meaning of subjects' scores. Furthermore, the statistics that can be calculated on the data are partially dependent on the measurement scale in the data. Thus, before we can select and compute statistics, we need to know the scale on which the data were measured.

Nominal Scale

A **nominal** scale classifies cases into categories. For that reason, it is also sometimes called a categorical scale. Here are some examples:

m = *male*, f = *female*

1 = *married*, 2 = *divorced*, 3 = *separated*, 4 = *never married*

tel = *owns a telephone*, notel = *does not own a telephone*

The nominal scale is the lowest level of measurement. It does not measure how much of a measured trait a person possesses. It merely categorizes the person as a "this" or a "that." Even when numbers are used to represent the categories, they are designations only, much like social security numbers or the numbers on a football jersey. Because the designations have no numeric meaning, it makes no sense to perform arithmetic operations on them. For example, you would learn nothing by knowing the average telephone number for the members of your statistics class or the average jersey number for the members of a football team.

There are two groups of people in the world: Those who believe that the world can be divided into two groups of people and those who don't.

Most of the statistics that you will use throughout this textbook require that the numbers have numeric meaning. Thus, most of the procedures that you will learn will not be appropriate for data that are measured on only a nominal level. However, certain statistical procedures have been developed to analyze data in a nominal form. For example, the chi-square statistic, which you will learn about later in the course, is computed on nominal data.

Ordinal Scale

An **ordinal** scale ranks people according to the degree to which they possess some measured trait. Persons are first measured on some attribute (e.g., height). Then, they are assigned ranks according to how much of the attribute they possess. Here are some examples:

1 = *tallest*, 2 = *second tallest*, 3 = *third tallest*, and so on.

1 = *highest grade point average (GPA)*, 2 = *second highest* GPA, 3 = *third highest* GPA, and so on.

1 = *fastest runner*, 2 = *second fastest runner*, 3 = *third fastest runner*, and so on.

To create ranks, place the scores in order by value (usually descending) and then assign the highest score a rank of 1, the second highest score a rank of 2, and so on. Table 2.1 presents scores and ranks for 10 students on a 100-item statistics exam given as a pretest before the course began.

With an ordinal scale, the difference in test scores between two adjacent ranks may not be the same as the difference in test scores between any other two adjacent ranks. Compare the test scores between adjacent ranks in Table 2.1. The test score difference between the first two ranks is one point (73 – 72), but between the next two ranks, the difference in test scores is 12 points (72 – 60). Clearly, certain aspects of relative performance are lost when scores are expressed as ranks.

Most of the statistics used throughout this textbook require a scale in which adjacent intervals on the measurement scale imply equal intervals between adjacent scores throughout the scale. Because this is not the case for ordinally scaled data, most of the statistics will not be appropriate for data that are measured on an ordinal scale. However, ranks do tell us more than nominal classifications. A higher-ranked person does have more of the measured

Table 2.1 Statistics Test Scores and Ranks

Score	*Rank*
73	1
72	2
60	3
59	4
57	5
56	6
52	7
48	8
36	9
28	10

trait, even if distance on the underlying scores is inconsistent. Therefore, certain statistical procedures have been developed to analyze data when they are in ordinal form. For example, a Spearman rho coefficient measures the degree of relationship between two sets of ranks.

Interval Scale

With an **interval** scale, the distances between adjacent scores are equal and consistent throughout the scale. Equal intervals on the scale imply equal amounts of the variable being measured. For this reason, the interval scale is sometimes referred to as the equal-interval scale. Here are some examples:

Scores on the final exam in this course

Scores on an intelligence test

Degrees Fahrenheit or Celsius

Scores on certain personality or career interest tests

Because interval scales are consistent throughout the scale, it makes sense to compare scores by adding or subtracting them. For example, an intelligence score of 120 is 10 points higher than an intelligence score of 110, just as an intelligence score of 60 is 10 points higher than an intelligence score of 50. In either case, the higher-scoring person has 10 more points of scaled intelligence than the lower-scoring person.

However, interval scales have no absolute zero point—the point at which a person would have none of the measured attribute. That is, even if a person scored zero on an intelligence test, we would not say that the person has no intelligence. This is because the test's starting point is fixed and arbitrary, whereas the starting point of the actual trait itself—in this case, intelligence—is unknown. This is typically the case in psychological or educational measurement.

Because of the lack of an absolute zero point in an interval measurement scale, it makes no sense to compare interval scores by multiplying or dividing them. Although a person who scores 120 on an intelligence test has scored twice as high as the person who scores 60, the person who scores 120 does not have twice as much intelligence as the person who scores 60.

Most of the statistics that you will learn throughout this textbook require at least an interval measurement scale. Fortunately, most data in social science research are also interval scaled. Therefore, the statistics that you will learn are ideally suited for most educational and psychological data.

One controversy in educational and psychological research is whether personality and career interest tests are interval scaled or only ordinal scaled. This is because individual test items typically ask test takers to use a noninterval scale. For example, test takers might be asked to rate the frequency with which they feel that "life has no meaning" with the following scale: *never, occasionally, often*, or *always*. Or the scale might ask test takers to rate the degree to which they agree with the statement "I enjoy working out-of-doors" with the following scale: *strongly agree, slightly agree, neither agree nor disagree, slightly disagree*, or *strongly disagree*. We cannot say that the difference in frequency between *never* and *occasionally* is the same as the difference in frequency between *often* and *always*. Similarly, we cannot say that the difference in degree between *strongly agree* and *slightly agree* is the same as the difference in degree between *slightly disagree* and *strongly disagree*. Therefore, the scale appears to be ordinal.

On the other hand, at the completion of the personality or career interest test, the test taker receives a score, much like the score you might receive on a test in this statistics course. Moreover, the scores are typically normed against a large number of test takers. This gives any one person's score both position and distance when compared with the known distribution of all scores. These features are typical of an interval scale, not an ordinal scale. Because the scores contain features of both ordinal and interval scales, the debate over the appropriate level of measurement, and hence the appropriate statistics to use to describe the scores, is ongoing.

Ratio Scale

A **ratio** scale is like an interval scale, in that the distance between adjacent scores is equal throughout the distribution. However, unlike an interval scale, in a ratio scale there is an absolute zero point. That is, there is a point at which a person does not have any of the measured traits. Because the trait's starting point is known, the scale reflects that zero point.

Ratio measurement typically applies to measures in the physical sciences. Here are some examples:

Height

Weight

Distance

Time

Because there is an absolute zero point, it makes sense to compare scores by multiplying or dividing them. For example, a person who weighs 110 lb not only weighs 55 lb more than a person who weighs 55 lb but also is twice as heavy. A person who makes a standing broad jump of 3 ft not only jumps 3 ft less than one who jumps 6 ft but also jumps only half as far.

Although most data in psychology and education are interval scaled, some are ratio scaled. For example, measures of reaction time or physical performance are ratio scaled. Any statistic that is appropriate for interval-scaled data is also appropriate for ratio-scaled data.

Finally, it is possible to measure data on more than one scale. That is, higher-level data can be measured on a lower-level scale. For example, your interval score on a statistics test can be nominally categorized as either pass or fail. However, lower-level data cannot be measured on a higher-level scale. For example, your sex (male or female) cannot be ranked. It can only be nominally categorized.

CHECK YOURSELF!

Give examples of nominal, ordinal, interval, and ratio data. Then, convert your ratio data example into interval, ordinal, and nominal scales.

PRACTICE

1. Indicate the first letter (N, O, I, R) of the *highest* possible scale for each of the following measures, where N is lowest and R is highest:

Measure	*Highest Scale*
a. Feet of snow	_____
b. Brands of carbonated soft drinks	_____
c. Class rank at graduation	_____
d. GPA	_____
e. Speed of a baseball pitch	_____

2. Indicate the first letter (N, O, I, R) of the *highest* possible scale for each of the following measures, where N is lowest and R is highest:

Measure	*Highest Scale*
a. Eye color	_____
b. Time taken to solve a puzzle	_____
c. Genre of favorite television program	_____
d. Level of depression	_____
e. Position in a starting lineup	_____
f. Number of angels that can fit on the head of a pin (Oops . . . angels may not be subject to earthly measurement constraints, so you may skip this one.)	_____

3. Indicate the first letter (N, O, I, R) of the *highest* possible scale for each of the following measures, where N is lowest and R is highest:

Measure	*Highest Scale*
a. Hair length	_____
b. Species of tree	_____
c. Military rank	_____
d. Political party membership	_____
e. Yearly income	_____

4. Indicate the first letter (N, O, I, R) of the *highest* possible scale for each of the following measures, where N is lowest and R is highest:

Measure	*Highest Scale*
a. Favorite radio station	_____
b. How much sleep you had last night	_____
c. How many calories you ate today	_____
d. Athletic ability	_____
e. Position in a graduation processional	_____

5. Indicate the first letter (N, O, I, R) of the *highest* possible scale for each of the following measures, where N is lowest and R is highest:

Measure	*Highest Scale*
a. Literature genres	_____
b. Relative academic position in one's graduating class	_____
c. Hair length	_____
d. Job salary	_____
e. Self-confidence	_____

6. Indicate the first letter (N, O, I, R) of the *highest* possible scale for each of the following measures, where N is lowest and R is highest:

Measure	*Highest Scale*
a. Type of disease	_____
b. Business profit	_____
c. Pass or fail status	_____
d. Conscientiousness	_____
e. Car brands	_____

Continuous Versus Discrete Variables

Continuous variables are variables whose values theoretically could fall anywhere between adjacent scale units. The data that are measured on a ratio scale are always continuously scored. A person's height or weight or even the time a person spends talking on the phone can fall anywhere between the scale units.

Some interval scale data are continuous variables. For example, although people do not score fractional points on IQ tests, depression inventories, or measures of talkativeness, they theoretically could score fractional points if the tests were so graded.

Bumper sticker: Statisticians do it both continuously and discretely.

Discrete data, on the other hand, are values that cannot even theoretically fall between adjacent scale units. Some interval scale data are discrete. Examples are the number of blue ribbons won, number of children in a family, or number of photographs taken. With discrete interval data, we can perform all the arithmetic operations on the data that we would with any other interval-scaled set of scores. That is, we can speak of twice as many blue ribbons won, the average number of children in families, or half as many photographs taken. At the same time, we recognize that individual scores cannot fall between the scale units. That is, no single person can earn a partial ribbon, have a partial child, or take a partial photograph.

CHECK YOURSELF!

Is the number of courses students register for in a semester a discrete variable or a continuous variable? Is the GPA that students earn in those courses a discrete variable or a continuous variable?

PRACTICE

7. Are the following variables discrete or continuous? Mark "D" or "C" to indicate your answer:

Measure	*Variable Type*
a. Inches of rainfall	_____
b. GPA	_____
c. Speed of a baseball pitch	_____
d. Time taken to solve a puzzle	_____
e. Level of depression	_____
f. Number of angels that can fit on the head of a pin (Oops . . . how'd that get in here again? You may skip this one.)	_____

8. Are the following variables discrete or continuous? Mark "D" or "C" to indicate your answer:

Measure	*Variable Type*
a. Hair length	_____
b. Yearly income	_____
c. How much sleep you had last night	_____
d. How many calories you ate today	_____
e. Athletic ability	_____

9. Are the following variables discrete or continuous? Mark "D" or "C" to indicate your answer:

Measure	*Variable Type*
a. Number of TVs in the household	_____
b. Level of extraversion	_____
c. Color saturation level	_____
d. How tired you feel right now	_____
e. How many awards you won as a child	_____

10. Are the following variables discrete or continuous? Mark "D" or "C" to indicate your answer:

Measure	*Variable Type*
a. Amount of irritation you felt today	_____
b. Number of times you felt irritated today	_____
c. How many different ice cream flavors you have tasted	_____
d. How much weight you have gained or lost in the past year	_____
e. How badly you need a vacation right now	_____

Real Limits

For continuous variables, scores can theoretically fall anywhere between scale units, even if actual scores fall only at specified locations along the scale. Because of this, the boundaries of any scaled score for a continuous variable are contiguous with the next adjacent scaled score. In lay terms, this means that adjacent scores butt right up against one another. To make that happen, mathematicians speak of the score's **real limits** as opposed to the actual observed score.

The value of the real limit is set at half the scale's unit. Thus, the real limits of any particular score are the score plus or minus (symbolized as ±) one-half scale unit. For example, if intelligence is measured to the nearest one point and if Claude scores 116, the real limits of Claude's score are 116 ± (0.5) (1 point), which is 116 ± 0.5, which is 115.5 and 116.5. The lower real limit (**LL**) is one-half unit below the score (in this case, 115.5), and the upper real limit (**UL**) is one-half unit above the score (in this case, 116.5). No scores actually fall at the real limits.

CHECK YOURSELF!

For a test measured on a scale of 10 to 100 in 10-point intervals, what is the value of the real limit? What, then, are the real limits (LL and UL) for a score of 70?

Real limits will be important in calculating medians and percentile ranks from tabulated data and for constructing histograms for continuous data. You will learn about these in the next few modules.

PRACTICE

11. What are the real limits of the following scores?

Score	*Size of Each Scale Interval*	*Real Limits*
a. IQ = 116	1 point	_______ and _______
b. Height = 66.5	1/2 in.	_______ and _______
c. Age = 20	10 years	_______ and _______
d. Driving speed = 60	5 mph	_______ and _______
e. Olympic performance = 9.7	1/10 point	_______ and _______

12. What are the real limits of the following scores?

Score	*Size of Each Scale Interval*	*Real Limits*
a. GPA = 3.2	1/10 point	_______ and _______
b. Test score = 83	1 point	_______ and _______
c. Miles to work = 20	10 miles	_______ and _______
d. Feet of snow = 3.25	1/4 ft	_______ and _______
e. Shoe size = 11	whole size	_______ and _______

13. What are the real limits of the following scores?

Score	*Size of Each Scale Interval*	*Real Limits*
a. GPA = 3.20	1/100 point	_______ and _______
b. Test score = 83.4	1/10 point	_______ and _______
c. Miles to work = 20	5 miles	_______ and _______
d. Feet of snow = 3	whole foot	_______ and _______
e. Shoe size = 11.0	half size	_______ and _______

14. What are the real limits of the following scores?

Score	*Size of Each Scale Interval*	*Real Limits*
a. IQ = 116.0	1/2 point	_______ and _______
b. Height = 66.5	1/10 in.	_______ and _______
c. Age = 20	whole years	_______ and _______
d. Driving speed = 60	10 mph	_______ and _______
e. Olympic performance = 9.70	1/100 point	_______ and _______

Visit the study site at www.sagepub.com/steinberg2e for practice quizzes and other study resources.

PART II

Tables and Graphs

"On Display"

3

Frequency and Percentile Tables

Terms: frequency, cumulative frequency, relative frequency, percentage, cumulative relative frequency, cumulative percentages, score interval, grouped frequency, percentile rank, percentile

Symbols: X, f, i, UL, LL, PR, X_{PR}

Learning Objectives:

- Convert scores to frequencies—simple, cumulative, relative, cumulative relative, and grouped
- Display scores in frequency tables
- Find percentiles and percentile ranks from tabled data

Why Use Tables?

Assume that you just received the results of your first statistics test. You scored 87. To better judge whether that is a high or low performance, you probably want to know how the other students in your class did. How many other students also got an 87? How many did better? How many did worse? Assume that there are 50 students in your class. The scores of all the students are listed in Table 3.1. Your score of 87 is highlighted in boldface in the table.

> I am ill at these numbers.
>
> —William Shakespeare

How helpful is that information? Not very. The information you need is there, but it certainly isn't easy to find. The information you are seeking would be a lot easier to find if the scores were arranged in score order from highest to lowest. That way, you could get a better idea of where in the set of scores your own score falls. Table 3.2 gives the same list of scores arranged in descending order.

Table 3.1 Statistics Test Scores

92	83	98	76	84	**87**	66	82	93	85
86	**87**	56	94	89	73	80	82	85	88
80	84	72	88	79	92	89	84	**87**	85
73	84	96	88	62	75	79	**87**	91	90
90	86	85	89	76	69	83	84	86	94

Table 3.2 Statistics Test Scores in Descending Order

98	89	86	84	76
96	89	86	84	76
94	89	86	83	75
94	88	85	83	73
93	88	85	82	73
92	88	85	82	72
92	**87**	85	80	69
91	**87**	84	80	66
90	**87**	84	79	62
90	**87**	84	79	56

From this list of arranged scores, you can see that you did well, although not superior. When the scores are divided into fifths, as they are in this list, your score fell within the second highest fifth of the class. You can also see that several other students scored the same as you.

A simple list of scores in descending order is sufficient for analysis in this case, because with only 50 students, you are able to manually count the number of scores falling above and below yours, as well as the number scoring the same as you. But what if the class had a large number of students—say, 300? A manual count would be tedious. Even more tedious, let's change the example. Suppose the data were not statistics test scores but maze running times for 264 rats in a laboratory experiment. Or levels of depression for 532 clients given a new antidepressant drug. Or reaction times for 148 participants in a perceptual task. The point is, research often involves many cases—far too many to efficiently list every case. How can you examine a set of data when there are many cases?

Frequency Tables

For large data sets, it is more efficient to place the data in a frequency table. A **frequency** table lists each observed score, along with the number of cases falling at each score. Table 3.3 shows the same statistics test data arranged in a frequency table. The columns on the left show all possible values, even those that no one scored. The columns on the right show only those values that someone scored.

SOURCE: Peanuts: © United Feature Syndicate, Inc.

Table 3.3 Frequency for Statistics Test Scores

Score	*Frequency*	*Score*	*Frequency*	*Score*	*Frequency*
98	1	73	2	98	1
97	0	72	1	96	1
96	1	71	0	94	2
95	0	70	0	93	1
94	2	69	1	92	2
93	1	68	0	91	1
92	2	67	0	90	2
91	1	66	1	89	3
90	2	65	0	88	3
89	3	64	0	**87**	**4**
88	3	63	0	86	3
87	**4**	62	1	85	4
86	3	61	0	84	5
85	4	60	0	83	2
84	5	59	0	82	2
83	2	58	0	80	2
82	2	57	0	79	2
81	0	56	1	76	2
80	2			75	1
79	2			73	2
78	0			72	1
77	0			69	1
76	2			66	1
75	1			62	1
74	0			56	1

The frequency tables make it easy to see that exactly 4 people scored 87 (shown in boldface), which was your score. You can also see that the most frequent scores were in the mid- to high 80s.

However, if you wanted to know exactly how many students scored below or above you, you would still need to manually count the cases above or below yours. A more efficient table for figuring the number of scores above or below yours is a **cumulative frequency** table. A cumulative frequency table shows how many scores are at or below (or at or above) any given score.

To create a cumulative frequency table, start with the bottom score and add the number of additional cases at each next higher score. On the next page is a demonstration for 8 subjects' shoe sizes. The numbers in bold, taken from the frequency column, are added at each step. Thus, 1 person wears a Size 9 shoe; 1 + 4 people wear a Size 10 or below; 5 + 0 people wear a Size 11 or below; and so on.

Shoe Size	*Frequency*	*Cumulative Frequency*
13	1	7 + **1** = 8
12	2	5 + **2** = 7
11	0	5 + **0** = 5
10	4	1 + **4** = 5
9	1	= **1**

Alternatively, you can start with the top score and add the number of additional cases at each next lower score. The numbers in bold, taken from the frequency column, are added at each step:

Shoe Size	*Frequency*	*Cumulative Frequency*
13	1	= **1**
12	2	1 + **2** = 3
11	0	3 + **0** = 3
10	4	3 + **4** = 7
9	1	7 + **1** = 8

CHECK YOURSELF!

Assume that three people are added to the shoe size data. Two wear a Size 8, and 1 wears a Size 14. Construct a frequency table in descending score order, showing frequency and cumulative frequency.

Returning to the 50 statistics test scores, Table 3.4 (see page 26) presents the data showing cumulative frequency.

From the cumulative frequency table, you can see that exactly 34 students out of 50 scored the same or lower than you (shown in boldface). Thus, the remaining 16 students (50 – 34) scored higher than you.

PRACTICE

1. Forty college juniors were surveyed regarding the number of dates they had during the previous 30 days. Here are the data. Create a table, in descending score order, showing frequency and cumulative frequency.

   4, 3, 0, 0, 1, 0, 2, 5, 0, 3

   2, 0, 1, 0, 0, 3, 8, 0, 2, 2

   0, 6, 3, 2, 0, 0, 0, 2, 3, 0

   1, 0, 0, 1, 1, 3, 1, 2, 5, 4

2. A college instructor logs attendance throughout the semester for a class of 32 students that meets on M-W-F. Here is the number of classes each student missed during the

semester. Create a table, in descending score order, showing frequency and cumulative frequency.

0, 1, 4, 2, 8, 1, 2, 4

5, 0, 2, 2, 1, 2, 1, 2

1, 0, 3, 1, 1, 2, 1, 2

2, 4, 0, 1, 3, 0, 3, 3

3. Twenty pet store customers were asked how many pets they currently own. Here are the data. Create a table, in descending score order, showing frequency and cumulative frequency.

   3, 0, 1, 4, 3, 2, 2, 1, 3, 0, 2, 4, 5, 3, 2, 4, 7, 1, 1, 2

4. Twenty-five families were asked how many TV sets they currently have in their home. Here are the data. Create a table, in descending score order, showing frequency and cumulative frequency.

   2, 3, 1, 2, 3, 0, 2, 4, 1, 2, 4, 3, 2, 1, 1, 3, 0, 2, 1, 1, 2, 3, 2, 5, 2

Table 3.4 Cumulative Frequency for Statistics Test Scores

Score	*Frequency*	*Cumulative Frequency*
98	1	50
96	1	49
94	2	48
93	1	46
92	2	45
91	1	43
90	2	42
89	3	40
88	3	37
87	**4**	**34**
86	3	30
85	4	27
84	5	23
83	2	18
82	2	16
80	2	14
79	2	12
76	2	10
75	1	8
73	2	7
72	1	5
69	1	4
66	1	3
62	1	2
56	1	1

SOURCE: Dana Fradon, New Yorker Cartoon 5/17/1976 © Dana Fradon/Condé Nast Publications/www.cartoonbank.com

Relative Frequency or Percentage Tables

According to the scores in Table 3.4, exactly 4 students out of 50 scored the same on the statistics test as you—that is, 8 students out of 100, or 8%. Also, 34 students out of 50 scored the same as or lower than you—that is, 68 students out of 100, or 68%.

For this set of data, a frequency table is probably sufficient for interpreting your score because a sample size of 50 is easy to manipulate arithmetically. But what if your class had 27 students, or 68 students, or 112 students? Admittedly, those changes would still not pose severe calculation difficulty. However, interpreting the results would certainly take more time. It seems that a percentage column would be helpful in interpreting the data. Such a table is called a **relative frequency** or **percentage** table, because it gives each score's frequency relative to 100%. Because relative frequencies are percentages, they can take values only between 0 and 100.

To calculate the relative frequency for a score, divide the frequency of that score by the total number of scores and then multiply by 100. Recall that you did this in Module 1. To demonstrate that process, let's look again at the shoe size data. There are 8 subjects in the shoe size study. For each shoe size, divide the frequency of that shoe size by 8 to find the proportion of subjects with that shoe size. Then, multiply by 100 to get the relative frequency or percentage.

Shoe Size	*Frequency*	*Relative Frequency or Percentage*
13	1	/8 × 100 = 0.125 × 100 = 12.5
12	2	/8 × 100 = 0.250 × 100 = 25.0
11	0	/8 × 100 = 0.000 × 100 = 00.0
10	4	/8 × 100 = 0.500 × 100 = 50.0
9	1	/8 × 100 = 0.125 × 100 = 12.5

Using the same process, you can calculate the relative frequency for the 50 statistics test scores, as shown in Table 3.5. From this table, you can easily see that 8% of the students scored the same as you (shown in boldface).

Table 3.5 Relative Frequency for Statistics Test Scores

Score	*Frequency*	*Cumulative Frequency*	*Relative Frequency or Percentage*
98	1	50	2.0
96	1	49	2.0
94	2	48	4.0
93	1	46	2.0
92	2	45	4.0
91	1	43	2.0
90	2	42	4.0
89	3	40	6.0
88	3	37	6.0
87	**4**	**34**	**8.0**
86	3	30	6.0
85	4	27	8.0

(Continued)

Table 3.5 (Continued)

Score	*Frequency*	*Cumulative Frequency*	*Relative Frequency or Percentage*
84	5	23	10.0
83	2	18	4.0
82	2	16	4.0
80	2	14	4.0
79	2	12	4.0
76	2	10	4.0
75	1	8	2.0
73	2	7	4.0
72	1	5	2.0
69	1	4	2.0
66	1	3	2.0
62	1	2	2.0
56	1	1	2.0

As in a frequency table, sometimes it is even more helpful to know the percentage above or below a given score, rather than merely the percentage at that given score. Such a table is called a **cumulative relative frequency** or **cumulative percentage** table. For that, we must add yet another column.

To create a cumulative relative frequency table, start at the bottom (or top) and add the percentage at each next higher (or lower) score. The process is the same as for relative frequencies. The statistics test data showing cumulative relative frequency are given in Table 3.6. Note that, because all cases are included, the percentages must add to 100%.

Table 3.6 Cumulative Relative Frequency for Statistics Test Scores

Score	*Frequency*	*Cumulative Frequency*	*Relative Frequency or Percentage*	*Cumulative Relative Frequency or Cumulative Percentage*
98	1	50	2.0	100.0
96	1	49	2.0	98.0
94	2	48	4.0	96.0
93	1	46	2.0	92.0
92	2	45	4.0	90.0
91	1	43	2.0	86.0
90	2	42	4.0	84.0
89	3	40	6.0	80.0
88	3	37	6.0	74.0
87	**4**	**34**	**8.0**	**68.0**
86	3	30	6.0	60.0
85	4	27	8.0	54.0
84	5	23	10.0	46.0
83	2	18	4.0	36.0
82	2	16	4.0	32.0
80	2	14	4.0	28.0
79	2	12	4.0	24.0

Score	*Frequency*	*Cumulative Frequency*	*Relative Frequency or Percentage*	*Cumulative Relative Frequency or Cumulative Percentage*
76	2	10	4.0	20.0
75	1	8	2.0	16.0
73	2	7	4.0	14.0
72	1	5	2.0	10.0
69	1	4	2.0	8.0
66	1	3	2.0	6.0
62	1	2	2.0	4.0
56	1	1	2.0	2.0

From Table 3.6, you can see that 68% of the students scored the same as or lower than you (shown in boldface). Thus, 32% (100% – 68%) must have scored higher than you.

CHECK YOURSELF!

In the previous *Check Yourself!* box, 3 people were added to the shoe size data: 2 wearing a size 8 and 1 wearing a size 14. You constructed a table showing frequency and cumulative frequency. Now add columns for relative frequency and cumulative relative frequency.

PRACTICE

5. Return to Exercise 1 from earlier in this module. Add columns for relative frequency (percentage) and cumulative relative frequency (cumulative percentage).
6. Return to Exercise 2 from earlier in this module. Add columns for relative frequency (percentage) and cumulative relative frequency (cumulative percentage).
7. Return to Exercise 3 earlier in this module. Add columns for relative frequency (percentage) and cumulative relative frequency (cumulative percentage).
8. Return to Exercise 4 earlier in this module. Add columns for relative frequency (percentage) and cumulative relative frequency (cumulative percentage).

Grouped Frequency Tables

With large numbers of scores, it is usually preferable to group the data into **score intervals.** Although this does not allow you to focus on a particular score, grouping the data often gives a clearer sense of the underlying trend in the data than does inspection of individual scores.

Here are the steps in creating a **grouped frequency** table:

1. Select the size of the score interval into which you want to group the data—for example, 2 points, 5 points, 10 points.
2. Divide the entire range of scores into intervals of the selected size.
3. Count and record the number of scores in each interval.

Note that each interval will be the same size, and each score can fall in one and only one interval. That is, there can be no overlapping intervals.

When should you group data? How wide should the intervals be? There is no hard-and-fast rule for when to group data or for how many groups to form. The guiding principle is clarity.

Let's group the statistics score data to see whether it helps or hinders data interpretation. In Table 3.7, the data on the right are grouped into 5-point intervals. In Table 3.8, the data on the right are grouped into 20-point intervals. In both tables, the initial data are on the left, with breaks indicating which scores were grouped.

Table 3.7 Grouped Frequency for Statistics Scores: 5-Point Intervals

		5-Point Intervals	
Score	*Frequency*	*Score*	*Frequency*
98	1	95–99	2
96	1	90–94	8
94	2	85–89	17
93	1	80–84	11
92	2	75–79	5
91	1	70–74	3
90	2	65–69	2
89	3	60–64	1
88	3	55–59	1
87	4		
86	3		
85	4		
84	5		
83	2		
82	2		
80	2		
79	2		
76	2		
75	1		
73	2		
72	1		
69	1		
66	1		
62	1		
56	1		

Notice that you cannot see the trend in the data in the initial frequency tables to the left because there are too many intervals to compare and too few cases in each interval. This is a good reason for grouping the data. However, the trend in the data is again lost in the 20-point interval table because there are too few intervals to compare and too many scores

Table 3.8 Grouped Frequency for Statistics Scores: 20-Point Intervals

		20-Point Intervals	
Score	*Frequency*	*Score*	*Frequency*
98	1	85–104	27
96	1	65–84	21
94	2	45–64	2
93	1		
92	2		
91	1		
90	2		
89	3		
88	3		
87	4		
86	3		
85	4		
84	5		
83	2		
82	2		
80	2		
79	2		
76	2		
75	1		
73	2		
72	1		
69	1		
66	1		
62	1		
56	1		

in each interval. The trend is readily apparent in the 5-point interval table. In general, interval widths that yield 5 to 12 groups are the best. However, you reach clarity for a specific set of data only by trial and error.

CHECK YOURSELF!

Why would it not be useful to group the shoe size data presented earlier in this module?

PRACTICE

9. The following 36 scores are the distances (to the nearest foot) that middle school boys threw a discus in a recent cross-country meet:

   69, 108, 70, 59, 86, 67, 54, 82, 95

   56, 74, 66, 74, 86, 82, 52, 87, 90

   89, 60, 83, 50, 76, 86, 101, 93, 98

   102, 79, 63, 87, 88, 93, 58, 83, 65

a. Create a grouped frequency table in 5-ft intervals.
b. Create a grouped frequency table in 10-ft intervals.
c. Create a grouped frequency table in 20-ft intervals.
d. Which table gives the clearest picture of the data?

10. Students in an experimental psychology lab time how long it takes rats to run a maze. Ten students run the same 20 rats. Here is the average running time (to the nearest second) for each rat:

Rat #	*Seconds*	*Rat #*	*Seconds*	*Rat #*	*Seconds*	*Rat #*	*Seconds*
1	32	6	106	11	78	16	38
2	25	7	46	12	112	17	68
3	67	8	66	13	63	18	36
4	45	9	52	14	76	19	96
5	64	10	43	15	45	20	103

a. Create a grouped frequency table in 10-s intervals.
b. Create a grouped frequency table in 15-s intervals.
c. Create a grouped frequency table in 20-s intervals.
d. Which table gives the clearest picture of the data?

11. Thirty freshmen living on-campus were asked how many times they skipped breakfast during the previous month. A check of their meal card swipes reveals the following number of skips:

2, 4, 0, 7, 5, 8, 4, 4, 12, 4, 6, 1, 6, 4, 3, 3, 6, 7, 15, 5, 6, 2, 8, 10, 4, 8, 5, 5, 1, 6

a. Create a grouped frequency table in 2-meal intervals.
b. Create a grouped frequency table in 5-meal intervals.
c. Create a grouped frequency table in 10-meal intervals.
d. Which table gives the clearest picture of the data?

12. Thirty computer users were asked to record the number of spam e-mails they received in the past week. After a week, they reported the following data:

24, 3, 18, 27, 13, 5, 2, 10, 7, 21, 5, 3, 1, 0, 23, 15, 4, 3, 9, 4, 2, 15, 6, 2, 17, 8, 4, 5, 17, 2

a. Create a grouped frequency table in 2-spam intervals.
b. Create a grouped frequency table in 5-spam intervals.
c. Create a grouped frequency table in 10-spam intervals.
d. Which table gives the clearest picture of the data?

Percentile and Percentile Rank Tables

When cumulative relative frequency is presented in descending order, as it is in Table 3.9, it is often called a **percentile rank** table because it indicates the percentage of cases falling at or below a given score. Note that a percentile rank, symbolized PR, does not indicate the percentage of cases falling below a given score. Rather, it indicates the percentage of cases falling *at or below* a given score.

Table 3.9 Cumulative Relative Frequency for Statistics Test Scores

Score	*Frequency*	*Cumulative Frequency*	*Relative Frequency or Percentage*	*Cumulative Relative Frequency or Percentile Rank*
98	1	50	2.0	100.0
96	1	49	2.0	98.0
94	2	48	4.0	96.0
93	1	46	2.0	92.0
92	2	45	4.0	90.0
91	1	43	2.0	86.0
90	2	42	4.0	84.0
89	3	40	6.0	80.0
88	3	37	6.0	74.0
87	**4**	**34**	**8.0**	**68.0**
86	3	30	6.0	60.0
85	4	27	8.0	54.0
84	5	23	10.0	46.0
83	2	18	4.0	36.0
82	2	16	4.0	32.0
80	2	14	4.0	28.0
79	2	12	4.0	24.0
76	2	10	4.0	20.0
75	1	8	2.0	16.0
73	2	7	4.0	14.0
72	1	5	2.0	10.0
69	1	4	2.0	8.0
66	1	3	2.0	6.0
62	1	2	2.0	4.0
56	1	1	2.0	2.0

Unfortunately, determination of percentile rank is not as simple as reading a number from a table. For "ballpark" purposes, the direct table reading works fine: The percentile rank for your score of 87 is about 68. However, 4 students scored 87. And recall from Module 2 that the true limits for any score extend from one-half unit of measurement below the score to one-half unit of measurement above the score. Therefore, to determine the exact percentile rank for your score, you must distribute the 4 students at your score between the upper real limit (UL) and lower real limit (LL) of that score.

> I am no prophet. I but calculate.

Calculation of the exact percentile rank for a score requires a formula. Here is the formula for percentile rank:

$$PR = \left(\frac{\text{cum } f_{LL} + (f_i/i)(X - LL)}{N}\right)100$$

where

cum f_{LL} = cumulative frequency of scores below the lower real limit of the interval containing the score X,

f_i = frequency of the interval containing the score X,

i = width of the interval,

X = score whose percentile rank is being determined,

LL = lower real limit of the interval containing the score X, and

N = total number of scores.

First, we must figure the real limits of the score containing the score X. The unit of measurement is 1 point; that is, scores are expressed in 1-point intervals. Therefore, the real limits for your score of 87 are 86.5 and 87.5. The real limits are shown in the relevant portion of Table 3.10. The row highlighted in boldface shows the interval containing your score.

Table 3.10 Real Limits of Score X

Score Interval	*Frequency*	*Cumulative Frequency*	*Percentage*	*Cumulative Percentage*
88.5–89.5	3	40	6.0	80.0
87.5–88.5	3	37	6.0	74.0
86.5–87.5	**4**	**34**	**8.0**	**68.0**
85.5–86.5	3	30	6.0	60.0
84.5–85.5	4	27	6.0	54.0

Taking information from the table, we can now find the percentile rank for your score of 87:

$$\begin{aligned}
\text{PR} &= \left(\frac{\text{cum } f_{\text{LL}} + (f_i/i)(X - \text{LL})}{N}\right)100 \\
&= \left(\frac{30 + (4/1)(87 - 86.5)}{50}\right)100 \\
&= \left(\frac{30 + (4)(0.5)}{50}\right)100 \\
&= \left(\frac{30 + 2}{50}\right)100 \\
&= \left(\frac{32}{50}\right)100 \\
&= (0.64)100 \\
&= 64
\end{aligned}$$

Note the difference between the ballpark percentile rank determined by reading from the table and the exact percentile rank determined by the formula. The ballpark percentile rank via the table was 68, but the percentile rank via the formula is 64. Why are they different? Which is correct?

Recall that the score of 87 includes all scores between 86.5 and 87.0 and between 87.0 and 87.5, although no such scores exist. If we assume that the four scores are evenly distributed between the upper and lower real limits of the score, then two of the four scores fall between 86.5 and 87.0, and the other two fall between 87.0 and 87.5. A score of exactly 87 falls halfway through the interval. According to the table excerpt, 60% of the scores fall below the interval, and 8% fall within the interval. Thus, 60% plus one half of the 8% within the interval equals 64%. So 64 is the correct percentile rank.

We could also ask what score falls at a given percentile rank. The score falling at a particular percentile rank is called a **percentile**, symbolized X_{PR}. The subscript indicates the particular percentile rank sought. Thus, X_{50} indicates the score at the 50th percentile rank, X_{25} indicates the score at the 25th percentile rank, and so on. Do not confuse percentiles and percentile ranks. Percentile ranks are percentages and can take values only between 0 and 100. Percentiles are scores and can take any value that the scores take.

For example, we could ask what score falls at the 64th percentile rank. What would that score be? By reading the table, we would say 86.5, because 64% falls halfway between 68% and 60% and, therefore, 86.5 falls halfway between the corresponding scores of 87 and 86.

89	3	40	6.0	80.0
88	3	37	6.0	74.0
87	4	34	8.0	**68.0**
86	3	30	6.0	**60.0**
85	4	27	4.0	54.0

By calculation using the formula, however, it will be exactly 87. Let's see how it works. The formula for finding a percentile is

$$X_{PR} = \text{LL} + (i/f_i)(\text{cum } f_{UL} - \text{cum } f_{LL})(0.5)$$

where

LL = lower real limit of the interval containing the percentile point,

i = width of the interval,

f_i = frequency of the interval containing the percentile point,

cum f_{UL} = cumulative frequency of scores below the UL of the interval containing the percentile point, and

cum f_{LL} = cumulative frequency of scores below the LL of the interval containing the percentile point.

Again, the real limits of the relevant portion of the table are given in Table 3.11. The row highlighted in boldface shows the score interval containing the percentile we are looking for.

Table 3.11 Selected Portion of Cumulative Relative Frequency Table

Score Interval	*Frequency*	*Cumulative Frequency*	*Percentage*	*Cumulative Percentage*
88.5–89.5	3	40	6.0	80.0
87.5–88.5	3	37	6.0	74.0
86.5–87.5	**4**	**34**	**8.0**	**68.0**
85.5–86.5	3	30	6.0	60.0
84.5–85.5	4	27	8.0	54.0

Taking information from Table 3.11, we can solve the formula for the percentile (score) for the 64th percentile rank:

$$\begin{aligned} X_{PR} &= LL + (i/f_i)(\text{cum } f_{UL} - \text{cum } f_{LL})(0.5) \\ &= 86.5 + (1/4)(34 - 30)(0.5) \\ &= 86.5 + (1/4)(4)(0.5) \\ &= 86.5 + (1/4)(2) \\ &= 86.5 + 0.5 \\ &= 87.0 \end{aligned}$$

Although percentile ranks and percentiles are somewhat cumbersome to compute, they are easy to interpret. For that reason, researchers commonly use them to communicate relative performance to the lay public. Throughout school, many of your standardized test scores were probably reported to you as percentile ranks. Conversely, percentiles (scores) were given for various percentile ranks.

Percentile ranks are useful for displaying data during the early stages of research. They give the researcher an initial picture of the data. However, they are seldom used in later analysis of research data because they do not easily answer the types of questions that much research asks. Therefore, you will not see percentile ranks again in this textbook.

PRACTICE

13. Select a score from Table 3.6. Using the appropriate formula, determine that score's percentile rank.
14. In Table 3.6, what exact score falls at the 75th percentile rank? Use the appropriate formula.
15. In Table 3.6, what exact score falls at the 25th percentile rank? Use the appropriate formula.
16. Select a score from Table 3.6. Using the appropriate formula, determine that score's percentile rank.

SPSS Connection

Download the file **data_statistic test scores.sav** from www.sagepub.com/steinberg2e. These data are used in the textbook example.

Alternatively, manually enter the data from Table 3.1 into the SPSS Data View spreadsheet and define the variables in the SPSS Variable View spreadsheet. Call the variable **statscor**, set the decimals at 0, and label the variable as **Statistics Test Score.**

If the file is not already in **Data View**, click that tab in the lower left portion of the screen.

On the tool bar on the top of the screen, click **Analyze**, then **Descriptive Statistics**, then **Frequencies.**

Highlight the variable **Statistics Test Score** in the left window and click on the **arrow** between the windows to send that variable into the right window.

Click the **Format** button, then click **Descending Values** (the default is Ascending Values).

Click **OK.** This is what you will see.

Frequencies

Statistics

Statistics Test Score

N	Valid	50
	Missing	0

Statistics Test Score

		Frequency	Percent	Valid Percent	Cumulative Percent
Valid	98	1	2.0	2.0	2.0
	96	1	2.0	2.0	4.0
	94	2	4.0	4.0	8.0
	93	1	2.0	2.0	10.0
	92	2	4.0	4.0	14.0
	91	1	2.0	2.0	16.0
	90	2	4.0	4.0	20.0
	89	3	6.0	6.0	26.0
	88	3	6.0	6.0	32.0
	87	4	8.0	8.0	40.0
	86	3	6.0	6.0	46.0
	85	4	8.0	8.0	54.0
	84	5	10.0	10.0	64.0
	83	2	4.0	4.0	68.0
	82	2	4.0	4.0	72.0
	80	2	4.0	4.0	76.0
	79	2	4.0	4.0	80.0
	76	2	4.0	4.0	84.0
	75	1	2.0	2.0	86.0
	73	2	4.0	4.0	90.0
	72	1	2.0	2.0	92.0
	69	1	2.0	2.0	94.0
	66	1	2.0	2.0	96.0
	62	1	2.0	2.0	98.0
	56	1	2.0	2.0	100.0
	Total	50	100.0	100.0	

Compare this output with the frequency tables in the textbook. It agrees with the textbook tables except that the final cumulative percentage column is ascending despite the descending scores in the first column. Recall that cumulative percentage (cumulative relative frequency) is synonymous with percentile rank. When cumulative percentage is displayed in descending order, as it is in the textbook and for most educational and psychological uses, the concordance between cumulative percentage and percentile rank is obvious. Unfortunately, that is not the case in SPSS. It certainly is not the case, for example, that 2% of the scores fall at or below a score 98!

Repeat the above steps, but change the default back to **Ascending Values**. After clicking **OK**, you will see this.

Frequencies

Statistics

Statistics Test Score

N	Valid	50
	Missing	0

Statistics Test Score

Valid	Frequency	Percent	Valid Percent	Cumulative Percent
56	1	2.0	2.0	2.0
62	1	2.0	2.0	4.0
66	1	2.0	2.0	6.0
69	1	2.0	2.0	8.0
72	1	2.0	2.0	10.0
73	2	4.0	4.0	14.0
75	1	2.0	2.0	16.0
76	2	4.0	4.0	20.0
79	2	4.0	4.0	24.0
80	2	4.0	4.0	28.0
82	2	4.0	4.0	32.0
83	2	4.0	4.0	36.0
84	5	10.0	10.0	46.0
85	4	8.0	8.0	54.0
86	3	6.0	6.0	60.0
87	4	8.0	8.0	68.0
88	3	6.0	6.0	74.0
89	3	6.0	6.0	80.0
90	2	4.0	4.0	84.0
91	1	2.0	2.0	86.0
92	2	4.0	4.0	90.0
93	1	2.0	2.0	92.0
94	2	4.0	4.0	96.0
96	1	2.0	2.0	98.0
98	1	2.0	2.0	100.0
Total	50	100.0	100.0	

Now the scores to the left are in ascending order. Unfortunately, the order of the Cumulative Percent column to the right never changes. This is a quirk in SPSS. Nevertheless, you can find Percentile Rank through this Ascending Values format. Simply read from the bottom of the table rather than from the top. Now you can see that 100% of the scores fall at or below a score of 98, 98% fall at or below a score of 96, and so on.

Visit the study site at www.sagepub.com/steinberg2e for practice quizzes and other study resources.

4 Graphs and Plots

Terms: stem-and-leaf, *X*-axis, abscissa, *Y*-axis, ordinate, histogram, continuous, frequency curve, normal curve, symmetric, skew, negative skew, positive skew, kurtosis, leptokurtic, platykurtic, bimodal, rectangular, bar graph, pie graph

Symbols: X, Y

Learning Objectives:

- Distinguish between normally distributed and nonnormally distributed data
- Select the best type of graph for data of a given scale
- Construct various types of graphs from data
- Apply graphing conventions so as not to distort the data

Why Use Graphs?

In Module 3, we were working with tabular data. Recall that there were 50 scores on a statistics test and that you scored 87. We had put the data into various frequency tables. As we saw in that module, tables are useful for interpreting data. Often, however, a graph provides a more understandable picture of the same data. In addition to providing numbers, graphs provide an "eyeball" test of the data. What is their shape—are they evenly distributed, lumpy, or lopsided? How spread out are the data—are they compact or widely dispersed? Where are most of the data—are they in one place or in more than one place? Are they generally high or low? Each of these aspects of the data is more visible in a graph than in a table.

> The purpose of models is not to fit the data but to sharpen the questions.
>
> —Samuel Karlin

Graphing Continuous Data

Before we can graph continuous data, we must list the data in score order. For many types of graphs, we must also first create frequency tables. Table 4.1 shows two previous ways we had arranged the 50 statistics test score data in Module 3.

Table 4.1 Ungrouped and Grouped Statistics Test Scores

		5-Point Intervals	
Score	*Frequency*	*Score*	*Frequency*
98	1	95–99	2
96	1	90–94	8
94	2	85–89	17
93	1	80–84	11
92	2	75–79	5
91	1	70–74	3
90	2	65–69	2
89	3	60–64	1
88	3	55–59	1
87	4		
86	3		
85	4		
84	5		
83	2		
82	2		
80	2		
79	2		
76	2		
75	1		
73	2		
72	1		
69	1		
66	1		
62	1		
56	1		

Stem-and-Leaf Displays

One type of graph that is a hybrid between a table and a graph is called a **stem-and-leaf** display. The column on the left indicates the first digit of a score—in this case, the tens place. This is the stem. The column to the right lists every instance of the next digit—in this case, the ones place. This is the leaf. To construct a stem-and-leaf display, you must first arrange the data in numerical order, as it appears in Table 4.1. By convention, the data in a stem-and-leaf display are displayed with the lowest scores on top. Table 4.2 shows a stem-and-leaf display for the 50 statistics test scores.

The first line of the display tells us that there was one score of 56. The second line tells us that there was one score of 62, one score of 66, and one score of 69. Note that in the third line, there was one score of 72 and 75 but two scores of 73, 76, and 79. The advantage of a stem-and-leaf display over a frequency table is that the shape of the distribution of

Table 4.2 Stem-and-Leaf Display for Statistics Test Scores

Stem	*Leaf*
5	6
6	2 6 9
7	2 3 3 5 6 6 9 9
8	0 0 2 2 3 3 4 4 4 4 4 5 5 5 5 6 6 6 7 7 7 7 8 8 8 9 9 9
9	0 0 1 2 2 3 4 4 6 8

scores is immediately apparent. We can see that there were very few low scores and also that scores were concentrated in the 80s. We can also see that the shape of the distribution is lopsided. We will meet a more technical term for this lopsidedness shortly.

When the number of scores is large, as it is in this example for scores between 80 and 89, it is acceptable to display each tens place twice and break the ones place into lower and higher digits. Note that you must do this for each tens place value and not just for the tens place where a large number of cases lie (in this case, the 8). Doubling rows for some tens place values and not others distorts the shape of the data. Table 4.3 shows the same data with the tens place properly doubled. For each tens place, the first row contains ones values of 0 to 4, and the second row contains ones value of 5 to 9.

Table 4.3 Stem-and-Leaf Display With Tens Place Doubled

Stem	*Leaf*
5	
5	6
6	2
6	6 9
7	2 3 3
7	5 6 6 9 9
8	0 0 2 2 3 3 4 4 4 4 4
8	5 5 5 5 6 6 6 6 7 7 7 7 8 8 8 9 9 9
9	0 0 1 2 2 3 4 4
9	6 8

This gives an even clearer picture of the data. Now we can see that the most frequent scores were not just anywhere in the 80s but in the high 80s. The distribution is still obviously lopsided, although less so.

Principles of Graph Construction

Although frequencies are not explicitly listed in a stem-and-leaf display, we get a visual sense of frequency via the length of the score lists from left to right. Most graphs, however, depict frequency via vertical height. If you rotate the textbook so that Tables 4.2 and 4.3 have their stems along the bottom of the page, you will see the vertical height of the data.

In a vertical graph, the horizontal axis, commonly called the ***X*-axis**, is the **abscissa**. The vertical axis, commonly called the ***Y*-axis**, is the **ordinate**. By convention, graphs indicate intervals of the measured variable on the *X*-axis and the frequency or proportion of cases on the *Y*-axis (see Figure 4.1).

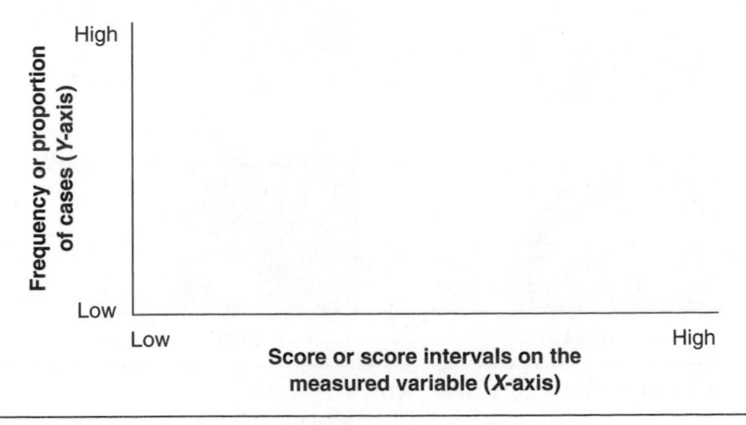

Figure 4.1 Template for a Vertical Graph

If the number of values on the *X*-axis is large, it is acceptable to group individual scores into score intervals. You did this in Module 3 when you created grouped frequency distributions. There is no single graph that is best for tabular data. However, the height, shape, and spread of any particular set of data look different depending on the number and width of the intervals on the *X*-axis and the relative height and width of the axes.

Figure 4.2 shows three graphs of the Scholastic Aptitude Test (SAT) scores of entering freshmen at BeachTown University over a 10-year period. Note how different the same data look in the three graphs. In the graphs in Figures 4.2a and b, the *X*-axis is the same, but the *Y*-axis differs. In the graphs in Figures 4.2b and c, the *Y*-axis is the same, but the *X*-axis differs. These differences in the *X*-axis and *Y*-axis make the data appear quite different in the three graphs. In Figure 4.2a, there appears to be little change in SAT scores. In contrast, in Figure 4.2b, there appears to be a steady and large increase in SAT scores. In Figure 4.2c, there appears to be a sudden and large increase in SAT scores.

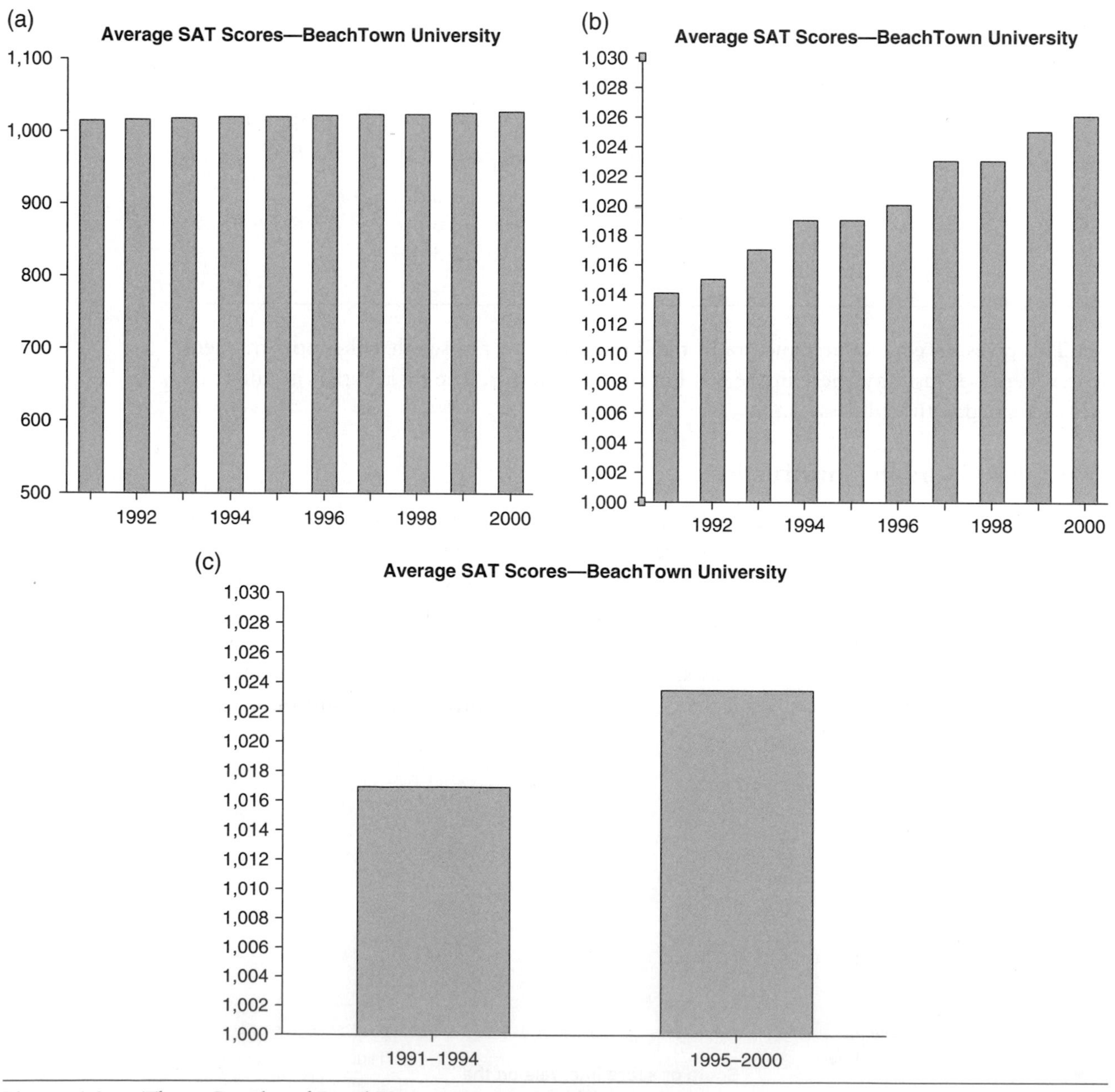

Figure 4.2 Three Graphs of BeachTown University SAT Scores

In order not to distort data, there are conventions for constructing graphs. These conventions are as follows:

1. The *Y*-axis should be about three fourths as tall as the *X*-axis is long.
2. When the number of score values on the *X*-axis is large, scores should be collapsed so that there are at least 5 intervals but no more than 12.
3. The width of each interval on the *X*-axis must be equal.
4. Frequency on the *Y*-axis must be continuous and regular.
5. Range on the *Y*-axis and *X*-axis must neither unduly compress nor unduly stretch the data.

Figure 4.3 shows BeachTown University's SAT data again, this time following graphing conventions. The increase in SAT scores is now appropriately conveyed as steady but modest.

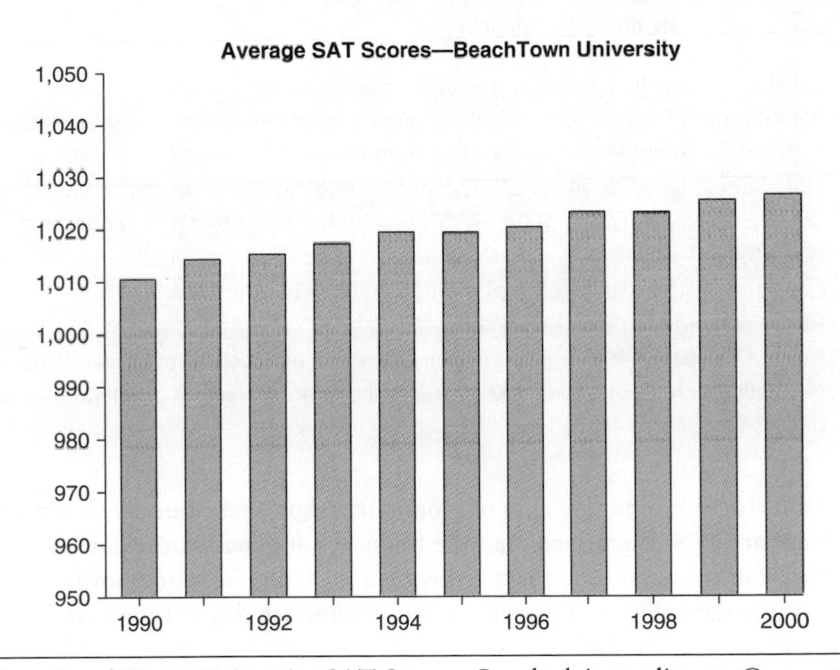

Figure 4.3 BeachTown University SAT Scores: Graphed According to Convention

Histogram

One type of graph for continuous data is the histogram. In a **histogram**, frequency is indicated by the height of the bar. The midpoint of the bar falls directly over the score value.

The bars of a histogram are connected rather than separate. The connection indicates that the data on the *X*-axis are **continuous**—that is, scores could theoretically fall anywhere between adjacent units. Ratio measures such as weight, distance, and time are continuous variables. Many interval measures, such as test scores, are also treated as continuous variables. Although people do not typically score fractional points, they could, if the test were so graded.

✓ CHECK YOURSELF!

The example below uses a simple set of data to illustrate the steps in constructing a histogram. Following the instructions, use the template in Figure 4.4 to draw the histogram for the data.

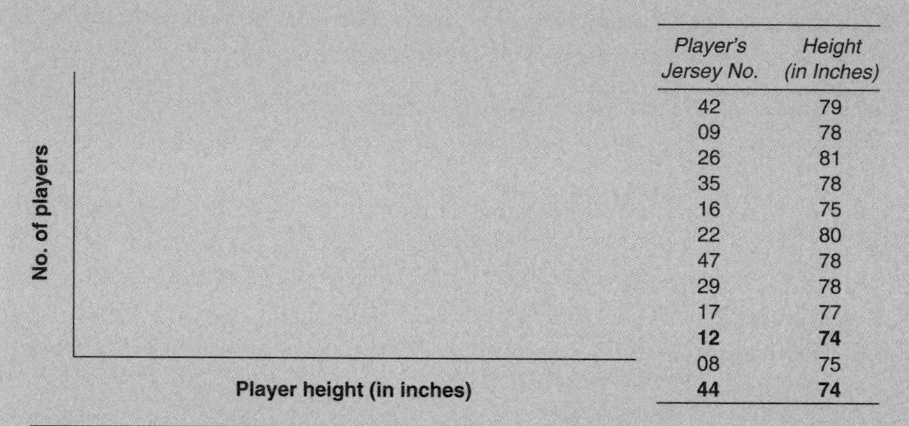

Player's Jersey No.	*Height (in Inches)*
42	79
09	78
26	81
35	78
16	75
22	80
47	78
29	78
17	77
12	**74**
08	75
44	**74**

Figure 4.4 Template for Creating a Histogram for Height of Basketball Team Members (in Inches)

1. Label the *X*-axis and *Y*-axis at appropriate intervals for the data. For these data, player height ranges from 74 to 81 in. and frequency at any one player height ranges from 0 to 4. Therefore, ranges just slightly larger than these are appropriate for each axis.
2. The shortest player is 74 in. tall. Two players (highlighted in the table above) are 74 in. tall. Draw a vertical bar over the 74-in. point on the *X*-axis to a point equal to a frequency of 2 on the *Y*-axis.
3. Continue this process for each player height. Note that no player is 76 in. tall; therefore, there will be no bar over that height.
4. Expand the bars left and right to their lower real limits and upper real limits. Adjacent bars will touch. Frequencies of 0 appear as a break between bars when a bar is missing (in this case, at 76 in.).

To read a histogram, find the score or score interval of interest on the *X*-axis, follow the corresponding bar to its height, and read the frequency for that bar height on the *Y*-axis.

Now that you know how to create a histogram, here is a histogram for the previous statistics test score data (Figure 4.5), with scores grouped in 5-point intervals.

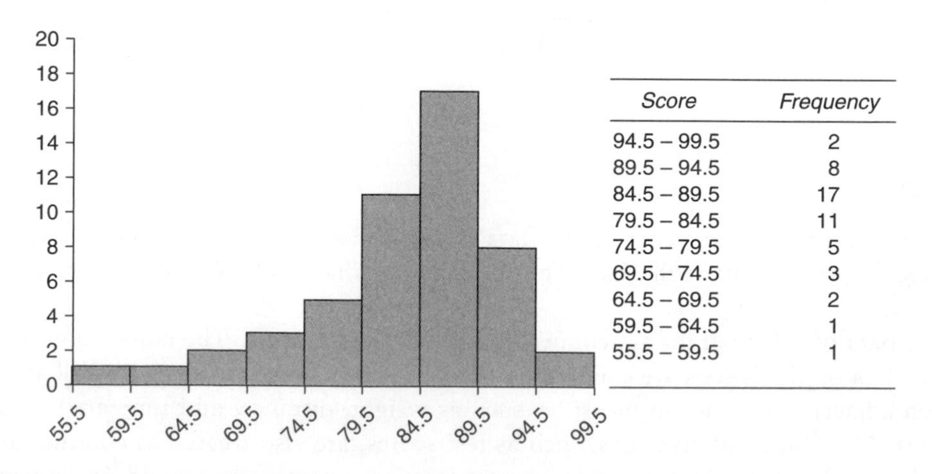

Score	*Frequency*
94.5 – 99.5	2
89.5 – 94.5	8
84.5 – 89.5	17
79.5 – 84.5	11
74.5 – 79.5	5
69.5 – 74.5	3
64.5 – 69.5	2
59.5 – 64.5	1
55.5 – 59.5	1

Figure 4.5 Histogram for Statistics Test Scores in 5-Point Intervals

Now here are the same data, with scores grouped in 10-point intervals (Figure 4.6). Note the need to increase the width of the intervals on the *Y*-axis due to the increased frequency in the broader score intervals on the *X*-axis.

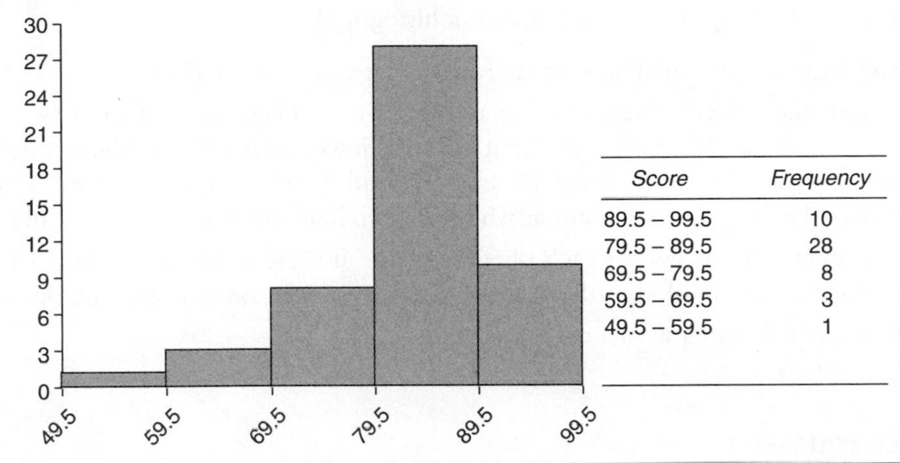

Score	*Frequency*
89.5 – 99.5	10
79.5 – 89.5	28
69.5 – 79.5	8
59.5 – 69.5	3
49.5 – 59.5	1

Figure 4.6 Histogram for Statistics Test Scores in 10-Point Intervals

Recall that you scored 87. On the 5-point interval histogram, 17 students scored in the same interval as you. On the 10-point interval histogram, 28 students scored in the same interval as you.

Both histograms show us that the data are lopsided, with more high scores than low scores, and with the most frequent scores in the 80s. The 5-point histogram is clearer than the 10-point histogram, however: It tells us that the most frequent scores are not just anywhere in the 80s but in the high 80s.

Frequency Curve or Line Graph

Suppose we connect the bars of the histogram at their midpoints and then eliminate the bars. This would give us another type of graph—a **frequency curve or line graph**. Here are the statistics test score data again, first with the histogram midpoints connected and then with the histogram bars eliminated (Figure 4.7).

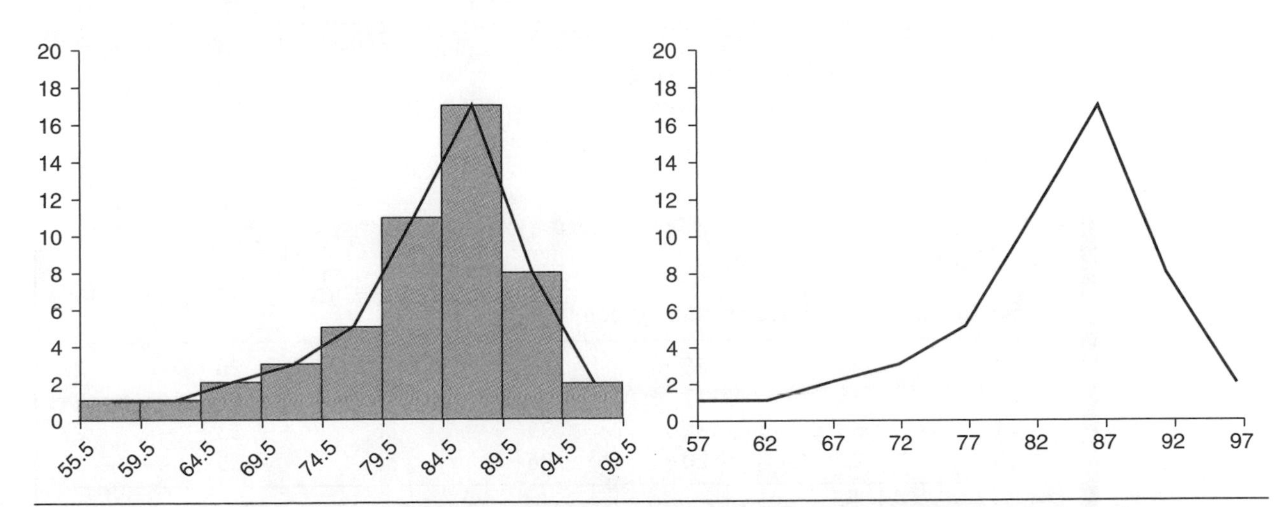

Figure 4.7 Statistics Score Histogram With Frequency Curve Overlay and With Overlay Alone

Frequency curves are commonly used for continuous score data, and they are easy to interpret. The height of the curve is the frequency or proportion. To interpret the curve, find the point on the curve directly above the score of interest, then read the height for that point on the *Y*-axis. For the score interval in which your score of 87 falls, the frequency is the same as in the histogram: It is 17.

To construct a frequency curve (without a histogram):

1. Label the *X*-axis and *Y*-axis in appropriate intervals for the data.
2. Locate the lowest obtained score on the *X*-axis. (This point should be midway between the score's real upper limit and real lower limit.) Locate the frequency for that score on the *Y*-axis. Mentally draw perpendicular lines from the axes to the interior of the graph. At the point at which the two lines intersect, place a small mark.
3. Continue this process for each obtained score. If there is no subject at a particular score, place the mark directly on the *X*-axis to represent no frequency on the *Y*-axis.
4. Connect the marks with a continuous line.

CHECK YOURSELF!

Here, again, are the basketball player data. Use the template in Figure 4.8 to create a frequency curve for the data.

Player's Jersey No.	*Height (in Inches)*
42	79
09	78
26	81
35	78
16	75
22	80
47	78
29	78
17	77
12	74
08	75
44	74

No. of players

Player height (in inches)

Figure 4.8 Template for Creating a Frequency Curve for Height of Basketball Team Members (in Inches)

PRACTICE

1. For the following 28 scores on a 40-point test,

27	25	25	23	32	26	27
24	25	17	35	34	37	34
21	38	31	39	26	33	27
36	27	20	36	30	35	35

   a. construct a stem-and-leaf display,

   b. construct a histogram using 4-point score intervals, and

   c. draw a frequency curve using 4-point score intervals.

2. Here are the Fahrenheit temperatures (to the nearest degree) at 3:00 p.m. each day during June for a town in the western part of the United States:

Date	*Temperature*	*Date*	*Temperature*	*Date*	*Temperature*
June 1	72	June 3	78	June 5	60
June 2	74	June 4	66	June 6	72
June 7	79	June 15	60	June 23	84
June 8	85	June 16	60	June 24	87
June 9	86	June 17	71	June 25	90
June 10	88	June 18	76	June 26	91
June 11	88	June 19	76	June 27	84
June 12	73	June 20	63	June 28	78
June 13	62	June 21	63	June 29	78
June 14	58	June 22	78	June 30	67

a. Construct a stem-and-leaf display.
b. Construct a histogram using 3-point score intervals.
c. Draw a frequency curve using 3-point score intervals.

3. For the following number of college credits being taken by 20 members of a college fraternity,

15, 12, 17, 13, 9, 12, 18, 15, 15, 14, 16, 15, 15, 12, 13, 13, 15, 17, 13, 21

a. construct a stem and leaf display
b. construct a histogram using 2-point intervals
c. draw a frequency curve using 2-point intervals

4. Here are the number of pounds lost in a 3-month period by 25 members of a weight-loss club

5, 7, 12, 4, 6, 11, 3, 8, 14, 10, 24, 8, 6, 15, 8, 12, 7, 11, 13, 12, 9, 4, 5, 18, 22

a. construct a stem and leaf display
b. construct a histogram using 3-point intervals
c. draw a frequency curve using 3-point intervals

Symmetry, Skew, and Kurtosis

As mentioned earlier in this module, one of the advantages of displaying data in graphs rather than in tables is that the shape and concentration of the data are more readily apparent. This is especially the case with frequency curves.

One type of frequency curve is the normal curve. A **normal curve** is a bell-shaped curve having special properties. The special properties of the normal curve make it very important in statistical analysis; therefore, I will have much more to say about the properties of a normal curve in future modules. For now, we will focus on its shape and symmetry. Examine the normal curve in Figure 4.9.

Like any other frequency curve, score values are located along the *X*-axis, and the frequencies of those scores are located along the *Y*-axis. In a normal curve, the score with the greatest frequency is the middle score. The frequencies of higher and lower scores are progressively lower as the scores are farther and farther from the middle. A normal curve is also perfectly **symmetric**: The frequency of a score at any given distance above the middle score is

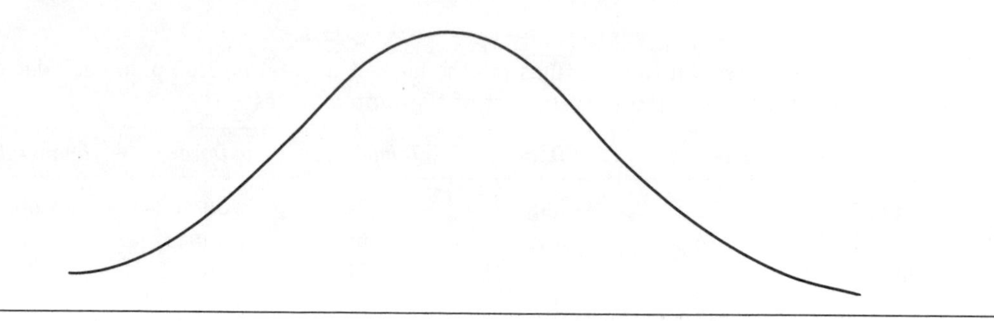

Figure 4.9 Normal Curve

exactly the same as the frequency of the corresponding score the same distance below the middle score. Thus, if we folded it in half, the right half would exactly overlap the left half.

Sometimes a distribution of scores is not symmetric but is instead lopsided: There are more high scores than low scores, or more low scores than high scores. We met this lopsidedness in the stem-and-leaf displays and histograms for the statistics test data, but now I am giving it a name: *skew*. A **skewed** distribution is an asymmetric distribution with a single peak. Imagine that the scores along the *X*-axis are on a number line that goes from negative infinity on the left to positive infinity on the right. Skew is then named according to the direction of the long tail. If there are many high scores, it is **negatively skewed** because the long tail goes to the left. We would expect this when a test is very easy. If there are many low scores, it is **positively skewed** because the long tail goes to the right. We would expect this when a test is very difficult (Figure 4.10).

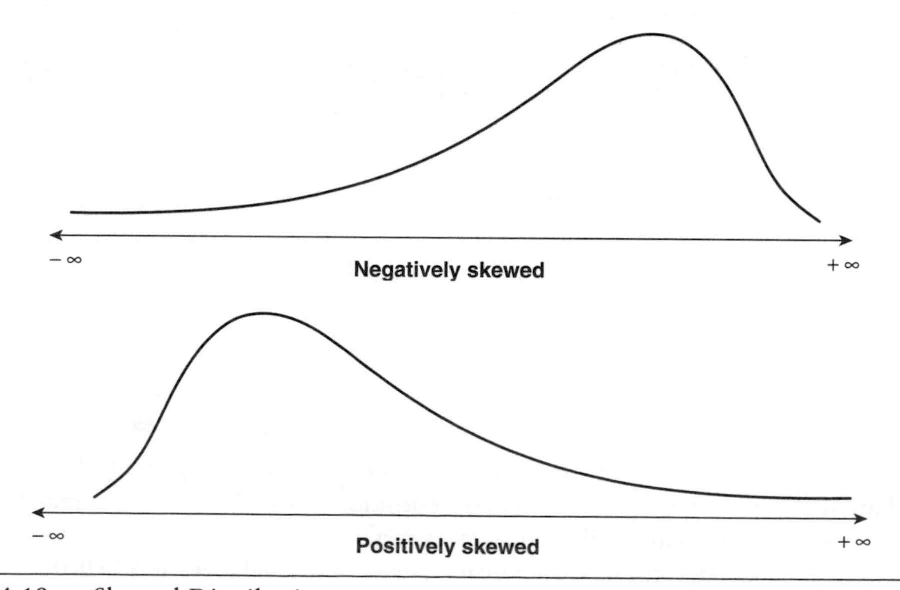

Figure 4.10 Skewed Distributions

It is tempting to label skew according to the direction in which the bulk of the scores lie, but that is incorrect. In a negatively skewed distribution, most of the scores are high, and in a positively skewed distribution, most of the scores are low. Yes, this labeling is counterintuitive.

SOURCE: Bizarro © Dan Piraro. King Features Syndicate.

Skew has an impact on central tendency and dispersion, which will be discussed in Modules 5 and 6, as well as the selection of correct test statistics (discussed in many subsequent modules). Therefore, skew is one of those concepts you should master now.

Sometimes a distribution is symmetric, but there are too many middle scores or too few middle scores. This, too, will affect the value of some of the later statistics we will compute. When there are too many or too few middle scores, the height of the distribution is affected. We refer to this as **kurtosis**. A distribution with many middle scores is called **leptokurtic.** To remember the prefix *lepto*, think of the scores as "leaping." A distribution with few middle scores is called **platykurtic**. To remember the prefix *platy*, think of the distribution as a "plate." Figure 4.11 shows what a leptokurtic and a platykurtic distribution might look like.

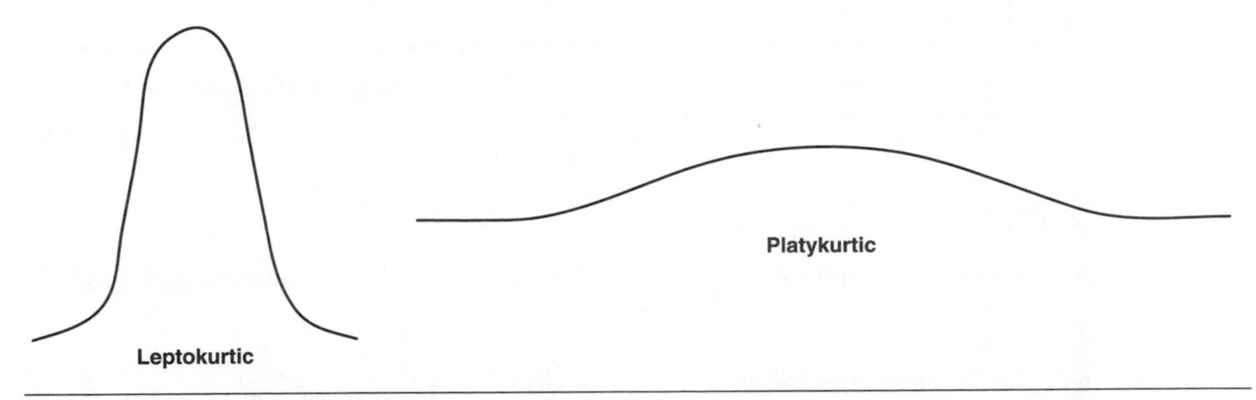

Figure 4.11 Kurtotic Distributions

At other times, a distribution has two or more peaks rather than just the one that you see in a normal curve. A distribution with two peaks is said to be **bimodal**. This is usually an indication that the sample of cases contains two distinct subgroups—one subgroup with a lot of the measured trait and one subgroup with only a little of the measured trait. We might see a bimodal distribution if we measured the weightlifting ability of men and women. Figure 4.12 shows what a bimodal distribution might look like.

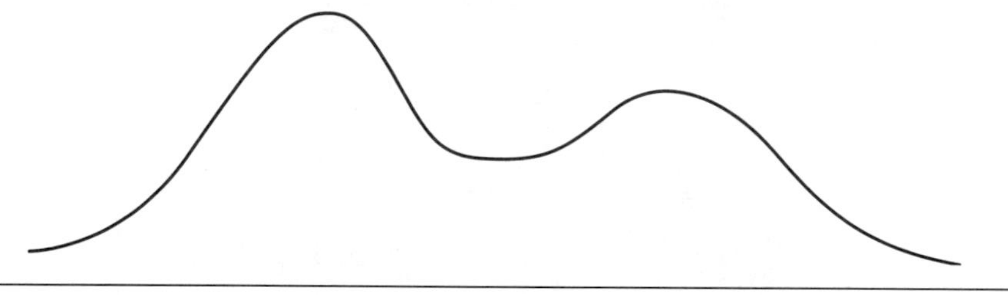

Figure 4.12 Bimodal Distribution

Other variables produce a distribution that is uniform, or **rectangular**. A multiple-choice test on which all students merely guessed would produce a nearly uniform distribution across scores. Ranks, too, produce rectangular distributions because, regardless of the actual scores, there is only one case at each rank (assuming no tied ranks). For this reason, it makes little sense to graph ranks (Figure 4.13).

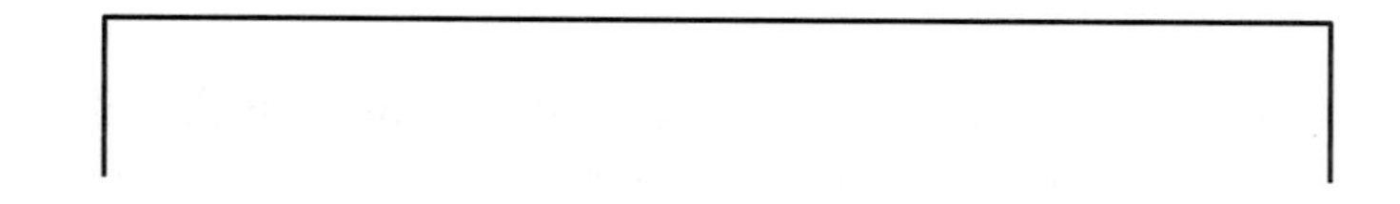

Figure 4.13 Rectangular Distribution

Most variables in psychology and education approximate a normal distribution. However, sometimes a distribution is so platykurtic that it approaches rectangularity. In such a case, the researcher would probably want to determine why the data were so evenly distributed.

CHECK YOURSELF!

Sketch the following distributions: normal, positively skewed, negatively skewed, leptokurtic, platykurtic, bimodal, and rectangular.

PRACTICE

5. Assume a frequency curve for the following sets of data. What shape will the curve probably take? Use the following key:

   NO = normal

   NS = negatively skewed

   PS = positively skewed

   BI = bimodal

   RE = rectangular

Data	*Shape of Graph*				
a. SAT scores for graduate students	NO	NS	PS	BI	RE
b. Age at which babies start to walk	NO	NS	PS	BI	RE
c. Self-esteem of Olympic medal winners	NO	NS	PS	BI	RE
d. Age of enlisted military personnel	NO	NS	PS	BI	RE
e. Running speed for sprinters in the Olympics and in the Special Olympics	NO	NS	PS	BI	RE
f. Number of ways to leave your lover (oops . . . the song says there are exactly 50 ways, so you may skip this one)	NO	NS	PS	BI	RE

6. Assume a frequency curve for the following sets of data. What shape will the curve probably take? Use the following key:

   NO = normal

   NS = negatively skewed

   PS = positively skewed

   BI = bimodal

   RE = rectangular

Data	*Shape of Graph*				
a. Number of children in American families	NO	NS	PS	BI	RE
b. Shoe size for a random sample of men	NO	NS	PS	BI	RE
c. Age of residents of a nursing home	NO	NS	PS	BI	RE
d. Number of crimes committed by a random sample of citizens	NO	NS	PS	BI	RE
e. Percentage of applicants admitted to a highly selective college	NO	NS	PS	BI	RE

7. For Exercise 1 in Module 3, you created a frequency table for the number of dates 40 college juniors reported during the previous 30 days. Review the table. Now describe the data's shape: Are they normally distributed, positively skewed, negatively skewed, rectangular, or multimodal?

8. For Exercise 2 in Module 3, you created a frequency table for the attendance log of a college instructor. Review the table. Now describe the data's shape: Are they normally distributed, positively skewed, negatively skewed, rectangular, or multimodal?

9. In Exercise 3 in Module 3, you created a frequency table for number of pets owned by pet store customers. Review the table. Now describe the data's shape: Are they normally distributed, positively skewed, negatively skewed, rectangular, or multimodal?

10. In Exercise 4 in Module 3, you created a frequency table for number of TV sets in homes. Review the table. Now describe the data's shape: Are they normally distributed, positively skewed, negatively skewed, rectangular, or multimodal?

Graphing Discrete Data

Nominal variables are not continuous. Measures of nominal variables are discrete because a case falls at one classification or another but cannot fall in between. Gender and political affiliation are examples of nominal variables. For nominal data, we need graphs that appropriately convey the noncontinuous nature of the data.

Bar Graph

One graph that is appropriate for nominal data is the bar graph. The *Y*-axis still indicates frequency, but the *X*-axis reflects categories instead of scores. A **bar graph** looks like a histogram except that the bars are separated by spaces. The space between the bars indicates that the categories are not continuous.

Because you already know how to construct a histogram, constructing a bar graph should be a snap. Here are the steps:

1. List the categories along the *X*-axis. Because there is no numeric value to the categories, the order in which you place the categories is irrelevant. Leave enough space between the categories so that bars drawn above the categories in the interior of the graph will not touch. Some researchers like to place the categories according to their frequencies—ascending or descending—for ease of interpretation. Spacing the categories evenly across the axis adds to the graph's attractiveness but is not necessary for interpretation.
2. Draw a vertical bar over the first category name to a height equal to its frequency.
3. Continue this process for each category.

Figure 4.14 shows a bar graph for the reported religious affiliation among students in a large class in a northeastern university.

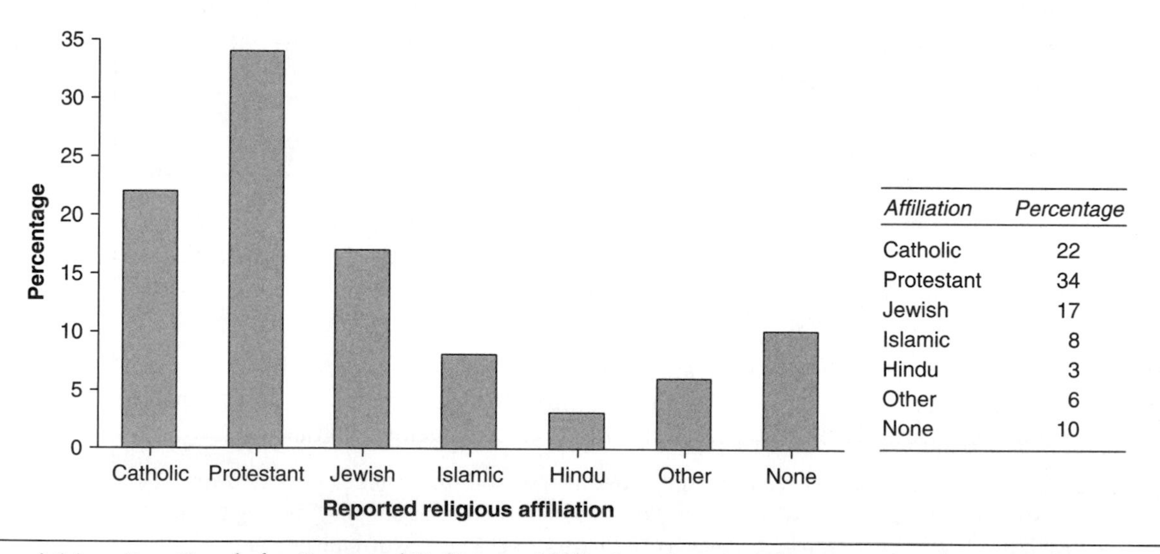

Affiliation	*Percentage*
Catholic	22
Protestant	34
Jewish	17
Islamic	8
Hindu	3
Other	6
None	10

Figure 4.14 Bar Graph for Reported Religious Affiliations in a Northeastern University Class

To read a bar graph, select a category along the *X*-axis and read the height of its bar on the *Y*-axis. In Figure 4.14, the largest number of students are Protestant, at 34%.

Pie Graph

Another type of graph commonly used for nominal data is the pie graph. A **pie graph** is a circle in which the "slices" represent percentages. To create a pie graph, use the following steps:

1. Calculate the percentage of cases falling in each category.
2. A circle contains 360°. Calculate the number of degrees in a category, proportional to its percentage. For example, in Figure 4.15, 34 out of 100 students, or 34%, are Protestant. 34% of 360° = 0.34 × 360° = 122.4°. Repeat this process for each category.
3. Draw slices within the circle for each category's calculated degree size. (You will need a compass for this.)
4. Label the slices for both category and percentage.

The pie graph in Figure 4.15 depicts the religious affiliation data from the previous bar graph.

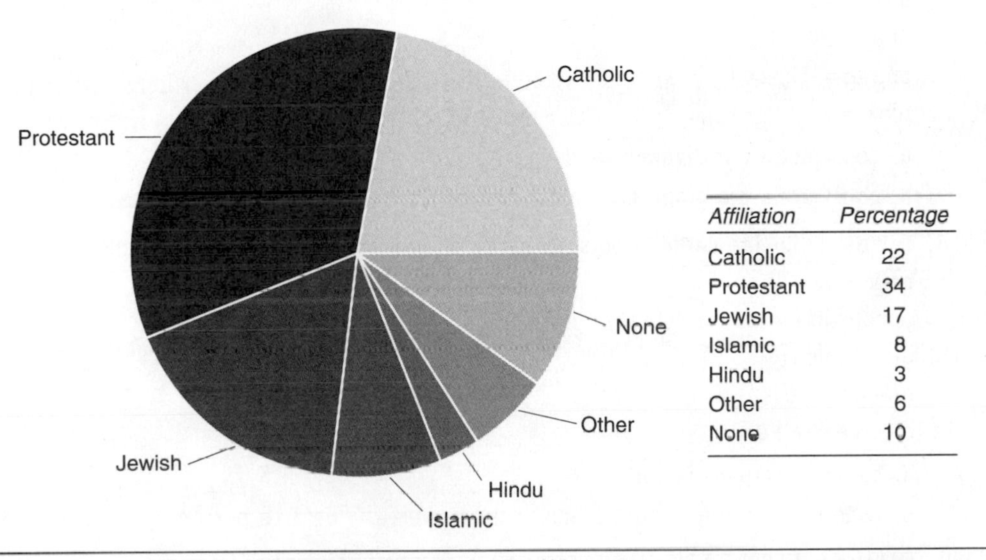

Affiliation	*Percentage*
Catholic	22
Protestant	34
Jewish	17
Islamic	8
Hindu	3
Other	6
None	10

Figure 4.15 Pie Graph for Reported Religious Affiliations in a Northeastern University Class

Reading a pie graph is straightforward. Simply read the category and percentage from the slice or from the accompanying key.

Pie graphs are commonly used for reporting survey data or for making summary statistics clear to the lay public. However, they are rarely used for reporting the results of experimental research.

CHECK YOURSELF!

How do a histogram and a bar graph differ in appearance and why? Which type of graph gives the same information as a frequency curve? Which type of graph must show every score individually, without grouping? For which graphs are there no scores but only frequencies within categories?

PRACTICE

11. For the following political party registrations in a voting district,

    Republican = 5,924,

    Democrat = 3,622,

Conservative = 840, and
Liberal = 614,

a. construct a bar diagram, and
b. construct a pie graph (you must first convert the data to percentages).

12. For the following color preferences of third-grade children at an elementary school,
red = 18,
orange = 4,
yellow = 3,
green = 15,
blue = 26,
purple = 12, and
pink = 7,

a. construct a bar diagram, and
b. construct a pie graph (you must first convert the data to percentages).

13. For the following carbonated soft drink preferences of 100 college students,
cola = 52
lemon-lime = 16
ginger ale = 4
orange = 10
root beer = 18

a. construct a bar diagram, and
b. construct a pie graph (you must first convert the data to percentages).

14. For the following declared concentrations of 100 college students,
social sciences = 32
business = 16
humanities = 28
science = 10
allied health fields= 14

a. construct a bar diagram, and
b. construct a pie graph (you must first convert the data to percentages).

15. What types of graphs would be appropriate for the following data? Circle as many types as are appropriate and would give clarity to the data. Use the following key:
SL = stem-and-leaf
H = histogram
F = frequency curve
B = bar diagram
P = pie chart

Data	*Types of Graphs*				
a. Feet of snowfall per storm last year	SL	H	F	B	P
b. Brands of carbonated soft drinks drunk	SL	H	F	B	P
c. High school class rank of entering college freshmen	SL	H	F	B	P
d. GPA of 500 freshmen	SL	H	F	B	P
e. Speed of a baseball pitch per team member	SL	H	F	B	P

16. What types of graphs would be appropriate for the following data? Circle as many types as are appropriate and would give clarity to the data. Use the following key:

SL = stem-and-leaf
H = histogram
F = frequency curve
B = bar diagram
P = pie chart

Data	*Types of Graphs*				
a. Eye color of students in your statistics class	SL	H	F	B	P
b. Types of tools owned by 50 heads of households	SL	H	F	B	P
c. Amount of water drunk by a dog	SL	H	F	B	P
d. Miles run by racers	SL	H	F	B	P
e. Relative positions in a contestant lineup	SL	H	F	B	P

17. What types of graphs would be appropriate for the following data? Circle as many types as are appropriate and would give clarity to the data. Use the following key:

SL = stem-and-leaf
H = histogram
F = frequency curve
B = bar diagram
P = pie chart

Data	*Types of Graphs*				
a. Consumer preference for 2-door, 4-door, or hatchback style cars	SL	H	F	B	P
b. Number of miles rowed by 30 campers	SL	H	F	B	P
c. Number of votes cast per candidate in a televised talent contest	SL	H	F	B	P
d. Number of ribbons won in a series of athletic contests	SL	H	F	B	P
e. Perfume fragrance preferences of 50 female shoppers	SL	H	F	B	P

18. What types of graphs would be appropriate for the following data? Circle as many types as are appropriate and would give clarity to the data. Use the following key:

SL = stem-and-leaf
H = histogram
F = frequency curve
B = bar diagram
P = pie chart

Data	*Types of Graphs*				
a. A,B,C, or D answers on a multiple-choice test answer key	SL	H	F	B	P
b. Diagnoses for 100 visitors to a medical clinic	SL	H	F	B	P
c. Number of months people dated their future spouses prior to marriage	SL	H	F	B	P
d. Number of pictures a digital camera will take before the battery is exhausted	SL	H	F	B	P
e. Preferred brand of laundry detergent	SL	H	F	B	P

SPSS Connection

Download the file **data_statistic test scores.sav** from www.sagepub.com/steinberg2e. These data are used in the textbook example.

Alternatively, manually enter the data from Table 3.1 into the SPSS Data View spreadsheet and define the variables in the SPSS Variable View spreadsheet. Call the variable **statscor**, set the decimals at 0, and label the variable as **Statistics Test Score.**

If the file is not already in **Data View**, click that tab in the lower left of the screen.

In the toolbar at the top of the screen, click on **Graphs,** then **Legacy Dialogs,** then **Histogram.** Highlight the variable **Statistics Test Score** in the left window and click on the **arrow** between the windows to send that variable into the right window. Click **OK.** This is what you will see.

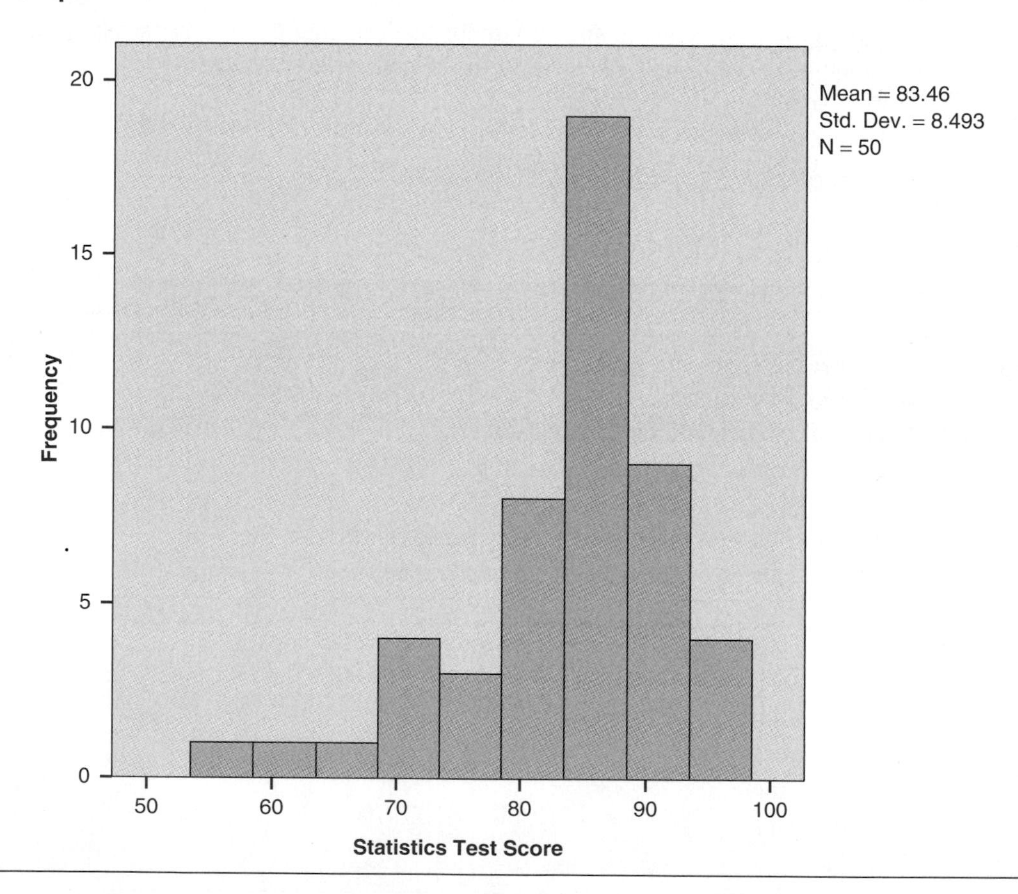

Although the variable in a histogram is continuous and so is properly defined by interval limits as discussed in the textbook, SPSS drops the interval limits in favor of whole numbers. The score groupings also differ from those in the textbook. The IBM® SPSS® Statistics Student Version 18.0 does not have an option to define axis score ranges. The program chooses the best fit for the graph.

Return to the Data View tab. In the toolbar at the top of the screen, click on **Graphs,** then **Legacy Display,** then **Line.** Click on the top example, **Simple.** Click on **Define.** Highlight the variable **Statistics Test Score** in the left window and click on the **arrow** between the windows to send that variable into the Category Axis window. Click **OK**. This is what you will see.

Graph

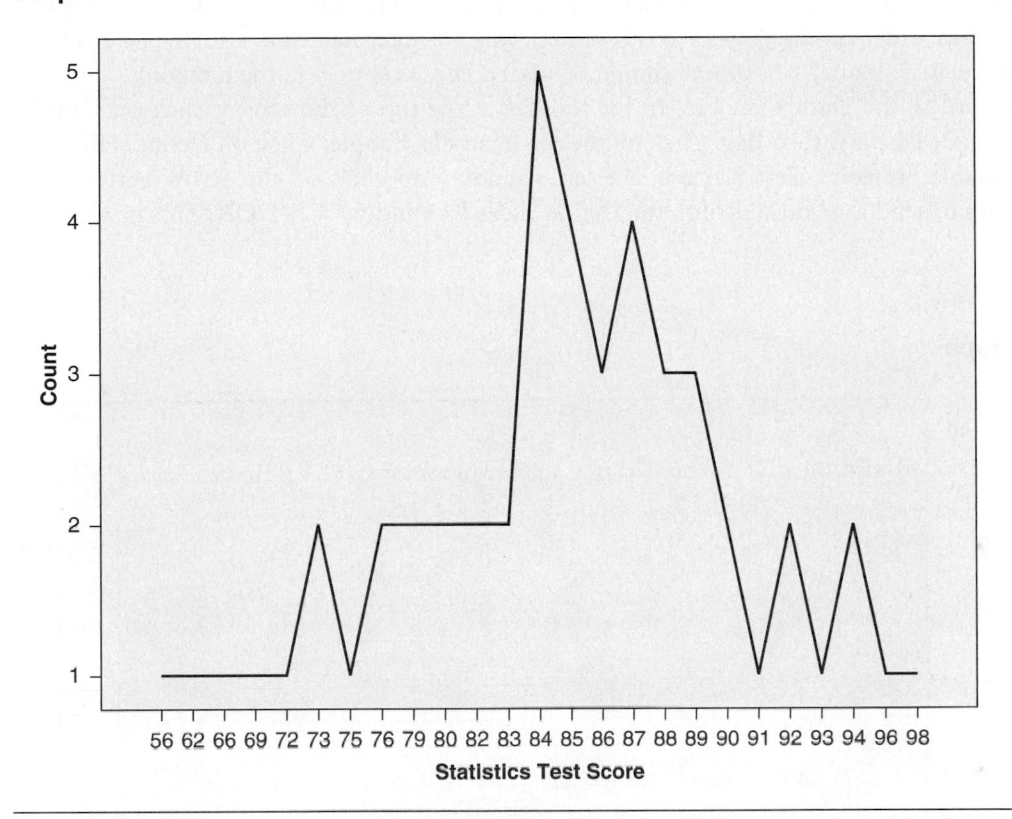

The line graph is more jagged than the one on the textbook because it shows individual scores rather than the grouped scores shown in the textbook. Again, the Student Version of SPSS does not have an option to define axis score ranges.

Download the file **data_religious affiliation.sav** from **www.sagepub.com/Steinberg2e.** These data are used in the textbook example. Alternatively, manually enter the following data into the SPSS Data View spreadsheet and define the variables in the SPSS Variable View spreadsheet.

p	c	i	j	p	p	c	p	n	o
c	j	p	c	p	p	h	j	i	c
c	p	j	o	p	n	p	p	c	j
p	o	i	c	p	p	n	p	j	c
j	n	p	c	i	p	p	h	c	p
j	J	p	o	c	p	j	i	c	h
j	p	c	i	p	c	j	p	n	j
c	n	p	p	c	o	p	j	p	n
c	n	j	p	o	c	j	c	p	n
i	p	p	n	c	p	j	c	i	p

Call the variable **relaffil,** change the Type to **string** (this is the SPSS name for nominal-level data), and label the variable as **Religious Affiliation.** Label the seven category values as follows: value = c, label = **Catholic;** value = h, label = **Hindu;** value = i, label = **Islamic;** value = **j,** label = **Jewish;** value = **n,** label = **None;** value = **o,** label = **Other;** value = **p,** label = **Protestant.** Value labels will appear on your graph output, making it more interpretable. You can enter the value labels in any order. However, SPSS output will display string variables in alphabetical order on the graph axis, regardless of how you enter them at the definition stage. This differs from the order shown in the textbook. I have left the textbook order different from SPSS's alphabetical order to highlight the fact that nominal variables have no relative relationships and so their order on the graph's axis is not relevant for interpretation. Usually, catchall categories such as "none" or "other" should be placed last, as shown in the textbook.

Return to the Data View tab. In the toolbar at the top of the screen, click on **Graphs,** then **Legacy Display**, then **Bar**. Click on the top example, **Simple**. Click on **Define**. Highlight the variable **Statistics Test Score** in the left window and click on the **arrow** between the windows to send that variable into the Define Slices By window. Click **OK**. This is what you will see.

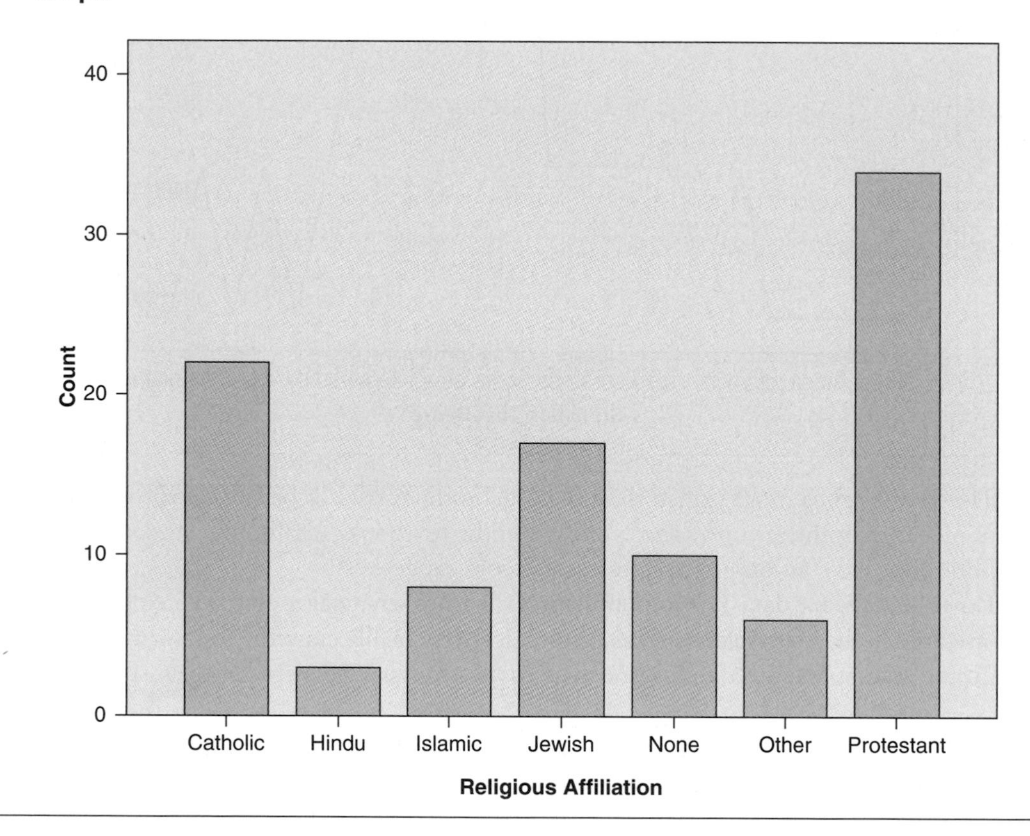

Return to the Data View tab. In the toolbar at the top of the screen, click on **Graphs,** then **Legacy Display**, then **Pie**. Click on the top example, **Simple**. Click on **Summaries for Groups of Cases** and **Define**. Highlight the variable **Statistics Test Score** in the left window and click on the **arrow** between the windows to send that variable into the Define Slices By window. Click **OK**. This is what you will see.

Graph

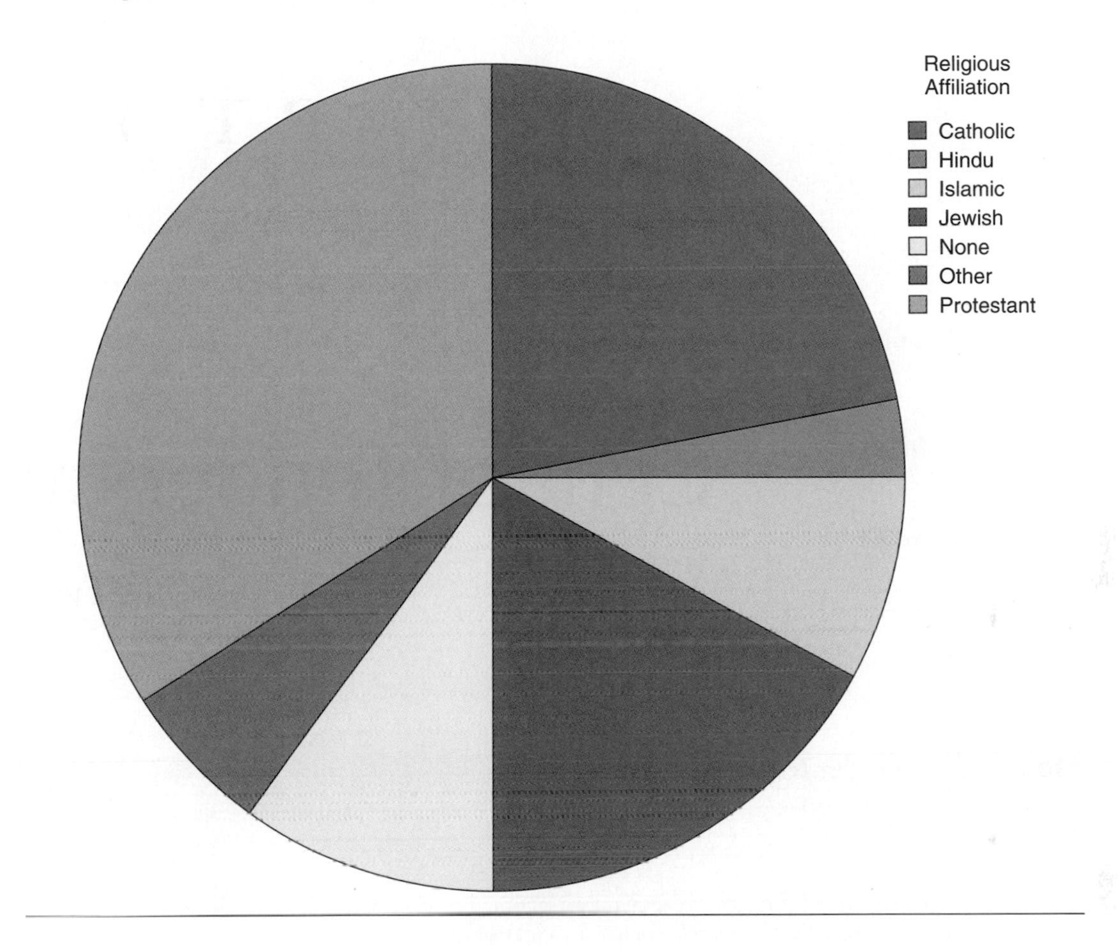

With a pie chart, as with a bar chart, the order of the slices is not relevant to graph interpretation.

Visit the study site at www.sagepub.com/steinberg2e for practice quizzes and other study resources.

PART III

Central Tendency

"Bull's-Eye"

5 Mode, Median, and Mean

Terms: central tendency, mode, median, mean, outlier

Symbols: *Mo, Mdn, M*, μ, Σ, *N*

Learning Objectives:

- Calculate various measures of central tendency—mode, median, and mean
- Select the appropriate measure of central tendency for data of a given measurement scale and distribution shape
- Know the special characteristics of the mean that make it useful for further statistical calculations

What Is Central Tendency?

You have tabulated your data. You have graphed your data. Now it is time to summarize your data. One type of summary statistic is called central tendency. Another is called dispersion. In this module, we will discuss central tendency.

Measures of **central tendency** are measures of location within a distribution. They summarize, in a single value, the one score that best describes the centrality of the data. Of course, there are lots of scores in any data set. Nevertheless, one score is most representative of the entire set of scores. That's the measure of central tendency. I will discuss three measures of central tendency: the mode, the median, and the mean.

Q: How are the mean, median, and mode like a valuable piece of real estate?

A: Location, location, location!

Mode

The **mode**, symbolized *Mo*, is the most frequent score. That's it. No calculation is needed.

Here we have the number of items found by 11 children in a scavenger hunt. What was the modal number of items found?

14, 6, 11, 8, 7, 20, 11, 3, 7, 5, 7

If there are not too many numbers, a simple list of scores will do. However, if there are many scores, you will need to put the scores in order and then create a frequency table. Here are the previous scores in a descending order frequency table.

Score	*Frequency*
20	1
14	1
11	2
8	1
7	3
6	1
5	1
3	1

What is the mode? The mode is 7, because there are more 7s than any other number.

Note that the number of scores on either side of the mode does not have to be equal. It might be equal, but it doesn't have to be. In this example, there are three scores below the mode and five scores above the mode.

Nor does the numerical distance of the scores from the mode on either side of the mode have to balance. It could balance, but it doesn't have to balance. Finding the distance of each score from the mode, we get the following values on each side of the mode. As shown below in boldface, –7 does not balance +29.

Distances Below Mode	*Distances Above Mode*
3 – 7 = –4	8 – 7 = +1
5 – 7 = –2	11 – 7 = +4
6 – 7 = –1	11 – 7 = +4
	14 – 7 = +7
	20 – 7 = +13
Σ = **–7**	Σ = **+29**

The mode is the least stable of the three measures of central tendency. This means that it will probably vary most from one sample to the next. Assume, for example, that we send these same 11 children on another equally difficult scavenger hunt. Now let's assume that every child in this second scavenger hunt finds the same number of items (a very unlikely occurrence in the first place), except that one child who previously found 7 items now finds 11. Compare the two sets of scores below. Only a single score (highlighted in boldface) differs between the two hunts, and yet the mode changes dramatically.

3, 5, 6, 7, 7, **7**, 8, 11, 11, 14, 20	first scavenger hunt
3, 5, 6, 7, 7, **11**, 8, 11, 11, 14, 20	second scavenger hunt

What is the new mode? It is 11, because there are now three 11s and only two 7s. This is a very big change in the mode, considering that most of the scores in the two hunts were the same. Furthermore, the two hunts would almost certainly be more different than I made them. This, of course, further increases the likelihood that the mode will change.

Because of its simplicity, the mode is an adequate measure of central tendency to report if you need a summary statistic in a hurry. For most purposes, however, the mode is not the best measure of central tendency to report. It is simply too subject to the vagaries of the cases that happen to fall in a particular sample. Also, for very small samples, the mode may have a frequency only one or two higher than the other scores—not very informative. Finally, no additional statistics are based on the mode. For these reasons, it is not as useful as the median or the mean.

Median

The **median**, symbolized *Mdn*, is the middle score. It cuts the distribution in half, so that there are the same number of scores above the median as there are below the median. Because it is the middle score, the median is the 50th percentile.

Here's an example. Seven basketball players shoot 30 free throws during a practice session. The numbers of baskets they make are listed below. What is the median number of baskets made?

22, 23, 11, 18, 22, 20, 15

To find the median, use the following steps:

1. Put the scores in ascending or descending order. If you do not first do this, the median will merely reflect the arrangement of the numbers rather than the actual number of baskets made. Here are the scores in ascending order.

   11, 15, 18, 20, 22, 22, 23

2. Count in from the lowest and highest scores until you find the middle score.

What is the median number of baskets? The median number of baskets is 20 because there are three scores above 20 and three scores below 20.

Here's another example. Twelve members of a gym class, some in good physical condition and some in not-so-good physical condition, see how many sit-ups they can complete in a minute. Here are their scores.

2, 3, 6, 10, 12, 12, 14, 15, 15, 15, 24, 25

What is the median number of sit-ups? Is it 12? 14? The median is 13, because there are six scores below 13 and six scores above 13. Note that the median does not necessarily have to be an existing score. In this case, no one completed exactly 13 sit-ups.

Here is the rule: With an odd number of scores, the median will be an actual score. But with an even number of scores, the median will not be an actual score. Instead, it will be the score midway between the two centermost scores. To get the midpoint, simply average the two centermost scores. In our example, this is (12 + 14)/2, which is 26/2, which is 13.

While the number of scores on each side of the median must be equal, the numerical distance of the scores on either side of the median will not necessarily be equal. It might be equal, but it doesn't have to be. Finding the distance of each score from the median, we get the following values. As shown in boldface, −33 does not balance +30.

Q: Where does a statistician park his car?

A: Along the median.

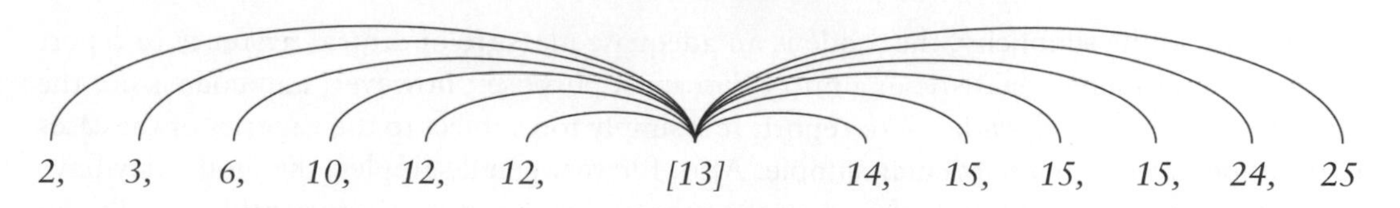

Distance Below Median	*Distance Above Median*
2 − 13 = −11	14 − 13 = +1
3 − 13 = −10	15 − 13 = +2
6 − 13 = −7	15 − 13 = +2
10 − 13 = −3	15 − 13 = +2
12 − 13 = −1	24 − 13 = +11
12 − 13 = −1	25 − 13 = +12
Σ = **−33**	Σ = **+30**

One nice feature of the median is that it can be determined even if we do not know the value of the scores at the ends of the distribution. In the following set of seven pop quiz scores (oops—it looks like the students weren't prepared!), we know that there is a score above 70 but do not know what that score is. Likewise, we know that there is a score below 30 but not what that score is:

>70
70
60
50
40
30
<30

Nevertheless, we can determine the median by counting up (or down) half the number of scores. In this case, the median is 50, because it is the fourth score from either direction. It does not matter whether the top score was 90, 100, or even 1,076 or whether the bottom score was 20, 10, or even −173. The median is still 50.

It is also possible to compute a median from a large number of scores when there are many duplicate scores. Suppose the pop quiz were given not just to 7 students but to 90 students. Because of the large number of students at each score, it is easier to interpret the data if they are arranged in a frequency table. Table 5.1 gives the scores and their frequencies.

Table 5.1 Pop-Quiz Scores for 90 Students

Score	*Frequency*	*Cumulative Frequency*
>70	3	90
70	7	87
60	19	80
50	31	61
40	14	30
30	12	16
<30	4	4
	$\Sigma = 90$	

There are 90 scores in all. Thus, the median will have 44.5 scores above it and 44.5 scores below it. To get an estimate for the median, start at the bottom and count upward: 4 + 12 = 16 cases (proceed); 16 + 14 more = 30 cases (proceed); 30 + 31 more = 61 cases (stop). The score of 50 is the median because we reach the middle case at that score.

Because there are many cases at each score, a reading from a frequency table gives, as we saw in Module 3, only a ballpark figure. Because the median is the 50th percentile, the median score out of 90 cases should be the score of the 44.5th case out of the 90 cases. But note that there are only 30 cases below our ballpark median of 50 (14 + 12 + 4), not 44.5 cases. Thus, if we take the bottommost case of the students who scored 50, that person's score is the 31st case from the bottom, and if we take the topmost case of the students who scored 50, that person's score is the 61st case from the bottom (31 + 14 + 12 + 4). We want the 44.5th case, not the 31st or the 61st case. Obviously, the 44.5th case falls somewhere within the 31 students who scored 50. We already know that 30 students scored below 50; thus, we need only 14.5 additional cases of the 31 cases at the score of 50 to reach the 44.5th case.

Settling for 50 as the median is like throwing darts at a dartboard. If it hits anywhere in the bull's-eye, we say we've hit the bull's-eye. But even within the bull's-eye, some points are more central than others. To determine the exact bull's-eye, we'd need to get out a measuring instrument and find the precise center of the bull's-eye.

And so it is with the median. For precision, we must use a formula. Here is the formula for a median:

$$Mdn = \text{LL} + (i)\left(\frac{0.5n - \text{cum } f_{\text{below}}}{f}\right)$$

where

LL = lower real limit of the score containing the 50th percentile,
i = width of the score interval,
$0.5n$ = half the cases,
cum f_{below} = number of cases lying below the LL, and
f = number of scores in the interval containing the median.

First, we determine the LL of the score containing the median. Recall that the real limits of a score extend from one half the unit of measurement below the score to one half the unit of measurement above the score. In our quiz example, the unit of measurement is 10 points; that is, scores are expressed to the nearest 10 points (40, 50, 60, etc.). Therefore, the real limits of a score are ±5 points. Table 5.2 is the frequency table with scores reexpressed in real-score limits.

Table 5.2 Real Limits of Pop-Quiz Scores for 90 Students

Score	*Real Limits*	*Frequency*
>70	>75	3
70	65–75	7
60	55–65	19
50	45–55	31
40	35–45	14
30	25–35	12
<30	<25	4
		$\Sigma = 90$

We already determined that the median falls in the score interval of 45 to 55. The LL of that interval is 45. Plugging the LL into the formula, we get the following:

$$\begin{aligned}
Mdn &= \text{LL} + (i)\left(\frac{0.5n - \text{cum } f_{\text{below}}}{f}\right) \\
&= 45 + (10)\left(\frac{(0.5)(90) - 30}{31}\right) \\
&= 45 + (10)\left(\frac{45 - 30}{31}\right) \\
&= 45 + (10)\left(\frac{15}{31}\right) \\
&= 45 + (10)(0.4838) \\
&= 45 + 4.838 \\
&= 49.838
\end{aligned}$$

The ballpark median was 50, but the exact mathematical median is 49.838. The mathematical median is a bit different from what we proposed by counting up cases from the bottom. Why? Remember that we needed only 14.5 of the 31 cases at score 50 to bring our case count up to 44.5. Out of 31 cases, 14.5 is just less than half. Remember also that the real limits for a score of 50 are 45 and 55. If we assume that the 31 scores at the score of 50 are evenly distributed between 45 and 55 and we go just less than halfway into the 45 to 55 range, what do we get? We get a little less than 50—or 49.838, our calculated median!

PRACTICE

1. One hundred and thirty-six breast cancer survivors participate in a community walk to raise money for fighting the disease. The number of women who walked various numbers of miles is listed below:

Miles Walked	*Number of Women*
5	18
4	23
3	54
2	19
1	8

   a. Counting up from the bottom of the table, what is the ballpark median number of miles walked?
   b. Using the formula for a median, find the exact median number of miles walked by these women.

2. Morbidly obese women attending the Healthy Weigh diet clinic are weighed at program entry, to the nearest 10 lb. To join, the women must weigh at least 200 lb. Here are the women's weights.

Weight (lb)	*Number of Women*
290	1
280	4
270	0
260	4
250	9

Weight (lb)	*Number of Women*
240	10
230	14
220	18
210	11
200	9

a. Counting up from the bottom of the table, what is the ballpark median weight of clinic clients?

b. Using the formula for a median, find the exact median weight of clinic clients.

3. Fifty clients of an outpatient mental clinic take an anxiety inventory. Scores range from 1 to 10. Here are the scores.

Anxiety Score	*Number of Clients*
10	2
9	8
8	13
7	10
6	7
5	4
4	2
3	2
2	1
1	1

a. Counting up from the bottom of the table, what is the ballpark median anxiety score?

b. Using the formula for a median, find the exact median anxiety score.

4. Twenty autistic children in a communication therapy program are scored on the number of times in a given session that they initiate eye contact or direct a comment toward the primary caretaker. Here are their scores.

Number of Contacts	*Number of Children*
8	1
6	1
5	1
4	2
3	4
2	7
1	4

a. Counting up from the bottom of the table, what is the ballpark median number of contacts?

b. Using the formula for a median, find the exact median number of contacts.

Mean

The **mean**, symbolized *M* (for samples) or μ (for populations), is the average score. You already know how to calculate an average. If you want to know the average score on a class test, you add up all students' scores and divide by the number of students in the class, right? In statistics, we make that process explicit with a formula. Here is the formula for a mean:

$$M = \frac{\sum X}{N}$$

where

$\sum$ = sum,

X = raw score, and

N = number of cases.

The formula says to add up ($\sum$) all the raw scores (X) and divide by the number of cases (N). Whenever you divide anything by the number of cases, you get an average of the thing you were dividing. The "thing" in this case is raw scores. Thus, the formula gives the average raw score.

Of the three measures of central tendency, the mean is the most stable. That is, if we drew many samples of the same size from the same population and calculated the mean of each sample, the mean would not likely vary much from sample to sample.

"The poor are getting poorer, but with the rich getting richer it all averages out in the long run."

SOURCE: Joseph Mirachi, New Yorker Cartoon 9/26/1988 © Joseph Mirachi/Condé Nast Publications/www.cartoonbank.com

Most important, the mean is the place where the numerical distances of scores on one side of the mean balance the numerical distance of scores on the other side of the mean. To demonstrate this principle, first find the mean of the following Fahrenheit temperatures for five winter days in Maine: −5, −5, +2, +4, +4. You should be able to do it in your head. The mean is 0:

$$M = \frac{\sum X}{N} = \frac{(-5)+(-5)+(+2)+(+4)+(+4)}{5} = \frac{0}{5} = 0$$

Finding the distance of each score from the mean, we get the following values. As shown by the boldface in this example, –10 exactly balances +10.

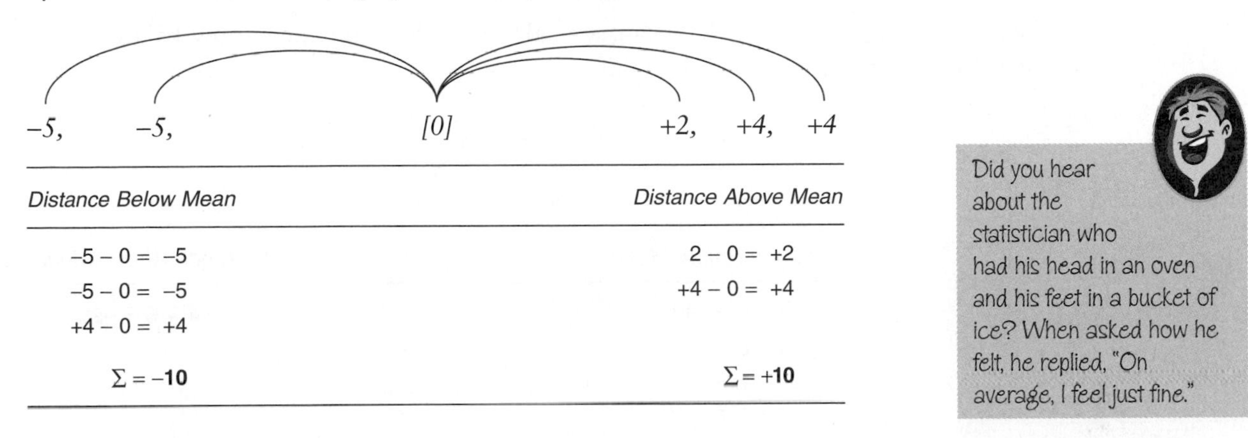

Distance Below Mean	*Distance Above Mean*
–5 – 0 = –5	2 – 0 = +2
–5 – 0 = –5	+4 – 0 = +4
+4 – 0 = +4	
Σ = **–10**	Σ = **+10**

Let's place these scores on a balance board such as a teeter-totter or seesaw. Imagine that each block is a score, and the scores are placed at distances equal to their numeric values. There are two scores of –5, one score of +2, and two scores of +4. As you can see in Figure 5.1, the scores balance at the mean, which in this case is 0.

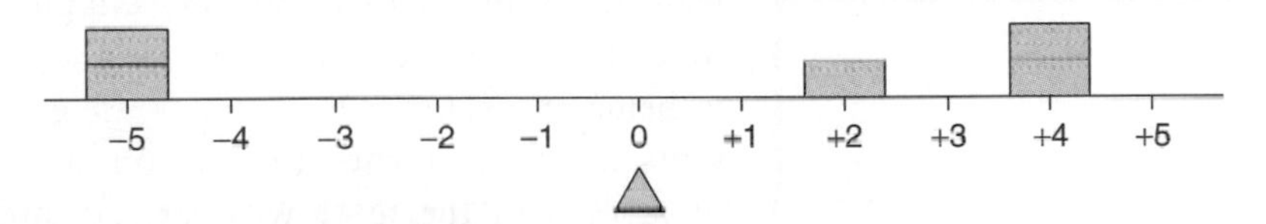

Figure 5.1 Scores Balancing at Mean of 0

The teeter-totter demonstrates other important characteristics of the mean. First, there do not necessarily have to be the same number of scores on each side of the mean. In this case, there are two scores below the mean and three scores above the mean.

Second, the numeric value of each score, X, is included in the calculation of the mean. Thus, the value of each score matters. The mean is not dependent on score frequencies, as was the case with the mode. Nor is the mean dependent on position within a distribution, as was the case with the median. But the mean is dependent on the value of each score.

This leads to a third important characteristic of the mean. Because the value of each score matters in the calculation of the mean, the mean is the most sensitive of the measures of central tendency to score aberrations. That is, a single extreme score has a marked effect on the mean's value.

To demonstrate this, let's consider the preceding set of temperatures again but replace one of the +4 temperatures with a temperature of +34. The new set of temperatures is –5, –5, +2, +4, +34. Here is the new mean. It has jumped from 0 to 6.

$$M = \frac{\sum X}{N} = \frac{(-5)+(-5)+(+2)+(+4)+(+34)}{5} = \frac{30}{5} = 6$$

Let's put this new set of scores on the teeter-totter as well. The break in the line to the right of Figure 5.2 indicates that a portion of the teeter-totter is missing. Without the break, the score of +34 would be somewhere off the right-hand side of the page.

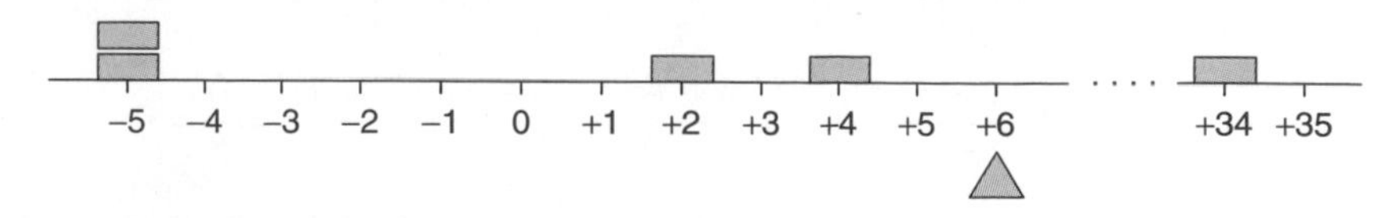

Figure 5.2 Scores Balancing at Mean of +6

Because of the increase in the mean, a single score above the mean now balances four scores below the mean, as shown in boldface below.

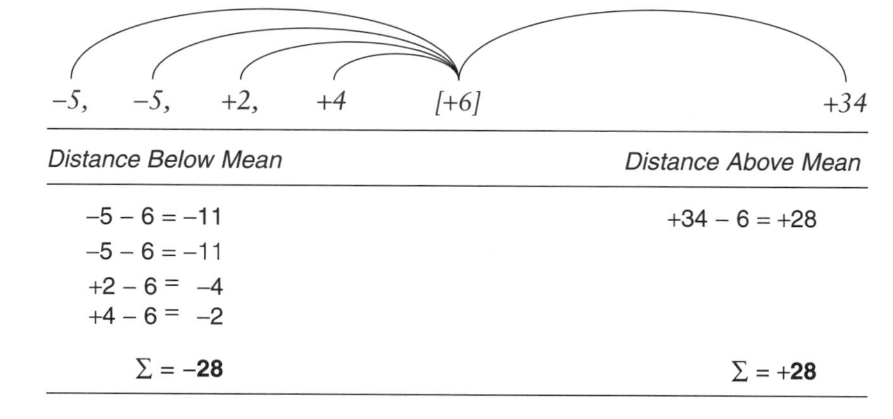

Distance Below Mean	*Distance Above Mean*
–5 – 6 = –11	+34 – 6 = +28
–5 – 6 = –11	
+2 – 6 = –4	
+4 – 6 = –2	
Σ = **–28**	Σ = **+28**

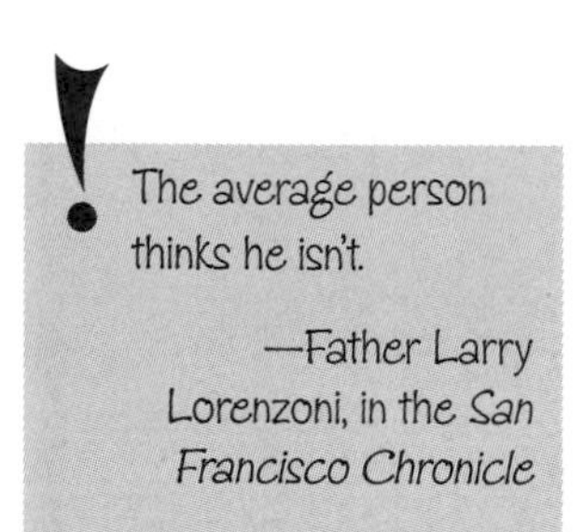

HERMAN®

"How do you expect me to average 55 miles an hour if I don't speed?"

SOURCE: HERMAN® is reprinted with permission from LaughingStock Licensing, Inc., Ottawa, Canada.

A score that is way out of line with the rest of the data is called an **outlier.** Sometimes outliers are legitimate—one person in the sample is simply much faster, smarter, or better along whatever scale is being measured. Other times an outlier represents a clerical error—the person was measured incorrectly or the score was entered into the data set incorrectly. Because outliers markedly affect the mean, researchers need to be especially alert for them so that they can determine whether the score legitimately belongs in the data set. Simply knowing the value of the mean does not, in itself, tell us that there is an outlier. Only visual inspection of the data tells us that. This is another reason why competent researchers always look at the data before calculating any statistic.

If there are several outliers, the median is a more appropriate measure of central tendency to report than the mean because the median is not influenced by outliers. In most cases, however, the mean is the preferred measure of central tendency to report. This is because further statistical analyses build on the mean. Sample means, for example, play an important role in the population-based statistics found throughout the remainder of this textbook.

CHECK YOURSELF!

Summarize the features of the mode, median, and mean:

	Mode	*Median*	*Mean*
Ease of calculation			
Stability over time			
Distance of scores above and below			
Number of scores above and below			

PRACTICE

5. Here, again, is the set of statistics test scores we worked with in Modules 3 and 4.

98	89	86	84	76
96	89	86	84	76
94	89	86	83	75
94	88	85	83	73
93	88	85	82	73
92	88	85	82	72
92	**87**	85	80	69
91	**87**	84	80	66
90	**87**	84	79	62
90	**87**	84	79	56

Find the (a) mode, (b) median (by formula), and (c) mean for these data.

6. Exercise 2 in Module 3 gave the following number of classes each of 32 students missed during the semester for a class meeting on M-W-F.

0, 1, 4, 2, 8, 1, 2, 4
5, 0, 2, 2, 1, 2, 1, 2
1, 0, 3, 1, 1, 2, 1, 2
2, 4, 0, 1, 3, 0, 3, 3

Find the (a) mode, (b) median (by formula), and (c) mean for these data.

7. Exercise 3 in Module 3 gave the following number of pets owned by customers of a pet store.

   3, 0, 1, 4, 3, 2, 2, 1, 3, 0, 2, 4, 5, 3, 2, 4, 7, 1, 1, 2

   Find the (a) mode, (b) median (ballpark), and (c) mean for these data.

8. Exercise 4 in Module 3 gave the following number of TV sets in homes.

   2, 3, 1, 2, 3, 0, 2, 4, 1, 2, 4, 3, 2, 1, 1, 3, 0, 2, 1, 1, 2, 3, 2, 5, 2

   Find the (a) mode, (b) median (ballpark), and (c) mean for these data.

Skew and Central Tendency

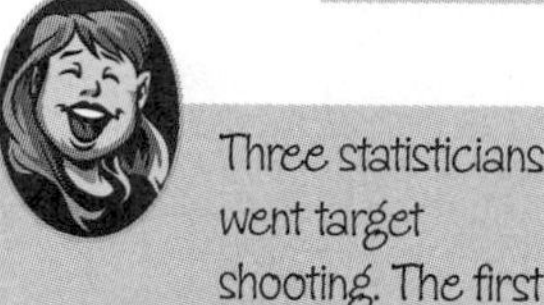

Three statisticians went target shooting. The first one took aim, shot, and missed by a foot to the left. The second one took aim, shot, and missed by a foot to the right. Whereupon the third one exclaimed, "We got it!" and walked away.

Recall from Module 4 that skew is a measure of asymmetry in a set of data. Skew affects the location of the mode, median, and mean. In a symmetric distribution such as a normal distribution, the three measures of central tendency coincide. That is, the most frequent score (mode) equals the midpoint (median), which equals the average (mean) (Figure 5.3).

This is not the case in a skewed distribution. Scores in the tail of a skewed distribution are outliers. And we already saw via the teeter-totter what happens when an outlier is introduced: The mean moves in the direction of the extreme score—that is, toward the tail. Because the mean is the most sensitive of the three measures of central tendency to extreme scores, the mean is pulled most toward the tail. The mode, which is simply the most frequent score, remains where it was. The median falls between the mean and the mode. This happens in both negatively and positively skewed distributions (Figure 5.4).

Because of the known relationship of the mode, median, and mean in normal versus skewed distributions, a researcher can tell from the calculated values whether a distribution is normally distributed or skewed. From Figures 5.3 and 5.4, we see that if the mean is lower than the mode, the distribution is negatively skewed. Conversely, if the mean is higher than the mode, the distribution is positively skewed. Similarly, a researcher can tell from the shape of the distribution where the mean, median, and mode will fall. If a distribution is negatively skewed, the mean must be lower than the mode. Conversely, if a distribution is positively skewed, the mean must be higher than the mode.

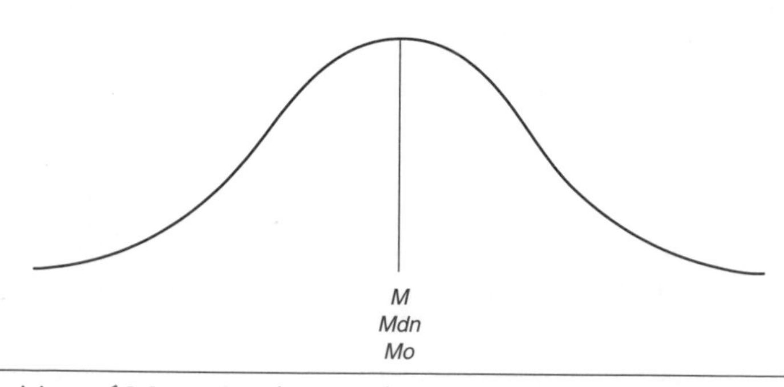

Figure 5.3 Position of Mean, Median, and Mode in Normally Distributed Data

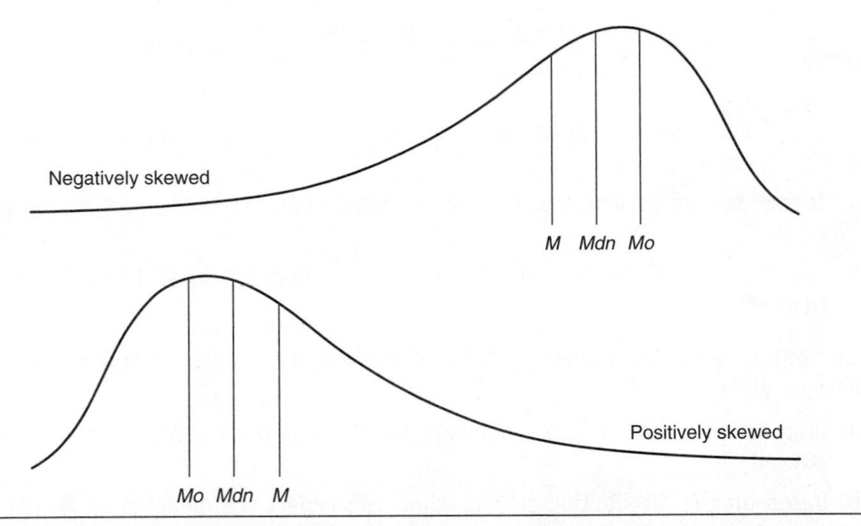

Figure 5.4 Position of Mode, Median, and Mean in Skewed Score Distributions

Comparing distribution shape with central tendency values is another way in which researchers check for clerical errors as they analyze their data. Minor deviations from expectation (say, a median that exceeds the mode or is lower than the mean) are usually due to "lumpiness" (additional lesser modes) in the data. But if the graphs and the numbers differ markedly, there is probably a calculation error.

Putting all this information together, which measure of central tendency should we report for a given set of data?

- Always report the mean if it is appropriate to do so, because it is the most useful measure for further statistical analysis.
- If we don't know the exact value of every score, we cannot report the mean. We may, however, be able to report the median.
- The median is a better choice when a distribution is known to be seriously skewed. In that case, the mean would be misleading.
- The mean is not appropriate for describing a bimodal or multimodal distribution. This is shown in Figure 5.5, in which the mean seriously misrepresents both the higher and the lower clusters of scores.

Note that multimodality is another instance in which a graph tells a story that a summary statistic cannot. As you can see from the graph, when a sample consists of two or more subgroups with very different performance levels, reporting any single measure of central tendency seriously misrepresents the performance of either group. In such a case, it is best to report two modes, one for each group.

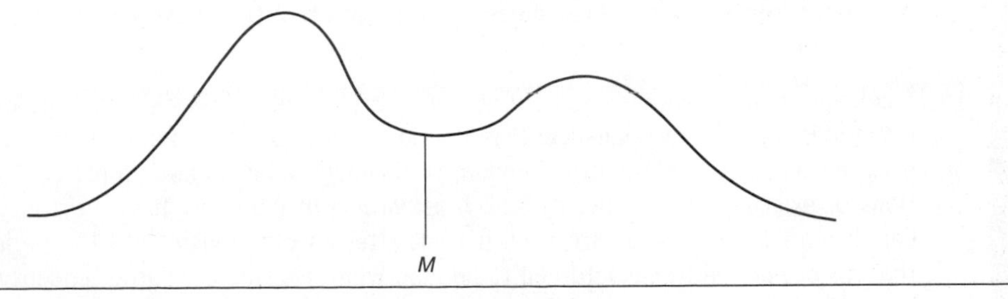

Figure 5.5 Position of the Mean in a Bimodal Distribution

PRACTICE

9. In Exercise 3 in this module, you found the median anxiety score for 50 mental health clinic clients.
   a. Is the set of anxiety scores normally distributed, positively skewed, or negatively skewed?
   b. Based on the distribution shape, would you expect the mean to be higher or lower than the median?
10. In Exercise 4 in this module, you found the median number of contacts made by 20 autistic children.
    a. Is the number of contacts normally distributed, positively skewed, or negatively skewed?
    b. Based on the distribution shape, would you expect the mean to be higher or lower than the median?
11. In Exercise 5 in this module, you found the mode, median, and mean for a set of 50 statistics test scores. Table 3.7 in Module 3 presents a grouped frequency table for that same data.
    a. Is the set of scores normally distributed, positively skewed, or negatively skewed?
    b. Do the relative magnitudes of the mode, median, and mean agree with the shape of the distribution? If not, can you explain why not?
12. In Exercise 6 in this module, you found the mode, median, and mean for an attendance log for a college class. In Exercise 2 in Module 3, you created a frequency table for that same data.
    a. Is the set of scores normally distributed, positively skewed, or negatively skewed?
    b. Do the relative magnitudes of the mode, median, and mean agree with the shape of the distribution? If not, can you explain why not?
13. You plan to try out for a track team. Every morning for a month, you run a mile and time yourself. What measure of central tendency best summarizes your running speed for the month? Why?
14. Your instructor gives a 60-item pretest on the first day of your statistics class. Because there has not yet been any statistics instruction, most students score quite low on the test. However, a small number of students have had statistics instruction as part of another course. Those students do very well on the test. What measure of central tendency best summarizes the whole class's statistics knowledge at the start of the course? Why?
15. A social psychologist wants to know the physical distance at which people are comfortable when in face-to-face conversation. She engages 20 participants in conversation on a neutral topic. Later, using a stationary video of the conversation taken against a 1 in. × 1 in. background pattern, she measures the number of inches each person stood from her when speaking. What measure of central tendency best summarizes the speaking distance of the 20 participants? Why?
16. A college's health center sees about 50 students per week. Most visits fall into one of these categories, in order of frequency: (1) infectious diseases, such as strep throat, mononucleosis, and influenza; (2) muscle, tendon, or ligament sprains and strains; or (3) complications of existing chronic disorders such as asthma or diabetes. It is the policy of Health Center staff to follow up with each patient after an office visit until the patient reports that his or her health has returned to normal. What measure of central tendency best summarizes patient recovery time for all patients seen during a typical week? Why?

SPSS Connection

Download the file **data_score set for central tendence.sav** from www.sagepub.com/steinberg2e. These data are used in the textbook example.

Alternatively, manually enter the following five scores into the SPSS **Data View** spreadsheet: –5, –5, +2, +4, +4. Click on the **Variable View** tab to define the variable. Name the variable **scr4mean,** set the decimals at **0**, and label the variable as **Scores.**

If the file is not already in **Data View,** click that tab in the lower left of the screen.

In the toolbar at the top of the screen, click on **Analyze,** then **Descriptive Statistics,** then **Descriptives.** Highlight the variable **Scores** in the left window and click on the **arrow** between the windows to send that variable into the right window. Click on **Options.** Several boxes are already checked by default. Keep the check mark in the box for **Mean. Remove checkmarks** in the three boxes in the Dispersion section by clicking on them. Click **Continue** and then **OK.** This is what you will see.

Descriptives

Descriptive Statistics

	N	Mean
Scores	5	.00
Valid N (listwise)	5	

Visit the study site at www.sagepub.com/steinberg2e for practice quizzes and other study resources.

PART IV

Dispersion

"From Here to Eternity"

6 Range, Variance, and Standard Deviation

Terms: dispersion, range, variance, deviation score, standard deviation, average absolute deviation, descriptive, inferential

Symbols: s^2, σ^2, s, σ

Learning Objectives:

- Calculate various measures of dispersion—range, variance, and standard deviation
- Select the appropriate measure of dispersion for data of a given distribution shape and for a given purpose
- Know the special characteristics of the standard deviation that make it useful for further statistical calculations
- Distinguish between descriptive and inferential formulas for the variance and standard deviation

What Is Dispersion?

In Module 5, we looked at measures of central tendency. Now, we will look at another type of summary statistic—dispersion. Measures of **dispersion** are measures of spread or variability within a set of scores. They summarize, in a single value, how spread out the data are. As with central tendency, there are several measures of dispersion. I will discuss three: the range, the variance, and the standard deviation.

Range

Range is a term that is part of our everyday language. We talk about salary range, age range, range of ability, range of opinion, and so on. The concept is easy to grasp. A **range** is simply the difference between the lowest and highest scores in a data set.

What is the age range of the following 13 children taking swimming lessons on a given day?

4, 5, 6, 7, 8, 8, 8, 12, 12, 13, 13, 14, 15

The range is 11, because 15 – 4 = 11.

Now, let's take those same children, but schedule the elementary school children (aged 5–11 years) for swimming lessons on a different day from the other children. That is, we will omit the second through seventh scores. What is the age range of the remaining 7 children?

4, 12, 12, 13, 13, 14, 15

The range is still 11, because 15 – 4 = 11.

Note that the range says nothing about the number of scores in the data set. The first set has 13 scores, and the second set has 7 scores, yet in both sets the range is 11.

Now, let's drop the preschool-age child from swimming lessons on either day. That is, we will drop the "4" from both sets of ages:

5, 6, 7, 8, 8, 8, 12, 12, 13, 13, 14, 15

12, 12, 13, 13, 14, 15

Only one score has changed, but it was the critical lowest score. Now, the ranges are 10 and 3, respectively. From this example, you can see that the range, like the mode in central tendency, is the least stable of the three measures of dispersion. This shows that the range will probably vary most from one sample to the next.

Because of its simplicity, the range is a good statistic to report if you need a summary statistic in a hurry. For most purposes, however, the range is not the best measure of dispersion to report. It is simply too subject to the vagaries of the cases that happen to fall at either end of a particular sample. Also, no other statistics are based on the range. For that reason, it is not often used in experimental research.

Variance

The **variance**, symbolized s^2 (for samples) or σ^2 (for populations), is the average area distance from the mean. Said another way, it is the average of the squared deviation scores.

Unlike the range, variance is not an everyday term. Therefore, its meaning is not intuitively obvious. You can learn what it means by studying its formula. The formula for the variance is as follows:

$$s^2 = \frac{\sum(X - M)^2}{N}$$

where

X = a raw score,

M = the mean, and

N = sample size.

Look carefully at the formula. The "$X - M$" portion in the numerator is called a **deviation score**, which is the amount by which a raw score deviates from the mean. What the formula does, then, is take each score's deviation from the mean, square those values, sum those squared deviation scores, and then take the average of that sum (i.e., divide by the number of cases). Hence, the variance is a measure of average squared dispersion.

Before we calculate the variance, let me briefly mention another formula:

$$s^2 = \frac{\sum (X - M)^2}{n - 1}$$

Note that this alternative formula's denominator is "$n - 1$" rather than "N". A discussion of the two formulas and their uses can be found in the final section of this module, "Controversy: N Versus $n - 1$". You might want to read that now. For the remainder of this module, I will use the "N" formula but will place answers using the "$n - 1$" formula in parentheses immediately following the answers. The correct answer will depend on which formula you use.

Now let's calculate a variance, and then I'll talk about what we use it for. For ease of calculation, our data set will consist of the following six scores: 90, 80, 80, 70, 60, 40. Real research, of course, would consist of many more than six cases.

1. According to the formula, we first find the mean:

Score
90
80
80
70
60
40
$\Sigma = 420$

$$M = \frac{\sum X}{N} = \frac{420}{6} = 70$$

2. Next, we find the deviation scores by subtracting the mean (70) from each score:

Score	*X – M*
90	90 – 70 = 20
80	80 – 70 = 10
80	80 – 70 = 10
70	70 – 70 = 0
60	60 – 70 = –10
40	40 – 70 = –30

3. Next, we square those deviation scores:

Score	*X – M*	*(X – M)²*
90	20	$20^2 = 400$
80	10	$10^2 = 100$
80	10	$10^2 = 100$
70	0	$0^2 = 0$
60	–10	$-10^2 = 100$
40	–30	$-30^2 = 900$

4. Next, we sum the squared deviation scores:

Score	$X - M$	$(X - M)^2$
90	20	400
80	10	100
80	10	100
70	0	0
60	10	100
40	30	900
		$\Sigma = 1{,}600$

5. Finally, we divide this sum of squared deviation scores by the number of cases. This gives us the average squared deviation score:

$$s^2 = \frac{1600}{6}$$
$$= 266.667 \text{ (320.0, using } n - 1)$$

We are done. The variance is 266.667 (or 320.0). But what does that mean?

That's a reasonable question because a variance is hard to interpret, in and of itself. Although I promised you an explanation after we had completed the calculation, the truth is, the variance has very little meaning for our current purposes. In later modules (when you learn about analysis of variance, for example), we will use the variance directly. But for our current purposes, it is merely a preliminary step in computing the next measure of dispersion—the standard deviation, which we can interpret directly.

Why isn't the variance useful at this point? Well, go back to the definition of a variance. What were we trying to find? We were trying to find the average dispersion from the mean. But the scores in our sample ranged from only 40 to 90, while the calculated variance is in the hundreds. How can that be an average dispersion? It can't. And the reason it can't is that we squared each deviation score, which increased each deviation score exponentially. So the variance doesn't give us the average deviation; rather, it gives us the average squared deviation.

So why did we square each deviation score? If we wanted the average dispersion, why didn't we simply take the average of the unsquared deviation scores? Well, let's see what happens when we find the average of the unsquared deviation scores.

Score	$X - M$
90	20
80	10
80	10
70	0
60	–10
40	–30
	$\Sigma = 0$

Black holes are where God divided by zero.

—comedian Stephen Wright

$$M = \frac{\sum X}{N}$$
$$= \frac{0}{6}$$
$$= 0$$

SOURCE: FRANK & ERNEST: © Thaves/Dist. by United Features Syndicate, Inc.

So the average deviation score is 0. Are you surprised by that? You shouldn't be. You have not already forgotten the teeter-totter illustration from our discussion of central tendency in Module 5, have you? The teeter-totter showed that the mean is the point at which the deviation scores balance. That is, the values of the deviation scores on either side of the mean are the same. Thus, the average of the deviation scores around the mean will *always* be 0, no matter what the actual scores are in the data set. So the average of the deviation scores isn't a very informative measure of dispersion, is it?

And that's why we squared the scores—to get rid of the negative values. Because a negative times a negative is a positive, all the squared deviation scores become positive. When we then sum those positive squared deviation scores, we get a positive number whose value varies directly with the actual amount of dispersion within the set of scores.

The problem is, although we now have a nonzero number, it is no longer a measure of linear dispersion because squared numbers are area measures, not linear measures. Our calculated variance tells us that the average distance from the mean within the set of scores is 266.667 (or 320.0) area score points. Does that make any sense? No. What we want is a measure of linear distance, the type you would find on a number line—that is, just plain, old-fashioned score points.

And how might we get that? If we got ourselves into this mess by squaring, how might we get ourselves back out of it?

If you said we should take the square root of the variance, you are correct. The square root of the variance is the standard deviation. For our current purpose, then, the variance was nothing more than a preliminary step toward calculating the standard deviation.

Standard Deviation

The **standard deviation**, symbolized s (for samples) or σ (for populations), is the average linear distance from the mean. It is simply the square root of the variance. The formula is as follows:

$$s = \sqrt{\frac{\sum(X - M)^2}{N}}$$
$$= \sqrt{s^2}$$

Oxymoron: a figure of speech in which contradictory words are brought together. Examples include a jumbo shrimp and a standard deviation.

Continuing with our example,

$$s = \sqrt{266.667}$$
$$= 16.330 \ (17.889, \text{using } n - 1)$$

That sounds like a more reasonable dispersion value. Now, we can say that the average amount of dispersion in this set of data is 16.33 (or 17.889) points.

CHECK YOURSELF!

Summarize the features of the range, variance, and standard deviation:

	Range	*Variance*	*Standard Deviation*
Ease of calculation			
Stability over time			
Interpretability			

Note that the standard deviation is the average standardized dispersion. Why did I say "standardized"? I said "standardized" because the process of squaring and taking the square root gives a value that falls in known locations in a particular standard distribution—the normal curve. As you will see throughout the remainder of this textbook, because of the standard deviation's known relationship to the normal curve, the standard deviation is of tremendous use in score interpretation both on an individual basis and in group research.

You will be calculating standard deviations, as well as their cousins—standard errors of various sorts—frequently throughout the remainder of this textbook. Thus, you should memorize the steps for calculating a standard deviation.

CHECK YOURSELF!

What are the steps in calculating a standard deviation?

SOURCE: Reprinted with permission from Sydney Harris, ScienceCartoonsPlus.com.

Average Absolute Deviation

Not all measures of average linear dispersion are standardized. One type of average linear dispersion that is not standardized is the average absolute deviation. To calculate the **average absolute deviation**, we take the absolute value of each deviation score rather than the square of each deviation score. Then, as with the standard deviation, we sum those deviation scores and find their average. Conceptually, this is a more intuitive measure of average linear dispersion than the standard deviation is, and some statisticians find it helpful in describing score spread. However, the average absolute deviation does not fall in known locations on a normal curve; thus, its further use is limited. For that reason, it is not widely used in experimental research.

PRACTICE

1. Here is the same set of statistics test scores we worked with in Modules 3, 4, and 5.

92	83	98	76	84	87	66	82	93	85
86	87	56	94	89	73	80	82	85	88
80	84	72	88	79	92	89	84	87	85
73	84	96	88	62	75	79	87	91	90
90	86	85	89	76	69	83	84	86	94

Find the (a) range, (b) variance, and (c) standard deviation of the statistics test scores. Show your work.

2. Here are the golf scores of 10 golfers.

92	84	97	76	70	67	88	95	92	94

Find the (a) range, (b) variance, and (c) standard deviation of the golf scores. Show your work.

3. A behavior therapist is about to begin working with an ADHD (attention-deficit/hyperactivity disorder) child in a classroom setting to reduce the number of times he gets out of his seat, talks out of turn, or goes off-task. To establish a baseline, the therapist observes the child for an hour each morning and afternoon for 5 days, keeping a tally of the target behaviors. Here are the therapist's behavior tallies for the 10 observation periods.

4	8	7	5	12	9	8	5	7	4

Find the (a) range, (b) variance, and (c) standard deviation of the behaviors. Display your work.

4. Exercise 2 in Module 3 and Exercise 6 in Module 5 gave the following number of classes each of 32 students missed during the semester for a class meeting on M-W-F.

0, 1, 4, 2, 8, 1, 2, 4

5, 0, 2, 2, 1, 2, 1, 2

1, 0, 3, 1, 1, 2, 1, 2

2, 4, 0, 1, 3, 0, 3, 3

Find the (a) range, (b) variance, and (c) standard deviation for the number of absences. Show your work.

5. Exercises 3 and 7 in Module 3 required you to create frequency distributions for the number of pets owned by customers of a pet store. Exercise 7 in Module 5 required you to find the mode, median, and mean of those data. Here are those data again.

   3, 0, 1, 4, 3, 2, 2, 1, 3, 0, 2, 4, 5, 3, 2, 4, 7, 1, 1, 2

   Find the (a) range, (b) variance, and (c) standard deviations for number of pets owned.

6. Exercises 4 and 8 in Module 3 required you to create frequency distributions for the number of TV sets in households. Exercise 8 in Module 5 required you to find the mode, median, and mean number of those data. Here are those data again.

   2, 3, 1, 2, 3, 0, 2, 4, 1, 2, 4, 3, 2, 1, 1, 3, 0, 2, 1, 1, 2, 3, 2, 5, 2

   Find the (a) range, (b) variance, and (c) standard deviations for number of TV sets per household.

7. Exercise 3 in Module 5 required you to find the median anxiety score for 50 clients of a mental health clinic. Here are those data again.

Anxiety Score	*No. of Clients (Frequency)*
10	2
9	8
8	13
7	10
6	7
5	4
4	2
3	2
2	1
1	1

   Find the (a) range, (b) variance, and (c) standard deviation for anxiety score. NOTE: Don't forget to enter each score the number of times indicated by its frequency!

8. Exercise 4 in Module 5 required you to find the median number of contacts made by 20 autistic children. Here are those data again.

No. of Contacts	*No. of Children (Frequency)*
8	1
6	1
5	1
4	2
3	4
2	7
1	4

   Find the (a) range, (b) variance, and (c) standard deviation for number of contacts. NOTE: Don't forget to enter each score the number of times indicated by its frequency!

Controversy: N Versus $n - 1$

Throughout this module, I have divided by N to obtain an average. This is the proper procedure whenever we know all cases in the data set. Whenever we use a summary statistic such as a mean or a standard deviation to describe a set of data and only those data, we use the statistic **descriptively**.

There are times, however, when we want to say something about a larger set of scores from which the smaller set of scores was drawn. In Module 1, you learned that the smaller set of scores is called a sample and that summary scores in a sample are called statistics. You also learned that the larger set of scores is called a population and that summary scores in a population are called parameters. When we use a sample statistic to say something about a population parameter, we are using the statistic **inferentially**. Because the sample data do not include all cases from the population, we must infer the population parameter from the sample statistic.

Unfortunately, statistics based on sample data tend to underestimate population parameters. We correct for this underestimation by dividing by $n - 1$ instead of N. Here are the descriptive and inferential formulas for variance and standard deviation. Note that the only difference is in the denominator, given in bold.

Descriptive	*Inferential*
$s^2 = \dfrac{\sum(X-M)^2}{N}$	$s^2 = \dfrac{\sum(X-M)^2}{\mathbf{n-1}}$
$s = \sqrt{\dfrac{\sum(X-M)^2}{N}}$	$s = \sqrt{\dfrac{\sum(X-M)^2}{\mathbf{n-1}}}$

By reducing the size of the denominator in the inferential formula, we increase the size of the quotient. Try it for yourself by plugging in any trial numbers. For example, $8/4 = 2$, but $8/3 = 2.67$. The smaller denominator makes the sample variance and sample standard deviation larger. This, in turn, makes them better estimates of the population variance and population standard deviation.

The debate among statisticians is not whether or not it is appropriate to use $n - 1$ rather than N when sample statistics are used inferentially. We all agree that $n - 1$ is the proper formula in that case. Rather, the debate is whether sample data ever exist for noninferential use. That is, can we talk about the standard deviation of a sample without there being a population from which the sample was drawn? And if there is no overriding population, then doesn't the sample become the population and, hence, no longer a sample?

This is more than just semantics. You should be aware of the controversy because different answers to those questions lead to different calculation methods. Some statisticians treat samples (n) and populations (N) interchangeably until inferential statistics are presented. They use N in the denominator for the standard deviation for descriptive statistics and switch to $n - 1$ in the denominator when using the standard deviation inferentially. Other statisticians figure that there is no reason to speak of standard deviation in samples unless inference to a population is expected. They always use $n - 1$ in the denominator, whether for samples or for populations.

There is merit in both approaches. Statisticians (and, hence, textbook authors) differ in their choice. Of the numerous elementary statistics textbooks currently on the market, the choice is split about 50-50. In this textbook, I have chosen N as the divisor for descriptive statistics. Then, when we get to inferential statistics, I will switch to $n - 1$ in the denominator for the variance and standard deviation (as do all statisticians), and I will remind you of that. On this matter, however, it is important that you ask your instructor which approach he or she prefers and follow your instructor's directions. For the sake of the textbook's utility across courses taught by either algorithm, I give the answer using $/n - 1$ in parentheses following the answer using $/N$.

If you are using a statistical calculator with buttons for the standard deviation, you should determine which formula your calculator uses. You can find this information in the calculator's instruction manual. Many calculators offer both buttons, with $/N$ listed on one button and $/n - 1$ listed on the other button. The dual option again reflects the ongoing controversy.

CHECK YOURSELF!

Which denominator ($/N$ or $/n - 1$) does your instructor prefer that you use when calculating a descriptive standard deviation?

SPSS Connection

Download the file **data_score set for dispersion.sav** from www.sagepub.com/steinberg2e. These data are used in the textbook example.

Alternatively, manually enter the following six scores into the SPSS **Data View** spreadsheet: 90, 80, 80, 70, 60, 40. Click on the **Variable View** tab to define the variable. Name the variable **scr4disp**, set the decimals at 0, and label the variable as **Scores.**

If the file is not already in **Data View**, click that tab in the lower left of the screen.

In the toolbar at the top of the screen, click on **Analyze**, then **Descriptive Statistics**, then **Descriptives.** Highlight the variable **Scores** in the left window and click on the **arrow** between the windows to send that variable into the right window. Click on **Options**. Several boxes are already checked, by default. **Remove** the check mark in the box for **Mean,** by clicking on it. Keep the checkmark in the box for **Standard Deviation** in the Dispersion section but also enter checkmarks in the boxes for the **Range** and **Variance.** Click **Continue** and then **OK.** This is what you will see.

Descriptives

Descriptive Statistics

	N	Range	Minimum	Maximum	Std. Deviation	Variance
Scores	6	50	40	90	17.889	320.000
Valid N (listwise)	6					

Note that the standard deviation and the variance produced by SPSS are calculated using $n - 1$ as the divisor and so produce the answers that I placed in parentheses in the textbook. While former versions of SPSS, as with handheld calculators, permitted a choice between N and $n - 1$ denominators, recent versions of SPSS do not. As you will see in future modules, SPSS's use of the $n - 1$ standard deviation algorithm to produce z scores leads to incorrect z scores, which in turn produce an incorrect Pearson r correlation coefficient when calculated via its definitional z-score formula. These problems can be avoided by keeping N in the denominator of descriptive standard deviations and variances.

Visit the study site at www.sagepub.com/steinberg2e for practice quizzes and other study resources.

PART V

The Normal Curve and Standard Scores

"What's the Score?"

7 Percent Area and the Normal Curve

Terms: inflection points, asymptotic, normal, robust

Learning Objectives:

- Know the history of the normal distribution
- Know the conditions under which data will be distributed normally
- Know the number of standard deviations in a normal distribution
- Know the percentage of cases falling between whole standard deviation values in a normal distribution

What Is a Normal Curve?

The study of standard deviations that we completed in Module 6 leads naturally to a discussion of the normal curve. That's because a normal curve consists of known numbers of standard deviations. But first, what is a normal curve? We have had a lot to say about normal curves so far, without ever defining what a normal curve actually is.

Mathematically, a normal curve is a symmetric bell-shaped curve having **inflection points** (changes in the curve direction) at exactly 1 standard deviation above and 1 standard deviation below the mean. Look at Figure 7.1. At the points indicated, the steepness of the curve changes from steeply down to gently out.

As with all graphs we have looked at thus far, the *X*-axis shows scores on a variable, and the *Y*-axis shows frequency or percentage of cases. As we saw in Module 5, in a normal curve the mean, median, and mode all fall at the same point. Frequency decreases as the distance from the mean increases in either tail. The normal curve is also **asymptotic**, meaning that it never quite reaches the *X*-axis. This allows for additional cases to fall way out in the tails of the distribution.

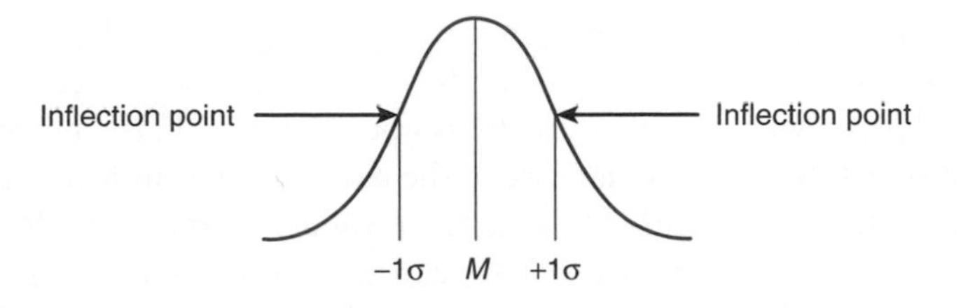

Figure 7.1 Inflection Points in the Normal Curve

A normal curve is only a theoretical distribution. It is how scores would distribute themselves if an infinite number of scores were collected. Of course, we never do have an infinite number of scores. Fortunately, however, even modest numbers of scores tend to distribute themselves approximately normally as long as that *variable* is normally distributed in the underlying population. Thus, in a sample of only 30 or 40 cases, a normal shape will typically emerge when the underlying population is normally distributed. With a sample of several hundred cases from the same normally distributed population, normality is practically assured. It is important to remember, however, that the normal curve and its percent areas that we will examine in this module are based on an infinite sample size. For that reason, the smaller the sample size, the greater the deviation from the percentages associated with a normal curve.

History of the Normal Curve

How was the normal curve discovered? In the early 1800s, the German mathematician Friedrich Gauss, when estimating the orbits of planets, noticed that the errors he made in orbit estimations distributed themselves symmetrically around the planets' true orbits. Furthermore, small errors close to the true orbits were much more frequent than large errors far from the true orbits. When he plotted the magnitude of the errors against the frequency of those errors, he found that the errors formed a symmetric bell-shaped curve. For that reason, the curve was initially called the curve of normal errors and was applied primarily to measures from the physical world. Later, the curve came to be called the Gaussian curve, after the man who discovered it. (Actually, Gauss rediscovered it: Laplace discovered it simultaneously, and DeMoivre had discovered it a century earlier.) Gauss and others derived a formula describing the normal curve. I will not present that formula here.

> ! Experimentalists think it is a mathematical theorem, while mathematicians believe it to be an experimental fact.
>
> —Gabriel Lippman, discussing the normal curve

Later, Sir Francis Galton, an Englishman studying the differences in intellectual capacity among people, noticed that measures of mental ability distribute themselves in the same bell-shaped curve. What Gauss, Galton, and numerous mathematicians since then have discovered is that many psychological, educational, and physical attributes are distributed in a Gaussian—now usually called **normal**—manner. Intelligence, for example, is normally distributed. So is achievement. So is height. Then again, so are shots at a target, time to a correct answer in a memory task, diameter of giant redwood trees, life span of the one-celled protist *Euglena gracilis*, and innumerable other natural and derived phenomena. This is because deviations from the mean for any continuously scaled variable whose occurrence is independent of other events distribute themselves normally. You will learn more about this symmetric tendency when you study the laws of probability in Modules 10 and 11.

Uses of the Normal Curve

Unfortunately, in real life not all events are independent of other events. Moreover, some events that appear to be independent are, in fact, not independent. Thus, although the normal curve remains useful for examining score distributions, its fit to real-world data, which is typically somewhat skewed or rather lumpy, is not perfect. Still, the normal curve is a useful model against which to evaluate data when the data's deviation from normality is small. In statistical terms, the normal curve is said to be **robust** to minor violations of its shape assumptions. In lay terms, this means that we can use the percentages associated with the normal curve to interpret our data even when our data are only approximately normally distributed. This, as we already saw, is usually the case as long as the underlying population data are normally distributed and the sample size is sufficiently large.

The normal curve is widely used in both physical and social sciences for interpreting individual scores. This is because a normal curve consists of known numbers of standard deviations—approximately 3 above the mean and 3 below the mean. This is demonstrated in Figure 7.2.

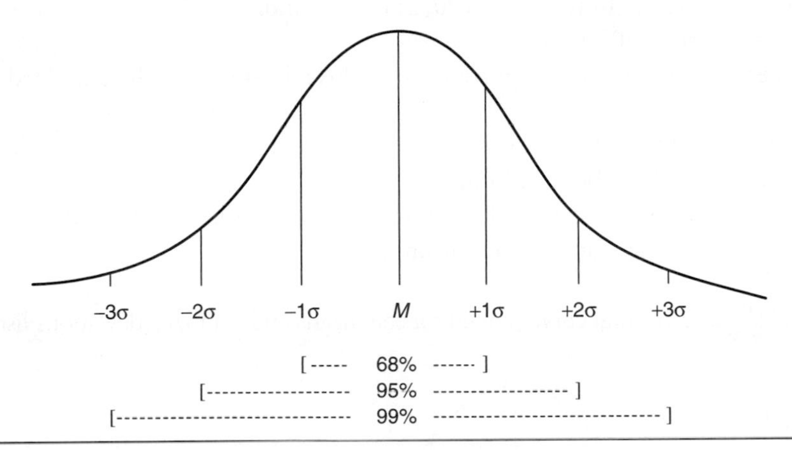

Figure 7.2 Standard Deviations in the Normal Curve

If we were to measure the area under the curve, we would find that fixed percentages fall between those standard deviations. Approximately 68% of the area falls within ±1 standard deviation from the mean, approximately 95% falls within ±2 standard deviations from the mean, and approximately 99% falls within ±3 standard deviations from the mean. Because area under the curve at any score location along the *X*-axis represents the percentage of cases on the *Y*-axis, it follows that approximately 68%, 95%, and 99% of actual scores fall within ±1, ±2, and ±3 standard deviations from the mean, respectively.

> Although this may seem a paradox, all exact science is dominated by the idea of approximation.
>
> —Bertrand Russell

These percentages are very important in hypothesis testing and will appear over and over throughout the remainder of this textbook. Therefore, you should memorize them.

CHECK YOURSELF!

What percentage of cases in a normal distribution fall within ±1, ±2, and ±3 standard deviations from the mean?

Exact percentages are somewhat higher. The more exact percentages, to two decimal places, are shown in Figure 7.3.

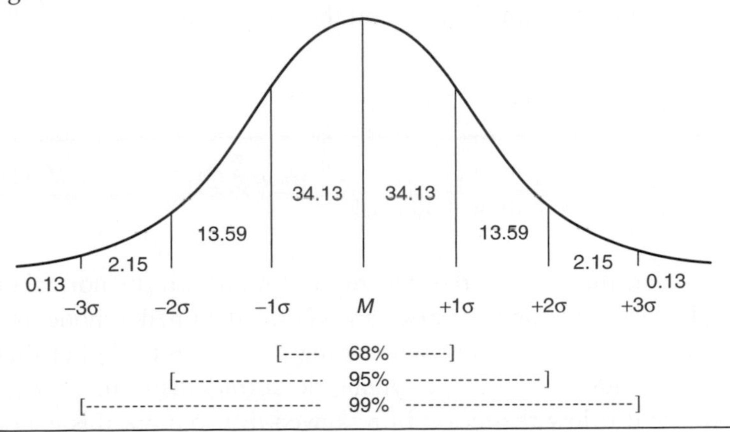

Figure 7.3 Normal Curve Percentages, to Two Decimal Places

Now let's add actual means and standard deviations to the normal curve. Here are the means and standard deviations for a variety of common scores, as well as for the 50 statistics test scores we worked with in Modules 3 through 6:

- The mean score on an IQ test is 100, and the standard deviation is 15 (or 16, depending on the test used).
- The mean score on a single portion of the SAT is 500, and the standard deviation is 100.
- The mean height of U.S. adult males is 69.50 in., and the standard deviation (this varies by source) is about 2.00 in.
- The mean score for the 50 statistics test scores we worked with in Modules 3 through 6 is 83.46, and the standard deviation is 8.41.

Figure 7.4 shows a normal curve plotted for the means and standard deviations listed above.

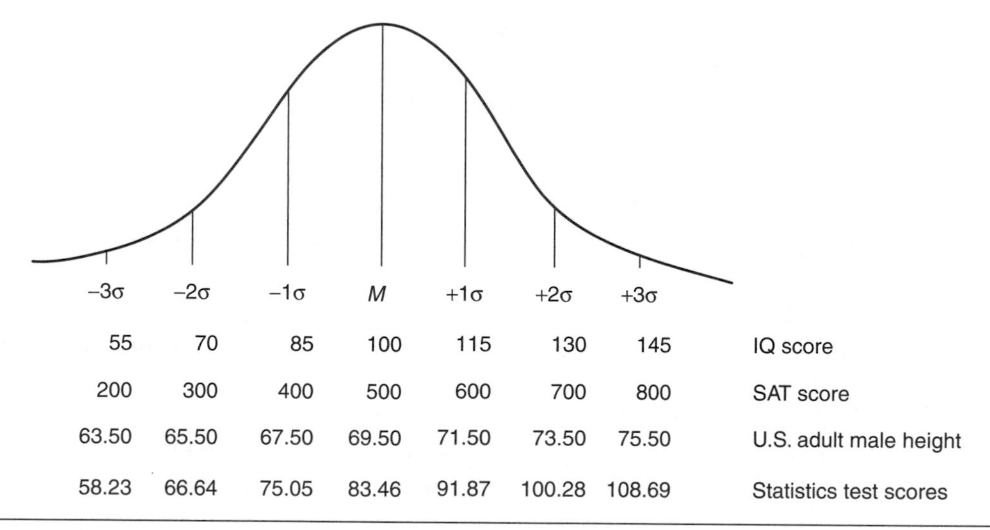

Figure 7.4 Normal Curve Showing Means and Standard Deviations for a Variety of Common Scores

Notice that each of the score distributions (IQ, SAT, U.S. male height, statistics test scores) has its own standard deviation value: 15 points, 100 points, 2.00 in., and 8.41 points, respectively. However, regardless of the standard deviation *value*, the standard deviations fall in the same location on the curve. Thus, if you score 1 standard deviation above the mean in IQ and 1 standard deviation above the mean in height, your scores on both variables will fall in the same location on the normal curve.

CHECK YOURSELF!

Dana scores 700 on one part of the SAT. If the SAT is a good indicator of intelligence, what would you expect Dana to score on the IQ test?

Finally, notice that the data for the statistics test do not fit the normal curve very well. This is because (1) there were only 50 cases and (2) the data had a moderate negative skew (high mean). Recall that the actual test scores ranged from 56 to 98 but that a majority of the scores were in the 80s and 90s. In a symmetric normal distribution, equal proportions of cases fall above and below the mean. In a skewed distribution, this is not the case. The small number of extreme scores on the lower end of the distribution inflated the size of the standard deviation. A normal curve allows for the possibility of scores at +3 standard deviations or higher. Given the standard deviation of 8.41, this implies scores of 108 or more. However, although most of the scores were, indeed, concentrated in the higher end of the distribution, no one did or could score above 100. This is another reason why a competent researcher graphs the data before calculating statistics. It is misleading to use percentages under the normal curve for interpreting seriously skewed data.

Once we know the mean and standard deviation for any normally distributed set of data, we can use the percentages under the curve to interpret individual scores. Using the means and standard deviations in Figure 7.4, along with the percentages in Figure 7.3, we can answer questions about the relative number of people scoring in certain portions of the curve, as in the following examples:

- What percentage of U.S. adult males are taller than 6 ft 3½ in.? Looking at the curve, that would be taller than 75.50 in., which is taller than 3 standard deviations above the mean, which is 0.13%. (That's *point* 13%, not 13%.)
- What percentage of people have IQs between 85 and 115? Looking at the curve, that would be IQs between ±1 standard deviation from the mean, which is 34.13% + 34.13% = 68.26%.

"When you lie about yourself, is it to appear closer to or farther away from the middle of the bell curve?"

SOURCE: William Haefeli, New Yorker Cartoon 8/18/2003 © William Haefeli/Condé Nast Publications/ www.cartoonbank.com

CHECK YOURSELF!

What percentage of people have higher SAT scores than Dana? (Recall that Dana scored 700.)

PRACTICE

Use the percentages in Figure 7.3 and the values in Figure 7.4 to answer the following questions:

1. What percentage of people have IQs between 70 and 130?
2. What percentage of people have IQs below 70? Above 130? Are the percentages the same?
3. What percentage of people have IQs between 85 and 100? Between 100 and 115? Are the percentages the same?
4. What percentage of SAT takers score below 600?
5. What percentage of SAT takers score between 500 and 600?
6. What percentage of SAT takers score above 600? Below 600? Are the percentages the same?
7. Someone with an average IQ and who completed a college preparatory program in high school with average grades should score about what on the SAT?
8. What percentage of U.S. adult males are between 5 ft 11½ in. and 6 ft 3½ in.?
9. What percentage of U.S. adult males are shorter than 5 ft 3½ in.? Taller than 6 ft 3½ in.? Are the percentages the same?
10. An adult American man who is 69.50 in. tall is taller than what percentage of other adult American men?

Looking Ahead

The symmetry and inflection points of the normal curve lead not only to the fixed percentages we have seen between whole standard deviation values but also to fixed percentages above and below *any* point along the curve. It is this feature that makes the normal curve so useful: by knowing where a score falls along the *X*-axis, we can determine the exact proportion of cases that fall above or below it. Because of the great importance of the normal curve in statistical analysis, we will continue our examination of the normal curve in the next module. There, we will see its use in individual score interpretation.

Moreover, sample statistics (e.g., means of repeated samples) also fall in a normal curve. As you will see in Module 15, means from repeated samples tend to distribute themselves normally even when the underlying population distribution is not normally distributed. For that reason, most of the inferential statistics you will come across later in this textbook make use of percentages under the normal curve. Thus, using the normal curve now to interpret individual scores sets the stage for its later inferential use.

Visit the study site at www.sagepub.com/steinberg2e for practice quizzes and other study resources.

8 *z* Scores

Terms: standard score, *z* score, rescale

Symbol: z

Learning Objectives:

- Know the advantages of standard scores over raw scores
- Understand the process of rescaling numerator units into denominator units
- Calculate a *z* score
- Use a normal curve table to determine percentage above, below, or between given *z* scores

What Is a Standard Score?

A **standard score** is a score expressed in terms of some standardized unit of measurement. Whereas a rank score is an ordinal measurement and can tell us only about the relative position of a score within a group of scores, a standard score is an interval measure. This means that it tells us that score's position relative to both the central tendency and the dispersion of scores within the group of scores. Thus, a standard score tells the value of a score relative to all other scores.

Benefits of Standard Scores

Let's start with an example. Suppose Bob earns a score of 72 on Test A and a score of 84 on Test B. On which test did he do better?

Did you say Test B? But what if Test A and Test B did not have the same number of possible points? Without that information, we don't know how well Bob did on the two tests. Raw scores alone do not give us that information.

Now, suppose I tell you that Test A had 95 possible points, and Test B had 105 possible points. With this information, Bob's performance on Test A is 72/95 = 76%. His performance on Test B is 84/105 = 80%. Now, can we tell on which test Bob did better?

Did you say Test B? On a percentage basis, it certainly does appear that he did better on Test B.

But wait. Maybe Bob's 80% on Test B was the lowest score among all test takers on that test, and his 76% on Test A was the highest score among all test takers on that test. Here we are, confused again. What is the missing piece of information? Yes, it's some indication of how Bob did compared with the other people who took the tests.

Here are the scores and the ranks of everyone, including Bob (see Table 8.1). Bob's score is indicated by an asterisk. His score and rank are given in boldface.

Table 8.1 Bob's Scores on Test A and Test B

Test A (of 95)		*Test B (of 105)*	
Score	*Rank*	*Score*	*Rank*
73	1	105	1
***72**	**2**	104	2
60	3	98	3
59	4	94	4
57	5	92	5
56	6	87	6
52	7	85	7
48	8	***84**	**8**
36	9	82	9
28	10	80	10
72/95 = 76%		84/105 = 80%	

We can see that Bob's score of 72 on Test A put him in the second-rank position out of 10 students, whereas his score of 84 on Test B put him in the eighth-rank position out of 10 students. Percentage-wise, he did better on Test B, but rankwise he did better on Test A. It seems that his score looks either good or bad depending on how we look at it.

Now, let's complicate things even further. Let's add Test C. Like Test A, Test C has 95 possible points. And as he did on Test A, Bob scores 72 on Test C. Also as on Test A, Bob's rank on Test C is second (see Table 8.2). On which test did Bob do best?

Table 8.2 Bob's Scores on Tests A, B, and C

Test A (of 95)		*Test B (of 105)*		*Test C (of 95)*	
Score	*Rank*	*Score*	*Rank*	*Score*	*Rank*
73	1	105	1	95	1
***72**	**2**	104	2	***72**	**2**
60	3	98	3	71	3
59	4	94	4	70	4
57	5	92	5	68	5
56	6	87	6	67	6
52	7	85	7	66	7
48	8	***84**	**8**	64	8
36	9	82	9	62	9
28	10	80	10	45	10
72/95 = 76%		84/105 = 80%		72/95 = 76%	

For both Tests A and C, Bob's score is 72, and his rank is second. Yet his performance is clearly different on the two tests. On Test A, Bob's score is very close to the top score, and the other scores are much lower than his. On Test C, in contrast, his score is a long way from the top score, and he did not score much higher than most of the other test takers. So on which test do you think Bob did best—Test A, B, or C?

As you can see, to fully evaluate Bob's performance we need to know more than how many points he scored (72, 84, 72). We also need to know more than what percentage he earned (76%, 80%, 76%). We also need to know more than his relative standing (second, eighth, second). We need to know how spread out the other scores are around his score and where his score falls within the spread of scores. A standard score gives us this information.

The most common type of standard score is a *z* score. A ***z* score** is the raw score reexpressed in standard deviation units. In other words, it is the number of standard deviation units that a score is above or below the mean. The formula for a *z* score is as follows:

$$z = \frac{X - M}{s}$$

where

X = any particular raw score,

M = the mean for the test, and

s = the standard deviation for the test.

Before we go on, let's take a look at the logic of the *z*-score formula. Let me ask you, how many feet is 18 in.? Can you answer that? Of course, you can. How did you figure out the answer? You learned how to answer that type of question in elementary school. To convert inches into feet, divide the inches by the value of a foot in inches:

$$\frac{18 \text{ in.}}{12 \text{ in./ft}}$$

The inches label in the numerator and the inches label in the denominator cancel each other, and you are left with 18/12 ft, or 1½ ft. You may have forgotten how to explain how to do it, but you intuitively knew what you needed to do, didn't you?

You may never have realized it, but whenever you want to convert one type of measurement (such as inches) into another type of measurement (such as feet), you **rescale** (change the scale of) the numerator by dividing it by the value of the measure you want to rescale it into.

That's a universal law. Inches into feet or, in the case of the *z*-score formula, distance from the mean (look at the numerator in the *z*-score formula) into standard deviation units (look at the denominator in the *z*-score formula). Voilà—the distance from the mean is rescaled into standard deviation units:

$$z = \frac{X - M}{s}$$

Now, recall the definition of a *z* score. A *z* score is the number of standard deviation units a score is above or below the mean. Because the definition of a *z* score is nothing more than the number of standard deviation units a score is above or below the mean, it is identical to the standard deviation row on our normal curve diagram. So let's add a row for *z* scores to the diagram (see Figure 8.1).

Q: Why did the tired man study the normal curve?

A: He was trying to catch up on his z's.

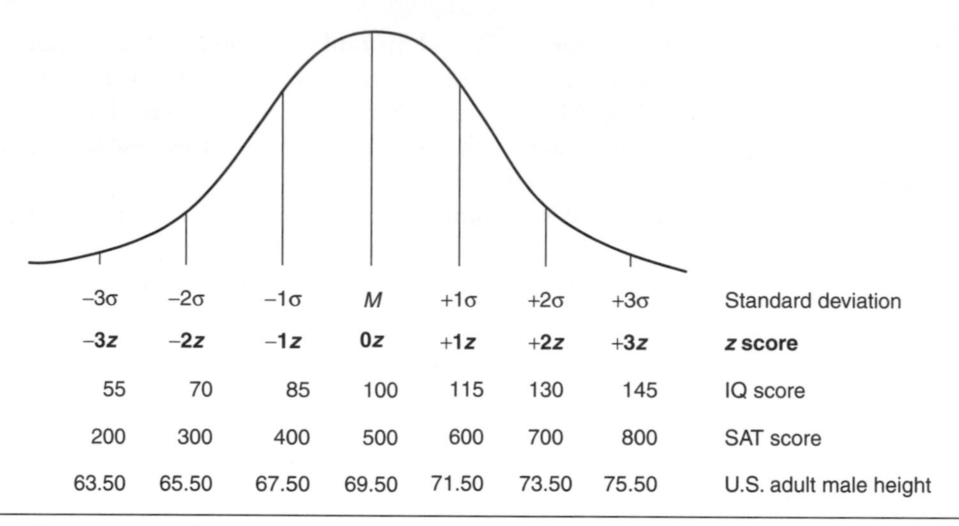

Figure 8.1 Normal Curve, With z Scores Indicated

Note that positive z scores fall above the mean, and negative z scores fall below the mean. When you calculate a z score, this is a good way to check yourself. If the raw score is above the mean, the z score must be positive; if the raw score is below the mean, the z score must be negative. If you find a negative z score for an above-average raw score or if you find a positive z score for a below-average raw score, you have made a calculation error.

CHECK YOURSELF!

If you score 1 standard deviation above the mean on a test whose scores are normally distributed, what will your z score be on that same test? Why?

Calculating z Scores

Let's do some examples, to see how this works.

Assume that your IQ is 120. What is your z score?

$$\frac{120 - 100}{15} = \frac{20}{15} = +1.33$$

Assume that you score 660 on the SAT. What is your z score?

$$\frac{660 - 500}{100} = \frac{160}{100} = +1.66$$

Assume that your friend Isaac scores 270 on the SAT. What is his z score?

$$\frac{270 - 500}{100} = \frac{-230}{100} = -2.30$$

The beauty of converting raw scores to z scores is that z scores distribute themselves in a normal curve; recall that a normal curve has known percentage areas above and below any point in the curve. Thus, once we know a z score, we also know exactly what percentage of a population scores above or below that score.

Let's see how common or rare your IQ of 120 is. Appendix A contains the normal curve table. A portion of the table is reproduced as Table 8.3.

Notice the diagram (Figure 8.2) at the top of the table. It shows a hypothetical z score, with the area at or below that z score shaded. The diagram is not meant to indicate that the z score you look up will fall at exactly that location in the curve. It is meant to indicate that whatever z score you look up, the tabled values will give the percentage *at or below* that z score.

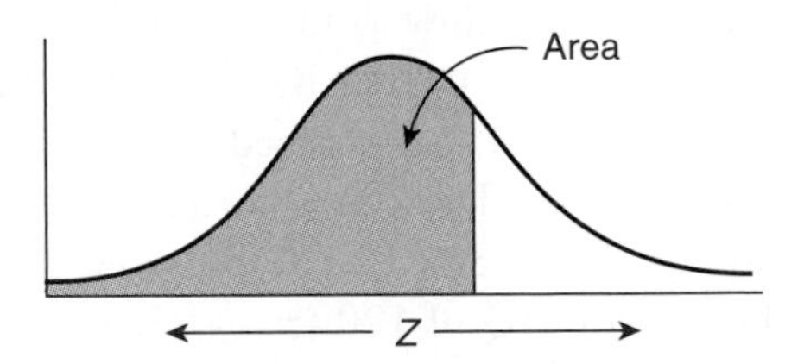

Figure 8.2 Shaded Area Below z

Table 8.3 A Portion of the Normal Curve Table

z	*Area*	*z*	*Area*	*z*	*Area*	*z*	*Area*
0.12	.5478	0.52	.6985	0.92	.8212	1.32	.9066
0.13	.5517	0.53	.7019	0.93	.8238	1.33	.9082
0.14	.5557	0.54	.7054	0.94	.8264	1.34	.9099
⋮	⋮	⋮	⋮	⋮	⋮	⋮	⋮

Now, look at the tabled values below the diagram. There are two columns repeated throughout the table: z score and percentage area. The z-score values range from −3.00 to +3.00 (which, as you learned in Module 7, accounts for more than 99% of all scores). The column next to each z score gives the percentage of scores falling at or below that score, as depicted in the diagram at the top of the table.

Let's finally look up your IQ score. Your IQ score was 120, which, you will recall, is a z score of +1.33. Now, find a z score of +1.33 in the normal curve table. Then, look at the column to the right of that z score to find the percentage of people having IQ scores at or below yours. What percentage of people score at or below a z score of +1.33? Yes, 90.82% have an IQ the same or lower than yours.

Now, if a normal curve includes all people (100%), and we just found the percentage of people having the same IQ as or a lower IQ than you, then what percentage of all people have an IQ higher than yours? Yes, of course, 100% − 90.82%, or 9.18%, have a higher IQ than you.

SOURCE: FRANK & ERNEST: © Thaves/Dist. by United Features Syndicate, Inc.

CHECK YOURSELF!

Recall that your friend Isaac scored 270 on the SAT. What percentage of people scored higher than Isaac? What percentage scored lower?

Finally, by locating your IQ *z* score within a normal curve, we know not only the percentage of people having an IQ at or below yours and the percentage having IQs above yours, but also the percentage of people having IQs between the mean and yours. How? Recall that in a normal curve, the mean is also the median and that 50% of scores are above a median and 50% are below a median. Thus, in a normal curve, 50% are above the mean and 50% are below the mean. In Figure 8.3, I have indicated those percentages, as well as the percentages above and below your IQ of 120 ($z = +1.33$).

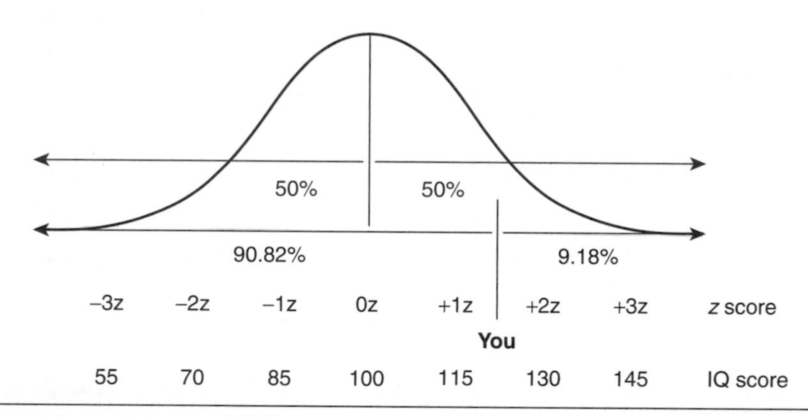

Figure 8.3 Normal Curve, With Percent Area Indicators

As you can see, if 50% of all people have IQs above the mean and 9.18% have IQs above your IQ, then 40.82% (50.00% – 9.18%) must have IQs between the mean IQ and your IQ.

Here's another way of looking at it. If 90.82% of all people have IQs the same as or lower than your IQ and 50.00% have IQs the same as or lower than the mean, then 40.82% (90.82% – 50.00%) must have IQs between the mean and your IQ.

PRACTICE

1. Refer to Figure 8.1. Jerry's IQ is 136. What percentage of people have IQs (a) the same as or lower than Jerry's, (b) higher than Jerry's, and (c) between the mean and Jerry's?
2. Refer to Figure 8.1. Luz's IQ is 123. What percentage of people have IQs (a) the same as or lower than Luz's, (b) higher than Luz's, and (c) between the mean and Luz's?
3. Refer to Figure 8.1. Drescher scored 430 on one portion of the SAT. What percentage of people have SAT scores (a) the same as or lower than Drescher's, (b) higher than Drescher's, and (c) between the mean and Drescher's?
4. Refer to Figure 8.1. Tara scores 570 on one portion of the SAT. What percentage of people have SAT scores (a) the same as or lower than Tara's, (b) higher than Tara's, and (c) between the mean and Tara's?

5. Refer to Figure 8.1. Shuai stands 68.00 in. tall. What percentage of people are (a) the same height as or shorter than Shuai, (b) taller than Shuai, and (c) between the mean and Shuai's height?
6. Refer to Figure 8.1. Joe stands 67.00 in. tall. What percentage of American men are (a) the same height or shorter than Joe, (b) taller than Joe, and (c) between the mean and Joes' height?
7. A college is located in the geographic center of the mainland United States. The admissions officer examines the east-west distance of the college from the hometown of each incoming resident student. Incoming resident students living to the east of the college are scored as + distances, and incoming resident students living to the west of the college are scored as − distances. The admissions officer finds distance to be normally distributed around a mean of 0.00 miles (the college's location) and a standard deviation of 163 miles. Incoming resident student Amy Brescia lives 82 miles east of the college. What percentage of incoming resident students live (a) in the same town as or west of Amy, (b) east of Amy, and (c) between the college and Amy?
8. Refer to the information in Exercise 7 above. Incoming resident student Eddie Franz lives 107 miles west of the college. What percentage of incoming resident students live (a) in the same town as or west of Eddie, (b) east of Eddie, and (c) between the college and Eddie?
9. Refer to the information in Exercise 7 above. Incoming resident student Belinda lives 236 miles east of the college. What percentage of incoming resident students live (a) in the same town as or west of Belinda, (b) east of Belinda, and (c) between the college and Belinda?
10. The mean human gestation period is 266 days. Actual time to delivery is approximately normally distributed with a standard deviation of about 12 days. Marcelle goes into labor and gives birth to a child 245 days after conception. What percentage of women give birth (a) at the same point in pregnancy as or earlier than Marcelle, (b) at a later point in pregnancy than Marcelle, and (c) between their due date and the point at which Marcelle gave birth?
11. Refer to the information in Exercise 10 above. Portia goes into labor and gives birth to a child 270 days after conception. What percentage of women give birth (a) at the same point in pregnancy as or earlier than Portia, (b) at a later point in pregnancy than Portia, and (c) between their due date and the point at which Portia gave birth?
12. Refer to the information in Exercise 10 above. Bo goes into labor and gives birth to a child 234 days after conception. What percentage of women give birth (a) at the same point in pregnancy as or earlier than Bo, (b) at a later point in pregnancy than Bo, and (c) between their due date and the point at which Bo gave birth?

Comparing Scores Across Different Tests

z Scores are useful not only for determining relative performance on a particular test but also for comparing performance across different tests. This is because the formula for a *z* score contains a value for the mean and also a value for the standard deviation (see Figure 8.4).

$$Z = \frac{X - M \leftarrow}{s \leftarrow}$$

Moreover, the formula for the mean itself contains a value for the sample size.

$$M = \frac{\Sigma X}{N \leftarrow}$$

Thus, a *z* score accounts for central tendency (*M*), dispersion (*s*), and sample size (*N*). These features of a *z* score permit us to compare someone's scores on different tests even when the standard deviations, means, and sample sizes of the tests differ. Once raw scores are converted into *z* scores, they can be interpreted via the normal curve table regardless of the actual means, standard deviations, and sample sizes of the various tests.

To see how this works, let's return to Bob's scores on Test A, Test B, and Test C. We finally have a mechanism to definitively answer the question "On which test did Bob do best?" We will calculate his *z* score on each test and then look up those *z* scores in the normal curve table.

Here are the data again. Bob's raw scores on each test are in bold and marked with asterisks. Below each test score distribution, I have provided that test's mean and standard deviation (see Table 8.4).

Table 8.4 Raw Scores, Means, and Standard Deviations for Tests A, B, and C

Test A (of 95)		*Test B (of 105)*		*Test C (of 95)*	
Score	*Rank*	*Score*	*Rank*	*Score*	*Rank*
73	1	105	1	95	1
***72**	**2**	104	2	***72**	**2**
60	3	98	3	71	3
59	4	94	4	70	4
57	5	92	5	68	5
56	6	87	6	67	6
52	7	85	7	66	7
48	8	***84**	**8**	64	8
36	9	82	9	62	9
28	10	80	10	45	10
72/95 = 76%		84/105 = 80%		72/95 = 76%	
M = 54.10		*M* = 91.10		*M* = 68.00	
s = 13.41		*s* = 8.53		*s* = 11.59	
(14.138, using *n* – 1)		(8.987, using *n* – 1)		(12.220, using *n* – 1)	

Here are Bob's *z* scores for the three tests and the associated percentages from the normal curve table:

Test A

$$z = \frac{72 - 54.10}{13.41} = +1.33(+1.27, \text{using}/n - 1)$$

$$\text{Percentage below} = 90.82(89.80, \text{using}/n - 1)$$

Test B

$$z = \frac{84 - 91.10}{8.53} = -0.83(-0.79,\ \text{using}/n - 1)$$

$$\text{Percentage below} = 20.33(21.48,\ \text{using}/n - 1)$$

Test C

$$z = \frac{72 - 68.00}{11.59} = +0.35(+0.33,\ \text{using}/n - 1)$$

$$\text{Percentage below} = 63.68(62.93,\ \text{using}/n - 1)$$

Bob outscored a greater percentage of theoretical test takers on Test A. Therefore, he did, as we had suspected earlier in this module, best on Test A.

CHECK YOURSELF!

If you plan to compare scores across different tests, why should you prefer that the scores be reported as z scores rather than as raw scores, percentage scores, or ranks?

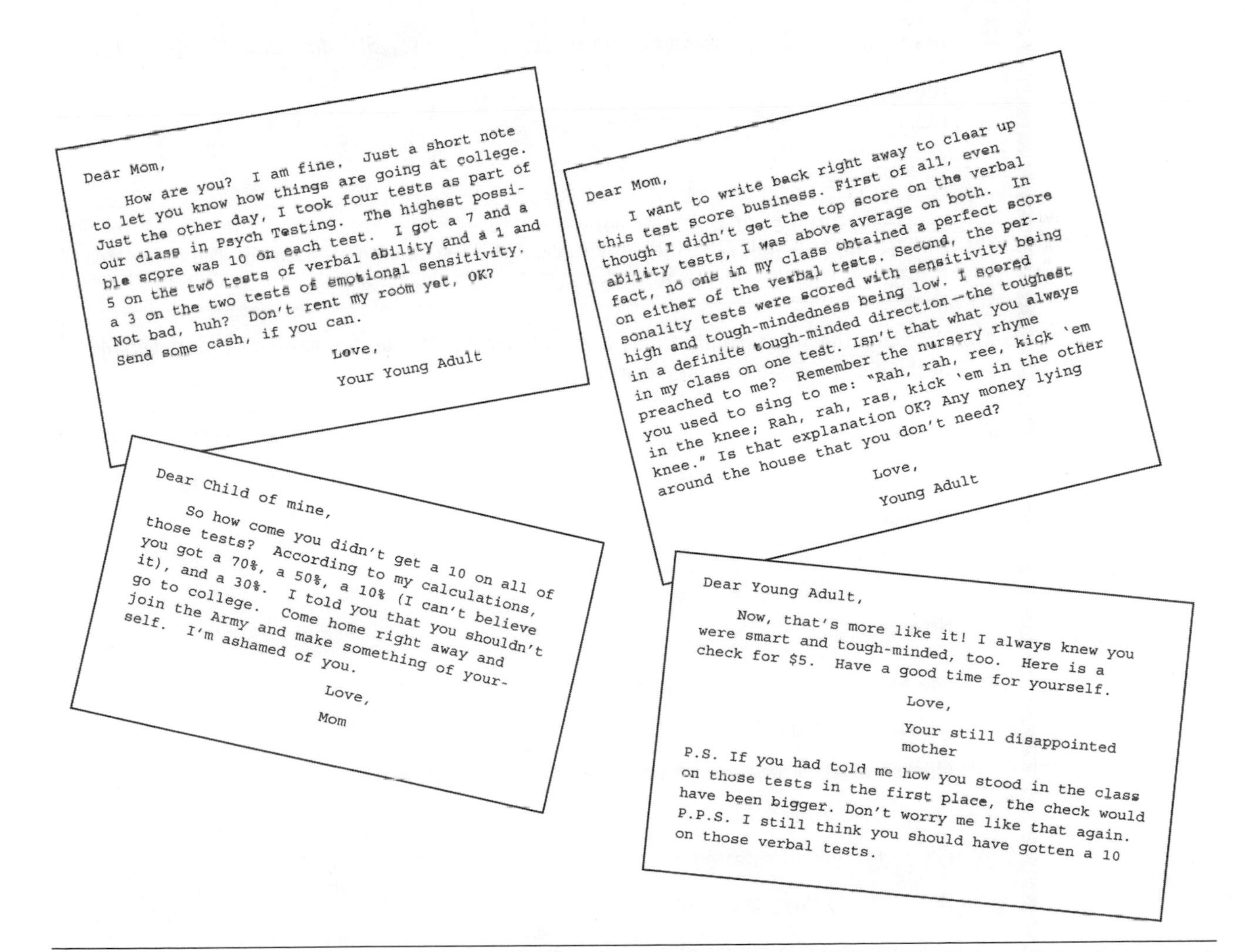

Dear Mom,

How are you? I am fine. Just a short note to let you know how things are going at college. Just the other day, I took four tests as part of our class in Psych Testing. The highest possible score was 10 on each test. I got a 7 and a 5 on the two tests of verbal ability and a 1 and a 3 on the two tests of emotional sensitivity. Not bad, huh? Don't rent my room yet, OK? Send some cash, if you can.

Love,
Your Young Adult

Dear Child of mine,

So how come you didn't get a 10 on all of those tests? According to my calculations, you got a 70%, a 50%, a 10% (I can't believe it), and a 30%. I told you that you shouldn't go to college. Come home right away and join the Army and make something of yourself. I'm ashamed of you.

Love,
Mom

Dear Mom,

I want to write back right away to clear up this test score business. First of all, even though I didn't get the top score on the verbal ability tests, I was above average on both. In fact, no one in my class obtained a perfect score on either of the verbal tests. Second, the personality tests were scored with sensitivity being high and tough-mindedness being low. I scored in a definite tough-minded direction—the toughest in my class on one test. Isn't that what you always preached to me? Remember the nursery rhyme you used to sing to me: "Rah, rah, ree, kick 'em in the knee; Rah, rah, ras, kick 'em in the other knee." Is that explanation OK? Any money lying around the house that you don't need?

Love,
Young Adult

Dear Young Adult,

Now, that's more like it! I always knew you were smart and tough-minded, too. Here is a check for $5. Have a good time for yourself.

Love,

Your still disappointed mother

P.S. If you had told me how you stood in the class on those tests in the first place, the check would have been bigger. Don't worry me like that again.
P.P.S. I still think you should have gotten a 10 on those verbal tests.

PRACTICE

13. During the past year, the daily price of gasoline per gallon at four local gas stations has been approximately normally distributed. The four gasoline stations' yearly data and current selling prices are listed below:

Station	*Mean Price*	*Standard Deviation of Price*	*Current Price*
Elm Street	1.45	0.08	1.64
Third Avenue	1.53	0.11	1.67
Devon Drive	1.56	0.12	1.65
Morton Lane	1.49	0.13	1.62

A citizen's advocate accuses one of the stations of current price gouging (charging an unreasonably high price). Based on current selling prices, which station appears to be price gouging, relative to its annual mean price?

14. The clerks of the traffic courts at three neighboring towns keep track of the number of miles by which convicted speeders drive over the speed limit. The data, which are approximately normally distributed, are listed below:

Town	*Mean mph Over Limit*	*Standard Deviation Over Limit*
Hillsboro	22	9
Lowville	18	10
Media	19	6

Decia was just issued a citation for driving 27 mph over the speed limit. If each town sets its fines according to how fast the speeder was driving relative to other convicted speeders in that town, in which town will Decia's fine be lowest?

15. Here are the means and standard deviations for various animals' gestation periods. Gestation times are normally distributed.

Animal	*Mean Days*	*Standard Deviation Days*
Human	266	12.00
Chicken	21	0.50
Coral snake	70	1.00
Pig	114	2.50
Horse	340	17.00

The following animals were just born:

Human at 251 days' gestation

Chicken at 20.5 days' gestation

Coral snake at 68.5 days' gestation

Pig at 110 days' gestation

Horse at 323 days' gestation

Which animal was born most prematurely?

SPSS Connection

Download the file data_scores on tests a b c.sav from www.sagepub.com/steinberg2e. These data are used in the textbook example.

Alternatively, click on the Data View tab at the bottom of the SPSS screen and manually enter the following data into the first four columns of the spreadsheet.

Abe	73	104	95
Bob	72	84	72
Carl	60	105	68
Don	59	98	71
Earl	57	92	66
Fred	56	87	70
Guy	52	94	62
Hal	48	82	64
Ike	36	80	67
Jay	28	85	45

Click on the Variable View tab to define the variables. Name first variable name, select string as the type of variable, and label the variable as Name. Name the second variable testa, set the decimals at 0, and label the variable as Score on Test A. Name the third variable testb, set the decimals at 0, and label the variable as Score on Test B. Name the fourth variable testc, set the decimals at 0, and label the variable as Score on Test C.

If the file is not already in Data View, click that tab in the lower left of the screen.

In the toolbar at the top of the screen, click on Analyze, then Descriptive Statistics, then Descriptives. Highlight all three of the numeric variables (Score on Test A, Score on Test B, and Score on Test C) in the left window and click on the arrow between the windows to send those variables into the right window. Click in the box in the bottom left, labeled Save Standardized Values as Variables. Click on Options. Several boxes are already checked, by default. Keep the checkmark in the box for Mean. In the Dispersion section, keep the checkmark in the box for Standard Deviation, but remove the checkmarks in the boxes for all other statistics. Click Continue and then OK. This is what you will see.

Descriptives

Descriptive Statistics

	N	Mean	Std. Deviation
Score on Test A	10	54.10	14.138
Score on Test B	10	91.10	8.987
Score on Test C	10	68.00	12.220
Valid N (listwise)	10		

This is not the output of interest. Select the Data View tab. Notice that z scores have been added to the data file. SPSS does not print the z scores on the output. It simply adds them to the data file as new variables. They are now available as variables for further analyses.

Recall the formula for a z score: $z = (X - M)/s$. Recall also that SPSS calculates the standard deviation using $n - 1$ in its denominator. This is the standard deviation printed on the output, and it is the one that SPSS uses to calculate z scores. As mentioned in Module 6, these z scores will produce an incorrect Pearson coefficient (r) when calculated using the definitional formula. This problem can be avoided by calculating the standard deviation with N, rather than $n - 1$, in the denominator.

Visit the study site at www.sagepub.com/steinberg2e for practice quizzes and other study resources.

9 Score Transformations and Their Effects

Term: transformation

Symbols: IQ, T, CEEB

Learning Objectives:

- Know the effect of score transformations on the percentage of cases falling above, below, or between various transformed scores
- Know the effect of score transformations on the mean and standard deviation of the set of scores
- Know the values of the mean and standard deviation for common standardized scores
- Convert scores from one type of standardized score to another

Why Transform Scores?

Sometimes it is desirable to adjust a set of scores so that the scores are higher, lower, more spread out, or less spread out than they originally were. When an adjustment is applied equally to all scores in a set, the adjustment is called a **transformation**. For example, perhaps a test was a little more difficult than the instructor intended. In that case, the instructor might give all students a few extra points on their final score (in student language, "curve" it). Or perhaps the instructor wanted the test to be worth twice as many points as it was actually worth. In that case, the instructor might make every question worth 2 points rather than 1 point. Is it OK to make these kinds of score transformations? What effect do they have on the central tendency and dispersion of the set of scores? What effect do they have on individual scores?

Effects on Central Tendency

Let's assume that your statistics course includes a lab section that meets once a week. There are 18 students in your lab. Your instructor gives a 60-item quiz. Here are the scores.

46
44
44
43
40
40
39
37
36
35
32
32
31
26
21
20
19
17
$\Sigma X = 602$

For this quiz, $\Sigma X = 602$ and $N = 18$. Therefore, the mean is 602/18 = 33.444.

Now assume that the lab instructor is chagrined to find that the highest score is only 46 and that the mean is only 33.444. To offset the quiz difficulty, the lab instructor decides to add 10 points to each student's score. Let's see what happens to the mean if we add a constant (+10 points) to each student's score:

Original Score	*10 Points Added*
46	56
44	54
44	54
43	53
40	50
40	50
39	49
37	47
36	46
35	45
32	42
32	42
31	41
26	36
21	31
20	30
19	29
17	27
	$\Sigma X = 782$

For this transformed set of scores, $\Sigma X = 782$ and $N = 18$. Therefore, the mean is 782/18 = 43.444.

After increasing each score by 10 points, did the mean increase, decrease, or stay the same? If it changed, by how much did it change? Yes, the mean increased by 10 points, from 33.444 to 43.444. Note that this 10-point increase in the mean is the same increase as was applied to each score.

Now assume that the lab instructor finds that it was a 50-item quiz, after all, not a 60-point quiz. In that case, the scores weren't too bad. However, assume that the instructor would like to have given a 100-point quiz, had there been time to give such a long one. So he or she decides to double each student's score on the 50-point quiz, making the 50-point quiz worth 100 points. Let's see what happens to the mean if we multiply every original score by a constant (×2).

Original Score	*Score Doubled*
46	92
44	88
44	88
43	86
40	80
40	80
39	78
37	74
36	72
35	70
32	64
32	64
31	62
26	52
21	42
20	40
19	38
17	34
	$\Sigma X = 1{,}204$

For this transformed set of scores, $\Sigma X = 1{,}204$ and $N = 18$. Therefore, the mean is 1204/18 = 66.888.

After doubling each score, did the mean increase, decrease, or stay the same? If it changed, by how much did it change?

Yes, the mean increased by twice its original value, from 33.444 to 66.888. Note that this doubling of the mean is the same increase as was applied to each score.

CHECK YOURSELF!

What is the effect on central tendency of adding or subtracting a constant to every score? What is the effect on central tendency of multiplying or dividing every score by a constant?

Effects on Dispersion

Now you know the effects of score transformations on central tendency. But what are the effects on dispersion? Let's make those same transformations on the original data set and see what happens.

Recall that the original mean was 33.444. First, we must find the standard deviation.

Score	$X - M$	$(X - M)^2$
46	12.556	157.6541
44	10.556	111.4291
44	10.556	111.4291
43	9.556	91.3171
40	6.556	42.9811
40	6.556	42.9811
39	5.556	30.8691
37	3.556	12.6451
36	2.556	6.5331
35	1.556	2.4211
32	−1.444	2.0851
32	−1.444	2.0851
31	−2.444	5.9731
26	−7.444	55.4131
21	−12.444	154.8531
20	−13.444	180.7411
19	−14.444	208.6291
17	−16.444	270.4051
		$\Sigma = 1490.445$

$$s^2 = \frac{\Sigma(X-M)^2}{N} = \frac{1490.445}{18} = 82.802$$

$$s = \sqrt{\frac{\Sigma(X-M)^2}{N}} = \sqrt{s^2} = \sqrt{82.802} = 9.10$$

> One merit of mathematics few will deny: It says more in fewer words than any other science.
>
> —David Eugene Smith

Recall that the lab instructor was chagrined to find that the highest score on the 60-point quiz was only 46 and that the mean was only 33.444. To offset the quiz difficulty, she added a constant +10 points to each student's score. This made the new mean 43.444. Now let's see what happens to the standard deviation when we add that same constant (+10 points) to each score.

Original Score	*10 Points Added*	$X - M$	$(X - M)^2$
46	56	12.556	157.6541
44	54	10.556	111.4291
44	54	10.556	111.4291

Original Score	*10 Points Added*	*X – M*	*(X – M)²*
43	53	9.556	91.3171
40	50	6.556	42.9811
40	50	6.556	42.9811
39	49	5.556	30.8691
37	47	3.556	12.6451
36	46	2.556	6.5331
35	45	1.556	2.4211
32	42	–1.444	2.0851
32	42	–1.444	2.0851
31	41	–2.444	5.9731
26	36	–7.444	55.4131
21	31	–12.444	154.8531
20	30	–13.444	180.7411
19	29	–14.444	208.6291
17	27	–16.444	270.4051
			Σ = 1490.4450

$$s^2 = \frac{\sum(X-M)^2}{N} = \frac{1490.445}{18} = 82.802$$

$$s = \sqrt{\frac{\sum(X-M)^2}{N}} = \sqrt{s^2} = \sqrt{82.802} = 9.10$$

After increasing each score by 10 points, did the standard deviation increase, decrease, or stay the same? If it changed, by how much did it change?

It stayed the same, didn't it? Are you surprised? Did you expect it to go up, just as central tendency had gone up? Well, look again at how a standard deviation is calculated. We subtract each score from the mean. Now recall that while each score went up (say, from 46 to 56), the mean went up by that same amount (from 33.444 to 43.444). Thus, the net subtraction from the mean is the same, and so the standard deviation remains unchanged.

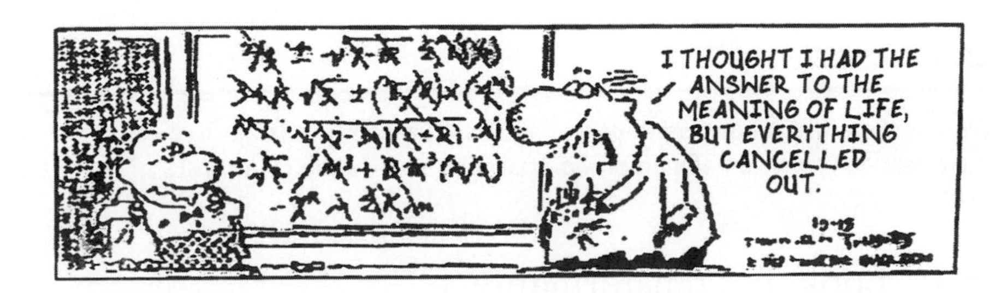

SOURCE: FRANK & ERNEST: © Thaves/Dist. by Feature Syndicate, Inc.

One more time, recall that the lab instructor discovered after the fact that it was a 50-point quiz, not one worth 60 points. In that case, the scores weren't too bad. However, she then decided to double each student's score on the 50-point quiz, making the 50-point quiz worth 100 points. Recall that this also doubled the mean—to 66.888. Now let's see what happens to the standard deviation if we multiply every score by that same constant (×2).

Original Score	*Score Doubled*	*X – M*	*(X – M)²*
46	92	25.112	630.6125
44	88	21.112	445.7165
44	88	21.112	445.7165
43	86	19.112	365.2685
40	80	13.112	171.9245
40	80	13.112	171.9245
39	78	11.112	123.4765
37	74	7.112	50.5805
36	72	5.112	26.1325
35	70	3.112	9.6845
32	64	–2.888	8.3405
32	64	–2.888	8.3405
31	62	–4.888	23.8925
26	52	–14.888	221.6525
21	42	–24.888	619.4125
20	40	–26.888	722.9645
19	38	–28.888	834.5165
17	34	–32.888	1081.6205
			Σ = 5961.7770

$$s^2 = \frac{\sum(X-M)^2}{N} = \frac{5961.777}{18} = 331.2908$$

$$s = \sqrt{\frac{\sum(X-M)^2}{N}} = \sqrt{s^2} = \sqrt{331.2908} = 18.20$$

After doubling each score, did the standard deviation increase, decrease, or stay the same? If it changed, by how much did it change?

The standard deviation increased by twice its original value, from 9.10 to 18.20. Note that this doubling is the same increase as was applied to each score.

CHECK YOURSELF!

What is the effect on dispersion of adding or subtracting a constant to every score? What is the effect on dispersion of multiplying or dividing every score by a constant?

A Graphic Look at Transformations

Let's consider one more example. This time, however, we will look at the effects graphically. The normal curve in Figure 9.1 has a mean of 100 and a standard deviation of 10.

On the same line, draw new normal curves according to the following instructions:

- Create a new normal curve to the right of the original curve by adding 50 to every score. Here's how it is calculated: The bottom score in the original curve is 70. If you add 50 to that score, the bottom score for the new curve will be 120. The middle score in the original curve is 100. If you add 50 to that score, the middle score for the

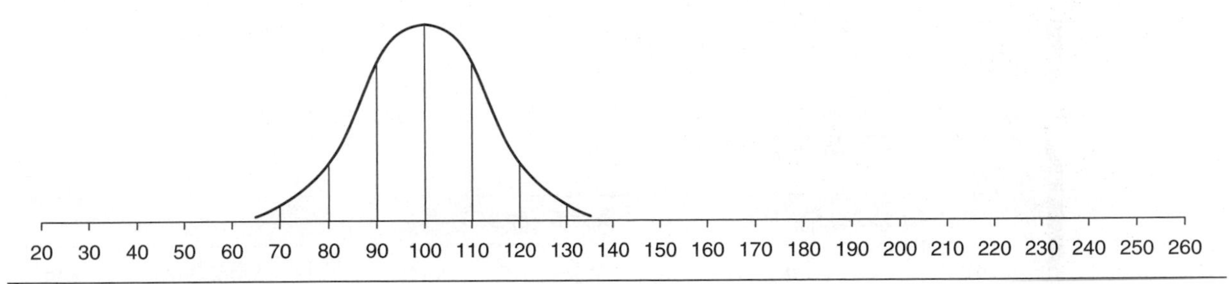

Figure 9.1 Effects of Score Transformations

new curve will be 150. The top score in the original curve is 130. If you add 50 to that score, the top score in the new curve will be 180. Keep the frequencies (the height of the curve) the same.

- o Now draw in the bars for the standard deviations on the new curve. Note that, regardless of their width, there must be seven bars—one for the mean, three to the left of the mean, and three to the right of the mean.
- o When you have completed drawing the new normal curve, label it "#1."

- Create another curve to the left of the original curve by subtracting 50 from every score. Use the same procedure as above. Label this second curve "#2."
- Create another curve to the right of the original curve by multiplying every score by 2. Use the same procedure as above. Label this third curve "#3."
- Create a fourth curve to the left of the original curve by dividing every score by 2. Use the same procedure as above. Label this fourth curve "#4."

Summary of Transformation Effects

Review both the calculation and the graphic examples. What is the effect on the mean of adding or subtracting a constant from every score? Yes, it goes up or down by that same amount.

What is the effect on the mean of multiplying or dividing every score by a constant? Yes, it is multiplied or divided by that same amount.

What is the effect on the standard deviation of adding or subtracting a constant from every score? Yes, it stays the same.

What is the effect on the standard deviation of multiplying or dividing every score by a constant? Yes, it is multiplied or divided by that same amount.

Now look again at the curves you drew in Figure 9.1. Find a score that falls 1 standard deviation above the mean in each of the five distributions. Note that regardless of the actual value of the mean or standard deviation, +1 standard deviation falls at the same place relative to other scores in that distribution. That is, whether the mean is 25, 50, or 100, and whether the standard deviation is 5, 10, or 20, the same proportion of scores fall above and below +1 standard deviation. It follows that the same proportion of scores fall above and below any given standard deviation. That is, once we locate a score within a standard normal curve, we can draw conclusions about the score's position relative to the other scores regardless of the distribution's actual mean or standard deviation. This is because, regardless of the value of the mean or the size of the standard deviation, the relative positions of the transformed scores remain the same.

The more things change, the more they stay the same.

PRACTICE

Reproduced below from Table 8.4 of Module 8 are Bob's scores and the mean and standard deviation for Tests A, B, and C.

Test A (of 95)		*Test B (of 105)*		*Test C (of 95)*	
Score	*Rank*	*Score*	*Rank*	*Score*	*Rank*
73	1	105	1	95	1
***72**	**2**	104	2	***72**	**2**
60	3	98	3	71	3
59	4	94	4	70	4
57	5	92	5	68	5
56	6	87	6	67	6
52	7	85	7	66	7
48	8	***84**	**8**	64	8
36	9	82	9	62	9
28	10	80	10	45	10
$M = 54.10$		$M = 91.10$		$M = 68.00$	
$s = 13.41$		$s = 8.53$		$s = 11.59$	

1. On Test A, add 2 points to every score. Compute the new mean and standard deviation.
   a. What effect did this change have on the mean? Does this fit with your understanding about the effect of score transformations on central tendency?
   b. What effect did this change have on the standard deviation? Does this fit with your understanding about the effect of score transformations on dispersion?

2. On Test A, add 5 points to every score. Compute the new mean and standard deviation.
   a. What effect did this change have on the mean? Does this fit with your understanding of the effect of score transformations on central tendency?
   b. What effect did this change have on the standard deviation? Does this fit with your understanding of the effect of score transformations on dispersion?

3. On Test B, multiply each score by 2 and then subtract 10 points from each score. Compute the new mean and standard deviation.
   a. What effect did these changes have on the mean? Does this fit with your understanding about the effect of score transformations on central tendency?
   b. What effect did these changes have on the standard deviation? Does this fit with your understanding about the effect of score transformations on dispersion?

4. On Test B, multiply each score by 3 and then subtract 5 points from every score.
   a. What effect did these changes have on the mean? Does this fit with your understanding about the effect of score transformations on central tendency?
   b. What effect did these changes have on the standard deviation? Does this fit with your understanding about the effect of score transformations on dispersion?

5. On Test C, add 5 points to each score and then divide each score in half. Compute the new mean and standard deviation.
   a. What effect did these changes have on the mean? Does this fit with your understanding about the effect of score transformations on central tendency?
   b. What effect did these changes have on the standard deviation? Does this fit with your understanding about the effect of score transformations on dispersion?

6. On Test C, add 10 points to each score and then divide each score by 3. Compute the new mean and standard deviation.
   a. What effect did these changes have on the mean? Does this fit with your understanding of the effect of score transformations on central tendency?
   b. What effect did these changes have on the standard deviation? Does this fit with your understanding of the effect of score transformations on dispersion?

Some Common Transformed Scores

Recall from Module 8 that +1 standard deviation is the same as $+1z$. Thus, a z score is one type of transformed score. This is because a z score rescales raw scores into standard deviation units. The distribution in Figure 9.2 should be familiar to you.

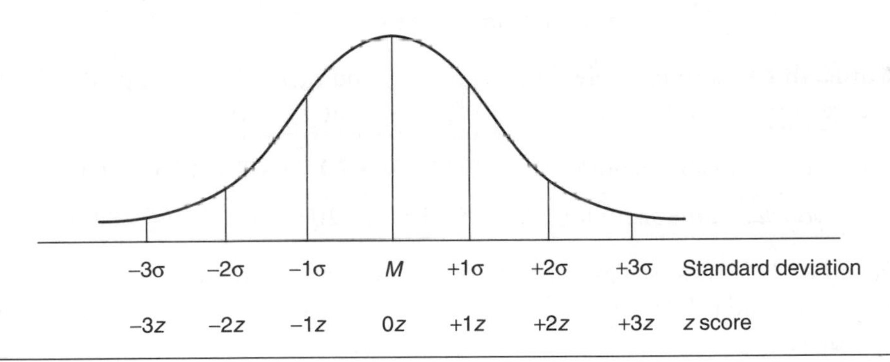

Figure 9.2 Normal Curve With Standard Deviations and z Scores

What you didn't know then, but you do now, is that IQ scores and SAT scores are merely transformed scores. Each transformed score first converts raw scores to z scores (standard deviation units) and then multiplies each score by a constant and adds a constant to each score. These are the types of transformations we have made in this module.

Let's look at these transformed scores more closely. The formula for converting a z score to an IQ score is

$$IQ = 15z + 100$$

> Mathematics is the art of giving the same name to different things.
>
> —J. H. Poincaré (1854–1912)

In words, this says to multiply the z score by 15 and then add 100. In Module 8, we assumed your IQ to be $+1.33z$. Applying the formula to your z score, your IQ is

$$15(+1.33) + 100 = 19.95 + 100 = 119.95 \approx 120$$

College Entrance Examination Board (CEEB) scores are another type of transformed score. CEEB scores are used to report scores on the Scholastic Aptitude Test (SAT) and the Graduate Record Exam (GRE). The formula for converting a *z* score to a CEEB score is

$$\text{CEEB} = 100z + 500$$

In words, this says to multiply the *z* score by 100 and then add 500. In Module 8, we assumed your SAT score to be +1.60*z*. Applying the formula to your *z* score, your SAT score is

$$100(+1.60) + 500 = 160 + 500 = 660$$

Another common transformed score is a *T* score. Scores on the California Psychological Inventory (CPI) and the Minnesota Multiphasic Personality Inventory (MMPI) are reported as *T* scores. The formula for converting a *z* score to a *T* score is

$$T = 10z + 50$$

In words, this says to multiply the *z* score by 10 and then add 50. Applying the formula to the *z* scores for your IQ and for your SAT, we get the following:

$$T \text{ score for your IQ score} = 10(+1.33) + 50 = 13.3 + 50 = 63.3$$

$$T \text{ score for your SAT score} = 10(+1.60) + 50 = 16.0 + 50 = 66.0$$

We can even create a transformed score of our own. Let's call it the ________________ (*yourlastname*) score. To obtain a *yourlastname* score, multiply the *z* score by 5 and then add 20. The formula for converting a *z* score to a *yourlastname* score is

$$\textit{yourlastname} \text{ score} = 5z + 20$$

In words, this says to multiply the *z* score by 5 and then add 20. Applying the formula to the *z* scores for your IQ and for your SAT, we get the following:

$$\textit{yourlastname} \text{ score for IQ} = 5(+1.33) + 20 = 6.65 + 20 = 26.65$$

$$\textit{yourlastname} \text{ score for SAT} = 5(+1.60) + 20 = 8.00 + 20 = 28.00$$

Let's add each of these scales to the normal curve (see Figure 9.3). Now you can see the positions of the standard deviations for *z* scores, IQ scores, SAT scores, *T* scores, and *yourlastname* scores.

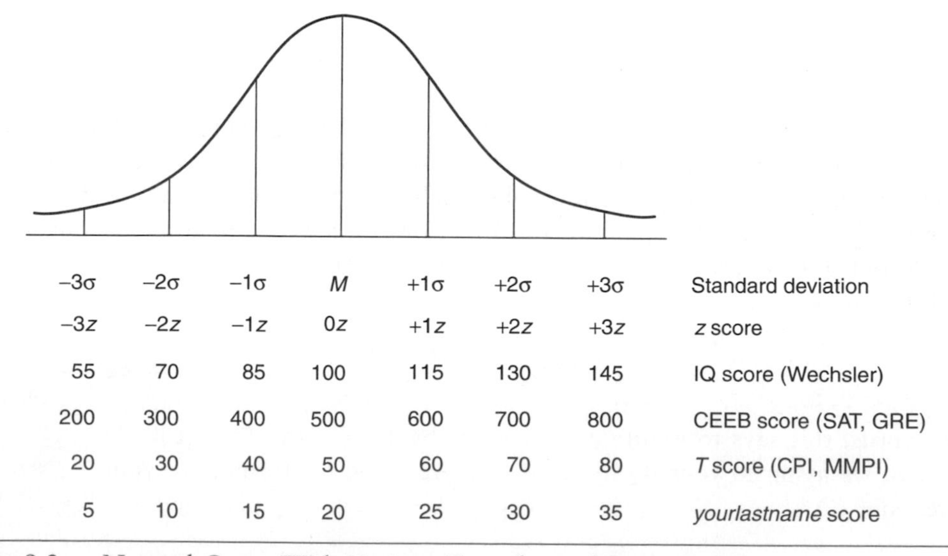

Figure 9.3 Normal Curve With Various Transformed Score Scales

Why all the different scales? Most scales were designed merely to eliminate negative numbers. After all, who wants to hear that their child's IQ is +0.63 or that their SAT score is –0.78? The scales themselves, however, are quite arbitrary. Notice from the diagram that rescaling a score into one of these scales does not change the location of the score within the distribution. Hence, score transformations have no effect on a score's interpretation relative to other scores.

Looking Ahead

This principle of location constancy applies to sample means as well as to individual scores. That is, once we find a sample mean within a distribution of sample means, the sample's actual mean and standard deviation become irrelevant. What matters is the sample's position within the distribution. This understanding underlies many of the inferential techniques you will learn in future modules.

CHECK YOURSELF!

Create a *yourfirstname* scale with the formula $25z + 30$. Place the *yourfirstname* scale beneath the normal curve diagram. (1) What are the mean and the standard deviation for the *yourfirstname* scale? (2) Convert the *z* score for your IQ (+1.33) to the *yourfirstname* scale. What is your IQ score on this new scale? Where does it fall in the normal curve relative to the original *z* score?

PRACTICE

7. Refer to Figure 9.3. Ludwig scores $+1.8\sigma$ on the SAT. What is his SAT score?
8. Refer to Figure 9.3. Macey scores -0.50σ on the SAT. What is her SAT score?
9. Refer to Figure 9.3. Bip scores $+1.4\sigma$ on the SAT. Bop scores -0.3σ on the SAT. By how many SAT points do Bip and Bop differ?
10. Refer to Figure 9.3. Burt scores -0.20σ on the SAT. Curt scores –.40 on the SAT. By how many SAT points do Burt and Curt differ?
11. Refer to Figure 9.3. Luigi scores -0.3σ on the extraversion scale of the CPI. What is his extraversion T score?
12. Refer to Figure 9.3. Catherine scores $+1.30\sigma$ on the extraversion scale of the CPI. What is her extraversion *T* score?
13. Refer to Figure 9.3. Barry scores $+1.5\sigma$ on the responsibility scale of the CPI. Harry scores -0.2σ on the responsibility scale of the CPI. By how many *T*-score points do Barry and Harry differ?
14. Refer to Figure 9.3. Svetlana scores -0.70σ on the responsibility scale of the CPI. Dymitri scores -0.30σ on the responsibility scale of the CPI. By how many *T*-score points do their scores differ?

Visit the study site at www.sagepub.com/steinberg2e for practice quizzes and other study resources.

PART VI

Probability

"Odds Are"

10

Probability Definitions and Theorems

Terms: probability, trial, proportion, outcome, equally likely model, mutually exclusive outcome, independent outcome, addition theorem, multiplication theorem, theoretical, empirical

Symbols: $p(A)$, $p(A \text{ or } B)$, $p(A \text{ and } B)$

Learning Objectives:

- Distinguish between mutually exclusive, embedded, and overlapping outcomes on a single trial
- Distinguish between independent and dependent outcomes in a series of trials
- Know the conditions under which the addition theorem applies
- Know the conditions under which the multiplication theorem applies
- Distinguish between theoretical and empirical outcomes
- Understand the use of empirical versus theoretical data in making inferences

Why Study Probability?

In a mathematical statistics course, formulas and the graphs of their functions are derived and proved. For example, the area under the normal curve would be derived via integral calculus. However, such instruction is both unnecessary and counterproductive in most applied social science statistics courses. For that reason, I have kept mathematical theory to a minimum in this textbook.

Nevertheless, even in applied statistics courses, certain questions arise. For example, what causes many natural events to distribute themselves normally over a period of time? What are the probabilities associated with the various outcomes over the long run? What are the implications of those probabilities for inferential statistics and hypothesis testing?

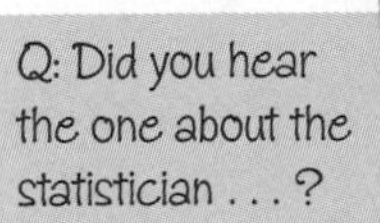

A course in statistics is not complete without some explanation of the frequency distributions on which the statistics depend. And understanding frequency distributions requires some understanding of the probability theory through which the distributions arise. Therefore, in this module and the next, I will discuss those aspects of probability theory most important for understanding the one distribution on which many inferential statistics are based—the normal curve. The current module is devoted to definitions and theorems. The next module presents examples and calculations.

Probability as a Proportion

Probability refers to the relative frequency of a particular outcome occurring over an infinite number of **trials** or occasions. Thus, probability is expressed as a **proportion**, which varies between .00 and 1.00. If the probability of an outcome is .00, it is certain that it will not occur. If the probability of an outcome is 1.00, it is certain that it will occur.

Bumper sticker: Statisticians probably do it better.

A good example is the probability of getting a "head" when tossing a coin. If we were to toss the coin from now to eternity, the number of heads we expect would be some proportion of the total number of tosses. The formula is written like this:

$$p(A) = \frac{\text{Number of outcomes classifiable as } A}{\text{Total number of possible outcomes}}$$

where $p(A)$ indicates the probability of outcome A—in this case, getting a head.

There are two possible **outcomes**, or results, in a coin toss—either a head or a tail. Only one of the outcomes is classifiable as a head. Thus, the probability of getting a head is calculated as follows:

$$p(A) = \frac{1}{2} = .50$$

Probability applies to a single toss as well. That is, when the probability of getting a head is .50 over the long term, the probability of getting a head on a single toss is also .50.

CHECK YOURSELF!

If someone tells you that the probability of an event occurring is 1.25, how should you respond?

CHECK YOURSELF!

If the probability of a particular outcome is .75 in the long term, what is $p(A)$ on the next trial?

Equally Likely Model

In an **equally likely model**, the generating event (e.g., coin tossing) yields possible outcomes that are each equally likely. Coin tossing fits the definition of an equally likely model because heads and tails, which are the only possible outcomes, each have a probability of 1/2, or .50. Card drawing also fits the definition of an equally likely model because each card has an equal probability of being drawn. The probability that any particular card will be drawn is 1/52, or ≈.0192. Each suit also has an equal probability of being drawn. The probability that any particular suit will be drawn is 1/4, or .25. Die rolling also fits the definition of an equally likely model. Because a die has six equal sides with a number on each side, each number has an equal probability of being thrown. The probability that any particular number will be thrown on a single toss of a single die is 1/6, or ≈.1667.

SOURCE: Peanuts: © United Feature Syndicate, Inc.

Not all outcomes are equally likely. If we shuffle together a regular deck of cards with a deck of pinochle cards (which has no cards under nine and two of every card from nine on up) and then draw a card at random, some cards in the mixed deck are more likely to be drawn than others. If a coin or a die is weighted, some outcomes are less likely than others. Even in a fair die, although the probability of each of the six possible outcomes is equal, the probabilities of, say, getting a four or not getting a four are not equal. The probability of the first outcome is 1/6, and the probability of the second outcome is 1 – 1/6, or 5/6.

It is possible to calculate probabilities when outcome likelihoods vary. However, the probability distributions produced from such models are not useful for the examples we will examine next or for most of the statistics we will study later. Most of our examples and later statistics assume that each outcome has an equal chance of occurring.

Mutually Exclusive Outcomes

Mutually exclusive outcome means that the outcome of a particular trial precludes any other outcome on that same trial. Still using the coin-tossing example, if a coin comes up heads on a toss, it cannot in that same toss come up tails. Note that *mutually exclusive outcomes* always refer to *a single trial*. They do not refer to a series of trials.

When I presented this definition in a class, one of my students responded aloud with, "Well, DUH!" And I guess the definition does have a "duh" aspect to it. After all, on a single toss, can there be any outcome *other* than a head or a tail? No, there cannot. Yet not all situations have mutually exclusive outcomes. Sometimes one outcome is embedded within another outcome. An example would be college students and sophomores. This embedded outcome is shown in Figure 10.1.

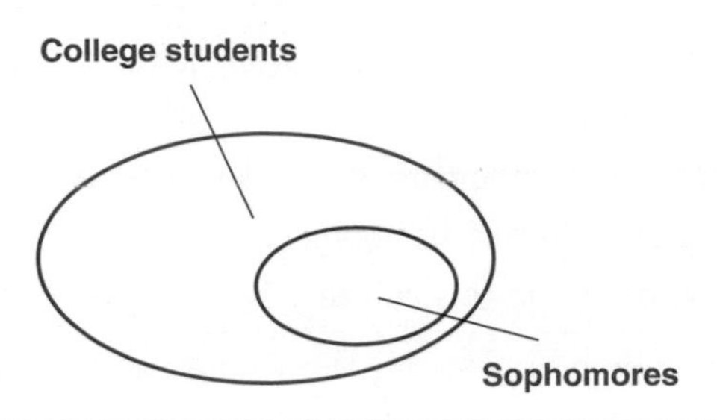

Figure 10.1 Embedded Outcomes

Other outcomes, although not embedded, overlap. For example, although you cannot be a sophomore and a junior at the same time, you can be a college student and an employee at the same time. This overlapping outcome is shown in Figure 10.2.

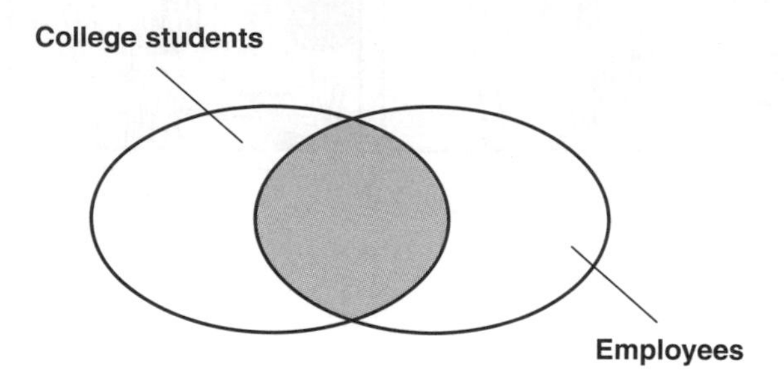

Figure 10.2 Overlapping Outcomes

It is possible to calculate the probabilities when outcomes are not mutually exclusive. However, the probability distributions produced from such models are not useful for the examples we will examine or for the statistics we will study later. All our examples and statistics assume that the possible outcomes are mutually exclusive.

PRACTICE

1. For which of these events are the outcomes mutually exclusive? Circle "Yes" if the outcomes are mutually exclusive or "No" if the outcomes are not mutually exclusive.

Event	*Mutually Exclusive?*	
a. Drawing a nickel or a dime from a bag of mixed coins	Yes	No
b. Drawing a heart or a diamond from a standard deck of playing cards	Yes	No
c. Being a basketball player or a football player	Yes	No
d. Being a man or being a woman	Yes	No
e. Being a man or being an athlete	Yes	No

2. For which of these events are the outcomes mutually exclusive? Circle "Yes" if the outcomes are mutually exclusive or "No" if the outcomes are not mutually exclusive.

Event	*Mutually Exclusive?*	
a. Being gifted in verbal tasks or being gifted in spatial tasks	Yes	No
b. Being short or being tall	Yes	No
c. Liking hotdogs or liking hamburgers	Yes	No
d. Preferring hotdogs or preferring hamburgers	Yes	No
e. Passing this course or failing this course	Yes	No

3. For which of these events are the outcomes mutually exclusive? Circle "Yes" if the outcomes are mutually exclusive or "No" if the outcomes are not mutually exclusive.

Event	*Mutually Exclusive?*	
a. Being born prematurely or being born on time	Yes	No
b. Being depressed or being anxious	Yes	No
c. Liking to go hiking or liking to watch TV	Yes	No
d. Having cancer or having diabetes	Yes	No
e. Being employed or being unemployed	Yes	No

4. For which of these events are the outcomes mutually exclusive? Circle "Yes" if the outcomes are mutually exclusive or "No" if the outcomes are not mutually exclusive.

Event	*Mutually Exclusive?*	
a. Wearing sneakers to today's statistics class or wearing sandals to today's statistics class	Yes	No
b. Being nervous or being afraid	Yes	No
c. Being valedictorian of your high school class or being salutatorian of your high school class	Yes	No
d. Doing your homework or not doing your homework	Yes	No
e. Having a headache or having a stomachache	Yes	No

Addition Theorem

When outcomes are mutually exclusive, the addition theorem applies. The **addition theorem** says that the probability of any of the possible outcomes occurring on a particular trial is the sum of their individual probabilities. Note that the addition theorem applies only when outcomes are mutually exclusive. Also, because it presumes mutually exclusive outcomes, the addition theorem applies only to a *single trial*. It does not apply to a series of trials. The formula is written like this:

$$p(A \text{ or } B) = p(A) + p(B)$$

where $p(A \text{ or } B)$ indicates the probability of Outcome A or Outcome B.

Hint: Look for the word "or." Still using the coin-tossing example, the probability of throwing either a head *or* a tail on a single toss is calculated as follows:

$$p(A \text{ or } B) = .50 + .50 = 1.00$$

> The greatest unsolved theorem in mathematics is why some people are better at it than others.
>
> —Adrian Mathesis

You can see that the probabilities of all possible outcomes will sum to 1.00.

The addition theorem is also useful for calculating subsets of possible outcomes. For example, we know that the probability of drawing a spade, a heart, a club, or a diamond in a standard deck of cards is 1.00 because

$$p(A \text{ or } B \text{ or } C \text{ or } D) = .25 + .25 + .25 + .25 = 1.00$$

We can then use the addition theorem to calculate the probability of drawing either a spade or a heart. In that case, Outcomes C and D are ignored:

$$p(A \text{ or } B) = .25 + .25 = .50$$

Independent Outcomes

Independent outcome means that the outcome of one trial has no relation to the outcome of another trial. For example, getting either a head or a tail on the first toss has no bearing on whether we get a head or a tail on the second toss. Note that *independent outcome* always refers to *a series of trials*. It does not refer to a single trial.

Lottery: A tax on people who are bad at math.

Not all outcomes are independent. As I write this, my son is out running—something his soccer coach has asked each team member to do every day so that they will be in good physical shape when the soccer season starts. His running speeds on successive days are not independent, however. Every day his speed is a little faster, just as the coach had hoped. That's because every day of running puts my son in a little better cardiovascular condition and muscular tone than he was on the previous day. The outcome on one day is not independent of the outcome on previous days. Rather, the outcomes are dependent.

Let's look at another example. If we stopped a married couple on the street and asked them which candidate they intend to vote for in an upcoming election, we would not expect their answers to be independent. Even if they have never discussed the issue, married couples tend to have similar values and beliefs. The outcomes would not be independent. Rather, the outcomes would be dependent.

Some of the statistics we will study later in the textbook assume that the outcomes are independent. Others assume that the outcomes are dependent. Thus, your study of statistics will include both independent and dependent outcomes. The formulas for the two types of situations differ in ways that take into account independent and dependent probabilities.

PRACTICE

5. For which of these events are the outcomes independent? Circle "Yes" if the outcomes are independent or "No" if the outcomes are not independent.

Event	*Independent Outcomes?*	
a. Getting a six on the fourth throw of a die, given known outcomes on the first through third throws	Yes	No
b. How much you love your boyfriend or girlfriend from one day to the next	Yes	No
c. How much two people chosen at random love their respective boyfriend or girlfriend on any given day	Yes	No
d. The average amount of weight lost by patients in successive weeks of a diet	Yes	No
e. The brand of soft drink favored by 100 people randomly selected from among shoppers in a retail mall	Yes	No

6. For which of these events are the outcomes independent? Circle "Yes" if the outcomes are independent or "No" if the outcomes are not independent.

Event	*Independent Outcomes?*	
a. Your chance of winning the lottery today and your chance of also winning the lottery tomorrow (assuming that you buy one ticket each day)	Yes	No
b. The amount it rains today and the amount it rains tomorrow	Yes	No
c. The color of the shirt you wear today and the color of the shirt you wear 16 days from today	Yes	No
d. How many points you are able to immediately meld in the first rummy hand you are dealt and how many points you are able to immediately meld in the next rummy hand you are dealt	Yes	No
e. Your final grade in this course and your final grade in any other course you are taking this semester	Yes	No

7. For which of these events are the outcomes independent? Circle "Yes" if the outcomes are independent or "No" if the outcomes are not independent.

Event	*Independent Outcomes?*	
a. Eating lunch today and eating lunch tomorrow	Yes	No
b. Rolling a 3 on a die and rolling another 3 on the next roll on the same die	Yes	No
c. How fast you run in two successive days of marathon training	Yes	No
d. Your score on a test of conscientiousness today and your score on a test of conscientiousness tomorrow	Yes	No
e. Your body temperature on Day 1 of an illness and your body temperature on Day 2 of the illness	Yes	No

8. For which of these events are the outcomes independent? Circle "Yes" if the outcomes are independent or "No" if the outcomes are not independent.

Event	*Independent Outcomes?*	
a. Whether or not you do your homework this week and whether or not you do your homework next week	Yes	No
b. Drawing a spade (with replacement) from a deck of cards and drawing another spade on the next draw from the same deck	Yes	No
c. Air temperature today and air temperature tomorrow	Yes	No
d. Liking what the college dining hall serves for dinner tonight and liking what it serves for dinner tomorrow night	Yes	No
e. Your IQ score this week and your IQ score next week	Yes	No

Multiplication Theorem

The multiplication theorem applies only when outcomes are independent. The **multiplication theorem** says that the probability of a series of particular outcomes occurring on successive trials is the product of their individual probabilities. Because the multiplication theorem assumes independent outcomes, it applies only to a *series of trials*. It does not apply to a single trial. The formula is written like this:

$$p(A \text{ and } B) = p(A) * p(B)$$

where $p(A \text{ and } B)$ indicates the probability of Outcome A and Outcome B.

Hint: Look for the word "and." Still using the coin-tossing example, the probability of getting two heads (i.e., getting a head *and* another head) on two successive tosses is calculated as follows:

$$p(A \text{ and } B) = .50 * .50 = .25$$

A Brief Review

This module has included many new terms. This is a good time to review and differentiate some of the more important terms.

- *Mutually exclusive outcomes* refer to outcomes on a single trial. Any one outcome precludes all other outcomes on that same trial.
- The *addition theorem* applies only to outcomes on a single trial. Look for the word "or." The outcomes must be mutually exclusive. The sum of the outcomes is 1.00.
- *Independent outcomes* refer to outcomes on a series of trials. The outcome on one trial has no bearing on the outcome on another trial.
- The *multiplication theorem* applies only to outcomes on a series of trials. Look for the word "and." The outcomes must be independent. The probability of a series of outcomes is the product of their individual probabilities.

CHECK YOURSELF!

Which terms and theorems apply to a single trial? Which apply to a series of trials?

PRACTICE

9. In which of these outcomes does the addition theorem apply? In which does the multiplication theorem apply? Circle "AT" for addition theorem or "MT" for multiplication theorem.

Outcome	*Applicable Theorem*	
a. Throwing a four, a five, and a two on three successive throws of a standard die	AT	MT
b. Throwing a four, five, or two on one throw of a standard die	AT	MT
c. Getting five red cards on five separate draws (with replacement) from a standard deck of playing cards	AT	MT
d. Getting either a red card (heart, diamond) or a black card (spade, club) when drawing a card from a standard deck of playing cards	AT	MT

10. In which of these outcomes does the addition theorem apply? In which does the multiplication theorem apply? Circle "AT" for addition theorem or "MT" for multiplication theorem.

Outcome	*Applicable Theorem*	
a. Winning the game of chance you are currently playing and winning the next game of chance you play	AT	MT
b. Winning or losing the game of chance you are currently playing	AT	MT
c. The probability that you will eat breakfast tomorrow morning and that you also will eat breakfast the next morning	AT	MT
d. The probability that you will either eat breakfast or not eat breakfast tomorrow morning	AT	MT

11. In which of these outcomes does the addition theorem apply? In which does the multiplication theorem apply? Circle "AT" for addition theorem or "MT" for multiplication theorem.

Outcome	*Applicable Theorem*	
a. Getting either a club or a spade when drawing a card from a standard deck of playing cards	AT	MT
b. Throwing a six or a five on one throw of a standard die	AT	MT
c. Getting two black cards on two successive draws (with replacement) from a standard deck of playing cards	AT	MT
d. Throwing a six and a five on two successive throws of a standard die	AT	MT

12. In which of these outcomes does the addition theorem apply? In which does the multiplication theorem apply? Circle "AT" for addition theorem or "MT" for multiplication theorem.

Outcome	*Applicable Theorem*	
a. Liking the dinner the college dining service offers today and liking the dinner it serves tomorrow	AT	MT
b. Liking the dinner the college dining service serves today and not liking it	AT	MT
c. Getting enough sleep tonight and not getting enough sleep tonight	AT	MT
d. Being well rested today and being well rested tomorrow	AT	MT

Probability and Inference

Note that with a finite number of trials, there is no guarantee that the observed outcome will be what you expect it to be. For example, even when a coin is fair, there is no guarantee that the number of heads we obtain in a series of tosses will be exactly half the number of tosses and the number of tails will be exactly half the number of tosses. That's because probabilities are **theoretical** (what is expected to happen, based on probability theory), whereas results in a series of tosses are **empirical** (what actually does happen). As they say in ads for various products, "actual results may vary."

This tension between actual results and theoretical results is an important principle in inferential statistics, where we will compare the actual outcome we observe in an experiment with the theoretical outcome we expected. For that reason, in inferential statistics results can never be definitively known; they can only be probabilistically known.

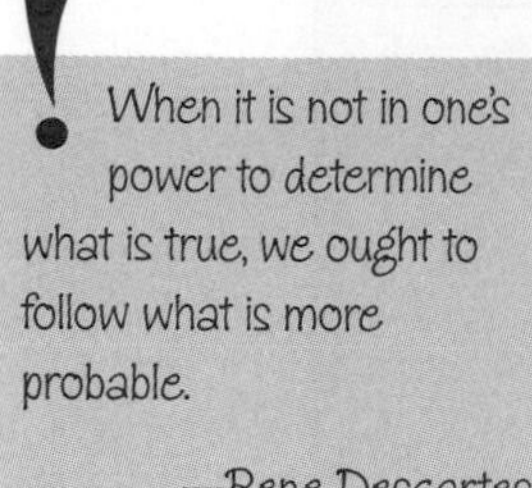

Nevertheless, as the number of tosses increases (say, from 2 tosses to 10,000 tosses to 50 million tosses), the actual outcome more closely approximates the theoretical outcome. It would be rare, indeed, for us to toss a fair coin even 100 times and get, say, 99 heads and one tail. Likewise, in a well-designed experiment with sufficient sample size, it would be rare, indeed, for our observed results to vary grossly from our expected results.

He who has heard the same thing told by 12,000 eyewitnesses has only 12,000 probabilities, which are equal to one strong probability, which is far from certain.

—Voltaire

Thus, theoretical distributions are the standard against which observed outcomes are judged. Under an equally likely model in which we expect 50 heads and 50 tails, if a coin comes up 99 heads and 1 tail, we have grounds for concluding that the coin is not fair. We conclude that the coin is most likely weighted and that its weightedness accounts for the observed outcome. Similarly, in an experiment testing the effectiveness of two different treatments, an equally likely model says that the two treatments are equally effective. That is, we expect no difference in observed outcome as a result of the treatment received. If we then conduct our study and find that people who receive one treatment do much better (or worse) than those who receive the other treatment, we have grounds for concluding that the two treatments are not equally effective. We conclude that the treatments' differential effects account for the observed outcome.

Indeed, in most experiments this is what we hope to find: results that lead us to reject the hypothesis of treatment equivalence. Nevertheless, our conclusions are never certain. Just as a fair coin *could* come up 99 heads and one tail, so, too, an ineffective treatment *can* appear to be effective. Thus, conclusions drawn from inferential statistics are probabilistic only. They are never certain.

CHECK YOURSELF!

It has been said that scientists look for proof where none exists. In what way is that a true statement? On what, then, does the scientist rely for evidence?

Visit the study site at www.sagepub.com/steinberg2e for practice quizzes and other study resources.

11 The Binomial Distribution

Terms: dichotomous, hit, miss, binomial expansion

Symbols: *p*, *q*

Learning Objectives:

- Calculate probability by listing all possible outcomes
- Calculate probability by expanding the binomial formula
- Find probability by using a binomial table
- Understand the relationship between likelihood in a dichotomous model and the shape of the outcome distribution

What Are Dichotomous Events?

Recall from Module 10 that when an outcome on a trial precludes any other outcome on that same trial, the outcomes are mutually exclusive. Recall also that when the outcome on one trial has no bearing on the outcome on any other trial, the outcomes are independent. Now I will introduce another term. When an event is restricted to only two possible outcomes, the outcomes are said to be **dichotomous.** That is, the outcome can be either one or the other but nothing in between. In this module, I will talk about the special case when there are only two possible outcomes, and those two outcomes are mutually exclusive and independent. I will begin with the case where the two outcomes are equally likely. Later in the module, I will discuss cases where the outcomes are not equally likely.

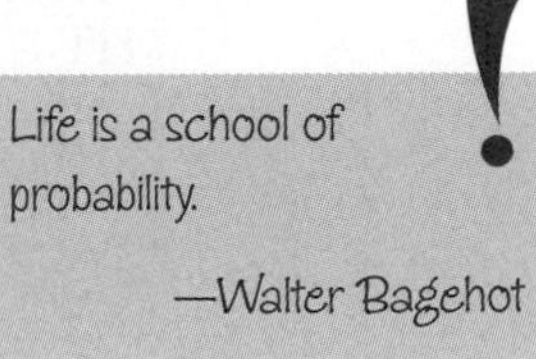

Many events in life are dichotomous. Coin tossing, for example, is dichotomous: Any given toss can come up either heads or tails but nothing else. Colors in a standard deck of cards are also dichotomous: A card can be either red or black but no other color. The direction a rat can turn at the end of a maze is also dichotomous: It can go either left or right but no other direction. Your score on a true-false or multiple-choice item is also dichotomous: You can get the item either correct or incorrect, but there is no partial credit.

As it turns out, outcomes for dichotomous events distribute themselves in particular patterns. For dichotomous events with outcome probabilities that are both .50 (an equally likely model), that pattern is symmetric and unimodal—quite like a normal curve.

Let's use a coin-toss example to calculate the probabilities for all possible outcomes in a series of trials (tosses). We will compute the probabilities by three methods: (1) list and calculation, (2) binomial formula expansion, and (3) binomial table. In each method, you will see how closely the results approximate a normal curve. Of course, it's not really a normal curve, because the measure on the *X*-axis of a normal curve is continuous, whereas outcome measures for dichotomous variables such as coin tosses are discrete. Nevertheless, the similarities in the graphs are striking. Their essential similarity is what allows us to use continuous-curve tables such as the normal curve table when making probability predictions, even in experiments whose possible outcomes are discrete.

As we work through the coin-toss example, try to keep the big picture in mind rather than focus merely on memorizing each calculation step. That big picture includes the following:

- All outcomes are mutually exclusive and independent.
- Probabilities are calculated with the addition and multiplication theorems.
- The theoretical frequencies of hits and misses (e.g., heads and tails) in a series of trials will follow a distinct pattern.
- For dichotomous events whose two probabilities are each .50, that pattern looks a lot like a normal curve.

Finding Probabilities by Listing and Counting

Let's assume that a head is the outcome of interest. In that case, we are said to make a **hit** when we obtain a head (H) and a **miss** when we obtain a tail (T). In probability notation, the symbol p is used to indicate the probability of a hit, and the symbol q is used to indicate the probability of a miss. Thus, in our coin-toss example, p = a head and q = a tail.

> Lest man suspect your tale untrue, keep probability in view.
>
> —John Gay

In this dichotomous example, there are only two outcomes—a head or a tail. From the addition theorem in Module 10, you know that the sum of all possible outcomes equals 1.00. You also know that $p = .5$ on any single toss. Therefore, $q = 1 - .5 = .5$ on that same toss.

If we toss the coin twice (in other words, make a series of two trials), the possible outcomes for the two tosses are the following:

TT

TH

HT

HH

From the multiplication theorem, the probabilities of each of these multi-trial outcomes are as follows:

Outcome	*Probability*
TT =	(.5)(.5) = .25
TH =	(.5)(.5) = .25
HT =	(.5)(.5) = .25
HH =	(.5)(.5) = .25

We can then view each possible outcome pattern in terms of the number of hits (i.e., heads) it contains.

Outcome	*No. of Hits*
TT =	0
TH =	1
HT =	1
HH =	2

Next, we look at how many ways there are to get each number of hits.

No. of Hits	*No. of Ways*
0 hits = (TT)	1
1 hit = (TH or HT)	2
2 hits = (HH)	1

Finally, we combine this information with the individual-outcome probability information. Now we can see that the probabilities for the number of hits in the two tosses are as follows:

No. of Hits	*No. of Ways*	*Probability for No. of Hits*
0 hits =	1 =	.25
1 hit =	2 =	.25 + .25 = .50
2 hits =	1 =	.25

Figure 11.1 shows a histogram of these probabilities. Note that, even with only two tosses, the graph of possible outcomes is unimodal and symmetric and is already beginning to look roughly like a normal curve.

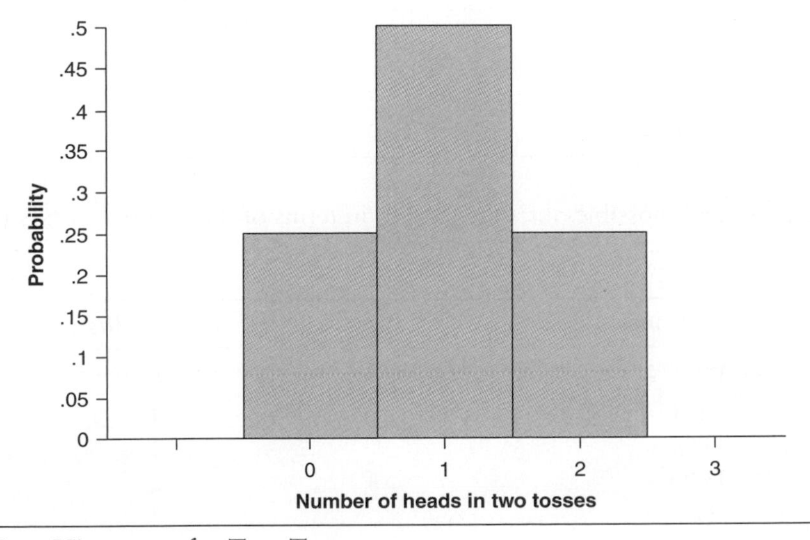

Figure 11.1 Histogram for Two Tosses

From the probabilities, we can answer questions about any possible outcome. For example, what is the probability that a person would toss a fair coin twice and get exactly two heads? Well, two heads is two hits, so the probability is .25. What is the probability of getting at least one head? Let's see . . . that's one hit or two hits, which is .50 + .25, which is .75.

CHECK YOURSELF!

What is the probability of getting fewer than two heads?

Now let's toss the same coin three times. As before, $p = .5$ and $q = .5$ on a single toss, but now the possible outcomes are the following:

TTT

TTH

THT

THH

HTT

HTH

HHT

HHH

From the multiplication theorem, the probabilities of each of these multi-trial outcomes are as follows:

Outcome	*Probability*
TTT =	(.5)(.5)(.5) = .125
TTH =	(.5)(.5)(.5) = .125
THT =	(.5)(.5)(.5) = .125
THH =	(.5)(.5)(.5) = .125
HTT =	(.5)(.5)(.5) = .125
HTH =	(.5)(.5)(.5) = .125
HHT =	(.5)(.5)(.5) = .125
HHH =	(.5)(.5)(.5) = .125

We reexpress each possible outcome pattern in terms of the number of hits (i.e., heads) it contains:

Outcome	*No. of Hits*
TTT =	0
TTH =	1
THT =	1
THH =	2
HTT =	1
HTH =	2
HHT =	2
HHH =	3

And then we look at how many ways there are to get each number of hits:

No. of Hits	*No. of Ways*
0 hits = (TTT) =	1
1 hit = (TTH, THT, or HTT) =	3
2 hits = (THH, HTH, or HHT) =	3
3 hits = (HHH) =	1

Finally, we combine this information with the individual probability information. Now we can see that the probabilities for the number of hits in the three tosses are as follows:

No. of Hits	*No. of ways*	*Probability for No. of Hits*
0 hits =	1 =	.125
1 hit =	3 =	.125 + .125 + .125 = .375
2 hits =	3 =	.125 + .125 + .125 = .375
3 hits =	1 =	.125

Figure 11.2 shows a histogram of these probabilities. The graph of possible outcomes continues to be unimodal and symmetric, and it continues to acquire the shape of a normal curve.

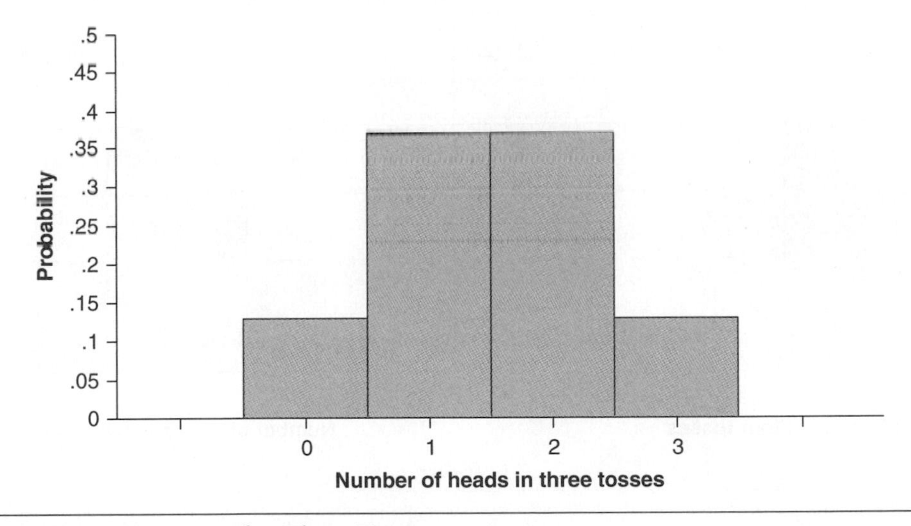

Figure 11.2 Histogram for Three Tosses

Again, from the probabilities, we can answer questions about any possible outcome. For example, what is the probability that a person would toss a fair coin three times and get exactly two heads? Well, two heads is two hits, so the probability is .375. What is the probability of getting at least one head? Let's see . . . that's one hit or two hits or three hits, which is .375 + .375 + .125, which is .875.

CHECK YOURSELF!

What is the probability of getting fewer than two heads?

Using the same calculation method, we could determine the probabilities for numbers of hits on 4 tosses, 5 tosses, 10 tosses, and any number of tosses. Note, however, the escalation in the number of possible outcomes. For 2 tosses the list of possible outcomes was 4, whereas for 3 tosses the list of possible outcomes was 8—twice as many as 4 but still manageable. For 4 tosses, the number of possible outcomes increases to 16, for 5 tosses it increases to 32, and so on. You can see that as the number of tosses increases arithmetically, the number of possible outcomes increases exponentially. Thus, a calculation method that requires listing all possible outcomes quickly becomes quite tedious.

SOURCE: Peanuts: © United Feature Syndicate, Inc.

Because of the tedium, I will skip the calculations and go straight to the histograms. Note how the graphs in Figure 11.3 look more and more similar to a normal curve as the number of tosses increases.

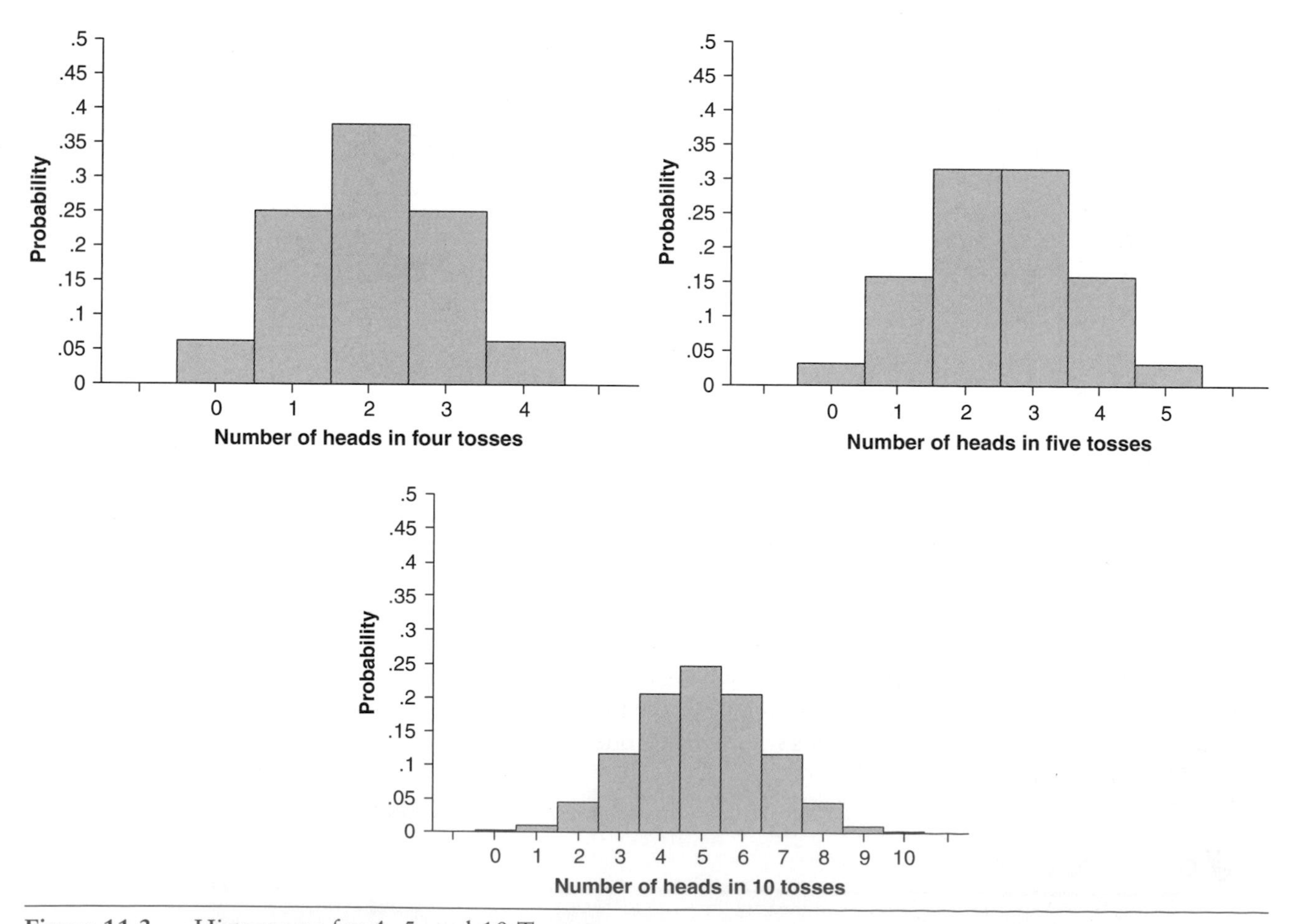

Figure 11.3 Histograms for 4, 5, and 10 Tosses

PRACTICE

Use the list and calculation method to answer the following questions. Show your work at each step.

1. You toss a coin three times. What is the probability that
   a. all three will be heads,
   b. fewer than three will be heads, and
   c. more than two will be heads?

2. A couple already has three children. All three children are boys. Assume that the probabilities of the man fathering a boy or a girl are .50 and .50, respectively. What is the probability that the fourth child will also be a boy?

3. A couple is planning on having four children. Assume that the probabilities of the man fathering a boy or a girl are .50 and .50, respectively. What is the probability that
   a. exactly two will be boys,
   b. fewer than two will be boys, and
   c. more than one will be a boy?

4. You draw three cards (with replacement) from a standard deck of cards. What is the probability that
   a. exactly one will be red,
   b. none will be red, and
   c. more than two will be red?

5. Assume that the probability of MyLady winning a horse race at a given track in given weather conditions against given horses is .50. If MyLady runs in three such races, what is the probability that MyLady will win
   a. exactly one race,
   b. exactly two races,
   c. all three races,
   d. no races, and
   e. at least one race?

6. You toss a coin four times. What is the probability that
   a. exactly three will be heads,
   b. fewer than three will be heads, and
   c. more than two will be heads?

7. You draw three cards (with replacement) from a standard deck of cards in which half the cards are red and half are black. Two of the three cards you draw are red and one is black. If you draw a fourth card, what is the probability that this fourth card will be red?

8. You draw three cards (with replacement) from a standard deck of cards in which half the cards are red and half are black. What is the probability that
   a. none will be black,
   b. all three will be black, and
   c. exactly two will be black?

9. A teacher grades in such a way that exactly 50% of the students earn either a B grade or higher. If a student drawn at random takes three unrelated courses with this teacher, what is the probability that the student will earn
   a. no grades of B or higher
   b. exactly two grades of B or higher, and
   c. one or more grades of B or higher
10. A student eats breakfast one out of every two class days (50% of the time). What is the probability that, in the next four class days, this student will eat breakfast
   a. no times
   b. at least twice, and
   c. exactly twice

Finding Probabilities by the Binomial Formula

Fortunately, there is an easier way to calculate the number of possible outcomes (4, 8, 16, 32, and so on), the number of possible hits (0, 1, 2, 3, and so on), and the probabilities associated with those numbers of hits. When events are dichotomous, we can calculate the probabilities for various numbers of hits via the binomial formula. Here is the binomial formula:

$$(p + q)^N$$

where

p = the probability of obtaining the favored hit,

q = the probability of obtaining the *un*favored miss, and

N = the number of independent trials or occasions.

To calculate probabilities using this formula, we use a process called **binomial expansion.** Expansion of the binomial takes the following form:

$$\begin{aligned}(p+q)^N = p^N &+ Np^{N-1}q + \frac{N(N-1)}{1 \times 2}p^{N-2}q^2 \\ &+ \frac{N(N-1)(N-2)}{1 \times 2 \times 3}p^{N-3}q^3 \\ &+ \frac{N(N-1)(N-2)(N-3)}{1 \times 2 \times 3 \times 4}p^{N-4}q^4 + \cdots + q^N\end{aligned}$$

The expanded formula looks horrid, I know. But let's take a look at its separate terms (each term is separated by an arithmetic operation). If p is a hit and q is a miss, the hit indicator decreases by 1 in each term (look at the exponents of p), and the miss indicator increases by 1 in each term (look at the exponents of q). The N coefficient in front of each term, once reduced, tells the number of ways those hits and misses can occur. Note the pattern within this N term as well: The numerator adds an increasingly smaller multiplier ($N - 1$, $N - 2$, $N - 3$, . . .), while the denominator adds a whole-digit multiplier (1×2, $1 \times 2 \times 3$, $1 \times 2 \times 3 \times 4$, . . .). Finally, the ". . ." at the end of the expansion indicates that these patterns continue until the number of trials, N, is met.

Let's do an example to see how it works. For the coin-tossing event, $p = .5$ and $q = 1 - .5 = .5$. Let's toss the coin five times. In that case, $N = 5$. Here, again, is the binomial formula:

$$(p+q)^N = p^N + Np^{N-1}q + \frac{N(N-1)}{1\times 2}p^{N-2}q^2 + \frac{N(N-1)(N-2)}{1\times 2\times 3}p^{N-3}q^3 + \frac{N(N-1)(N-2)(N-3)}{1\times 2\times 3\times 4}p^{N-4}q^4 + \cdots + q^N$$

Plugging our example values into the binomial formula, the binomial expansion becomes the following:

$$(.5+.5)^5 = (.5)^5 + 5(.5)^4(.5)^1 + \frac{5(4)}{1\times 2}(.5)^3(.5)^2 + \frac{5(4)(3)}{1\times 2\times 3}(.5)^2(.5)^3 + \frac{5(4)(3)(2)}{1\times 2\times 3\times 4}(.5)^1(.5)^4 + \frac{5(4)(3)(2)(1)}{1\times 2\times 3\times 4\times 5}(.5)^5$$
$$= 1(.5)^5 + 5(.5)^4(.5)^1 + 10(.5)^3(.5)^2 + 10(.5)^2(.5)^3 + 5(.5)^1(.5)^4 + 1(.5)^5$$

Substituting p and q back into the formula, the expanded binomial is

$$(p+q)^5 = 1(p)^5 + 5(p)^4(q)^1 + 10(p)^3(q)^2 + 10(p)^2(q)^3 + 5(p)^1(q)^4 + 1(q)^5$$

Note the exponents for each term. The exponents are 5, then 4 and 1, then 3 and 2, then 2 and 3, then 1 and 4, and then 5. Also note to which symbol (p or q) the exponents are attached. Recall that p is a hit and q is a miss. Thus, the first term, $(p)^5$, depicts five hits. The second term, $(p)^4(q)^1$, depicts four hits and one miss, and so on.

Also note the coefficients for each term. The coefficients are 1, 5, 10, 10, 5, and 1. These are the numbers of ways of getting the successive numbers of hits. For example, $1p^5$ says that there is one *way* to get five *hits*, $5p^4q^1$ says that there are five *ways* to get four *hits* and one *miss*, and so on. Note that the binomial expansion allowed us to get to this step without first listing all 32 ways (HHHTH, THHTH, etc.). We have gone straight to the numbers of hits and to the numbers of ways these numbers of hits can occur. From the binomial expansion, the numbers of hits and the numbers of ways can be listed as follows:

No. of Hits	*No. of Ways*
$0 = 0p, 5q =$	1
$1 = 1p, 4q =$	5
$2 = 2p, 3q =$	10
$3 = 3p, 2q =$	10
$4 = 4p, 1q =$	5
$5 = 5p, 0q =$	1
	32 possible outcomes

With 32 possible outcomes, each individual outcome (e.g., THHTH) is 1/32 of all possible outcomes. In decimal form, that individual probability is .03125. Multiplying the individual outcome probability by the number of ways the specified number of hits can occur, the probability of each specified number of hits is as follows:

No. of Hits	*No. of Ways*	*Probability for No. of Hits*
0	1	= (1)(.03125) = .03125
1	5	= (5)(.03125) = .15625
2	10	= (10)(.03125) = .31250
3	10	= (10)(.03125) = .31250
4	5	= (5)(.03125) = .15625
5	1	= (1)(.03125) = .03125

Figure 11.4 shows a histogram of these probabilities. Note that this is the same graph we produced previously by the list and calculation method.

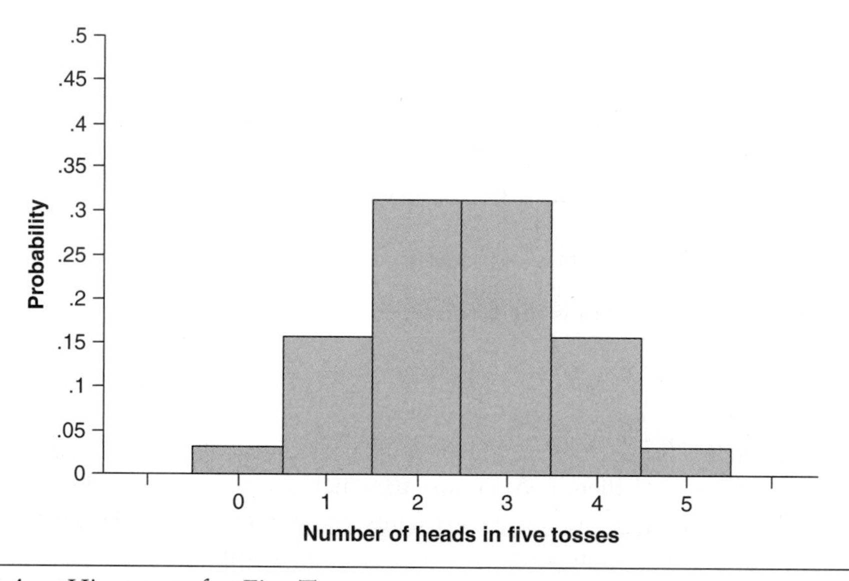

Figure 11.4 Histogram for Five Tosses

Once again, from the probabilities we can answer questions about any possible outcome. For example, what is the probability that a person would toss a fair coin five times and get exactly two heads? Well, two heads is two hits, so the probability is .31250. What is the probability of getting at least four heads? Let's see . . . that's four hits or five hits, which is .15625 + .03125, which is .1875.

CHECK YOURSELF!

What is the probability of getting either two heads or five heads?

PRACTICE

Use the binomial formula expansion method to answer the following questions. Show your work at each step.

11. You toss a coin three times. What is the probability that
    a. none will be heads,
    b. fewer than two will be heads, and
    c. more than one will be a head?

12. You draw six cards from a standard deck of cards. What is the probability that
    a. exactly three will be red,
    b. none will be red, and
    c. fewer than four will be red?

13. Assume that the probability of MyLady winning a horse race at a given track in given weather conditions against given horses is .50. If MyLady runs in four such races, what is the probability that MyLady will win
    a. fewer than three races,
    b. exactly two races,
    c. all four races,
    d. no race, and
    e. at least one race?

14. Assume a man has a .50 chance of fathering a boy and a .50 chance of fathering a girl. He fathers three children. What is the probability that
    a. all three will be girls,
    b. exactly two will be girls, and
    c. at least two will be girls?

15. A teacher grades in such a way that exactly 50% of the students earn either a B or higher grade. If a student drawn at random takes four unrelated courses with this teacher, what is the probability that the student will earn
    a. all four grades of B or higher
    b. exactly three grades of B or higher, and
    c. two or more grades of B or higher?

16. A student goes to breakfast one out of every two class days (50% of the time). What is the probability that, in the next five class days, this student will go to breakfast
    a. exactly three days
    b. at least three days, and
    c. all five days?

Finding Probabilities by the Binomial Table

The binomial expansion is less tedious than listing all possible outcomes (HHHTH, THHTH, etc.), then counting the number of ways each specified number of hits can occur, and then multiplying the number of ways by the individual outcome probabilities. However, the binomial formula method is also a bit tedious. Fortunately, statisticians have already performed the formula expansions for us and have created a table of the results.

> Everything should be made as simple as possible, but not simpler.
>
> —Albert Einstein

Appendix B contains a binomial table. A portion of the table is reproduced in Table 11.1.

Across the top of the table is the probability of a hit on a single trial. The leftmost column lists the number of trials, and the column next to that lists the number of hits in the multi-trial series. Table entries are the probabilities for the specified number of hits.

Table 11.1 A Portion of the Binomial Table

	No. of	*p* or *q*									
N	*p* or *q* Events	*.05*	*.10*	*.15*	*.20*	*.25*	*.30*	*.35*	*.40*	*.45*	*.50*
5	0	.7738	.5905	.4437	.3277	.2372	.1681	.1160	.0778	.0503	.0312
	1	.2036	.3280	.3915	.4096	.3955	.3502	.3124	.2592	.2059	.1562
	2	.0214	.0729	.1382	.2048	.2637	.3087	.3364	.3456	.3369	.3125
	3	.0011	.0081	.0244	.0512	.0879	.1323	.1811	.2304	.2757	.3125
	4	.0000	.0004	.0022	.0064	.0146	.0284	.0488	.0768	.1128	.1562
	5	.0000	.0000	.0001	.0003	.0010	.0024	.0053	.0102	.0185	.0312

For the coin-tossing example we just finished calculating via the binomial expansion, the probability of getting a hit on a single trial is .50, which is the rightmost column. This is the only column we need to look at. We tossed the coin five times, so we go down the leftmost column to $N = 5$. This is the only N we have to look at. The possible numbers of hits are listed next to that column. They are 0, 1, 2, 3, 4, and 5. The probabilities of getting those numbers of hits are listed in the column for the individual trial probability, which is the rightmost column of .50. Thus, the probabilities for each number of hits are as follows:

No. of Hits	*Probability*
0 hits =	.0312
1 hit =	.1562
2 hits =	.3125
3 hits =	.3125
4 hits =	.1562
5 hits =	.0312

Notice that these are the same values we calculated through the binomial expansion. One element not shown in the binomial distribution table is the number of ways we can get each number of hits (recall that the numbers of ways were 1, 5, 10, 10, 5, and 1, respectively). However, in most cases, we have no need for that information. It is but one step on the way to the tabled probabilities.

The data in the table can be used to answer questions about any combination of outcomes. For example, for the five-toss problem we just completed, what is the probability of getting exactly two hits? It is .31250. What is the probability of getting three or more hits? It is .31250 + .15625 + .03125, which is .50. What is the probability of getting either no hits or five hits? It is .03125 + .03125, which is .0625.

CHECK YOURSELF!

Assume that we toss the coin seven times instead of five times. What is the probability of getting exactly two hits? What is the probability of getting five or more hits? What is the probability of getting either no hits or seven hits?

Probability and Experimentation

In experimentation, we are seldom interested in coin tossing or card drawing. Instead, we are interested in the effectiveness of one treatment over another or the validity of one claim over another. This is where probability comes into play.

The frequency distribution for discrete outcomes is properly graphed as a bar graph or histogram (statisticians debate which is more appropriate). The midpoints of the bars give the numbers in the binomial distribution table we just consulted. A normal distribution, on the other hand, is a graph of continuously scaled data rather than discrete data. Scores can occur at any point along the axis. However, when the midpoints of the bars are connected, the line graph takes on a near-normal shape. This tendency increases as the number of trials increases. In the case of 10 tosses shown in Figure 11.5, a normal curve has been superimposed on the histogram for comparison. Note that the line graph connecting the midpoints of the bars is practically indistinguishable from a normal curve.

> What is research, but a blind date with knowledge.
>
> —William Henry

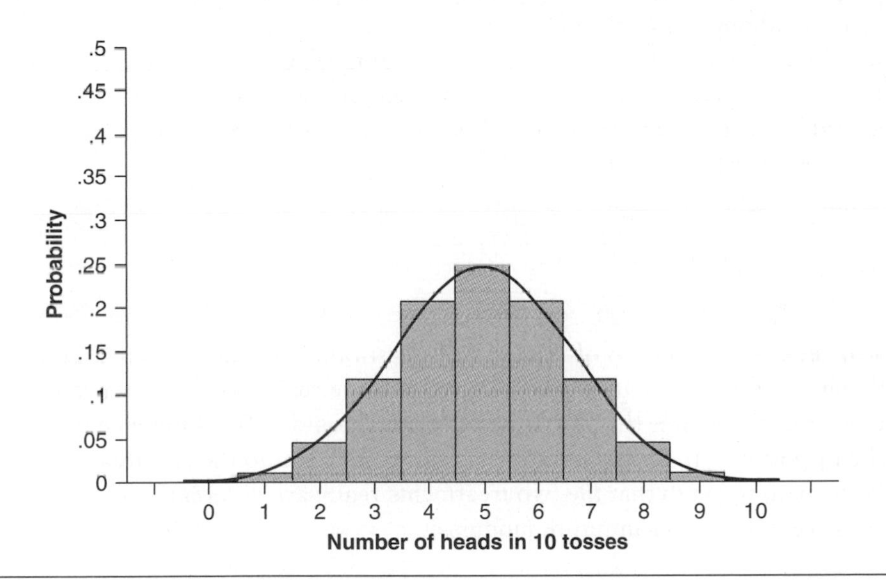

Figure 11.5 Histogram With a Line Graph Superimposed

This is good news for us because it allows us to use the probabilities listed in the normal curve table, rather than the binomial distribution table, to answer questions concerning the likelihood of random event outcomes. That is, the normal curve either will be explicitly consulted or will implicitly underlie the inferential statistics that we will look at during much of the remainder of this textbook.

Furthermore, just as we do not actually have to toss a coin thousands of times in order to know the expected probabilities of various outcomes, so also we will not have to conduct an experiment thousands of times in order to know the expected probabilities of various outcomes. Instead, we will rely on tables created by statisticians familiar with the underlying probabilities.

> There are a hundred things wherein we mortals must be content with probability, where our best light and reasoning will reach no farther.
>
> —Isaac Watts

For example, suppose your friend claims to have unusual psychic powers. She claims she can see in which of your two hands a penny is hidden. You decide to test the truth of her claim. Ten times in a row, and out of her sight, you hide a penny in either your left hand or your right hand. She then guesses which hand the penny is in. If she is a fraud and merely guessing, what is the probability that she will be right all 10 times?

Let's see. If she is guessing, then on any given trial she has a 50/50 chance of guessing the left hand or the right hand correctly. Therefore, $p = .50$, and we look in the column labeled .50. You tested her 10 times, so we look at the set of rows labeled $N = 10$. The number of hits (correct guesses) was 10, so we look at the single row showing the number of p events = 10. According to the table, the probability that she would guess correctly on all 10 trials is only .0010. In other words, the chance is very slim.

Now suppose she correctly identifies the penny's location eight times out of the 10 trials. Should you accept her claim that she has special psychic powers?

Well, if you are willing to accept her claim of special psychic powers if she gets 8 correct, then you will certainly also accept her claim if she gets either 9 or 10 correct. According to the table, the probability that she will get 8, 9, or 10 correct by mere guessing is $.0439 + .0098 + .0010 = .0547$. In other words, even without special psychic powers she could get 8 or more correct about 5% of the time.

So should you accept her claim? Statistics don't tell you that. They tell you only the probability that she performed well by chance alone, not what you should decide. Only you can decide whether a 5% chance of fraud is so little that you will accept her claim or too much for you to accept her claim.

Looking Ahead

It is easy to see the relationship of this type of question to the more familiar research question, "Which of two treatments works better?" If there really is no difference between two different treatments, what is the probability that you would find a difference by mere chance alone? The opposite form of that question is "How different do the two treatment outcomes have to be for you to claim that the two treatments really are different?" These are the types of questions we will answer in future modules.

CHECK YOURSELF!

When we use probability to answer a research question, what part of our answer is certain? What part is subjective?

PRACTICE

Use the binomial table to answer the following questions:

17. You toss a coin six times. What is the probability that
    a. none will be heads,
    b. fewer than three will be heads, and
    c. more than four will be heads?

18. You draw nine cards from a standard deck of cards. What is the probability that
    a. four or more will be red,
    b. exactly two or three will be red,
    c. two or fewer will be red, and
    d. exactly five will be red?

19. Assume that the probability of MyLady winning a horse race at a given track in given weather conditions against given horses is .50. If MyLady runs in five such races, what is the probability that MyLady will win
    a. fewer than two races,
    b. exactly four races,
    c. all five races,
    d. no race, and
    e. at least two races?

20. Assume that a man has a .50 chance of fathering a boy and a .50 chance of fathering a girl. He fathers six children. What is the probability that
    a. two or fewer will be girls,
    b. exactly four will be girls, and
    c. at least four will be girls?

21. A teacher grades in such a way that exactly 50% of the students earn either a B grade or higher. If a student drawn at random takes five unrelated courses with this teacher, what is the probability that the student will earn a B or higher grade
    a. in all five courses,
    b. in exactly two courses, and
    c. in either three or four courses?

22. A student eats breakfast one out of every two class days (50% of the time). What is the probability that, in the next eight class days, this student will eat breakfast
    a. seven days or fewer,
    b. all eight days, and
    c. exactly three days

Nonnormal Data

The binomial formula, as well as the binomial table derived from the binomial formula, is not restricted to cases where the probability of getting a hit on a single trial is .50. So far, I have discussed such cases only. But what if we toss a coin that is weighted in such a way that it comes up heads only 25% of the time? In that case, $p = .25$ and $q = 1 - .25 = .75$. Suppose we toss that coin six times. What is the probability of getting each possible number of hits (i.e., heads)? N is still six tosses, but the appropriate column in the table is now the .25 column, not the .50 column. Reading from the table, here are the probabilities of each number of hits:

No. of Hits	*Probability*
0 hits =	.1780
1 hit =	.3560

(Continued)

(Continued)

No. of Hits	*Probability*
2 hits =	.2966
3 hits =	.1318
4 hits =	.0330
5 hits =	.0044
6 hits =	.0002

Clearly this is *not* a normal distribution. Figure 11.6 shows a histogram of these probabilities. The distribution displays a strong positive skew.

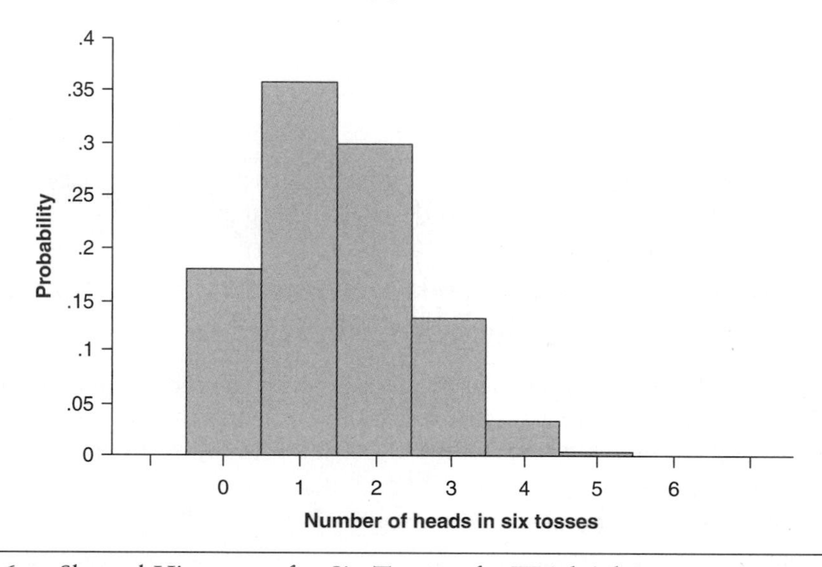

Figure 11.6 Skewed Histogram for Six Tosses of a Weighted Coin

With our weighted coin, over half the time the six tosses will yield two or fewer heads. The probability of getting five heads barely shows on the graph, and the probability of getting six heads is so small that it doesn't show at all.

Or how about your expected performance on a pop quiz in statistics—one for which you were completely unprepared and had to guess on every question? If there were 10 questions on the test and each question were in multiple-choice format with four choices from which to choose, what is the probability that you got at least 7 questions correct—that is, a score of at least 70%? Well, the probability of getting a single question correct on a four-choice item by mere guessing is 1/4, which is .25. Thus, you must look in the .25 column of the binomial table. There were 10 questions on the test, so $N = 10$. According to the table, the probability that you got at least 7 items correct by guessing is .0031 + .0004 + .0000 + .0000 = .0035. In other words, the probability that you will score at least a 70% on such a quiz by mere guessing is less than 1%. Don't let that happen to *you*!

Many events in life do not have equally likely outcomes. For example, although the probabilities of drawing a spade, club, heart, or diamond from a standard deck of playing cards are each .25, the probabilities of drawing a spade or not drawing a spade are .25 and .75, respectively. And although the probabilities of fathering a girl or a boy are roughly .50 and .50 for all men across all time, for any individual man the probabilities may be far different depending on the gender proportion among his sperm and even his age (older men produce more viable female sperm than male sperm). The proportions of right-handed and left-handed people are not equal. The probabilities of your graduating or not graduating from college once you have matriculated in college are not equal. The probability that a person will suffer from schizophrenia is not equal to the probability that he or she will not suffer from schizophrenia. In each of these cases, we may use binomial methods for calculating probabilities, but we must be careful to apply the appropriate values for p and q. We cannot assume that the probabilities are equal.

CHECK YOURSELF!

What will be the shape of the outcome distribution for a dichotomous event whose two possible outcomes are equally likely? What will be the shape of the outcome distribution for a dichotomous event whose two possible outcomes are *not* equally likely?

PRACTICE

Use the binomial table to answer the following questions:

23. You draw nine cards from a standard deck of cards. What is the probability that
    a. six or more will be diamonds,
    b. exactly two will be diamonds,
    c. all nine will be diamonds, and
    d. exactly four or five will be diamonds?

24. You draw six cards from a standard deck of cards. What is the probability that
    a. exactly three will be hearts,
    b. all six will be hearts,
    c. five or more will be hearts, and
    d. only one will be a heart?

25. You read an article that says that freshmen who live on campus have a 5% chance of contracting a contagious viral infection that requires them to miss at least one full week of classes. You follow the health of six randomly selected freshmen who neither live together nor are friends with one another. What is the probability that
    a. two or more of the six freshmen will contract a contagious viral infection that requires them to miss at least one full week of classes,
    b. exactly three of the six freshmen will contract a contagious viral infection that requires them to miss at least one full week of classes, and
    c. none of the six freshmen will contract a contagious viral infection that requires them to miss at least one full week of classes?

26. You read an article in your college newspaper that says that 30% of all cars on campus are blue. The next morning, you stand at the entrance to your college and log the color of the first 10 cars entering the campus. Four of them are blue. If it is true that 30% of the cars on your campus really are blue, what is the probability that exactly 4 out of the 10 cars you observed were blue?

27. Your friend Suzie forgets to read an assigned chapter. The next day, the instructor gives a pop quiz on the chapter. Suzie must guess at every question. The quiz consists of five questions. Each question is a multiple-choice question, with five possible choices. What is the probability that Suzie will get at least 80% on the pop quiz?

28. The Cook-A-Lot company advertises that 90% of its Crock-Pots are still working without need of repair 5 years after purchase. You purchase six Crock-Pots for friends as gifts. What is the probability that all six Crock-Pots will still be working without need of repair after 5 years?

29. Your friend Ike forgets to read an assigned chapter. The next day the instructor gives a pop quiz on the chapter. Ike must guess at every question. The quiz consists of four questions. Each question is multiple-choice, with four choices. What is the probability that Ike will get at least three of the four questions correct?

30. At a given college, 40% of the students graduate with cum laude honors. What is the probability that all five seniors selected at random will graduate with cum laude honors?

31. You roll a die six times and get a 2 exactly four of those times. Why can't you use the binomial table to calculate the probability of that happening?

32. Your friend Priyadarshi forgets to read an assigned chapter. The next day the instructor gives a pop quiz on the chapter. Priyadarshi must guess at every question. The quiz consists of six questions. Each question has three choices: "always," "sometimes," or "never." Priyadarshi hopes to score at least 50% by guessing. Why can't you use the binomial table to calculate the probability of that happening?

33. The same friend discussed in this module who claimed to have special psychic powers now claims to have an extraordinary palate as well. She claims that she can tell the difference between store-brand peanut butter and name-brand peanut butter by taste alone. You put her claim to the test. Ten days in a row, you give her identical-sized portions of four different kinds of peanut butter—one store brand and three name brands. If she is just guessing and really cannot tell the difference, what is the probability that she will be able to pick out the store brand from the name brands on at least 6 of those 10 days?

34. The same friend discussed in this module who claimed to have special psychic powers now claims to be able to detect disease by mere touch. Seven students in your dorm feel ill. Prior to the students going to a medical clinic for a diagnostic test, the friend with claimed diagnostic powers predicts the medical tests' results merely by touching the students. If she really does have special diagnostic ability, what is the probability that she will be able to correctly predict at least five of the seven students' test results (hit or miss)?

Visit the study site at www.sagepub.com/steinberg2e for practice quizzes and other study resources.

PART VII

Inferential Theory

"Of Truth and Relativity"

12

Sampling, Variables, and Hypotheses

Terms: descriptive, sample, inferential, population, parameter, representative sample, sampling, simple random sampling, systematic random sampling, stratified random sampling, cluster sampling, convenience sampling, independent variable, dependent variable, extraneous variable, confounding variable, post hoc study, hypothesis, research hypothesis, alternative hypothesis, directional hypothesis, nondirectional hypothesis, null hypothesis

Symbols: IV, DV, XV, CV, H_0, H_A, H_1

Learning Objectives:

- Distinguish between various sampling methods
- Distinguish between types of variables
- Distinguish between null and alternative hypotheses
- Distinguish between directional and nondirectional hypotheses
- Write various types of hypotheses
- Understand the relationship between a research question and the directionality of the hypothesis
- Understand the impact of extraneous variables on interpretation of a study's result

From Description to Inference

So far, we have been using statistics descriptively. With its **descriptive** use, a statistic says something about, or describes, the group of subjects in the study. For example, we have calculated the mean and the standard deviation of a group of subjects' scores. However, the only subjects described have been those actually observed—in other words, the **sample**.

We are now ready to begin using statistics inferentially. With its **inferential** use, a statistic is used to draw a conclusion about the characteristics of a larger group from which the sample was drawn. This larger group is called the **population**. A characteristic of the population—say, its mean—is called a **parameter**. For example, when a researcher draws a sample of subjects from a population, we might ask if the mean for the sample is similar to the mean of the population. That is, is the statistic similar to the parameter? This is an important question because researchers are seldom able to test or observe all subjects to which an experimental treatment might be applied. Rather, they treat and observe a subset of subjects—a sample—and then generalize their conclusions to the larger group from which the subjects were drawn—the population.

Sampling

For researchers to trust that the data from a sample will be similar to the data from the population about which they draw conclusions, the sample must be a representative sample. A **representative sample** is a sample whose relevant characteristics are similar to those of the population from which it was drawn. For example, if the population about which we want to draw a conclusion is men and women of middle and upper socioeconomic status, then the sample must include both men and women and these men and women must be of both middle and upper socioeconomic status. The sample cannot include children, and it cannot include men or women of lower socioeconomic status. If the population about which we want to draw a conclusion consists of golden retriever dogs, then the sample must consist only of golden retriever dogs, not poodles, terriers, or spaniels.

Sampling is the process of drawing a sample. Because the validity of an inference depends on whether or not the sample was representative, researchers take great care in drawing representative samples. The best way to ensure a representative sample is through simple random sampling.

> Anyone who considers arithmetical methods of producing random digits is, of course, in a state of sin.
>
> —John Von Neuman

In **simple random sampling,** every case in the population has an equal chance of being selected. The classic example is "drawing names from a hat." Each person is assigned a number, and his or her number is written on a slip of paper that is put into a jar, or "hat." Then a predetermined number of cases are drawn from the hat. Each person has the same chance as every other person of being selected. In real life, researchers rarely number slips of paper and draw cases from a hat. Instead, they start with a list—say, an alphabetical or numerical directory of all cases in the population—and use a table of random numbers to indicate the specific cases to be selected. A table of random numbers is, as the name implies, a table of randomly arranged numbers. The researcher enters the table at a random point and reads, from that point on, numbers for the desired number of cases—say, 62, 91, 24, 56, and so on—until he or she reaches the desired number of cases. The researcher then selects from the population the 62nd, 91st, 24th, and 56th cases, and so on, corresponding to the numbers read from the table of random numbers. Many statistical software programs now supply random digits, thus eliminating the need to use a table. The details on how to use a table or software program of random numbers is better left to a textbook on research methods. For our purposes, suffice it to say that the procedure results in a simple random sample.

Systematic random sampling, like simple random sampling, selects individual cases from within the entire population. However, instead of drawing names "from a hat," the researcher selects every *n*th case from a list of all cases. For example, the researcher may select every 10th person in an organization's membership list, or he might interview the head of every 4th house on the left-hand side of the street on every street in the village. Properly done, systematic random sampling will give samples as uncontaminated by extraneous variables as will simple random sampling. The caveat is "properly done." The list or pool from which the cases are selected must include all members of the population of interest, and it must not have an internal organization or grouping that would exclude certain members from being selected. If that is not the case, characteristics of the selected cases will be systematically different from the population. For example, if a directory consists only of couples, and if it lists their names in two lines each—first male and then female—any even *n*th number will select members of only one sex.

Sometimes, the types of subjects to which the researcher wishes to generalize his or her results are not equally represented in the population from which the sample is drawn. For example, at a given university, two thirds of the undergraduate students might be female and one third male. If we want to determine the attitudes of only this university's undergraduate students toward premarital sex, a simple random sample should approximate the university's gender ratio. However, if the number of undergraduate males and females nationwide is approximately equal and we want to draw a conclusion about the attitudes of students nationwide, then we would first need to classify the students at that university as either male or female and then draw equal-sized samples from within those groups. Such a procedure is called **stratified random sampling**, because the members of the population are first classified into strata or categories and then random sampling is done from within those strata. For the sample to be representative of undergraduates nationwide, we would also need to stratify on other relevant criteria, such as type of school (public, private independent, religious), geographic area of the country, size of the school, and so on. Again, a textbook in research methods would go into much more detail on sampling techniques. For our purposes, the result will be a sample that is representative of the population about which we draw our conclusions.

Unfortunately, in the social sciences, it is not always possible to draw a random sample—be it simple random, systematic random, or stratified random. The subjects in many social science studies are people, and it is not always practical to randomly select individual subjects. For example, in a study of reading achievement among children taught by two different curricula, it is not practical to teach different children in the same classroom with different curricula. Rather, whole classes are selected. Then, some classes are taught using one curriculum, and other classes are taught using the other curriculum. This whole-group sampling is called **cluster sampling**. Unfortunately, cluster sampling does not ensure that the subjects in the groups are equivalent prior to administration of the treatment. For example, in the reading achievement study, students in one class may be smarter or better prepared than students in another class. Or the teacher for one class may be more effective than the teacher for another class. Cluster sampling complicates the interpretation of a study's results.

The worst sampling technique is **convenience sampling**. In convenience sampling, the researcher uses whatever subjects are available—typically volunteers from a restricted population. For example, a clinical psychologist may study the effect of a particular treatment on his own clients. However, his own clients are probably not a representative sample of any particular population. Similarly, psychology professors often gather data for studies from among volunteers in freshman general psychology courses. However, freshman psychology students are representative only of other freshmen psychology students. The results cannot be generalized to all people, or even to all college students.

CHECK YOURSELF!

Tell how you would draw a sample of each of the following types. Tell when you would use each type of sample. What are its strengths and weaknesses?

Type of Sample	*How to Draw*	*Strengths*	*Weaknesses*
Simple random			
Systematic random			
Stratified random			
Cluster			
Convenience			

PRACTICE

1. Professor Knotso Swift wants to know whether students learn better with or without an accompanying textbook for his lectures. What sampling technique (simple random, stratified random, cluster, convenience) is Professor Swift using in each of the following studies?
   a. He uses the textbook for students in one of his classes and does not use the textbook for students in another of his classes.
   b. He announces the intent of his study and lets students who state that they prefer to use the textbook use it and lets students who state that they prefer not to use the textbook not use it.
   c. He classifies the students in each of his classes by class year, sex, and cumulative GPA. From each of these three groups, he selects students to take the course with the textbook and students to take the course without the textbook.
   d. Starting at an arbitrary point in his class roster of names, he assigns every other student either to receive a textbook or not to receive a textbook.

2. Assume that administrators at a university administered a satisfaction survey to a sample of students. To get a representative sample, they first determined the proportion of students who are freshman, sophomores, juniors, or seniors. Then, they classified courses as 100, 200, 300, or 400 level (corresponding to freshman through senior level). Then, assuming that most of the students in the 100-level course are freshman, in the 200-level course are sophomores, and so on, they randomly selected proportional numbers of 100- through 400-level courses in which to administer the surveys. All students in these courses were administered the surveys regardless of their actual class level. However, the results did closely approximate the desired freshman through senior proportions. Which two sampling techniques did the administrators use?

3. One day, while eating a prepackaged meal, my son bit down on something hard. It was a metal clip of some sort. We sent the object to the manufacturer, who determined that it was a clamp from one of their processing machines. In a letter of apology, the director of quality assurance indicated that all packages are screened by an X-ray machine prior to shipping to prevent incidents such as what happened to us. Comment on the manufacturer's quality assurance sampling:
   a. In what way does the sampling not meet the criteria of any sampling method discussed in this module?
   b. Why is inference unnecessary?

4. A market research company maintains an office in a shopping mall. Employees of the market research company stop every *n*th (i.e., some predetermined number, such as every 16th) shopper who passes by to ask a series of market preference questions. What sampling method are they using?

5. An organization is running a raffle to raise money. The fund-raisers place all of the purchased tickets into a bowl. When sufficient tickets have been sold, a volunteer sticks his hand into the bowl and draws three winning tickets. What sampling method are they using?

6. A state education department wants to try out a new fourth grade history curriculum. It assigns a number to every fourth-grade class in the State. Then it randomly selects 100 fourth-grade classes from among the numbered classes. Teachers of the selected fourth-grade classes teach with the new curriculum. Those not selected teach with the old curriculum. What sampling method are they using?

7. Researchers place an advertisement in a newspaper, offering $25 to anyone who participates in their study. Everyone who volunteers is accepted until the study's participant cap is reached. What sampling method are they using?

8. A polling organization asks every 10th person who exits the voting booth to say which candidate he or she voted for. What sampling method are they using?

Variables

When a researcher sets out to do a study, she has an idea in mind about what she is testing and what the effect may be. For example, she may believe that the type of food given to rats will affect the rats' running speeds. Or she may believe that the type of treatment given to clinically depressed patients will affect their recovery. Or she may believe that the type of reading curriculum children are taught with will affect their reading achievement.

Typically, the researcher expresses her idea in a particular form, so that it is clear to readers and other researchers exactly what the researcher did and what she expected. The researcher states her idea in terms of the expected effect of an independent variable on a dependent variable. So before we discuss the form of the statement, let's discuss these different types of variables.

The **independent variable** (IV) can be thought of as the cause. It is the treatment or condition that the researcher expects will make subjects perform either better or worse on some measure of behavior.

The **dependent variable** (DV) can be thought of as the effect. It is the measured outcome or behavior, which the researcher then assumes is attributable to the treatment.

Let's try picking out the IV and DV in hypothetical studies.

- A researcher wants to know the effect of sweetener (sweetened, unsweetened) added to drinking fluid on the amount of fluid drunk by participants. The IV is the presence or absence of sweetener, and the DV is the amount of fluid drunk.
- A researcher wants to know the effect of type of counseling (group, individual) on clients' level of emotional adjustment. The IV is the type of counseling, and the DV is the level of emotional adjustment.

CHECK YOURSELF!

A researcher wants to know the effect of students' study time (1–3 hr, 4–6 hr, 7–9 hr, 10+ hr) on exam grade. What is the IV? What is the DV?

Another type of variable warrants mentioning. An **extraneous variable** (XV), sometimes called a **confounding variable** (CV), is an unintended variable that coexists with the IV and may therefore explain the measured effect in the DV. I like to think of XVs as parasites or leeches because they function only by their association with the IV. Unfortunately, because they coexist with the IV, when we find a measured effect on the DV, we have no way of knowing whether the effect was due to the IV or due to the XV. For this reason, XVs are not good. We want to avoid them in our studies.

Let's consider a hypothetical study of the type of television show viewed and aggressive behavior in children. If we had randomly assigned the children to television-viewing conditions, then any preexisting relevant differences between the children (say, in aggressive tendencies) could be assumed also to be randomly distributed across the viewing conditions and, therefore, not associated with any particular television-viewing condition. That, indeed, is the advantage of random sampling: It randomizes XVs across treatment conditions. But what if the children had not been randomly assigned to the television-viewing conditions? Suppose that, instead, we found people who had already freely chosen to view certain types of television shows? Such a study is called a **post hoc** study, because the treatment of interest—type of television show watched—has already happened at the time the study begins. That is, in a post hoc study, we do not assign the IV to subjects; rather, we find subjects to whom the IV has already happened. What does that do to our conclusions?

It is possible that children who are more aggressive by nature would choose to watch violent action-adventure shows rather than educational or situation-comedy shows. Children's aggressive natures would then be confounded with the type of show watched. If a relationship is found between type of show watched and subsequent aggressive behavior, we would not know whether the aggressive behavior was due to the type of show watched or to the child's preexisting nature.

This type of situation is fairly common in social science research. For example, in a study of the effect of childhood abuse on adults' later depression, we would not actually abuse children to see what effect it has on them when they became adults. That would certainly be unethical! Instead, people who have already experienced abuse as children are "found." That is, the treatment of interest has already happened at the time the study begins. The study is post hoc.

Other variables, while not unethical to assign, cannot even theoretically be assigned. Socioeconomic status, educational level, sex, height, and many other variables preexist in the subjects being studied and are, hence, not assignable. However, sometimes such a variable is the IV of interest. For example, suppose we want to know if economically poor people and economically wealthy people differ in the type of books they prefer to read. Socioeconomic status is a found variable rather than an assigned variable. The study is post hoc.

The results of post hoc studies are difficult to interpret. Children who were abused may very well be more depressed as adults, but the depression might or might not be due to the abuse. It might be due to XVs that often coexist with childhood abuse, such as low achievement in school and on the job, few close friends throughout childhood, and even having the same genetic mood disorder as the abuser. Similarly, the poor may like romance novels and the wealthy may like political satires, but it is doubtful that this is due to differential income or differential cost of the two types of books. Rather, educational and lifestyle differences associated with income probably account for the differences in book preferences.

> The grand aim of all science is to cover the greatest number of empirical facts by logical deduction from the smallest number of hypotheses or axioms.
>
> —Albert Einstein

We are back to sampling methods again. The most effective way to rule out XVs is to (a) randomly select individual subjects for inclusion in the study and then (b) randomly assign the treatment conditions to the groups. Anything less confounds interpretation of the study's results.

CHECK YOURSELF!

How does a post hoc study differ from a true experiment? Why are the results of post hoc studies difficult to interpret?

PRACTICE

9. A psychologist studies the number of video games played per week by students who hold a job while attending school and students who don't hold a job while attending school. What are the IV and the DV in this study?
10. A psychologist assigns teenagers to listen to rap music or classical music, and then the psychologist studies the amount of aggression displayed by those teenagers in a simulated conflict situation. What are the IV and the DV in this study?
11. A study by De Belis and others (2000) found that the hippocampus of adolescents and young adults who were heavy drinkers was 10% smaller than the hippocampus of their peers who were not heavy drinkers. What are the IV and the DV in this study?
12. Olweus (1993) found that boys who had been bullied during childhood suffered more depression in their 20s than boys who had not been bullied during childhood. What are the IV and the DV in this study?
13. A camp director is investigating the number of recreational activities that boys versus girls sign up to participate in. What are the IV and DV in this study?
14. A school librarian is investigating whether children from lower socioeconomic neighborhoods sign out a different number of books than do children from higher socioeconomic neighborhoods. What are the IV and DV in this study?
15. An educational psychologist is investigating whether or not asking questions at the start of each class lecture, changes student retention of the subsequent lecture materials. In some classes, he starts the class lectures with questions, and in others he does not. What are the IV and DV in this study?
16. A clinical psychologist is studying whether scheduling clients for appointments prior to their workday or after their workday results in a better rate of showing up for their appointments. He randomly schedules half of his clients for early-morning appointments and half for late-afternoon appointments. What are the IV and DV in this study?
17. Which of the studies described in the exercises above are probably post hoc studies? Why does this complicate interpretation of the results?

Hypotheses

A prudent question is one-half of wisdom.

—Francis Bacon

The greatest challenge to any thinker is stating the problem in a way that will allow a solution.

—Bertrand Russell

When a researcher sets out to conduct a study, he or she has an idea in mind about what he or she is doing and what the effect may be. We call the statement of expected results a **hypothesis**. Hypotheses are expressed in a particular form, so that it is clear to readers and other researchers exactly what the researcher did and what he or she expected. That is, hypotheses are stated in terms of the expected effect of the IV on the DV.

Research hypotheses, also known as **alternative hypotheses**, state what the researcher expects to find. They are designated as H_1 or H_A. They can be either directional or nondirectional.

A **directional** research hypothesis states which IV treatment is expected to result in the better (or worse) effect on the DV. Look for words such as *higher than, better than, more than* or their reverses such as *lower than, worse than, less than*. Sticking with the television viewing and aggression example, a directional research hypothesis might state, *Children exposed to violent action-adventure shows will display more aggression than children exposed to either educational or situation-comedy shows.*

CHECK YOURSELF!

State the directional hypothesis for the study of the effect of sweetener on the amount of fluid drunk.

A **nondirectional** research hypothesis states that the IV is expected to have an effect on the DV, but it does not specify which IV treatment will have the better or worse effect on the DV. Look for the words *different than*. Still sticking with the television-viewing example, a nondirectional research hypothesis would state, *Children exposed to educational, situation-comedy, or violent action-adventure television shows will differ in the amount of aggression they display*.

CHECK YOURSELF!

State the nondirectional hypothesis for the study of the effect of sweetener on the amount of fluid drunk.

Now we turn to the null hypothesis. A **null hypothesis,** designated H_0, states that there is no expected effect on the DV due to the IV. Look for the words *no difference*. Sticking with the television-viewing example, the null hypothesis states, *Children exposed to either educational, situation-comedy, or violent action-adventure television will show no difference in the amount of aggression they display*.

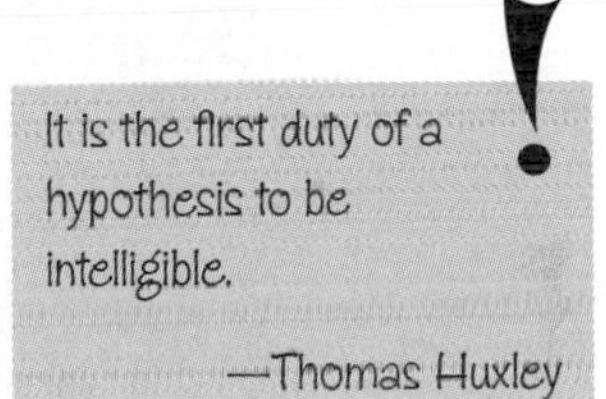

CHECK YOURSELF!

State the null hypothesis for the study of the effect of sweetener on the amount of fluid drunk.

At this point, you may be wondering why anyone would say something like that. Your wonder would be warranted because the null hypothesis usually is not what the researcher believes to be so. That is, if a researcher is conducting a study, it is generally because he or she believes that the treatment *does* have an effect, not that it *doesn't* have an effect. In the study of the type of television viewed and aggression, the researcher believes that viewing violent television affects children's aggression, not that it doesn't affect it. What the researcher expects to find is expressed in his or her research hypothesis (either directional or nondirectional), not in the null hypothesis.

The null hypothesis is used for statistical purposes only. It is not what the researcher really believes to be so, and it is not what he or she expects to find. Rather, it is the "straw man" that the researcher sets up, hoping to knock it down. Let me discuss that further.

The great tragedy of science is the slaying of a beautiful hypothesis by an ugly truth.

—Thomas Huxley

Because hypothesis testing is inferential, conclusions—no matter how probabilistically true—are merely an inference from sample data. For this reason, a researcher *can never prove a research hypothesis to be true*. The most he or she can do is disprove the null hypothesis and thereby gain support for the research hypothesis.

It works like this. Viewing violent television shows either makes children more aggressive or doesn't make them more aggressive. We may not be able to prove that viewing violent television makes children more aggressive (after all, we will not be able to observe all children; we can observe only a sample of children). However, if, for the children observed, we can prove that it is *not true* that viewing violence on television *does not* make them aggressive, we have gathered evidence that viewing violence on television *does* make them aggressive. What an argument! However, this backward reasoning is important in research because *it is only by disproving the null hypothesis that we gain support for the research hypothesis.*

> As far as the laws of mathematics refer to reality, they are not certain; and as far as they are certain, they do not refer to reality.
>
> —Albert Einstein

Yes, it's true: From this point on, we will leave certainty behind. We will be drawing only probabilistic conclusions, and these conclusions will be based only on *dis*proving the null hypothesis of no treatment effect. Fortunately for us, as we enter these murky inferential waters, probability is on our side.

CHECK YOURSELF!

Your friend tells you that a recent study proved beyond a shadow of doubt that AcheBeGone is more effective than AcheAway in relieving headache pain. How should you respond to your friend's assertion?

PRACTICE

For each of the following studies, write

a. a directional research hypothesis,

b. a nondirectional research hypothesis, and

c. the null hypothesis.

18. A researcher is studying whether men and women differ in their preferred house temperature.
19. A researcher is studying whether the number of video games played per week differs for students who hold a job while attending school and students who don't hold a job while attending school.
20. A researcher is studying whether the amount of aggression displayed in a simulated conflict situation differs in teens who listen to rap music and teens who listen to classical music.
21. A researcher is studying whether the size of the hippocampus of adolescents and young adults who are heavy drinkers differs from that of their peers who are not heavy drinkers.
22. A researcher is studying whether boys who had been bullied during childhood experience a different amount of depression in their 20s than boys who had not been bullied during childhood.
23. A camp director is investigating the number of recreational activities that boys sign up to participate in and the number of recreational activities that girls sign up to participate in.

24. A school librarian is investigating whether children from lower socioeconomic neighborhoods check out a different number of books than do children from higher socioeconomic neighborhoods.
25. An educational psychologist is investigating whether or not asking questions at the start of each class lecture changes student retention of the subsequent lecture materials.
26. A clinical psychologist is studying whether scheduling clients for appointments prior to their workday or after their workday results in a better rate of showing up for their appointments. He randomly schedules half of his clients for early-morning appointments and half for late-afternoon appointments.

Visit the study site at www.sagepub.com/steinberg2e for practice quizzes and other study resources.

13

Errors and Significance

Terms: sampling error, significant difference, Type 1 error, alpha, Type 2 error, beta

Symbols: α, β

Learning Objectives:

- Understand that probability, being uncertain, always includes error
- Distinguish between two types of error—Type 1 and Type 2

Random Sampling Revisited

In Module 12, you learned how to state a hypothesis in terms of the expected effect of the independent variable on the dependent variable. You also learned how to sample subjects to be included in a study. When we sample randomly from a population, we expect the sample to reflect the characteristics of the population. That is what allows us to infer the characteristics of the larger population from the characteristics of the sample.

Returning to the example of the effect of water's sweetness on the amount of water drunk, we expect that the difference we find in the amount of water drunk by participants given sweetened or unsweetened water is about the same difference we would see in any participants given sweetened or unsweetened water. We can then say that the difference we observe probably really exists in the populations from which the participants were drawn—that is, in the population of all participants given sweetened water and all participants not given sweetened water.

Sampling Error

Naturally, we do not expect sample statistics to exactly match the population parameters. To the degree that our sample statistics miss the true population parameters—indicating that our participants drank either more or less than what is true of the populations—any inference we make about the populations based on the samples will be in error. We call this type of error sampling error. **Sampling error** is any deviation due only to the particular cases falling within the samples. In other words, sampling error is deviation from expectation that is due to mere chance.

Ordinarily, sampling error is small. If we toss a fair coin 100 times, it probably won't come up exactly 50 heads and 50 tails, but it will probably be close to that—say, 46 heads and 54 tails. The difference from expectation is mere sampling error. Similarly, in the sweetened and unsweetened water example, we don't expect participants in the two treatment conditions to drink exactly the same amount of water. But under the null hypothesis, we do expect them to drink similar amounts of water. Minor differences in the amount drunk are mere sampling error.

Nevertheless, while sampling error is ordinarily small, it is possible to obtain a very large difference from expectation and yet still have it be due merely to sampling error. It's not likely, but it could happen. We will see that in a moment.

Significant Difference

How small does a difference from expectation have to be for it to be mere sampling error, and how big is too much for it to be mere sampling error?

Let's return to the coin that we toss 100 times. Under the null hypothesis, we expect no difference in the number of heads and tails. In other words, we expect 50 heads and 50 tails. Now, suppose the coin comes up 97 heads and 3 tails. That certainly doesn't seem like a fair coin, does it? Such an outcome probably isn't just a chance occurrence. Rather, such a large deviation from the expected number of heads and tails would lead us to conclude that the coin is probably weighted. With such an outcome, we would reject the null hypothesis that the coin produces no difference in the number of heads and tails. This is the backward reasoning—the "straw man" process—that you met in Module 12.

Similarly, in the study of the effect of sweetener on the amount of water drunk, the null hypothesis states that we expect no difference in the amount of water drunk by participants given sweetened versus unsweetened water. Now, suppose we find a very large difference in the amount of water drunk by participants given sweetened versus unsweetened water. As with the coin example, this outcome probably isn't just a chance occurrence. Rather, such a large difference in the amount of water drunk by participants in the two treatment conditions would lead us to conclude that the presence of sweetener in the water does, indeed, make a difference in the amount of water drunk. With such an outcome we would reject the null hypothesis that there is no difference in the amount of water drunk by participants given sweetened or unsweetened water.

When we reject the null hypothesis, it is because we have found a large difference from expectation. In statistics, we call this a significant difference. A **significant difference** is any difference that is greater than would be expected by mere chance. If two comparable samples that differ only in their independent variable (IV) treatment show a significant difference in their dependent variable (DV) outcome, we can assume that the difference in the DV outcome is due to the IV treatment. In other words, the treatment probably caused the observed difference.

Nevertheless, although large differences from expectation probably do reflect real treatment differences, it is possible to observe a very large difference from expectation and yet still have it be due merely to sampling error. It is not likely, but it could happen. We will see that in a moment.

Statisticians do it significantly better.

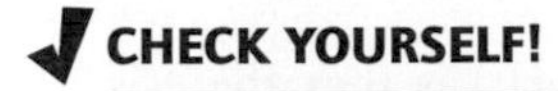

What is the relationship between *sampling error* and a *significant difference*?

The Decision Table

> You can find truth with logic only if you have already found truth without it.
>
> —G. K. Chesterton

Figure 13.1 shows a decision table. Perhaps you have seen one in a previous course in mathematics or logic. Across the top is the "real" truth. It's what is so in the entire population of cases, if only we had a way of knowing the real truth—which we don't. Down the left side is what we "think" to be the real truth. It's the judgment we make about the real truth based on the available sample data. Because we are making judgments about a population truth based on sample evidence, the decision table is a table of inference.

Your conclusion about the H_0: based on the samples	Real truth of the H_0: in the population(s): True	False
Retain	Hit	Miss (Type 2) (β)
Reject	Miss (Type 1) (α)	Hit

Figure 13.1 Elements of a Decision Table

> Contradiction is not a sign of falsity, nor the lack of contradiction a sign of truth.
>
> —Blaise Pascal

And therein lies the rub. Because we are making inferences about cases beyond those actually included in the study, our decision may or may not agree with what is really true. Notice that two of the cells in the decision table are labeled "Hit" and two are labeled "Miss." The "Miss" cells demonstrate that our decision might be wrong. Yes, that's right: We have left certainty behind and are now dealing with probability—educated guesses about what is so, but without certainty. That's the way it will be for the rest of this textbook. Fortunately, however, probability is on our side.

CHECK YOURSELF!

In the following situations, have you made a hit or a miss?

- The null hypothesis is really true and your sample data lead you to retain it.
- The null hypothesis is really false and your sample data lead you to reject it.

Type 1 Error

Continuing with the drinking water example, our null hypothesis (H_0) is that there is no difference in the amount of water drunk by participants given sweetened water and those given unsweetened water. Figure 13.2 is a diagram of the null hypothesis.

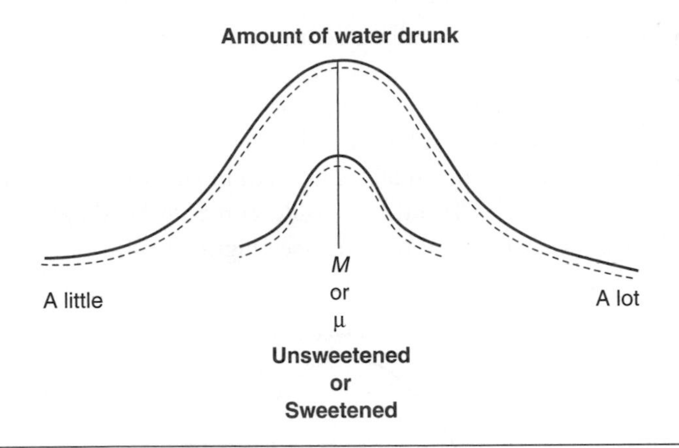

Figure 13.2 Null Hypothesis for Amount of Water Drunk

NOTE: The solid line curve represents the amount of sweetened water drunk, and the dashed line curve represents the amount of unsweetened water drunk; larger curves represent populations, and smaller curves represent samples; "μ" is the population mean, and "M" is the sample mean.

> But what is Truth? Is Truth unchanging law? We all have Truths. Are mine the same as yours?
>
> —Pontius Pilate, speaking to Jesus, in Webber and Rice's rock opera, *Jesus Christ, Superstar*

Look again at the decision table. A true null hypothesis is located in the left column. Now, if the null hypothesis is really true, when we draw a random sample of participants to whom we give sweetened water and we draw another random sample of participants to whom we give unsweetened water, we probably will not find much difference in the amount of water each sample's participants drink. And if, indeed, we find very little difference between the samples in the amount of fluid drunk, what would be the most logical decision to make regarding the null hypothesis? Should we retain it or reject it? Yes, of course, we should retain (not reject) the H_0.

Now, find the "Retain" decision point on the decision table. It's the top row. And now, find where our decision (top row) intersects with the real truth (left column). They intersect at the top left cell. Notice that the top left cell says "*Hit.*" This indicates that we have made a correct decision regarding the real truth of the null hypothesis (see Figure 13.3).

Your conclusion about the H_0: based on the samples	Real truth of the H_0: in the population(s): True	False
Retain	***Hit***	Miss (Type 2) (β)
Reject	Miss (Type 1) (α)	Hit

Figure 13.3 Correct Decision Under a True Null Hypothesis

Given that the null hypothesis is really true, most of the time we will not reject the null hypothesis. Most of the time, we will make a *hit*. But does it have to happen that way? No, it does not. It is possible, even when the null hypothesis is really true, and even when subjects are randomly sampled, that we will find a large difference in the amount of water drunk by the two groups of participants. It's not likely, but it could happen—in the same way that the fair coin we examined in Modules 10 and 11 could come up heads all the time or none of the time. We can diagram the situation like this (see Figure 13.4).

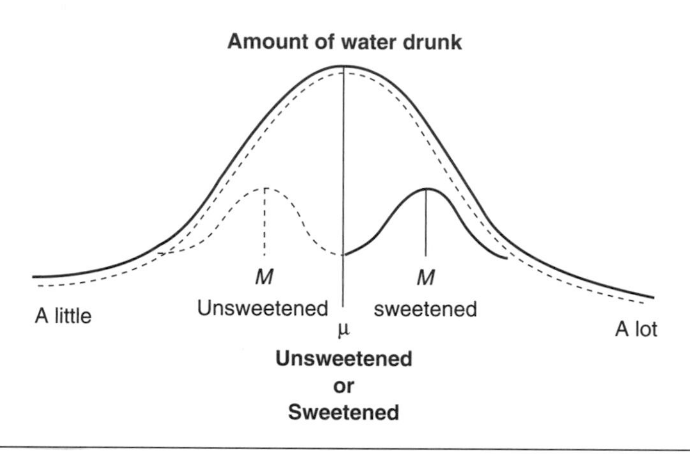

Figure 13.4 Disparate Samples Under a True Null Hypothesis

NOTE: The solid line curve represents the amount of sweetened water drunk, and the dashed line curve represents the amount of unsweetened water drunk; larger curves represent populations, and smaller curves represent samples; "μ" is the population mean, and "*M*" is the sample mean.

But remember, we never know the real truth regarding the null hypothesis. The only data available to us come from the subjects in our samples. So if we find a large difference between the samples in the amount of water drunk, as shown in Figure 13.4, what would be the most logical decision to make regarding the null hypothesis? Should we retain it or reject it? Yes, of course, we should reject the null hypothesis.

Now, find the "Reject" decision point on the decision table (see Figure 13.5). It's the bottom row. And find where our decision (bottom row) intersects with the real truth (left column). They intersect at the bottom left cell. Notice that the bottom left cell says "*Miss.*" This indicates that we have made an incorrect decision regarding the real truth of the null hypothesis.

Your conclusion about the H_0: based on the samples data	Real truth of the H_0: in the population(s): True	False
Retain	Hit	Miss (Type 2) (β)
Reject	***Miss (Type 1) (α)***	Hit

Figure 13.5 Decision Table Showing Type 1 Error

There is a name for this type of error. It's called a **Type 1 error**. It is also called **alpha** and is designated by the symbol α. Type 1 error, or α, occurs when the null hypothesis is really true, but we incorrectly reject it. Said another way, it's when there really is no effect, but we find one.

Type 2 Error

So far, we have looked only at the situation when the null hypothesis is really true. But what if the research hypothesis, rather than the null hypothesis, is really true?

Continuing with the drinking water example, the nondirectional research hypothesis (H_A) is that there really is a difference in the amount of water drunk by participants given sweetened water versus unsweetened water. We can diagram the research hypothesis like this (see Figure 13.6).

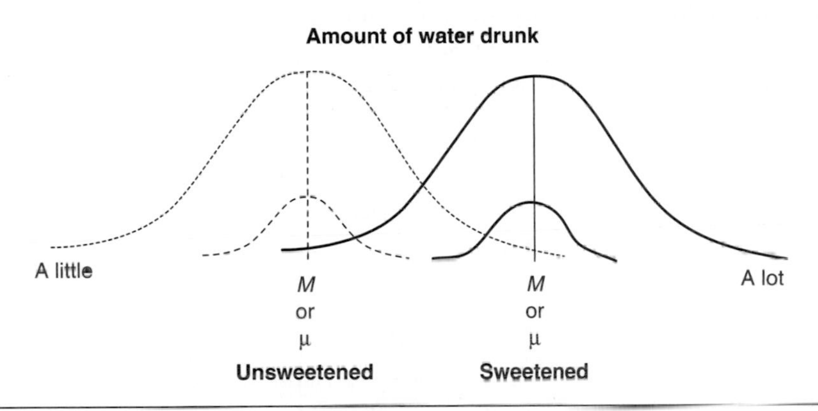

Figure 13.6 False Null Hypothesis for Amount of Water Drunk

NOTE: The solid line curve represents the amount of sweetened water drunk, and the dashed line curve represents the amount of unsweetened water drunk; larger curves represent populations, and smaller curves represent samples; "μ" is the population mean, and "*M*" is the sample mean.

> It is not certain that everything is uncertain.
>
> —Blaise Pascal

Look again at the decision table. A true research hypothesis is the same as a false null hypothesis. Therefore, the research hypothesis is located in the right column. Now, if the null hypothesis is really false, then when we draw a random sample of participants to whom we give sweetened water and we draw another random sample of participants to whom we give unsweetened water, probably we will find quite a bit of difference in the amount of water each sample's participants drink. And if, indeed, we do find quite a lot of difference between the samples in the amount of water drunk, what would be the most logical decision to make regarding the null hypothesis? Should we retain it or reject it? Yes, of course, we should reject the H_0.

Now, find the "Reject" decision point on the decision table. It's the bottom row. And now, find where our decision (bottom row) intersects with the real truth (right column). They intersect at the bottom right cell. Notice that the bottom right cell says "*Hit.*" We have made a correct decision regarding the real truth of the null hypothesis (see Figure 13.7).

Given that the null hypothesis is really false, rejecting the null hypothesis is the most likely outcome. Most of the time, we will make a *hit*. But does it have to happen that way? No, it does not. It is possible, even when the null hypothesis is really false, and even when

Real truth of the H_0: in the population(s)

Your conclusion about the H_0: based on the samples data	True	False
Retain	Hit	Miss (Type 2) (β)
Reject	Miss (Type 1) (α)	***Hit***

Figure 13.7 Correct Decision Under a False Null Hypothesis

participants are randomly sampled, that we would not find much of a difference in the amount of water drunk by the two groups. It's not likely, but it could happen—in the same way that a weighted coin could come up heads half the time and tails the other half of the time. We can diagram the situation like this (see Figure 13.8).

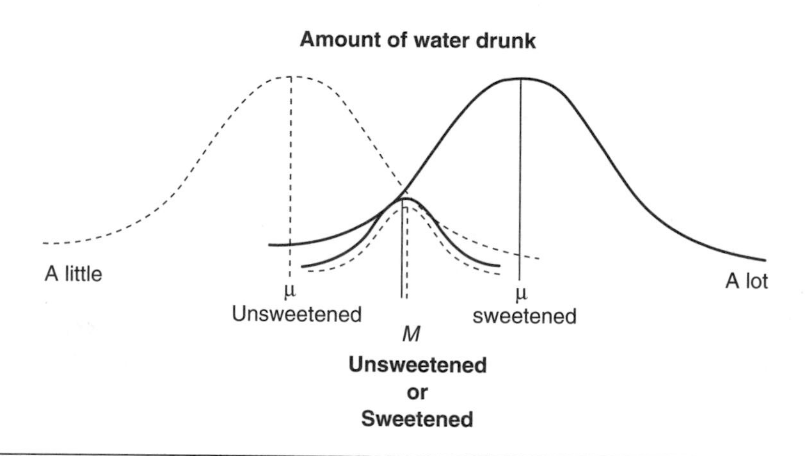

Figure 13.8 Homogeneous Samples Under a False Null Hypothesis

NOTE: The solid line curve represents the amount of sweetened water drunk, and the dashed line curve represents the amount of unsweetened water drunk; larger curves represent populations, and smaller curves represent samples; "μ" is the population mean, and "*M*" is the sample mean.

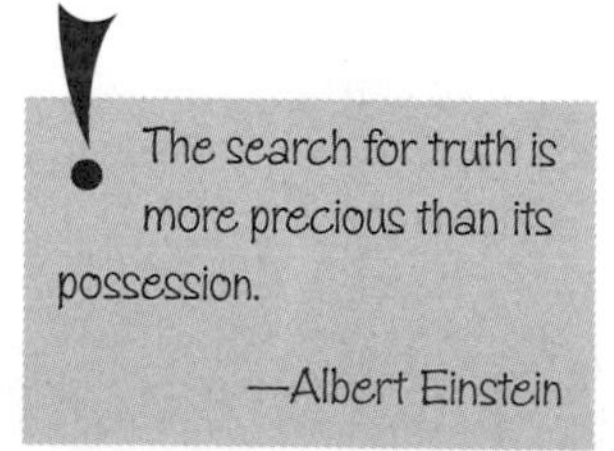

But remember, we never know the real truth regarding the null hypothesis. The only data available to us come from the subjects in our samples. So if we find very little difference between the samples in the amount of water drunk, as shown in Figure 13.8, what would be the most logical decision to make regarding the null hypothesis? Should we retain it or reject it? Yes, of course, we should retain (fail to reject) the null hypothesis.

Now, find the "Retain" decision point in the decision table. It's the top row. And find where our decision (top row) intersects with the real truth (right column). They intersect at the top right cell. Notice that the top right cell says "*Miss.*" This indicates that we have made an incorrect decision regarding the real truth of the null hypothesis (see Figure 13.9).

Your conclusion about the H_0: based on the samples data	Real truth of the H_0: in the population(s): True	False
Retain	Hit	***Miss (Type 2) (β)***
Reject	Miss (Type 1) (α)	Hit

Figure 13.9 Decision Table Showing Type 2 Error

There is a name for this type of error. It's called **Type 2 error**. It is also called **beta** and is designated by the symbol β. Type 2 error, or β, occurs when the null hypothesis is really false, but we incorrectly retain it. Said another way, it's when there really is an effect, but we don't find one.

Because hypothesis testing always involves the possibility of error, we will want to test our hypotheses with the least amount of error. Therefore, consideration of Type 1 and Type 2 errors underlies everything we will do for the remainder of this textbook.

CHECK YOURSELF!

Draw a decision table like the one in this module but with the two column headings reversed. Keep the two row headings where they were. Now, enter *hit* and *miss* designations in the cells according to the definitions in this module. In what cell does each designation fall? Are they the same cells as in this module? Why or why not?

Draw another decision table, this time leaving the two column headings as they were in this module but reversing the two row headings. Then, enter *hit* and *miss* designations in the cells according to the definitions in this module. In what cell does each designation fall? Are they the same cells as in this module? Why or why not?

PRACTICE

In answering the following exercises, it will be helpful to have before you the definitions of Type 1 and Type 2 errors that you have already learned. Here is a reminder:

Type 1: There really is *not* an effect (population truth), but you find one (your decision) in your study.

Type 2: There really *is* an effect (population truth), but you don't find one (your decision) in your study.

1. A man is accused of committing a crime. The jury finds the man not guilty. Create a decision table for this situation like the one in this module and answer the following questions:
   a. In which row of the table does the jury's action fall?
   b. Assume that the man really did commit the crime. In which column of the table does his guilt or innocence fall?
   c. By finding this guilty man not guilty, which type of error did the jury make—Type 1 or Type 2?

2. It is a population fact that men and women differ in height. You draw a random sample of men and a random sample of women, measure their heights, and find that the average height of the men and women do not differ. Given that men's and women's heights really are different in their respective populations, what type of error occurred in your study—Type 1 or Type 2? Explain your answer.

3. People who have infectious mononucleosis ("mono") test positive for the virus on a heterophile test of sheep cell agglutination. A man who does not have mono but is merely tired and nursing a cold takes the diagnostic test, and the test result is positive. What type of error is this false positive—is it Type 1 or Type 2? Explain your answer.

4. Stan applied to work at Great Company. Because Stan had a degree from a prestigious university, the company expected him to be a great employee. Given their expectation (i.e., given what they thought was the population truth regarding applicants from prestigious universities), managers at Great Company hired Stan. However, Stan turned out to be a very unproductive worker, and Great Company ended up firing him. What type of error did Great Company make in its initial decision to hire Stan—was it Type 1 or Type 2? Explain your answer.

5. Two fishermen, Abe and Gabe, are debating the relative merits of two lakes for fishing. Based on their experience in fishing in the two lakes, they decide that the two lakes are equivalent in terms of the number of fish. A researcher overhears their conversation and decides to test their conclusion. Using advanced sonar sensing, she counts the actual number of fish in each lake. She finds that one lake has many more fish than the other lake does. Assuming that her findings reflect the population truth about the two lakes, what type of error did Abe and Gabe make in their judgment—was it Type 1 or Type 2? Explain your answer.

6. In general, students who attend private universities come from higher socioeconomic families than those who attend public universities. A sociologist selects a random sample of students from each type of university and finds that the students from the public university sample have higher socioeconomic backgrounds than those from the private university sample. Given that this result is contrary to actual population fact, what type of error did this study lead to—Type 1 or Type 2? Explain your answer.

7. Teenage males tend to drive faster and get more speeding citations than teenage females do. A traffic officer reviews the traffic court records in his jurisdiction over the past month and finds no significant difference in speeding citation numbers for males and females. Given that this result is contrary to actual population fact, what type of error did this study lead to—Type 1 or Type 2? Explain your answer.

8. There is no significant difference between male and female infants in the age at which they first walk. A pediatrician reviews the age of walking for all infants in her practice and is surprised to find a significant difference between males and females in the age at which they first walked. Given that this result is contrary to actual population fact, what type of error did this study lead to—Type 1 or Type 2? Explain your answer.

Visit the study site at www.sagepub.com/steinberg2e for practice quizzes and other study resources.

14

The *z* Score as a Hypothesis Test

Term: prototype

Symbol: *z*

Learning Objective:

- Understand the logic underlying the numerator and denominator in most inferential test statistics

Inferential Logic and the *z* Score

In Module 8, you learned to calculate *z* scores with the following formula:

$$z = \frac{X - M}{s}$$

For calculation purposes, we pretended that your IQ was 120. The mean IQ score is 100, and the standard deviation for IQ is 15. Thus, we computed your IQ *z* score to be +1.33.

$$z = \frac{120 - 100}{15}$$
$$= +1.33$$

Then, we looked up your +1.33 *z* score in a normal curve table and found that 90.82% of all people have IQ scores the same or lower than yours, and 9.18% have IQ scores higher than yours.

Let's take a closer look at the logic of the *z*-score formula. The **prototype** (generic model) for any inferential test of statistical significance is

$$\frac{\text{What did you get?} - \text{What did you expect?}}{\text{Standardized random error}}$$

Note the numerator in this prototype formula: It is the difference, or deviation, from expectation. Also, note the denominator: It is random error. Now, recall that any time you divide a numerator by a denominator, you rescale the numerator into the denominator unit. Thus, the *z*-score formula rescales the deviation from expectation (numerator) into standardized error units (denominator). In other words, it tells how many standardized error units the observed score is away from the expected score.

The z-score formula follows the prototype for all inferential statistics. Here is the prototypical inferential formula and its z-score substitutions.

In the Prototype Formula	*In the z-Score Formula*
"What did you get?"	Raw score (X)
"What did you expect?"	Sample mean (M)
"Standardized random error"	Standard deviation (s)

In the case of our example, the expected score is the mean because the mean is the score that best summarizes the population. It is also your own most likely score if you are someone picked at random from the population.

With those substitutions, the prototype gives us the following formula:

$$z \text{ Score} = \frac{\text{Raw score} - \text{Sample mean}}{\text{Standard deviation}}$$

In symbols, this is

$$z = \frac{X - M}{s}$$

Once we know the population parameters—the mean and standard deviation—the obtained score can be interpreted via tabled, standardized percentages in the normal curve table. Thus, a z score works like an inferential statistic.

Constructing a Hypothesis Test for a z Score

If we are to interpret your score inferentially, we should first reexpress the problem as a test of a hypothesis. From Module 12, you know how to do that. Here is the problem reexpressed as a null hypothesis:

> There is no significant difference between your IQ and the IQ of the general public.

Recall that the null hypothesis is the "straw man" we set up to be knocked down with sufficient evidence. That is, is the deviation of your IQ from the expected IQ so little that it is just random variation and we should therefore retain the null hypothesis? Or is the deviation of your IQ from the expected IQ so great that your IQ is significantly different and we should therefore reject the null hypothesis?

If the null hypothesis were really true, there would be no significant difference between your IQ and the mean IQ of the general public. The numerator for calculating the z score would look like this:

$$\begin{aligned} z &= \frac{100 - 100}{15} \\ &= \frac{0}{15} \\ &= 0 \end{aligned}$$

However, your IQ was not 100. It was 120. Clearly, you are not average. The numerator for calculating your z score looks like this:

$$z = \frac{120 - 100}{15}$$
$$= +1.33$$

Thus, your IQ is 1.33 standard deviations above what the expected IQ would be under the null hypothesis. The question is, is that just a small and random deviation, or is it a large and statistically significant deviation? That is, what is the probability that you would have such a high IQ if the null hypothesis that you are not significantly different than average is really true?

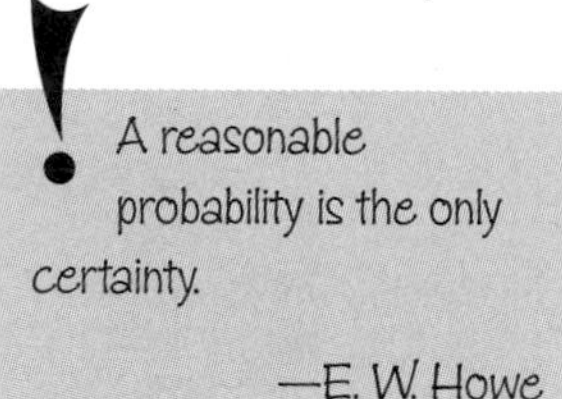

A reasonable probability is the only certainty.

—E. W. Howe

You already know the answer to that question. You found the answer when you looked up your z score in the normal curve table. When the average IQ is 100 and the standard deviation is 15, an IQ of 120 or higher would occur only 9.18% of the time. In other words, it's not common, but it's not rare either.

Is 9.18% rare enough to reject the null hypothesis and conclude that you come from some other type of population—say, a population of high-IQ Mensa members? Statistics cannot answer that question. Statistics provide probabilities on which decisions can be based, but decision making is a wholly judgmental task. As it turns out, however, there are guidelines for suggesting when a deviation is too small to reject the null hypothesis and when a deviation is too big to retain a null hypothesis. Under most guidelines, 9.18% would not be small enough to reject the null hypothesis. But the guidelines are only that—guidelines. I will have more to say about decision making in future modules.

Looking Ahead

Reconceptualizing a z score from a descriptive report of percentage above or below the score into an inferential test of the null hypothesis may seem like mere semantics. However, the reconceptualization is critical for what we will be doing in the remainder of this textbook. For the remainder of this textbook, the questions before us will be as follows:

- How different was the value we observed from the value we expected to get under the null hypothesis?
- Was the difference only a little, and therefore probably due to mere sampling error? Or was it a lot, and therefore probably due to a real treatment effect?

CHECK YOURSELF!

Compare the two bulleted questions above with the prototype formula for an inferential statistic. Which of the two questions is addressed by the numerator? Which is addressed by the denominator?

PRACTICE

This module did not introduce any new calculations. However, test your understanding of the concepts introduced with the following exercises.

1. Draw a normal curve showing ±3 standard deviations, and place the *z*-score scale along the *X*-axis (see Module 8 if you have forgotten the scale). Now, write the null hypothesis.
   a. Where will a *z* score fall in the distribution curve if the null hypothesis is true (and therefore should be retained)?
   b. Where will a *z* score fall in the distribution curve if the null hypothesis is false (and therefore should be rejected)?

2. The mean IQ for a population is 100, and the standard deviation is 15. Girard's IQ is 142.
   a. State the null hypothesis.
   b. State the research (alternative) hypothesis.
   c. If Girard is just a random person from the population, what is the probability that his IQ would be so high?
   d. You have not learned the standards for rejecting or retaining the null hypothesis. However, given the percentage of people with IQs as high or higher than Girard's in this population, does it seem like Girard's IQ is high enough to reject the null hypothesis? Justify your decision.

3. The average SAT score is 500. The standard deviation is 100. Monica scores 440.
   a. State the null hypothesis.
   b. State the research (alternative) hypothesis.
   c. If Monica is just a random SAT taker, what is the probability that she would score as low as she did or lower?
   d. You have not learned the standards for rejecting or retaining the null hypothesis. However, given the percentage of people who score at or below Monica's SAT score, does it seem like Monica scores low enough to reject the null hypothesis? Justify your decision.

4. Professor Laffing peppers his lectures with corny jokes. At the end of the course, he administers the Laugh-o-Meter test to see how funny his students found his lectures. The Laugh-o-Meter test has a nationwide mean of 40 and a standard deviation of 6. Professor Laffing's funniness rating by the members of his class is 54.
   a. State the null hypothesis.
   b. State the research (alternative) hypothesis.
   c. If Professor Laffing is just a random person from the population, what is the probability that he would be as funny as he is?
   d. You have not learned the standards for rejecting or retaining the null hypothesis. However, given the test's nationwide norms, does it seem that someone as funny as Dr. Laffing is funny enough to reject the null hypothesis? Justify your decision.

5. The mean height for U.S. adult males is 69.5 in. The standard deviation is 2 in. Dominic is 72.0 in. tall.
   a. State the null hypothesis.
   b. State the research (alternative) hypothesis.
   c. If Dominic is just a random male from the U.S. population, what is the probability that he would be as tall as he is?
   d. You have not learned the standards for rejecting or retaining the null hypothesis. However, given the percentage of U.S. adult males who are as tall or shorter than Dominic, does it seem like Dominic is tall enough to reject the null hypothesis? Justify your decision.

Visit the study site at www.sagepub.com/steinberg2e for practice quizzes and other study resources.

PART VIII

The One-Sample Test

"Are They From Our Part of Town?"

15 Standard Error of the Mean

Terms: central limit theorem, sampling distribution of the mean, standard error, standard error of the mean

Symbol: σ_M

Learning Objectives:

- Understand the difference between a raw score distribution and a sampling distribution of the mean
- Understand why any sampling distribution of the mean is normally distributed
- Understand the impact of sample size on the shape of a sampling distribution
- Understand the impact of sample size on the size of the standard error of the mean

Central Limit Theorem

Assume that there are 100 people in a required freshman class in college mathematics. Assume also that we are interested in that class as our population. That is, they are the group about which we want to draw conclusions. Now, suppose we administer each class member a test of mathematical ability. If we do, we will get a distribution of scores, one score from each class member. And from that score distribution, we can compute a mean score for the 100-person population.

OK, let's assume we do that. Assume that the mean score for the population is 40.00 and the standard deviation is 8.00.

Now suppose that we test only 49 students instead, selected randomly from among the 100 students in the class. The population still has 100 people, but the sample now has only 49 people. Suppose we compute the mean for this sample. Do you think that the mean score for this sample of 49 students will be exactly the same as the mean score for the entire population of 100 students?

No, probably not. Random error due to the particular students included in the sample will probably make this sample mean somewhat different from the population mean. But it should be close to the population mean, shouldn't it? Let's assume that the sample mean is 43.00.

Now suppose we include those 49 students back in the population of 100 students and draw a second sample of 49 students. We test those students and compute their mean. (For this example, we have to assume that having possibly been tested previously won't affect current performance on the test.) And suppose we put those students back into the pool and draw yet another sample of 49 students, test them, and compute their mean. Suppose we do this many, many times, each time computing the mean score for the sample of 49 students. The odds are, very few of the sample means will be exactly the same as the population mean. However, most of the sample means should be close to the population mean, with some being above the population mean and some being below the population mean.

> The infinite! No other question has ever moved so profoundly the spirit of man.
>
> —David Hilbert

Now suppose we draw an infinite number of samples of 49 students. (I am getting tired just thinking about it!) For each sample, we compute its mean. If we then plot the frequencies of these infinite number of sample means, what do you suppose the distribution of the sample means will look like?

As it turns out, the distribution will be normal. This is because sample means close to the population mean will be more common than sample means far away from the population mean; also, sample means above the population mean are just as likely as sample means below the population mean. With most of the sample means falling toward the middle, and with means farther and farther from the middle being less and less likely and being symmetric in either direction, the distribution of sample means forms a normal distribution around the population mean.

Of course, it makes intuitive sense that when a population of scores is normally distributed, the means of an infinite number of samples drawn from that population will also be normally distributed. After all, samples drawn from such a population should include many middle scores and only a few extreme scores, because that's what the population included. Hence, the means of those samples will more often be moderate than extreme.

What is *not* intuitive is that the means of an infinite number of samples will approximate a normal distribution *even when the underlying population from which the samples were drawn is not normally distributed*, as long as sample size is sufficient. The tendency of the distribution of sample means to be normal regardless of the shape of the underlying distribution is known as the **central limit theorem**.

The effect of the central limit theorem becomes pronounced as sample size increases. However, the distribution of sample means will tend toward normality even when the sample sizes are quite small—say, as few as 4 or 5 scores. By the time sample size reaches 25 to 30, deviations from normality are minimal—in the third and fourth decimal places of percent area under the normal curve. This is why you can trust the data in a distribution of sample means when sample size reaches about 25 to 30 cases.

CHECK YOURSELF!

Why are sample means usually close to the population mean rather than farther from it? Why does this cause the distribution of sample means to be normally distributed?

PRACTICE

1. Clients of an eating disorder clinic ($N = 64$) take a self-esteem test. The scores are positively skewed. What will be the shape of the sampling distribution of the mean for this test? Explain your answer.
2. This exercise demonstrates that the means of samples drawn from a population will approximate a normal curve even when the sample size is quite small and the population is not normally distributed.

Directions:

Step 1: You have the following population of scores: 1, 2, 3, 4, 5.

Step 2: Draw all possible samples of Size 2 from this population, each time replacing the element drawn before the next draw. I have started the draws, in the second column of the chart below. You complete the second column.

Sample ID	*Scores Drawn*	*Sample Mean*
#1	1, 1	1.0
#2	1, 2	1.5
#3	1, 3	2.0
#4	1, 4	2.5
#5	1, 5	3.0
#6	2, 1	1.5
#7	2, 2	2.0
#8	2, 3	2.5
#9		
#10		
#11		
#12		
#13		
#14		
#15		
#16		
#17		
#18		
#19		
#20		
#21		
#22		
#23		
#24		
#25	5, 5	5.0

Step 3: Now compute the mean for each sample. I have begun the calculations in the third column of the chart above. You complete the third column.

Step 4: Count the number (frequency) of means of each value. I have started the count in the chart below. You complete the chart:

Value of the Mean	*No. of Means at That Value*
1.0	1
1.5	2
2.0	3
2.5	
3.0	
3.5	
4.0	
4.5	
5.0	1

Add the frequencies in the second column. You should have 25, because that's how many samples there were. If you do not have 25, you have counted wrong.

Step 5: Here is a frequency graph of the original population (Figure 15.1). Recall that there was only one score at each value—a rectangular distribution. The population clearly was *not* normally distributed:

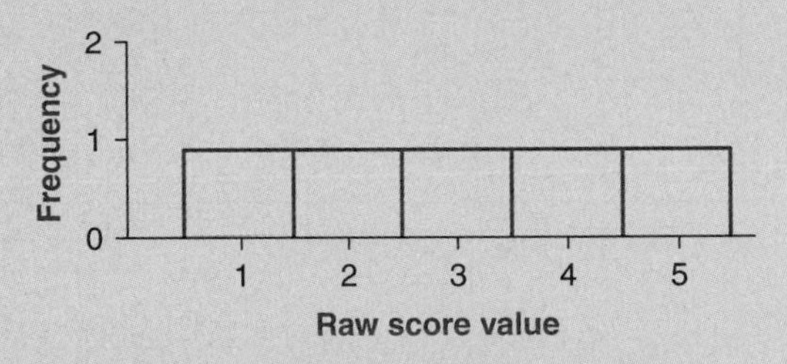

Figure 15.1 Frequency Graph of Original Population

Using the frequencies you entered in the chart in Step 4 above, graph the frequencies of the sample means (Figure 15.2).

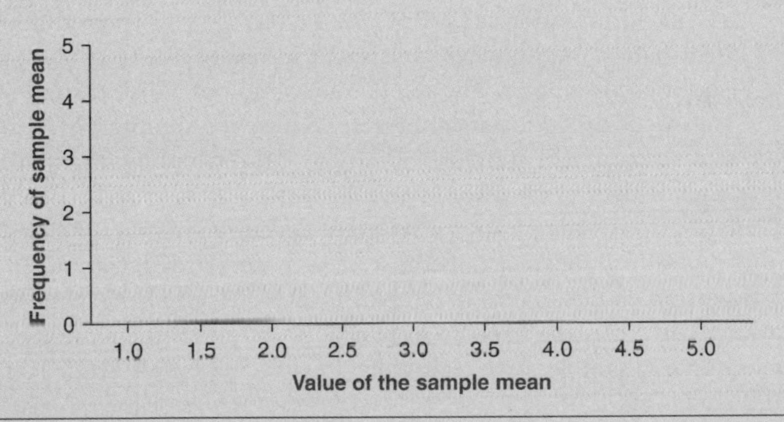

Figure 15.2 Template for Graphing Frequency of Sample Means

What is the shape of the distribution of sample means? It should look approximately normal. (If it does not, you have miscalculated one or more sample means in Step 3, or you have miscounted the number of those means in Step 4.) Note how closely the distribution of sample means approximates a normal curve despite the fact that the population was not normally distributed (it was rectangular) and the sample size was very small (it was Size 2).

Sampling Distribution of the Mean

The distribution of the means of an infinite number of samples of a given size from a population is called a **sampling distribution of the mean.** Of course, a sampling distribution of the mean is only theoretical because we cannot actually draw an infinite number of samples. Nevertheless, if we *could* draw an infinite number of samples of a given size from a population and plot those sample means, the sampling distribution of the mean would look like Figure 15.3.

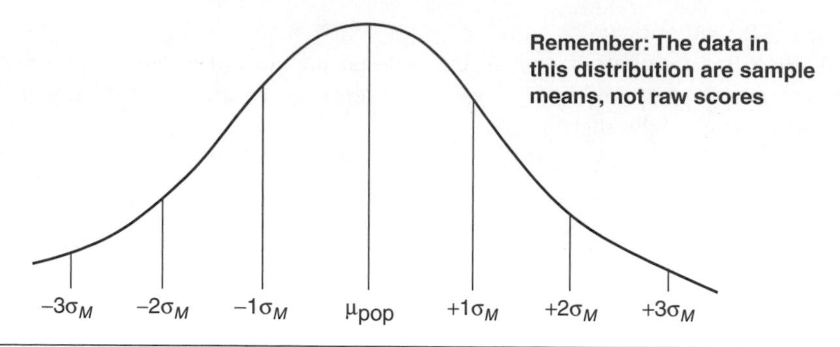

Figure 15.3 Sampling Distribution of the Mean

Look at Figure 15.3. What is the mean of this sampling distribution of means? Because the sample means distribute themselves symmetrically around the population mean, the mean of this sampling distribution of means is the population mean.

This distribution, like any normal distribution, also has a standard deviation. But it is not called a standard deviation. The term *standard deviation* is reserved for distributions of raw scores only, and a sampling distribution of the mean does not consist of raw scores. As you learned above, a sampling distribution of the mean consists of means—which are themselves summary statistics of samples. The standard deviation of a distribution of any summary statistic is called a **standard error**. Because the summary statistic in this distribution is the sample mean, the standard deviation of the sampling distribution of the mean is called the **standard error of the mean**, symbolized σ_M.

The standard error of the mean is a standard deviation; it's the standard deviation of a sampling distribution of the mean. Because it is a standard deviation, it is the average amount of dispersion we can expect in sample means from sample to sample. In the example we are working with, it is the average dispersion in mean mathematical ability we would expect from sample to sample if we repeatedly drew samples of 49 students from our population of 100 students.

Calculating the Standard Error of the Mean

Look again at Figure 15.3. The standard deviation of this distribution (now referred to as the standard error of the mean) is designated as σ_M. Its formula is

$$\sigma_M = \frac{\sigma}{\sqrt{n}}$$

Let's compute the σ_M for our example:

$$\sigma_M = \frac{8}{\sqrt{49}} = \frac{8}{7} = 1.14$$

Note how much smaller the standard error of the mean, σ_M, is than the population standard deviation, σ. It is 1.14 versus 8.00. The much smaller size of σ_M in comparison with σ makes sense. Recall that the data points in a sampling distribution of the mean are themselves means of samples. By plotting only the central tendency of each sample, the variability in each sample has already been condensed. Thus, any measure of the variability of the set of means must be smaller than the measure of variability within the original population.

CHECK YOURSELF!

Contrast a sampling distribution of the mean and a raw score distribution in terms of their shapes and the size of their standard deviations.

Sample Size and the Standard Error of the Mean

In addition, let's consider the effect of sample size on σ_M. Our sample size was 49. The sample size is the denominator in the formula for σ_M. If the sample size had been smaller—say, only 36 people—then σ_M would have been as follows:

$$\sigma_M = \frac{8}{\sqrt{36}} = \frac{8}{6} = 1.33$$

If the sample size had been larger—say, 81 people—then σ_M would have been as follows:

$$\sigma_M = \frac{8}{\sqrt{81}} = \frac{8}{9} = 0.89$$

These examples demonstrate that as sample size decreases, the size of σ_M increases; conversely, as sample size increases, the size of σ_M decreases. This is an important principle for later topics in hypothesis testing, so you will want to remember it.

The effect of sample size on the size of σ_M makes sense. As sample size approaches the population size, there are fewer population cases not included in the sample. Thus, we can put more faith in the sample's ability to estimate the true value of the parameter for that population. In that situation, the size of the random error (our σ_M) will be smaller. Conversely, as sample size becomes smaller, more population cases are left out of the sample. Thus, the estimate of the population parameter is less trustworthy. In that case, the size of the random error (our σ_M) will be larger.

CHECK YOURSELF!

State the relationship between sample size and the size of the standard error of the mean.

CHECK YOURSELF!

Why is a larger sample more trustworthy than a smaller sample as an estimate of population parameters?

PRACTICE

3. SAT scores are normally distributed with a mean of 500 and a standard deviation of 100. Using the formula for σ_M, what will σ_M be when
   a. the sample size is 25,
   b. the sample size is 121, and
   c. the sample size is 400?

4. Members of a freshmen orientation class take a time management test. Their average time management score is 82, and the standard deviation for their scores is 6. Using the formula for σ_M, what will σ_M be when
   a. the sample size is 25,
   b. the sample size is 36, and
   c. the sample size is 49?

5. Assume that the distribution for mile running times is normally distributed for members of college athletic teams nationwide with a mean of 6.3 min and a standard deviation of 0.4 min. Using the formula for σ_M, what will σ_M be when
   a. the local team size is 9,
   b. the local team size is 16, and
   c. the local team size is 25?

6. Assume that a personality trait is normally distributed in the general population, with a mean of 20 and a standard deviation of 5. Using the formula for σ_M, what will σ_M be when
   a. the research sample size is 49,
   b. the research sample size is 64, and
   c. the research sample size is 81?

7. Assume that the score distribution for the time management test mentioned in Exercise 4 is platykurtic. What shape will the sampling distribution of the mean take? Why?

8. Assume that a test is given to a large number of people but we do not yet know their scores or the shape of the score distribution. Can we be sure that the sampling distribution of the mean for this test will be normally distributed? Why or why not?

9. This exercise demonstrates that (a) the larger the sample size, the smaller the σ_M and (b) σ_M values calculated from a limited number of samples will closely approximate the values based on an infinite number of samples. The exercise requires a minimum of 15 students. Your instructor may want to assign the exercise, gather the data from all students, and calculate the actual σ_M for the samples drawn by members of your class. Or you may want to coordinate the data collection with other students and calculate σ_M yourself. Or you could pretend to be 15 or more students and complete the exercise once for each student you pretend to be.

Directions

On page 181 is a set of squares, numbered 1 through 9 (Figure 15.4). The average value of these numbers is 5:

$$\frac{1+2+3+4+5+6+7+8+9}{9 \text{ squares}} = \frac{45}{9} = 5$$

Step 1: Carefully cut out each of the nine numbered squares. Do not tear them out: Each square must be exactly the same size and must feel the same to the touch along its edges. Tightly and equally wad each square, and place all wads into a bowl. Be sure that no wad accidentally becomes stuck inside another wad.

Step 2: Mix the wads of paper well.

Step 3: Without looking, draw one wad of paper, unfold it, and mark down its value on the line in Step 5 below.

Step 4: Tightly rewad the slip of paper, and return it to the bowl.

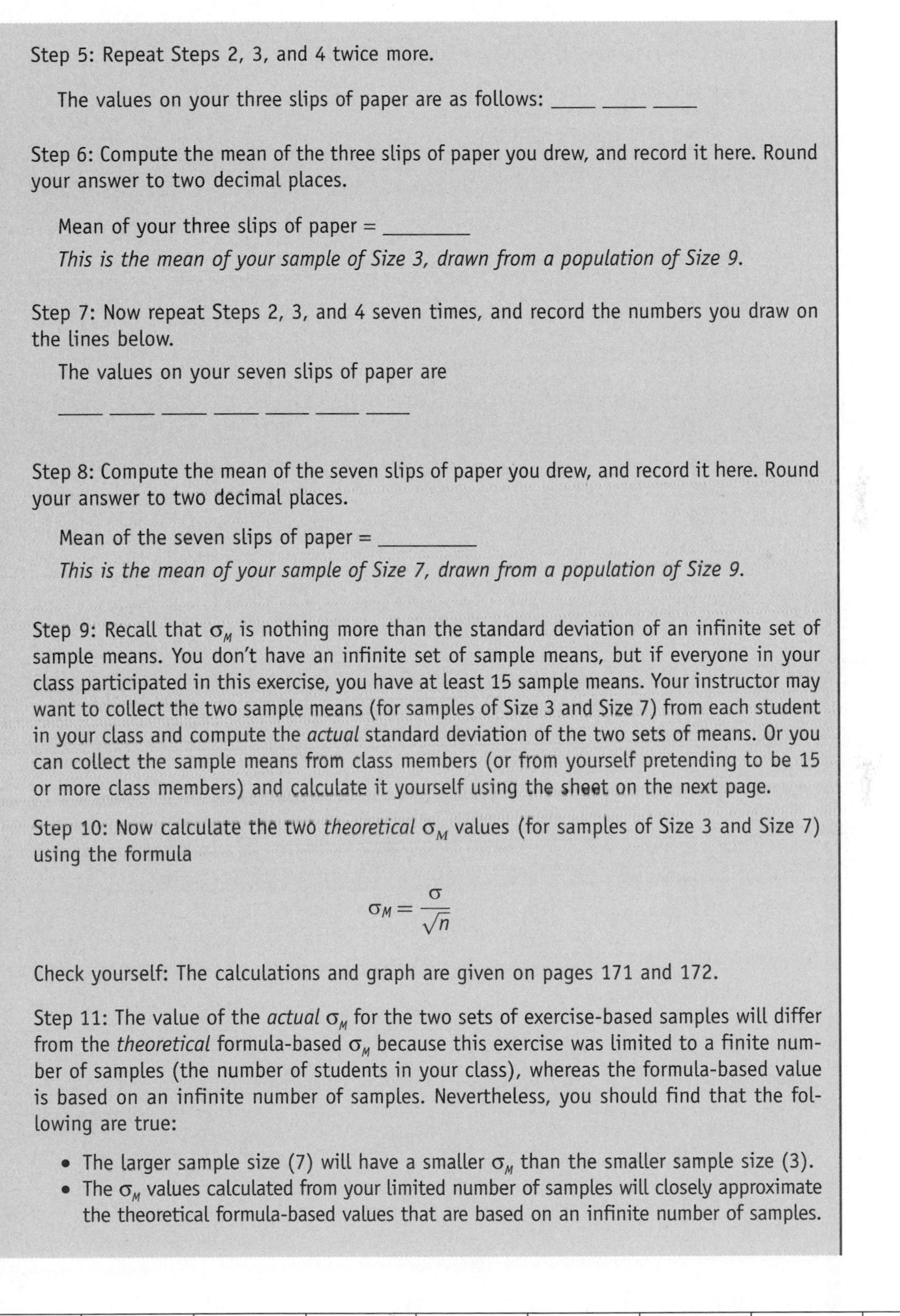

Step 5: Repeat Steps 2, 3, and 4 twice more.

The values on your three slips of paper are as follows: ____ ____ ____

Step 6: Compute the mean of the three slips of paper you drew, and record it here. Round your answer to two decimal places.

Mean of your three slips of paper = ________

This is the mean of your sample of Size 3, drawn from a population of Size 9.

Step 7: Now repeat Steps 2, 3, and 4 seven times, and record the numbers you draw on the lines below.

The values on your seven slips of paper are

____ ____ ____ ____ ____ ____ ____

Step 8: Compute the mean of the seven slips of paper you drew, and record it here. Round your answer to two decimal places.

Mean of the seven slips of paper = ________

This is the mean of your sample of Size 7, drawn from a population of Size 9.

Step 9: Recall that σ_M is nothing more than the standard deviation of an infinite set of sample means. You don't have an infinite set of sample means, but if everyone in your class participated in this exercise, you have at least 15 sample means. Your instructor may want to collect the two sample means (for samples of Size 3 and Size 7) from each student in your class and compute the *actual* standard deviation of the two sets of means. Or you can collect the sample means from class members (or from yourself pretending to be 15 or more class members) and calculate it yourself using the sheet on the next page.

Step 10: Now calculate the two *theoretical* σ_M values (for samples of Size 3 and Size 7) using the formula

$$\sigma_M = \frac{\sigma}{\sqrt{n}}$$

Check yourself: The calculations and graph are given on pages 171 and 172.

Step 11: The value of the *actual* σ_M for the two sets of exercise-based samples will differ from the *theoretical* formula-based σ_M because this exercise was limited to a finite number of samples (the number of students in your class), whereas the formula-based value is based on an infinite number of samples. Nevertheless, you should find that the following are true:

- The larger sample size (7) will have a smaller σ_M than the smaller sample size (3).
- The σ_M values calculated from your limited number of samples will closely approximate the theoretical formula-based values that are based on an infinite number of samples.

1	2	3	4	5	<u>6</u>	7	8	<u>9</u>

Figure 15.4 Numbered Slips of Paper

Observed (Exercise-Based) Standard Error of the Mean

M = sample mean

M_M = mean of the sample means

	Sample of Size 3			*Sample of Size 7*		
Student	M	$M - M_M$	$(M - M_M)^2$	M	$M - M_M$	$(M - M_M)^2$
1						
2						
3						
4						
5						
6						
7						
8						
9						
10						
11						
12						
13						
14						
15						
16						
17						
18						
19						
20						
Sum of the means =	______			______		
Mean of the means =	______			______		
Variance of the means =	______			______		
Standard deviation of the means (i.e., σ_M) =	______			______		

Standard Deviation of the Population (Numbered Slips of Paper)

M	$M - M_M$	$(M - M_M)^2$
1	−4	16
2	−3	9
3	−2	4
4	−1	1
5	0	0
6	+1	1
7	+2	4
8	+3	9
9	+4	16
$\sum = 45$		$\sum = 60$

$$M = \frac{45}{9} = 5$$

$$\sigma^2 = \frac{\sum(M - M_M)^2}{N} = \frac{60}{9} = 6.666$$

$$\sigma = \frac{\sum(M - M_M)^2}{N} = \sqrt{6.666} = 2.58$$

Theoretical (formula-based) standard error of the mean for samples of Size 3 and Size 7 from this population of nine scores are as follows (Figure 15.5):

Sample of Size 3

$$\sigma_M = \frac{\sigma}{\sqrt{n}} = \frac{2.58}{\sqrt{3}} = \frac{2.58}{1.732} = 1.490$$

Sample of Size 7

$$\sigma_M = \frac{\sigma}{\sqrt{n}} = \frac{2.58}{\sqrt{7}} = \frac{2.58}{2.646} = 0.975$$

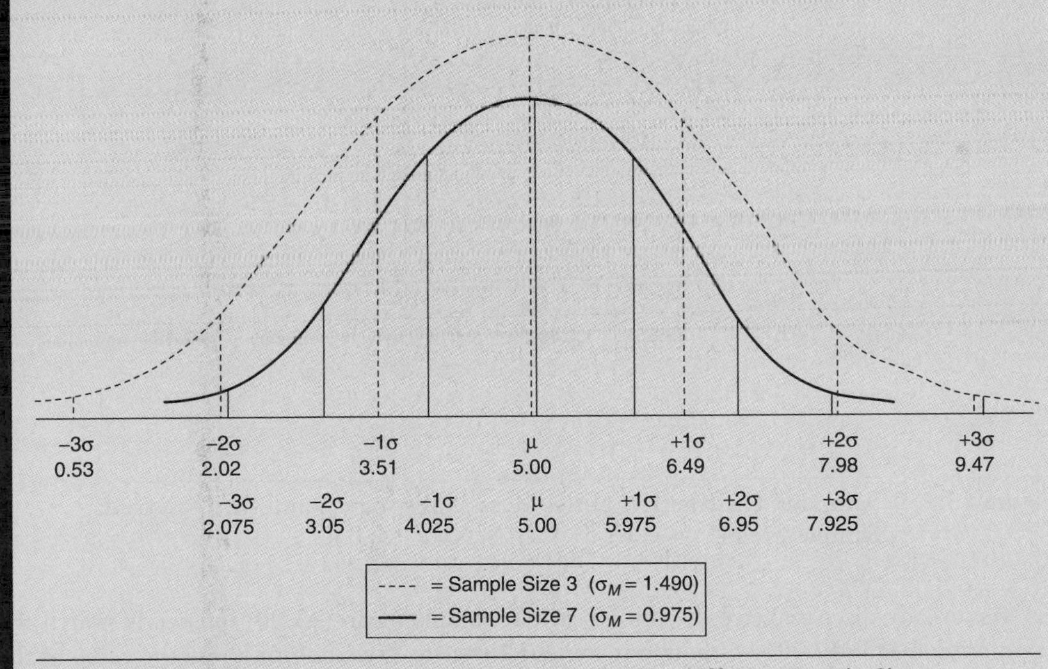

Figure 15.5 Size of Standard Error of the Mean for Two Different Sample Sizes

Looking Ahead

Recall that the value of σ_M for samples of Size 49 in our population of 100 students was 1.14. Let's add that value to our sampling distribution of the mean (Figure 15.6).

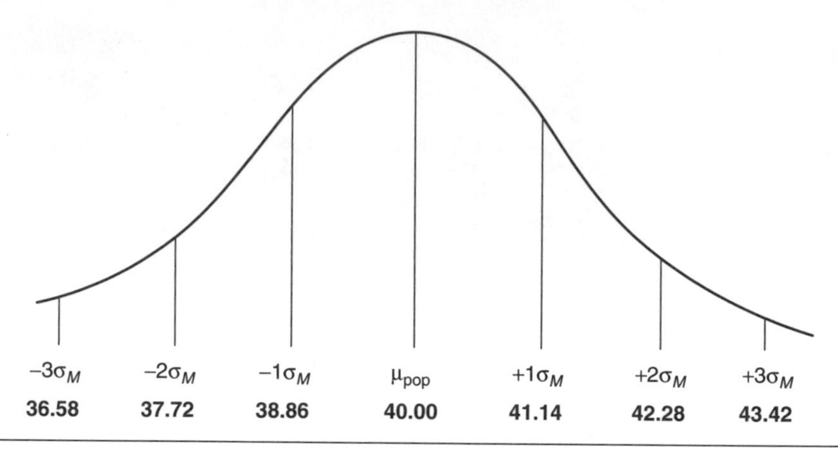

Figure 15.6 Sampling Distribution of the Mean, Showing the Value of σ_M

Recall also that the mean score on the test of mathematical ability for our sample of 49 students was 43.00. Let's add that sample mean to our diagram as well (Figure 15.7).

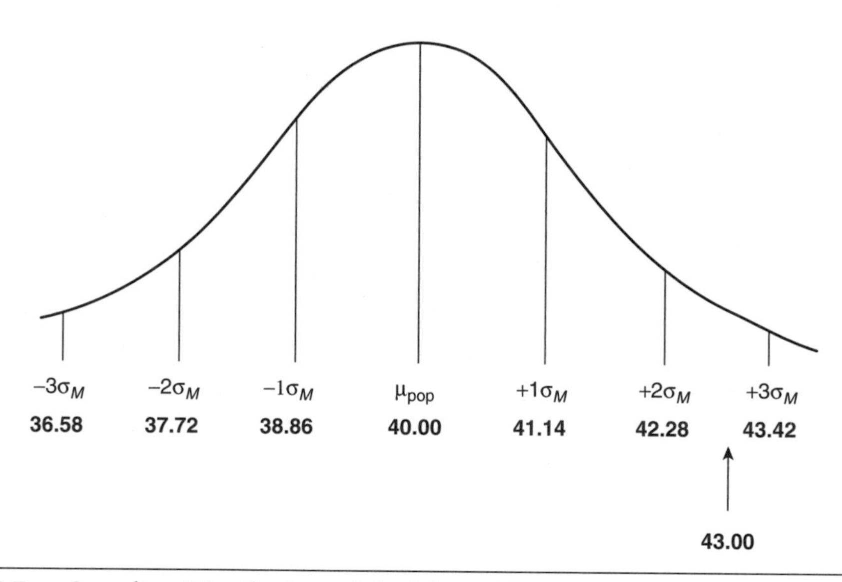

Figure 15.7 Sampling Distribution of the Mean, Showing σ_M and the Observed Sample Mean

As already discussed, we do not expect our sample mean (43.00) to exactly match the population mean (40.00)—although it could. However, if we did not know the value of the population mean, we might use our sample mean to make an inference about the value of the population mean. To the degree that our sample mean "misses" the true population mean, any inference we might make about the population mean based on the sample mean would be in error. As you learned in Module 13, we call this sampling error.

Most sample means will, as already discussed, fall quite close to the population mean. Thus, any inference we make about the population mean based on a sample mean should be reasonably accurate. However, we could—indeed, we sometimes will—obtain a sample mean that is extremely far from the population mean. In such a case, any inference we make about the population mean based on the unusual sample mean will be inaccurate.

Look again at our sample mean of 43.00 in the sampling distribution of the mean. Judging by distance on the X-axis, our sample mean appears to be very different from the population mean. But *how* different? Is it very different or only a little different? To answer this question, we need a statistic that gives us the probability of obtaining a sample mean falling this far from the known population mean. You will learn that statistic in the next module.

Visit the study site at www.sagepub.com/steinberg2e for practice quizzes and other study resources.

16

Normal Deviate *Z* Test

Symbol: $Z_{\text{norm dev}}$

Learning Objectives:

- Distinguish between a *z* score and a *Z* test
- Know the conditions under which it is appropriate to use a normal deviate *Z* test
- Calculate a normal deviate *Z* test
- Use a normal curve table to interpret a normal deviate *Z*

Prototype Logic and the *Z* Test

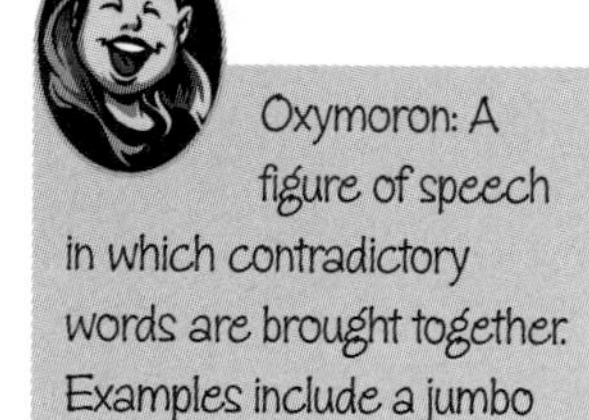

Oxymoron: A figure of speech in which contradictory words are brought together. Examples include a jumbo shrimp or a normal deviate.

In Module 15, we administered a test of mathematical ability to a population of 100 students. Then, we drew a sample of 49 students from that population. We knew the population mean: It was 40.00. And we knew the sample mean: It was 43.00. That the sample mean was not exactly the same as our population mean is not the point. The question is "Is the difference a lot or only a little?" As we ended the module, we realized that we needed a statistic to determine the probability of obtaining a sample mean that is so far from the known population mean. This statistic is called a normal deviate *Z* test.

The formula for the normal deviate *Z* test is

$$Z_{\text{norm dev}} = \frac{M - \mu}{\sigma_M}$$

Life is largely a matter of expectation.

—Horace, 65 BCE to 8 BCE

But wait! The normal deviate *Z* looks a lot like the *z* score we previously encountered, doesn't it? Yes, it does. That's because both are examples of the prototype for any test of statistical significance. Recall that a prototype is a generic model. Here is the prototype for a test of statistical significance:

$$\frac{\text{What did you get?} - \text{What did you expect?}}{\text{Standardized random error}}$$

When we studied z scores, the substitutions we made were as follows:

"What did you get?"	Raw score
"What did you expect?"	Sample mean
"Standardized random error"	Standard deviation

This gave us the formula

$$z \text{ Score} = \frac{\text{Raw score} - \text{Sample mean}}{\text{Standard deviation}}$$

In symbols, this was

$$z \text{ Score} = \frac{X - M}{s}$$

For the normal deviate Z test, the substitutions we now make are as follows:

"What did you get?"	Observed sample mean
"What did you expect?"	Population mean
"Standardized random error"	Standard error of the mean

This gives us the formula

$$Z_{\text{norm dev}} = \frac{\text{Sample mean} - \text{Population mean}}{\text{Standard error of the mean}}$$

In symbols, this is

$$Z_{\text{norm dev}} = \frac{M - \mu}{\sigma_M}$$

where (from Module 15)

$$\sigma_M = \frac{\sigma}{\sqrt{n}}$$

Calculating a Normal Deviate Z Test

Now that we have the appropriate test statistic, let's finally see how different our sample mean of 43.00 really is from the population mean of 40.00. Here, again, are the statistics and parameters from Module 15:

Population $N = 100$

Population $\mu = 40.00$

Sample $n = 49$

Sample $M = 43.00$

Population $\sigma = 8$

$\sigma_M = 1.14$

Calculating the normal deviate Z test, we get

$$Z_{\text{norm dev}} = \frac{M - \mu}{\sigma_M}$$
$$= \frac{43.00 - 40.00}{1.14}$$
$$= +2.63$$

The "2.63" tells us that our sample mean of 43.00 is 2.63 standard error units (i.e., 2.63 of the σ_M units) away from the population mean of 40.00. Let's add this statistic to the diagram we were working with in Module 15, which gives us Figure 16.1.

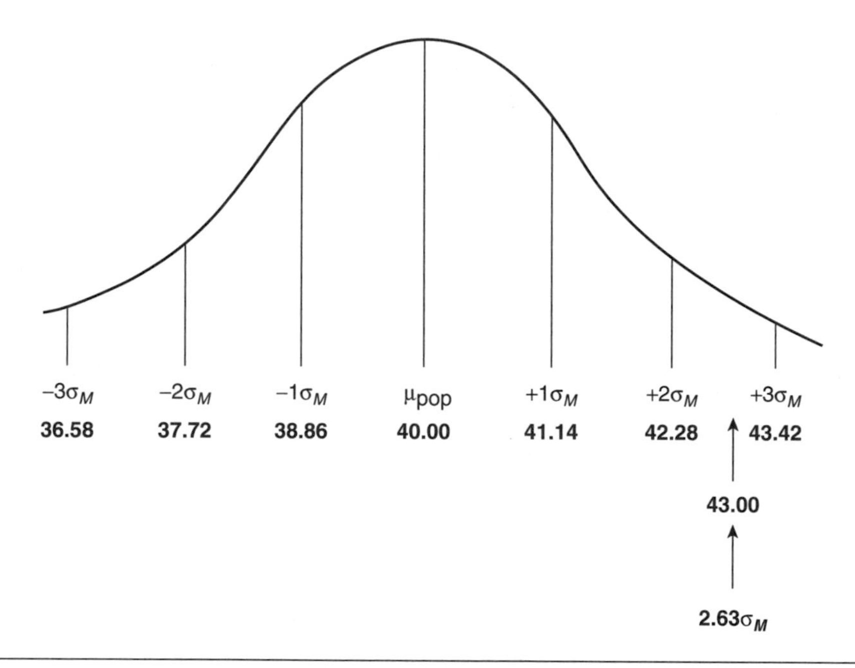

Figure 16.1 Sampling Distribution of the Mean, Showing the Value and Location of the Normal Deviate Z

This sample mean does seem very different from the population mean, based on its distance on the X-axis. But how different is it? What is the probability that we would draw a sample whose mean is 2.63 standard error units away from a population mean of 40.00?

Whether we are determining the probability of a raw score falling at a given distance from the mean of a normally distributed set of raw scores or the probability of a sample mean falling at a given distance from the mean of a distribution of normally distributed sample means, the logic is the same. Thus, all we have to do is look up the probability in a normal curve table.

Let's do that. Appendix A contains the normal curve table. A portion of the table is reproduced as Table 16.1.

Table 16.1 A Portion of the Normal Curve Table

z	*Area*	z	*Area*	z	*Area*	z	*Area*
1.92	.9726	2.27	.9884	2.62	.9956	2.97	.9985
1.93	.9732	2.28	.9887	2.63	.9957	2.98	.9986
1.94	.9738	2.29	.9890	2.64	.9959	2.99	.9986
⋮	⋮	⋮	⋮	⋮	⋮	⋮	⋮

Our normal deviate Z for our sample mean of 43.00 is +2.63. Looking at the table, what is the probability of having obtained a normal deviate Z of at least +2.63?

Recall that the percentages in the normal curve table indicate percentage *below* the given score. The probability of having obtained a sample of 49 people whose mean is 43.00 or below from a population whose mean is 40.00 is .9957. Therefore, the probability of having obtained a mean of 43.00 or higher is 1 – .9956, or only .0043. That's about 4/10 of 1%. Clearly, this is a very improbable sample to have drawn from this population.

In fact, this sample is so improbable that, if you didn't know better, you'd probably say that this sample almost certainly did not come from our known population at all, because it is just too unlikely. And you would be absolutely right.

I'm sorry to tell you this, but having done all that we just did, it was a rather silly thing to do. That's because, given the way I set up this problem, we already knew that the sample did come from the given population. After all, I told you that the sample of 49 students was drawn from the population of 100 students. So why would we bother figuring out the probability that the sample mean came from that population?

This is why: In most research, we don't know what population a sample comes from. Rather, that's precisely what we are trying to determine. We have only a sample mean, and from that sample mean, we must *infer* the population from which it came. That is, we hypothesize what that population mean might be and must determine the probability that the one obtained sample came from that hypothesized population.

Examples of Normal Deviate Z Tests

Normal deviate Z tests are common in social science research. Here are two examples for which a normal deviate Z test would be appropriate:

- A teacher gives his students a nationally standardized achievement test and wants to know whether or not his students' achievement is significantly different from the national norm. That the mean score for his students is not identical with the national mean score is not the issue. The question is "*How different* is it?" Is his students' achievement significantly different from the national norm? Or is it just a little different, no more than one would expect by mere chance?

  The teacher knows the population mean: It's the national average. And he knows the sample mean: It's the average score of his own students. So he does a normal deviate Z test to determine how different his students' average achievement is from the average achievement of the national population. If there is no significant difference between his students' achievement and the achievement of the national population, he can conclude that his students are well described by the population. However, if there is a significant difference between his students' achievement and the achievement of the national population, he will conclude that his students are not well described by the national population. Instead, they are more typical of a very different population—say, a gifted population or an academically challenged population.

- An experimental psychologist orders laboratory rats from a rat-breeding firm. She plans to test the effectiveness of a new food on the rats' running speed. However, prior to running the experiment, she wants to ensure that the rats are typical of their breed—neither unusually fast nor unusually slow runners. That the mean running speed for her rats is not identical with the breed's overall running speed is not the issue. The question is "*How different* is it?" Is her rats' speed significantly different from the overall breed's speed? Or is it just a little different, no more than one would expect by mere chance?

  She knows what the population running speed is for rats of that breed, because the breeder provides that information in the catalog. She also knows what the running speed is of her sample of three dozen rats, because she runs each of the rats through a trial run prior to beginning the experiment. So she does a normal deviate *Z* test to determine how different her rats' running speed is from the running speed of average rats of that breed. If there is no significant difference between her rats' average running speed and the running speed of the overall breed, she can conclude that her rats are well described by the breed's data. However, if there is a significant difference between her rats' speed and the speed of the overall breed, she will conclude that her rats are not well described by the breed's data. Instead, her rats are more typical of a very different population—say, a breed of very fast runners or very slow runners.

CHECK YOURSELF!

How is a normal deviate *Z* test different from a *z* score? How is it similar?

Decision Making With a Normal Deviate Z Test

Let's return to the example we calculated. Suspend belief for a minute and pretend that we do not know from what population our sample of 49 students was drawn. However, we hypothesize that it might have been drawn from a population whose parameters we do know. These population parameters are as already given: $\mu = 40.00$ and $\sigma = 8.00$. The question is "Did this sample probably come from that population?" That is, is this a typical sample from this population?

Our null hypothesis states that there is no significant difference in mathematical ability between this sample of students and a population with a mathematical ability score of 40.00. We test that null hypothesis to determine whether the difference between our observed sample mean (43.00) and our hypothesized population mean (40.00) is a lot, and therefore we can reject the null hypothesis, or only a little, and therefore we cannot reject the null hypothesis. Now, how would we do that? Why, we would conduct the normal deviate *Z* test as we have just done!

As we have already determined, the probability of obtaining a sample of 49 people whose mean is at least 43.00 from a population with a known mean of 40.00 is only .0043, which is only 4/10 of 1%. Now we can do more than merely remark on how far this sample mean "appears" to be from the population mean. We can state conclusively that such a sample would be drawn from this population less than 4/10 of 1% of the time. The other

.9958, which is 99 and 6/10 of 1% of the time, a sample with such a high mean would have been drawn from some other population—one with a higher mean mathematical ability.

Given these probabilities, we would almost certainly reject the null hypothesis that there is no significant difference in scores between our sample and this population. We would conclude, instead, that our sample probably comes from some other population—one with a higher population mean.

In summary, the normal deviate Z test is appropriate whenever you are comparing the mean of a single sample with a population whose mean and standard deviation are known.

PRACTICE

1. A biokinetics instructor measures the resting heart rate of a new class of students. Their average resting heart rate is 72. The instructor wants to know whether or not this is significantly different from that of the population at large. What *two* additional pieces of information does the instructor need to conduct the normal deviate Z test?

2. Below are the arithmetic test scores of Ms. Teachwell's fourth-grade class. The statewide average on this same test for all fourth graders is 77.0. The statewide standard deviation is 7.0. You want to know if the performance of Ms. Teachwell's class is typical of the state's fourth graders.

72	91	74	83	94	75	89	82	73	67	82	83	79	82	82
80	81	92	80	70	74	78	92	81	75	76	90	72	77	68

   a. State the null hypothesis.
   b. Calculate the normal deviate Z test. (*Note*: To do this, you will need to find the mean for Ms. Teachwell's class and the σ_M for a sample of this size.)
   c. What is the probability that Ms. Teachwell's class would score at least as high as it did if her class were just a random fourth-grade class from across the state?
   d. The principal commends Ms. Teachwell for the outstanding performance of her class. *Review*: What factors other than her exceptional teaching ability might explain the performance of Ms. Teachwell's class? Give explanations in terms of (i) random error and (ii) confounding variables.

3. According to the information on a granola raisin bran cereal box, the cereal is packaged in boxes containing 18.0 oz. of cereal, with a range (standard deviation) of 0.5 oz. Assume that a quality assurance investigator pulls out a random sample of 32 boxes of this cereal and weighs the boxes' contents. Here are the results:

17.3	16.9	18.2	17.4	17.3	18.0	16.8	17.4
18.2	17.8	18.4	16.6	17.3	17.8	17.3	17.7
18.6	17.9	16.5	16.9	17.5	18.3	17.0	17.0
17.5	18.3	16.8	17.2	17.6	18.2	17.1	18.1

   a. State the null hypothesis.
   b. Calculate the normal deviate Z test. (*Note*: To do this, you will need to find the mean weight for the 32 boxes of cereal and the σ_M for a sample of this size.)
   c. What is the probability that the average weight per box of this sample's cereal would be that far below the stated standard weight if the company's granola raisin bran cereal really does average 18.0 oz. per box?

4. According to a college guide, *selective* means that the average SAT score of incoming freshmen is 600, with a standard deviation of 50. A high school guidance counselor suspects that many colleges rated by the guide as selective do not meet the guide's standard. He randomly selects 30 colleges rated by the guide as selective. He then asks the admissions officer at each of the 30 colleges for the average SAT scores of the college's entering freshmen. The average SAT score of entering freshmen at the 30 schools turns out to be 562.
   a. State the null hypothesis.
   b. Calculate the normal deviate Z test.
   c. What is the probability that the average SAT score for these 30 colleges would be so far below the stated 600 average if the guide's claim is really true?

5. Each scale on the NEO-PR Personality Inventory is scaled with a mean of 50 and a standard deviation of 10. (Do you recall the name of this scale, from Module 9?) Professor Jovial administers the NEO-PR to a class of psychology majors ($N = 30$) and finds that the students' average score on the Gregarious (extraversion) scale is 72. If psychology majors are really no different in gregariousness from the population at large, what is the probability that Professor Jovial's students scored so high?

6. A ladies' clothing store chain reports an average sales increase of 7% over last year's sales for all stores nationwide, with a standard deviation of 2%. A district store manager pulls the shopping record of the 49 stores in his district and finds that the sales volume for this group of stores was up only 6%. If this district's stores are representative of stores throughout the company nationwide, what is the probability that this district's sales increase would be so low?

7. Affection Connection is a respected dating service for busy professionals seeking an alternative to singles' clubs. Nationwide, the company claims that the mean length for relationships they connect is 10 months, with a standard deviation of 2 months. The local franchise's office has a mean connection time of only 9 months. This length was calculated from a survey of 36 former clients. If this local franchise's clients are representative of clients throughout the company nationwide, what is the probability that this local franchise's relationship length was so short?

8. The average height of U.S. adult males is 69.5 in. and the standard deviation is 2.0. A physician thinks that this national mean seems a little short for his patients, so he records the height of the next 100 adult males to visit his office. Their average height is 69.75 in. If this physician's adult male patients are just a representative sample of nationwide adult males, what is the probability that his patients were so tall?

Looking Ahead

I have not yet discussed the criteria for accepting or rejecting the null hypothesis. However, I did note in Module 14 that 9.18% almost certainly is *not* a small enough probability to reject the null hypothesis that your IQ of 120 is about the same as the average IQ of 100. In this current module, on the other hand, I noted that 4/10 of 1% almost certainly *is* a small enough probability to reject the null hypothesis that a sample of 49 students with a mean mathematical ability of 43.00 came from a population with a mean mathematical ability of 40.00. These judgments should give you a clue that statisticians typically look for a value somewhere between these probabilities when determining statistical significance: 9.18% is too big to reject the null hypothesis, but 0.43% is too small to retain the null hypothesis. Keep considering what you think the criterion should be for rejecting the null hypothesis. Should it be 1%, 5%, or something else?

I also have not yet discussed practical importance. In the exercise regarding Ms. Teachwell's class, how many points was Ms. Teachwell's class's mean above the statewide mean? What is your sense of the practical importance of that difference? I will discuss more about this topic in a later module.

Finally, each of the examples in this module and its exercises compared the sample with a known population mean and standard deviation. However, more often than not in social science research, we do not know the population standard deviation. We can hypothesize a population mean, but the population standard deviation is not known. In that case, we must use the sample standard deviation to estimate the population standard deviation. Our statistic is then not a Z test but a t test. We will discuss t tests in the next several modules.

Visit the study site at www.sagepub.com/steinberg2e for practice quizzes and other study resources.

One-Sample *t* Test

17

Terms: estimated population standard deviation, degrees of freedom, biased indicator, correction factor, unbiased indicator, one-tailed, two-tailed, region of retention, region of rejection

Symbols: σ_{est}, *df*, $t_{1\text{-samp}}$
Learning Objectives:

- Distinguish between a normal deviate *Z* test and a one-sample *t* test
- Know the conditions under which it is appropriate to use a one-sample *t* test
- Understand the similar logic underlying various test statistics
- Understand the concept of degrees of freedom
- Distinguish between biased and unbiased estimates of a population parameter
- Find the regions of retention and rejection
- Understand the relationship between directionality of the hypothesis and tail in the sampling distribution
- Calculate a one-sample *t* test
- Use a table to interpret calculated *t*
- Report results in APA format

Z Test Versus *t* Test

In Module 16, you learned how to conduct and interpret a normal deviate *Z* test. In a normal deviate *Z* test, the investigator compares the mean of a single sample with a known mean of a population, to determine the probability that the sample came from that population.

A one-sample *t* test does the same thing. However, a one-sample *t* test can be used when a normal deviate *Z* test cannot. Do you remember the two criteria that had to be met for the normal deviate *Z* test to apply? The two criteria were as follows: (1) the sample size must be at least 25 to 30 and (2) the standard deviation for the population with which the sample is being compared must be known. When either of those two criteria is not met, the normal deviate *Z* test cannot be used. Instead, we must use the one-sample *t* test.

Comparison of *Z*-Test Versus *t*-Test Formulas

Before computing anything, let's compare the formulas for the two statistics:

Normal Deviate Z Test	*t Test*
$Z_{\text{norm dev}} = \dfrac{M - \mu}{\sigma_M}$	$t_{\text{1-samp}} = \dfrac{M - \mu}{\sigma_M}$
where	where
$\sigma_M = \dfrac{\sigma}{\sqrt{n}}$	$\sigma_M = \dfrac{\sigma_{\text{est}}}{\sqrt{n}}$
where	where
$\sigma = \sqrt{\dfrac{\sum (X - \mu)^2}{n}}$	$\sigma_{\text{est}} = \sqrt{\dfrac{\sum (X - M)^2}{n - 1}}$

The formulas for the normal deviate *Z* test and the *t* test are nearly identical. They differ only in the calculation of the standard error of the mean (σ_M), which is in the denominator of the *Z* and *t* formulas. For the normal deviate *Z* test, the numerator for the standard error of the mean is, as you learned in Module 16, the known standard deviation of the population. For the one-sample *t* test, in contrast, the numerator for the standard error of the mean is the **estimated population standard deviation**, symbolized σ_{est}. It is estimated because the standard deviation of the population is not known. This distinction is repeated here, with the difference shown in **boldface:**

$\sigma_M = \dfrac{\boldsymbol{\sigma}}{\sqrt{n}}$	$\sigma_M = \dfrac{\boldsymbol{\sigma_{\text{est}}}}{\sqrt{n}}$

Moreover, the formulas for the known population standard deviation (σ) and for the estimated population standard deviation (σ_{est}) differ. When calculating the known population standard deviation, the denominator is the sample size, as you learned in Modules 6 and 16. But when calculating the estimated population standard deviation, the denominator is the *sample size minus 1*. This distinction is repeated here, with the difference shown in **boldface:**

$\sigma = \sqrt{\dfrac{\sum (X - \mu)^2}{\boldsymbol{n}}}$	$\sigma_{\text{est}} = \sqrt{\dfrac{\sum (X - M)^2}{\boldsymbol{n - 1}}}$

Be careful about putting the $n - 1$ in the denominator of the correct formula. It goes in the formula for the estimated population standard deviation (σ_{est}), *not* in the formula for the standard error of the mean (σ_M). This is a common error you will want to avoid.

CHECK YOURSELF!

Write the formula for the standard error of the mean (σ_M) and the formula for the estimated population standard deviation (σ_{est}). Which one has *n* in the denominator? Which one has $n - 1$ in the denominator?

Degrees of Freedom

Did you hear about the statistician who was thrown in jail? He now has zero degrees of freedom.

The $n - 1$ in the denominator is the degrees of freedom (df). **Degrees of freedom** refers to how many numbers are free to vary in a calculation sequence. Recall that in calculating the standard deviation, the deviation scores (distance from the mean) must sum to 0. That restriction causes the loss of 1 df. See for yourself:

X	M	X – M
2	5	–3
4	5	–1
6	5	+1
8	5	?

For the deviations to sum to 0, the fourth deviation (where the "?" is) *must* be +3. Thus, as soon as we know $n - 1$ deviations, we also know the fourth deviation.

You will come across this concept often throughout the remainder of this book. Every statistic has a different formula for calculating the df, depending on the number of fixed parameters—that is, restrictions—in the calculation sequence for the statistic. For a t test, the formula for df is as follows:

$$df_t = n - 1 \; df \text{ for each sample.}$$

For a one-sample t test, there is only one sample. Therefore, the df for a one-sample t test is $n - 1$.

Biased and Unbiased Estimates

Let me say more about the $n - 1$ df in the denominator for the estimated population standard deviation. Recall that a standard deviation is a measure of dispersion. A sample is, by definition, smaller than the population from which it was drawn. In a normal distribution, scores are concentrated in the middle. Thus, when a sample is drawn from a normally distributed population, many more middle scores than extreme scores are typically included in the sample. As a result, the dispersion in scores—the standard deviation—tends to be smaller in the sample than in the population from which the sample was drawn. Because of this, if the sample standard deviation is used as an estimate of the population standard deviation—as is the case in the t-test formula—it will tend to systematically underestimate the true value of the population standard deviation. We call such a statistic a **biased indicator** because the tendency is in only one direction—smaller.

We need a way to correct for this tendency of the sample standard deviation to underestimate the population standard deviation. That is, we need to increase the size of the estimated population standard deviation. How might we do that? Look again at the formula for the known population standard deviation:

$$\sigma = \sqrt{\frac{\sum (X - \mu)^2}{n}}$$

If we want to increase the value of the σ on the left side of the equation, we have only two choices for actions on the right side of the equation. We must either (1) increase the numerator, which is the actual difference between the sample mean and the population mean $(X - \mu)$, or (2) decrease the denominator, which is the sample size. Either one will increase the value of the σ.

Because we have no control over the actual difference between the sample and population mean in the numerator, the approach we take is to decrease the sample size in the denominator. That's what the −1 in the denominator of the estimation formula does. For that reason, the −1 is called a **correction factor**.

No estimation formula will guarantee the same value as the actual formula. However, with the correction factor, the estimation formula no longer systematically underestimates the population standard deviation. That is, the estimated population standard deviation is now just as likely to be higher than the true population standard deviation as lower than the true population standard deviation. For that reason, the formula is referred to as an **unbiased indicator**.

Note, too, that the correction is greater for smaller samples than for larger samples. For example, assume a sample size of 10. Dividing by $n - 1$, we divide the numerator by 9 rather than by 10. Now, assume a sample size of 100. Dividing by $n - 1$, we divide the numerator by 99 rather than by 100. For a sample size of 10, removing one case takes away 1/10 of the sample size; for a sample size of 100, removing one case takes away only 1/100 of the sample size. Thus, there is a bigger reduction in the divisor, and hence a bigger correction in the estimated standard deviation for the population, when the sample size is smaller. This is as it should be. Because statistics based on smaller samples are less reliable predictors of population parameters than statistics based on larger samples, the greater correction for smaller samples makes sense.

CHECK YOURSELF!

Explain why we need a correction factor when estimating a population standard deviation from a sample standard deviation.

When Do We Reject the Null Hypothesis?

The null hypothesis says that there is no significant difference from expectation in the measured dependent variable. Let's use as our example the arithmetic scores of Ms. Teachwell's class. This example was introduced in Exercise 2 in Module 16. For this example, the null hypothesis says that there is no significant difference between Ms. Teachwell's class mean and the statewide mean. Now, if the sample and population means are not significantly different, then when we subtract one from the other—which is what the numerator of the *t* test does—we will get a value close to 0. Thus, if we do get a value close to 0, we know that the sample and population are not very different. It follows that when the test statistic is *close to 0*, we *do not reject* the null hypothesis.

On the other hand, if the sample and population means are significantly different from each other, then when we subtract one from the other we will get a value far from 0. Thus, if we do get a value far from 0, we know that the sample and population means are quite different. It follows that when the test statistic is *far from 0*, we *reject* the null hypothesis.

The test statistic, whether *Z* or *t*, is expressed in standardized dispersion units. For both the normal deviate *Z* test and the one-sample *t* test, that standardized dispersion unit is the standard error of the mean. Thus, a *Z* test or a *t* test tells how many standard errors of the mean (denominator) the sample mean is away from the population mean (numerator).

Recall that the sampling distribution of the mean is symmetric, with a mean of 0 and a standard deviation (i.e., a standard error of the mean) of 1. Thus, a Z or t statistic close to 0 falls in the central portion of the distribution. This is referred to as the **region of retention** for the null hypothesis. A t statistic far from 0 falls in either of the two tails of the distribution. Thus, the tails are referred to as the **regions of rejection** for the null hypothesis. The tail to the left is the *negative tail*, and the tail to the right is the *positive tail*. We can diagram the situation as shown in Figure 17.1.

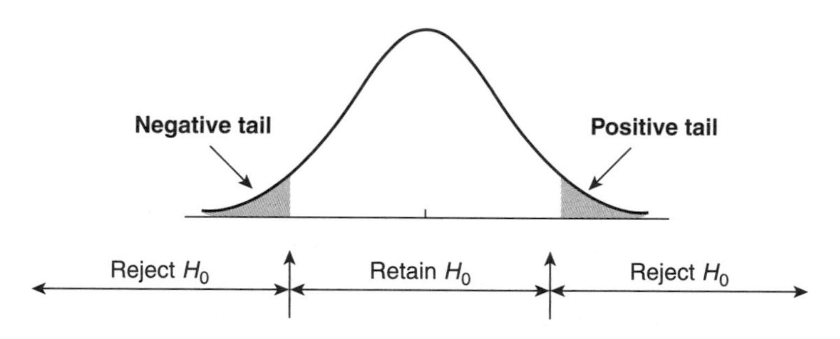

Figure 17.1 Regions of Rejection and Retention

We already know from Module 16 that Ms. Teachwell's class scored above average when compared with the statewide average. Thus, her class mean falls toward the right-hand side of the distribution. But is their performance far enough from 0 for us to reject the null hypothesis? Or is it still too close to 0 for us to reject the null hypothesis? Said another way, what is the probability that the null hypothesis really is true but her class just happened to perform so much better than average? These are the questions we will soon answer.

CHECK YOURSELF!

Draw a normal curve. Mark and label the area of retention and the areas of rejection when testing a null hypothesis.

One-Tailed Versus Two-Tailed Tests

Before examining the performance of Ms. Teachwell's class, let's discuss one-tailed versus two-tailed tests. Recall that there are two types of research hypotheses:

- A *nondirectional* hypothesis says that there is a difference between what is observed and what is expected under the null hypothesis, but it *does not specify the direction* of the difference. In our example, it says that the arithmetic scores of Ms. Teachwell's class are significantly different from those of the statewide students. Look for the words *different from*. A nondirectional hypothesis allows for the possibility that the scores of Ms. Teachwell's class may be either significantly higher or significantly lower than the statewide scores.
- A *directional* hypothesis says that there is a difference between what is observed and what is expected, but it also *specifies the direction* of the difference. In our example, it says that the arithmetic scores of Ms. Teachwell's class are significantly higher than those of the statewide students. (*Note*: It could also say that they are significantly lower than those of the statewide students.) Look for the words *higher than*, *lower*

than, or some similar comparative phrase. A directional hypothesis allows for the possibility that the scores of Ms. Teachwell's class are in the specified direction only. Should the scores fall in the opposite direction—even significantly so—the hypothesis would not be supported.

The relationship between directionality of the hypothesis and tails of the distribution is straightforward. If a hypothesis is *directional*, the test statistic is **one-tailed.** It says that the difference between what is observed (sample mean) and what is expected (population mean) will be in a particular tail of the distribution. The diagram can be seen in Figure 17.2.

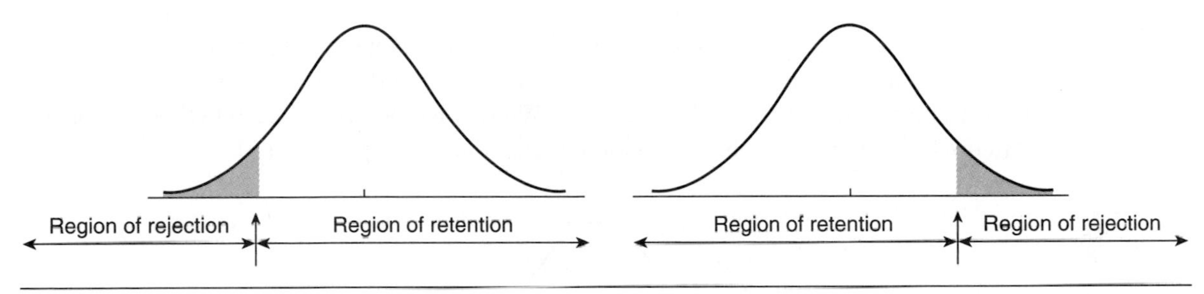

Figure 17.2 Region of Rejection in a One-Tailed Test

If a hypothesis is *nondirectional*, the test statistic is **two-tailed.** It says that the difference between what is observed and what is expected will be in either of the two tails of the distribution. The diagram can be seen in Figure 17.3.

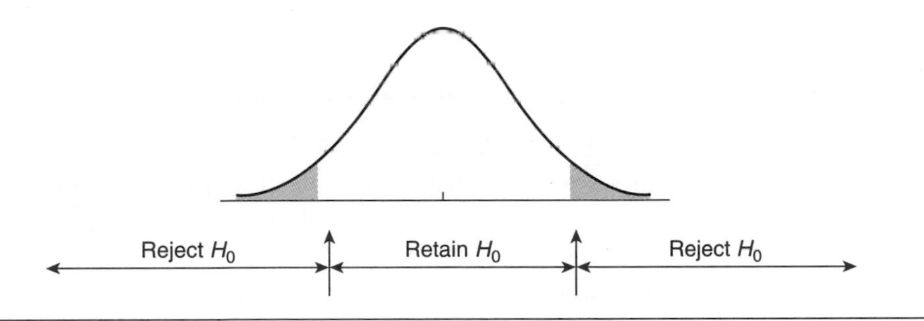

Figure 17.3 Regions of Rejection in a Two-Tailed Test

The researcher decides which type of hypothesis (directional or nondirectional), and hence which type of test statistic (one-tailed vs. two-tailed), is most appropriate for the study. If the arithmetic test had not already been given, the principal might propose a nondirectional hypothesis, which would then be evaluated with a two-tailed test. That is, does Ms. Teachwell's class perform either significantly higher than or significantly lower than statewide students? However, if the principal has reason to believe that Ms. Teachwell is a superior teacher whose students routinely outperform state norms, the principal might propose a directional hypothesis, which is then evaluated with a one-tailed test. That is, does Ms. Teachwell's class perform significantly higher than statewide students? Similarly, if the principal has reason to believe that Ms. Teachwell is an ineffective teacher or that her students are of particularly low ability, a directional hypothesis with a one-tailed test could determine whether Ms. Teachwell's class performs significantly lower than statewide students. In our example, we already know that Ms. Teachwell's class performed above average, so it makes no sense to propose a nondirectional hypothesis and conduct a two-tailed test. Instead, we will propose a directional hypothesis and conduct a one-tailed test.

The researcher also decides what level of Type 1 error he or she is willing to live with. If the consequences of falsely rejecting the null hypothesis are not very high, the researcher may be willing to accept a 10% error rate. However, if the consequences of falsely rejecting the null hypothesis are high—for example, people will die, be fired, or be denied admission—then the researcher would want a more stringent limit to error: say, 1% or even 1/10 of 1%. In the case of Ms. Teachwell, if we mistakenly conclude that her class is achieving at a significantly higher level than students statewide when in fact they are not, the worst that might happen is that she and her students will be showered with undeserved praise. To me, this does not seem like a high-risk consequence. However, others might see this as very undesirable. Again, evaluation of the consequences of wrong decisions is subjective.

Let's select a 5% Type 1 error, as that is a middle-ground decision. Figure 17.4 is a diagram for a one-tailed test at 5% Type 1 error. We will be looking to see whether or not our calculated *t*-test statistic falls in the region of rejection in the positive tail.

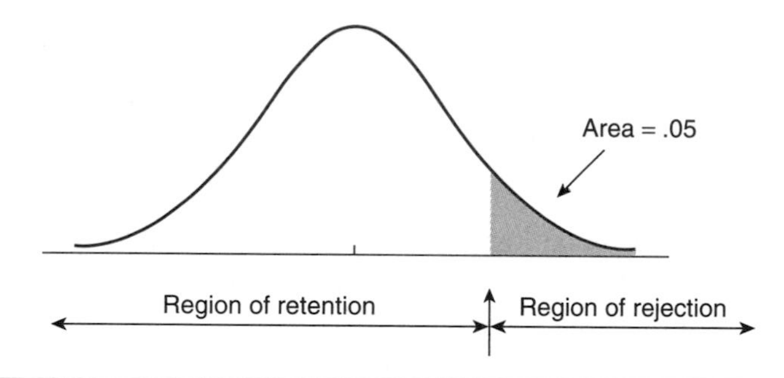

Figure 17.4 Region of Rejection for a One-Tailed Test of Ms. Teachwell's Class

The *t* Distribution Versus the Normal Distribution

Before we evaluate the arithmetic performance of Ms. Teachwell's class, let's look at an important difference between the normal deviate Z test and the one-sample t test—the shape of the distribution of the test statistic.

Recall that a normal distribution assumes an infinite sample size but that the sampling distribution of the mean will not differ much from normality as long as the sample size is at least 25 to 30. It follows that the smaller the sample size, the more the sampling distribution of the mean departs from normality. Although t tests can be conducted on small sample sizes, we cannot assume that the sampling distribution of the mean is normally distributed. Therefore, we cannot interpret the test statistic via the normal distribution. Instead, we interpret the test statistic via a t distribution. We also must use the t distribution whenever the population standard deviation is not known, regardless of the sample size. This is the case in our current example.

Like the normal distribution, the t distribution is unimodal and symmetric. However, it is somewhat leptokurtic (peaked in the center), and its tails are slightly raised compared with those of a normal curve. Furthermore, the t distribution, unlike the normal distribution, is not a single distribution but a family of distributions. That is, there is a different t distribution for each sample size, reflecting slight differences in area under the curve at different points along the curve for different sample sizes (see Figure 17.5).

Recall Exercise 2 in Module 15, in which you plotted the means of repeated samples of Size 2 drawn from a population of Size 5. Even with such a small sample size, the distribution of repeated sample means was nearly normal. Also, recall Exercise 7 in Module 15, in

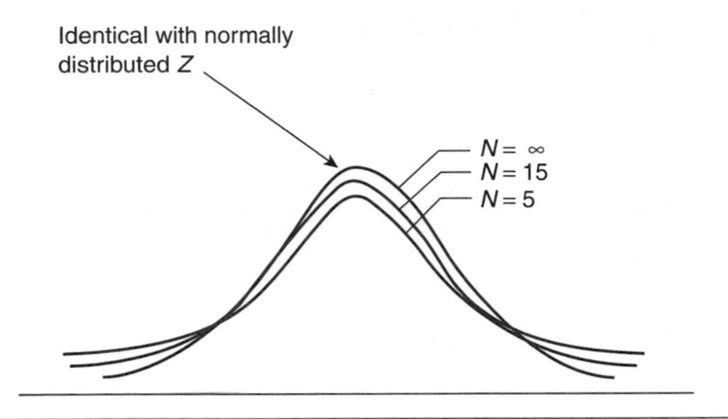

Figure 17.5 Change in *t* Distribution as a Function of Sample Size

which you plotted the size of the standard error of the mean for repeated samples of Size 3 versus Size 7 drawn from a population of Size 9. The standard error of the mean (dispersion among sample means) decreased as the sample size approached the population size. Now, in like manner for a *t* distribution, as sample size increases, the sampling distribution of the mean for a *t* statistic approximates a normal distribution.

For that reason, as the distributions in Figure 17.5 illustrate, deviation from normality is almost imperceptible when the sample size is greater than 25. This is why most statisticians agree that 25 to 30 cases are sufficient to assume a normally distributed test statistic. There is nothing magical about that number, however. Some statisticians say a sample size of 20 is sufficient; others prefer a sample size of 50 or even 100. Regardless of the sample size criterion, what the family of *t* distributions illustrates is that deviation from normality is greatest for smaller sample sizes and that the deviation is particularly pronounced in the tails—which is the critical area for deciding for or against the null hypothesis. Thus, it is important that in interpreting a *t* test, we use the more accurate *t* distributions rather than the generalized normal distribution.

The *t* Table Versus the Normal Curve Table

Just as a normal table displays area under a normal curve, a *t* table displays area found under the family of *t* distributions. Appendix C contains the *t* table. A portion of the *t* table is reproduced in Table 17.1 and a portion of the normal curve table is reproduced in Table 17.2.

Table 17.1 A Portion of the *t* Table

	Level of Significance for One-Tailed Test (%)			
	5	*2.5*	*1*	*.5*
	Level of Significance for Two-Tailed Test (%)			
df	*10*	*5*	*2*	*1*
1	6.314	12.706	31.821	63.657
2	2.920	4.303	6.965	9.925
3	2.353	3.182	4.541	5.841
4	2.132	2.776	3.747	4.604
⋮	⋮	⋮	⋮	⋮
40	1.684	2.021	2.423	2.704
60	1.671	2.000	2.390	2.660
120	1.658	1.980	2.358	2.617

Table 17.2 A Portion of the Normal Curve Table

z	*Area*
−3.00	.0013
−2.99	.0014
−2.98	.0014
⋮	⋮
−0.01	.4960
0.00	.5000
0.01	.5040
⋮	⋮
2.98	.9986
2.99	.9986
3.00	.9987

Compare the layout and information in the t table with that of the normal curve table. They do not look the same, do they? The format of the two tables is reversed. With a normal curve table, we look up the value of Z that we computed, and the table entries tell us the probability that we obtained such a deviant sample by chance alone—that is, a Type 1 error. It is the opposite with a t table. With a t table, we select and look up the probability that we committed a Type 1 error, and the table entries tell us the minimum value of t that we must obtain to reject the null hypothesis with the specified error level. Said another way, the table tells us the *critical* value that our calculated t must meet for us to reject the null hypothesis.

There is a good reason for this difference in table format. It is due to the t table's provision for various sample sizes. Because the normal curve assumes only one sample size (infinite), the table rows are dedicated to test statistic values, and the column entries give the probabilities associated with those test statistic values. But the shape of the t distribution differs slightly for every sample size (expressed in *df*). Thus, a t table can be set up like a Z table only if we have a compendium of t tables—one for each sample size. If we did have such a compendium, the rows in each table could be dedicated to t values, just as those in a normal curve table are dedicated to Z values, and the column entries would give the probabilities associated with each t value.

Although statisticians have in fact created a compendium of such tables (these are the types of tasks that give statisticians their boring reputations!), such an exercise seems like overkill. After all, when evaluating the truth of a null hypothesis, do we really care if the probability that we have made a Type 1 error is .0002 versus .0003? Isn't it enough to say it was less than 1/10 of 1% and let it go at that? And what about those huge Type 1 error probabilities? I might struggle over whether or not I should reject the null hypothesis when the probability that I am making a Type 1 error is 4% versus 2%. But what if the probability is 29%, 56%, or 89%? Would anyone reject a null hypothesis when the probability that they are doing so in error is so large? Of course not. Thus, much of the data in a compendium of t tables are simply unnecessary.

For that reason, statisticians have provided a condensed version of the t table. They give us only a few Type 1 error values—typically, .001, .01, .025, .05, and .10. These appear as the column headings. Values in between, should we be interested in them, must be interpolated. In return for this massive deletion of data, the table can then display the other two variables: rows reflect the *df* value (which is a function of sample size), and column entries display the minimum t value needed to reject the null hypothesis for the given sample size and specified Type 1 error level. Voilà—a single table is all we need.

CHECK YOURSELF!

What do *t* table entries tell us?

Calculating a One-Sample *t* Test

It is finally time to calculate a one-sample *t* test. Let's return to the arithmetic scores of Ms. Teachwell's class from Exercise 2 in Module 16:

72 91 74 83 94 75 89 82 73 67 82 83 79 82 82

80 81 92 80 70 74 78 92 81 75 76 90 72 77 68

We were given the statewide population mean: It was 77.0. The sample size is large enough to conduct a normal deviate *Z* test too. However, this time, let's pretend that the state did not release the statewide standard deviation. Therefore, we will need to estimate the statewide standard deviation from the standard deviation of Ms. Teachwell's class. Thus, we will do a one-sample *t* test instead of a normal deviate *Z* test. The formula is as follows:

$$t_{1\text{-samp}} = \frac{M - \mu}{\sigma_M}$$

First, we find the mean and standard deviation of the sample—that is, of Ms. Teachwell's class:

X	*X* – *M*	(*X* – *M*)2
72	–7.80	60.84
91	11.20	125.44
74	–5.80	33.64
83	3.20	10.24
94	14.20	201.64
75	–4.80	23.04
89	9.20	84.64
82	2.20	4.84
73	–6.80	46.24
67	–12.80	163.84
82	2.20	4.84
83	3.20	10.24
79	0.80	0.64
82	2.20	4.84
82	2.20	4.84
80	0.20	0.04
81	1.20	1.44
92	12.20	148.84
80	0.20	0.04
70	–9.80	96.04
74	–5.80	33.64
78	–1.80	3.24
92	12.20	148.84

(Continued)

(Continued)

X	$X - M$	$(X - M)^2$
81	1.20	1.44
75	−4.80	23.04
76	−3.80	14.44
90	10.20	104.04
72	−7.80	60.84
77	−2.80	7.84
68	−11.80	139.24
Σ = 2394		Σ = 1562.80

$$M = \frac{2394}{30} = 79.80$$

Note the "$n - 1$" in the denominator here:

$$\sigma^2_{\text{est}} = \frac{\sum (X - M)^2}{n - 1} = \frac{1562.80}{29} = 53.890$$

$$\sigma_{\text{est}} = \sqrt{53.890} = 7.341$$

Next, we calculate the standard error of the mean:

$$\sigma_M = \frac{\sigma_{\text{est}}}{\sqrt{n}} = \frac{7.341}{\sqrt{30}} = \frac{7.341}{5.477} = 1.34$$

Finally, we calculate the one-sample t test:

$$t_{\text{1-samp}} = \frac{M - \mu}{\sigma_M} = \frac{79.80 - 77.00}{1.34} = \frac{2.80}{1.34} = 2.09$$

We are finally done. The 2.80 points' difference in mean arithmetic score between students in Ms. Teachwell's class and students statewide is 2.09 standard errors of the mean. But what does that mean?

Interpreting a One-Sample *t* Test

> ! The only relevant test of the validity of a hypothesis is comparison of prediction with experience.
>
> —Milton Friedman, U.S. economist

Let's see where our observed sample mean falls within the sampling distribution of the mean (see Figure 17.6).

From the diagram, it certainly does look like Ms. Teachwell's class did quite well. But how well? As usual, the question is whether the class mean's distance from the population mean is *a lot*, and therefore probably due to a real difference in student achievement between Ms. Teachwell's students and statewide students, or the class mean's distance from the population mean is *only a little*, and therefore probably due to mere sampling error.

The t table answers that question. Appendix C contains the t table. A portion of the table is reproduced in Table 17.3.

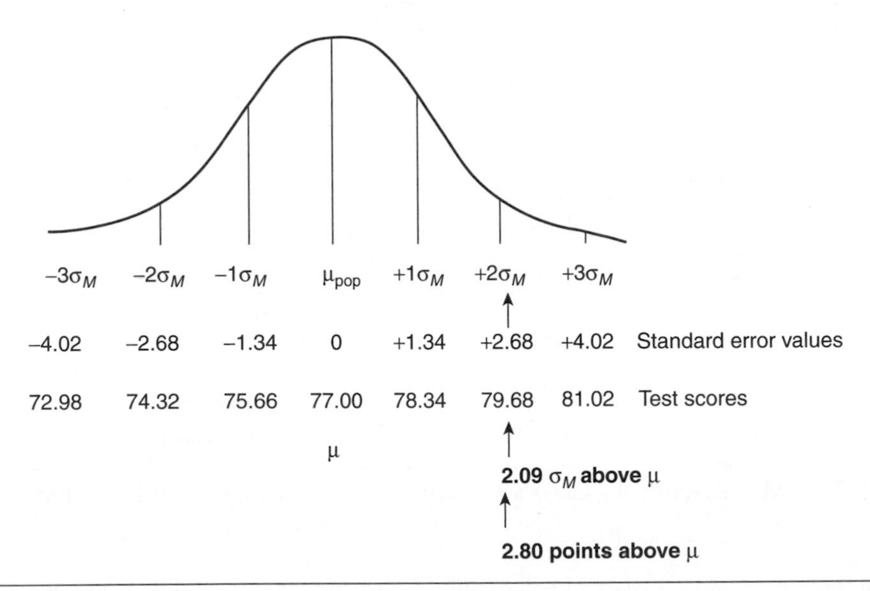

Figure 17.6 Position of Ms. Teachwell's Class Relative to the Statewide Population

Table 17.3 A Portion of the *t* Table

	Level of Significance for One-Tailed Test (%)			
	5	*2.5*	*1*	*.5*
	Level of Significance for Two-Tailed Test (%)			
df	*10*	*5*	*2*	*1*
28	1.701	2.048	2.467	2.763
29	1.699	2.045	2.462	2.756
30	1.697	2.042	2.457	2.750

We must enter the table at the correct *df*. Recall that $df = n - 1$ for each sample. We have one sample in our study, and the sample consists of 30 students. Thus, the *df* is $30 - 1 = 29$.

Find that row in the table. Now, look at the column headings. There are two sets of headings. The top set is for a one-tailed test, and the bottom set is for a two-tailed test. As discussed earlier, a directional hypothesis seems appropriate for our study. Therefore, we will limit ourselves to the one-tailed columns.

Also, as discussed earlier, 5% seems like about an appropriate level of Type 1 error (α). Looking at the table, what is the minimum *t* value that we must have to reject the null hypothesis for a one-tailed test at $\alpha = .05$? Can we reject the null hypothesis?

The minimum *t* needed to reject the null hypothesis is 1.699. (Note that all values in the chart are absolute values. That is, −1.699 and +1.699 are both represented by 1.699.) Yes, we can reject the null hypothesis because our calculated *t* of 2.090 meets or exceeds the one-tailed critical *t* of 1.699. That is, our calculated *t* falls within the region of rejection. This is shown in Figure 17.7.

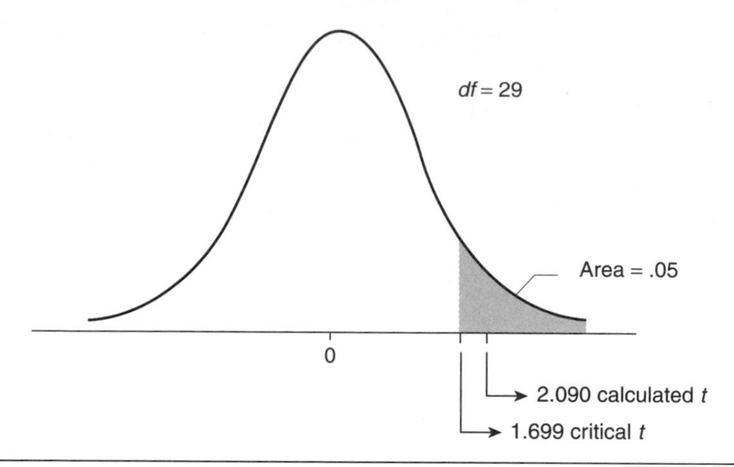

Figure 17.7 Ms. Teachwell's Class's *t* Relative to the .05 Region of Rejection

Now, let's assume that we had been willing to make a Type 1 error only 1% of the time rather than 5% of the time. Looking at the table, what is the minimum *t* value that we must have to reject the null hypothesis for a one-tailed test at $\alpha = .01$? Can we reject the null hypothesis?

The minimum *t* needed to reject the null hypothesis is 2.462. No, we cannot reject the null hypothesis because the calculated *t* of 2.090 does not meet or exceed the one-tailed critical *t* of 2.462. That is, the calculated *t* does not fall within the area of rejection. This is shown in Figure 17.8.

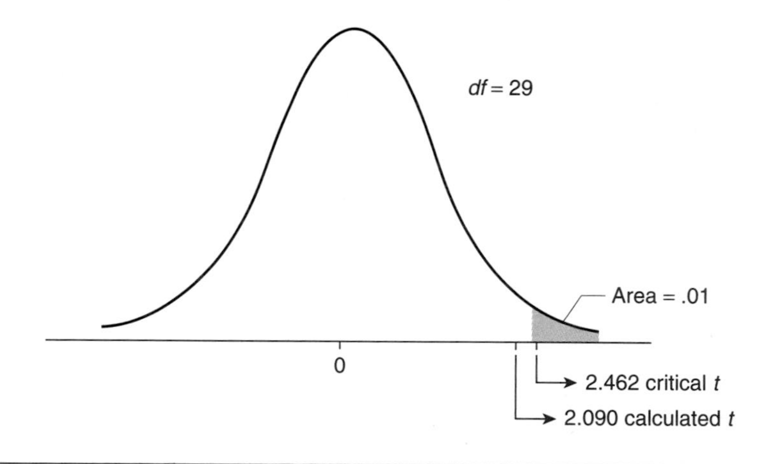

Figure 17.8 Ms. Teachwell's Class's *t* Relative to the .01 Region of Rejection

Remember, the researcher determines the maximum level of Type 1 error permissible. In our example, if you as the researcher had determined that 5% α were permissible, you could reject the null hypothesis. However, if you as the researcher had determined that α must be less than 1%, then you could not reject the null hypothesis.

CHECK YOURSELF!

How is a *t* test different from a normal deviate *Z* test? How is it similar?

PRACTICE

1. The average height of U.S. adult males is 69.5 in., and the standard deviation is 2.00 in. A nurse measures the height of the 62 male patients who visit a physician's office during one week and finds their average height to be 70.3 in. Assume that the nurse wants to know whether the height of these patients is significantly different from the nationwide height. Why should a Z test rather than a t test be used to test this hypothesis?

2. Betty, an employee of Shining Sun Daycare Center, read an article in Healthy Child Magazine saying that the average 3-year-old child is 37 in. tall. Betty works with 3-year-olds at Shining Sun, so later that week, she measured the height of each child who had just turned or was about to turn 3 years old. Here are their heights in inches: 41, 40, 36, 42, 39, 38, 33, 44, 39, 41.
   a. State the nondirectional hypothesis.
   b. Determine the critical t for $\alpha = .05$.
   c. Calculate t. Show your calculations.
   d. Is the height of 3-year-olds in Shining Sun Daycare Center significantly different from the height given in the magazine?

3. Clyde has recently opened a pizza restaurant. His main competitor, Lorraine's Pizza Palace, claims it can take a customer's order, prepare the pizza, and deliver it to the customer's table in an average of 8 min. Having eaten at Lorraine's many times before, Clyde suspects that the process takes longer than 8 min. Clyde and his friends record the time it takes to order and be served the next time each eats at Lorraine's. Here are their times, to the nearest minute, for 15 visits to Lorraine's: 6, 12, 9, 10, 8, 8, 7, 11, 13, 9, 8, 11, 7, 6, 14.
   a. State the directional hypothesis.
   b. Determine the critical t for $\alpha = .05$.
   c. Calculate t. Show your calculations.
   d. Has Lorraine's Pizza Palace significantly understated its service time?

4. A study reports that 26% of 4-year college graduates nationwide who majored in biology obtain a job related to biology within 1 year of graduation. Fifteen 4-year colleges in one state compare their placement rates for biology graduates with the national average at 1 year postgraduation. Their rates are 41%, 23%, 19%, 61%, 22%, 28%, 18%, 29%, 33%, 23%, 35%, 37%, 26%, 38%, and 21%.
   a. State the nondirectional hypothesis.
   b. Determine the critical t for $\alpha = .05$.
   c. Calculate t. Show your work.
   d. Is the placement rate for biology graduates in these 15 schools significantly different from the national placement rate?

5. Ms. DuGood is chairing a local fund-raising campaign for Helps-A-Lot charity. Literature from the charity's national headquarters indicates that the average local contributor gives \$27.32. Under Ms. DuGood's leadership, 632 donors contribute an average of \$28.19 to the local campaign. The standard deviation (using $n - 1$) for her donors is \$7.82.
   a. What additional information would you need to conduct a normal deviate Z test?
   b. State the directional hypothesis.
   c. Determine the critical t for $\alpha = .01$.

d. Calculate *t*, using the sample standard deviation as the estimated population standard deviation. Show your calculations.

e. Did the local campaign that Ms. DuGood chaired raise significantly more money than the nationwide average?

6. Assume a hypothetical survey reports that the average number of articles published over a 5-year period in refereed academic journals by faculty at 4-year liberal arts colleges nationwide is 3.2. The faculty at Almost Ivy college publish an average of 3.7 articles during that same 5-year period, with a standard deviation (using $n - 1$) of .80. There are 60 faculty at Almost Ivy.

   a. State the directional hypothesis.

   b. Determine the critical *t* for $\alpha = .01$.

   c. Calculate *t*, using the sample standard deviation as the estimated population standard deviation. Show your work.

   d. Did the faculty at Almost Ivy publish significantly more articles than their peers nationwide?

7. The average life expectancy of the migratory Canada goose is 7.2 months, but the first half year of life shows the fewest survivors. A wildlife biologist is studying the life span of a local flock. Through banding and radiomonitoring, she determines the life span of a local flock of 30 geese to be, in months: 2, 26, 3, 54, 9, 0, 2, 5, 3, 43, 4, 3, 7, 0, 3, 5, 12, 18, 3, 2, 0, 1, 3, 31, 2, 1, 7, 4, 10, 1.

   a. State the nondirectional hypothesis.

   b. Determine the critical *t* for $\alpha = .05$.

   c. Calculate *t*, using the sample standard deviation as the estimated population standard deviation. Show your work.

   d. Is the life span of this flock significantly different from that of the average Canada goose?

8. The average reading achievement score for second graders on a nationally standardized test is 76.2%. The average score for Merriweather school district's 46 second graders on that same test is 74.6%. The standard deviation (using $n - 1$) is 3.1%.

   a. State the directional hypothesis.

   b. Determine the critical *t* for $\alpha = .05$.

   c. Calculate *t*, using the sample standard deviation as the estimated population standard deviation. Show your work.

   d. Are Merriweather's second graders reading at a significantly lower level than second graders nationwide?

9. An annual race director records an average finish time of 12.3 min over many years of results. Runners are not permitted to run the same race twice, so each new year features a new crop of runners. This year's race by 51 runners completed the course in an average of 12.1 min, with a standard deviation (using $n - 1$) of 1.2 min.

   a. State the nondirectional hypothesis.

   b. Determine the critical *t* for $\alpha = .05$.

   c. Calculate *t*, using the sample standard deviation as the estimated population standard deviation. Show your work.

   d. Did this year's runners complete the race in a significantly different time from previous years' runners?

10. A manufacturing company reports average annual sales nationwide of 32.4 million per franchise store. The 10 franchise stores in one geographic area have annual sales of 33.6 million, with a standard deviation (using $n - 1$) of 1.2 million.
    a. State the nondirectional hypothesis.
    b. Determine the critical t for $\alpha = .01$.
    c. Calculate t, using the sample standard deviation as the estimated population standard deviation. Show your work.
    d. Was this franchise's sales volume significantly different from the national average?

11. You conduct a one-sample t test and find that $t = +1.69$. Your study has 41 participants. Your research hypothesis is nondirectional.
    a. Can you reject the null hypothesis at $\alpha = .01$?
    b. Can you reject the null hypothesis at $\alpha = .05$?
    c. Can you reject the null hypothesis at $\alpha = .10$?

12. You conduct a one-sample t test and find that $t = -1.83$. Your study has 81 participants. Your research hypothesis is directional. Sketch a normal curve and draw vertical lines for and label from $-3\sigma_M$ to $+3\sigma_M$, locate and label the tabled critical t, and locate and label the computed t.
    a. Can you reject the null hypothesis at $\alpha = .01$?
    b. Can you reject the null hypothesis at $\alpha = .05$?

Looking Ahead

The ability to reject the null hypothesis depends not only on how far the sample mean is from the population mean but also on what level of Type 1 error we are willing to accept. In the next module, we will look at this concept of error in more detail. As it turns out, dichotomous decisions (reject/do not reject) are less meaningful than reports of actual Type 1 error.

SPSS Connection

Download the file **data_mrs teachwell class arithmetic scores.sav** from www.sagepub.com/steinberg2e. These data are used in the textbook example.

Alternatively, manually enter the following 30 scores into the first column of the SPSS **Data View** spreadsheet: 72, 91, 74, 83, 94, 75, 89, 82, 73, 67, 82, 83, 79, 82, 82, 80, 81, 92, 80, 70, 74, 78, 92, 81, 75, 76, 90, 72, 77, 68. Click on the **Variable View** tab to define the variable. Name the variable **arithscr**, set the decimals at 0, and label the variable as **Arithmetic Score.**

If the file is not already in **Data View**, click that tab in the lower left of the screen.

In the toolbar at the top of the screen click on **Analyze**, then **Compare Means**, then **One-Sample T-Test.** Highlight the variable **Arithmetic Score** in the left window and click on the **arrow** between the two windows to send that variable into the right window. In the small box labeled Test Value enter **77**, which was Ms. Teachwell's class mean. Click **OK.** This is what you will see.

T-Test

One-Sample Statistics

	N	Mean	Std. Deviation	Std. Error Mean
Arithmetic Score	30	79.80	7.341	1.340

One-Sample Test

	Test Value = 77					
					95% Confidence Interval of the Difference	
	t	df	Sig. (2-tailed)	Mean Difference	Lower	Upper
Arithmetic Score	2.089	29	.046	2.800	.06	5.54

Visit the study site at www.sagepub.com/steinberg2e for practice quizzes and other study resources.

18

Interpreting and Reporting One-Sample *t*

Error, Confidence, and Parameter Estimates

Terms: confidence, parameter estimation, point estimate, interval estimate, confidence interval

Symbols: *p*, CI

Learning Objectives:

- Distinguish between tabled and incurred alpha
- Understand the relationship between error and confidence
- Estimate parameters—point and interval

What Is Confidence?

As we saw in Module 17, our calculated t of 2.090 met the tabled value for statistical significance at $\alpha = .05$ as well as at $\alpha = .025$ but did not meet the tabled value for statistical significance at $\alpha = .01$.

So if we reject the null hypothesis, how confident can we be that the difference in arithmetic achievement between Ms. Teachwell's class and statewide students is due to a real achievement difference between the two groups and not due to mere sampling error? Said another way, if α is the probability that we wrongly reject the null hypothesis, then what is the probability that we do not wrongly reject the null hypothesis?

Well, either we wrongly reject the null hypothesis or we don't. Thus, if α is the probability that we have made a Type 1 error, then the probability that we have *not* made a Type 1 error must be $1 - \alpha$. We call this probability that we have not made a Type 1 error our **confidence**. Said another way, confidence is the probability that the effect we found is real and not due to mere sampling error. Let's reexpress the tabled α levels as confidence levels.

> Logic is the art of going wrong with confidence.
>
> —Morris Kline

At $\alpha = .10$, confidence $= 1 - \alpha = .90$.

At $\alpha = .05$, confidence $= 1 - \alpha = .95$.

At $\alpha = .025$, confidence $= 1 - \alpha = .975$.

At $\alpha = .01$, confidence $= 1 - \alpha = .99$.

CHECK YOURSELF!

At $\alpha = .005$, what is the confidence?

In our study of Ms. Teachwell's class performance, we were able to reject the null hypothesis at $\alpha = .025$. Therefore, what is our confidence that we correctly rejected the null hypothesis and did not make a Type 1 error? Yes, it is $1 - \alpha$, or .975. That is, we can be 97.5% sure that the difference in performance between Ms. Teachwell's students and statewide students is real and not due to mere sampling error.

Refining Error and Confidence

The actual Type 1 error we incur in a study, and hence our actual confidence, is somewhat different from the values we just calculated. That's because a t table gives only a few α levels, not all the possible α levels. Our calculated t met the tabled value under the .025 column, so we know we incurred .025 or less Type 1 error. We also know that our incurred error is not as low as .01 because we did not meet the tabled t value for .01. Thus, our incurred α is somewhere between .01 and .025.

The level of α that we actually incur in a study is reported as a probability, symbolized p. In real research, we typically use a statistical software program such as SPSS to find t. This same software provides actual incurred error, or p. Without the software, the best we can do is interpolate (estimate the location of) p between the tabled α columns. Table 18.1 shows a portion of the t table.

Table 18.1 A Portion of the t Table

	Level of Significance for One-Tailed Test (%)			
	5	*2.5*	*1*	*.5*
	Level of Significance for Two-Tailed Test (%)			
df	*10*	*5*	*2*	*1*
28	1.701	2.048	2.467	2.763
29	1.699	2.045	2.462	2.756
30	1.697	2.042	2.457	2.750

Recall that Ms. Teachwell's class had 30 students. That's 29 *df*, so we look in the row labeled $df = 29$. At $df = 29$, here are the one-tailed critical t values for the two relevant α columns (.025 and .01):

At $\alpha = .025$, critical $t = 2.045$.

At $\alpha = .01$, critical $t = 2.462$.

Our calculated t of 2.090 meets the tabled critical t somewhere between α values .025 and .01. So now we interpolate the location of our calculated t between these two α columns.

- The tabled critical t for the .025 α column is 2.045. Our calculated t was 2.09. So the distance between the critical t and our calculated t is 2.045 – 2.09, or –.045.
- The tabled critical t for the .01 α column is 2.462. Our calculated t was 2.090. So the distance between the critical t and our calculated t is 2.462 – 2.09, or +.372.

Our calculated t falls much closer to the .025 α than to the .01 α. Thus, our incurred p is much closer to .025 than to .01. A p of about .023 is probably about right. That is, we are able to reject the null hypothesis with about a .023 Type 1 error level. This is the p that a software program such as SPSS will display.

CHECK YOURSELF!

What is the difference in meaning between α and p? Why would we use one rather than the other?

At the beginning of this module you learned that confidence = 1 – α. Now we have refined our preselected error, α, to actual incurred error, p. So if confidence = 1 – error and we have refined error from the preselected α to incurred p, then what is our actual confidence in rejecting the null hypothesis in our study? Yes, our actual confidence is 1 – p, which is 1 – .023, or .977. We can be 97.7% confident that in rejecting the null hypothesis, we have not made a Type 1 error. Said another way, we can be 97.7% confident that the observed arithmetic achievement difference between Ms. Teachwell's students and statewide students is real and not due to mere sampling error. Said yet another way, we can be 97.7% confident that Ms. Teachwell's class is like a wholly different type of population than the statewide population—say, like a population of mathematically gifted students.

In reporting the results of a study in an academic journal, it is acceptable to report either the precise p provided by the software program—in this case, p = .023—or the lowest tabled α column that our t met—in this case, α < .025. Here is the way the results would be reported in a journal. The results are written in APA style:

$$t(29) = 2.09,\ p = .023 \text{ or } t(29) = 2.09,\ \alpha < .025$$

This is read as follows: "t at 29 degrees of freedom is 2.09. The probability that the observed difference between our sample mean and the population mean was due to mere chance rather than to a real difference in achievement is 2.3%" (or, in the second statistical sentence, <2.5%).

PRACTICE

1. A researcher conducts a two-tailed one-sample t test on 15 subjects and finds that t = 2.38.
   a. How many degrees of freedom are in this study?
   b. About what is the actual Type 1 error for the study?
   c. How confident can the researcher be that the results are true and not due to mere sampling error?
   d. Report the results as they would appear in an academic journal.

2. A researcher conducts a one-tailed one-sample t test on 25 subjects and finds that $t = 1.89$.
   a. How many degrees of freedom are in this study?
   b. About what is the actual Type 1 error for the study?
   c. How confident can the researcher be that the results are true and not due to mere sampling error?
   d. Report the results as they would appear in an academic journal.

3. A researcher conducts a one-tailed one-sample t test on 31 subjects and finds that $t = 2.60$.
   a. How many degrees of freedom are in this study?
   b. About what is the actual Type 1 error for the study?
   c. How confident can the researcher be that the results are true and not due to mere sampling error?
   d. Report the results as they would appear in an academic journal.

4. A researcher conducts a two-tailed one-sample t test on 24 subjects and finds that $t = 1.89$.
   a. How many degrees of freedom are in this study?
   b. About what is the actual Type 1 error for the study?
   c. How confident can the researcher be that the results are true and not due to mere sampling error?
   d. Report the results as they would appear in an academic journal.

5. You read a study that reports $p = .08$. What confidence can you have that the results are due to the treatment rather than to mere sampling error?

6. You read a study that reports $p = .02$. You do not know the study's sample size. How confident can you be that the results are not due to mere sampling error?

Decision Making With a One-Sample t Test

To reject or not to reject, that is the question. Whether 'tis nobler in the eyes of man to take aim against . . . Oh, sorry, this is a statistics textbook, not a Shakespearean treatise. So let's get back to interpreting our results. We now have two choices:

- We can say that the achievement of Ms. Teachwell's students is really no different from that of all other statewide students but Ms. Teachwell's class just happened to score unusually high this one time. This scenario is diagrammed in Figure 18.1.

In this case, we will retain (fail to reject) the null hypothesis that there is no significant difference between Ms. Teachwell's class and statewide students. The probability of this scenario is, as we just calculated, about .023.

- Alternatively, we can say that the achievement of Ms. Teachwell's students is too much higher than that of statewide students for us to believe that it happened by chance alone and, thus, Ms. Teachwell's class really must be achieving at a higher level than statewide students. This scenario is diagramed in Figure 18.2.

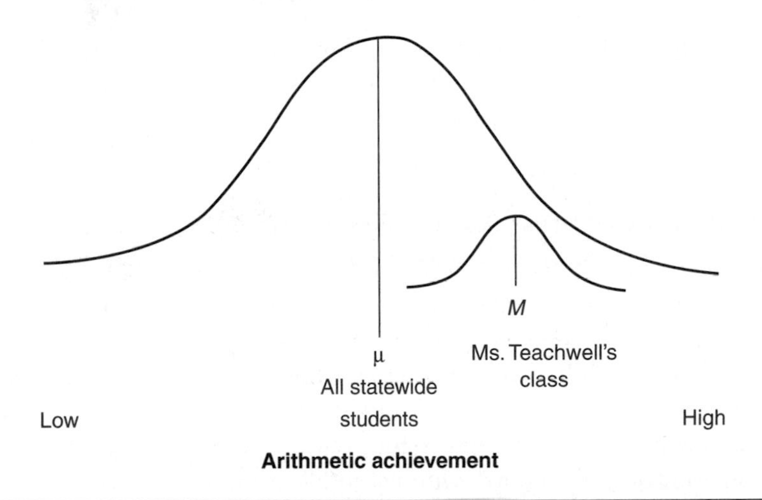

Figure 18.1 Ms. Teachwell's Class, Shown as an Unusual Sample From the Population

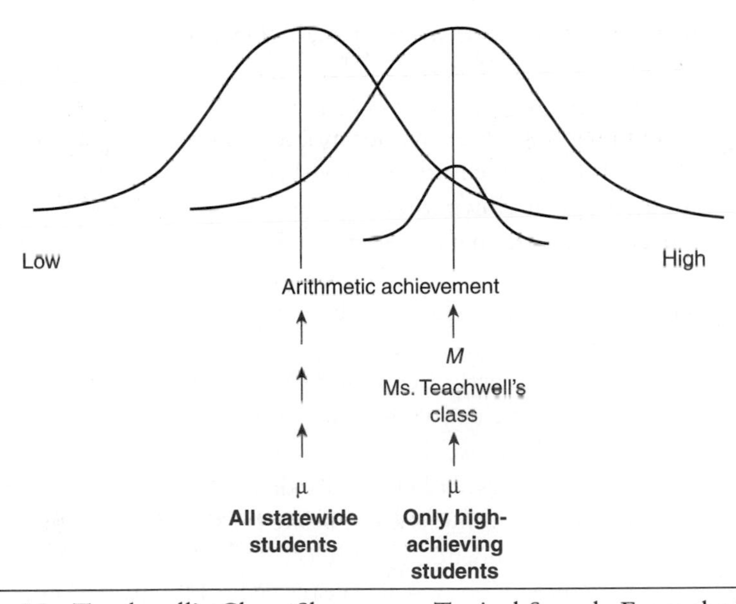

Figure 18.2 Ms. Teachwell's Class, Shown as a Typical Sample From the Population

In this case, we will reject the null hypothesis that there is no significant difference between Ms. Teachwell's class and statewide students. We conclude, instead, that the reason we found such a large difference between Ms. Teachwell's class and statewide students is because Ms. Teachwell's class really is achieving more highly than are students statewide. The probability of this scenario is, as we just calculated, about .977.

You may have noticed that the recommendation for whether or not we reject the null hypothesis, and hence whether or not the study's results are disseminated and put to use, depends wholly on the predetermined level of α. This is because, when we first learn how to calculate hypothesis tests such as t, either (a) our instructor tells us the value of α at which to conduct the test or (b) we preselect an appropriate value of α based on a subjective judgment of the risks involved, as discussed in Module 17. In the social sciences, this value is typically .05 for studies in which making a Type 1 error will not have serious consequences for people or programs. When the consequences will be more serious, .01 is adopted. After selecting an appropriate value of α, we let the tables make the rejection decision for us. In the case of our study of Ms. Teachwell's class, if we had conducted the study at $\alpha = .05$ or $\alpha = .025$, we would reject the null hypothesis. But if we had conducted the study at $\alpha = .01$, we would not reject the null hypothesis.

Dichotomous Decisions Versus Reports of Actual *p*

So if we conduct our study at the .025 error level or higher, we praise Ms. Teachwell's teaching methods because her students perform significantly higher than statewide students. But if we conduct our study at the .01 error level or lower, we do not praise Ms. Teachwell because her students do not perform significantly higher than statewide students. Something about this situation doesn't seem quite right. Why should the students' achievement be judged either superior or not superior, with no middle ground? And why should a predetermined α level determine the either/or decision? Are there not degrees of support for the null and alternative hypotheses? Yes, there are. Rosnow and Rosenthal (1989) expressed the issue well when they wryly observed,

> Can there be any doubt that God views the strength of evidence for or against the null as a fairly continuous function of α? Surely, God loves the .06 as much as the .05! (p. 1277)

The authors are commenting on the inappropriateness of making all-or-nothing, reject- or don't-reject-decisions. Is a study whose null hypothesis can be rejected with only 5% error very different from one whose null hypothesis can be rejected with 6% error? No, it is not. Yet a dichotomous decision based on a preselected .05 error level would be very different in the two studies.

As already discussed, decisions on how much error is permissible are made on the basis of subjective variables such as danger to affected people if programs are changed or not changed as a result of the study, costs of changing or not changing programs as a result of the study, and so on. Statistics give us probabilities, but they don't make decisions. People, usually program managers, make those decisions.

Furthermore, who are we as researchers to decide on behalf of all other users (of our data) what level of statistical significance is appropriate for their own programs? Let's return to the study of Ms. Teachwell's students. If we select a stringent α level for our study, say $\alpha = .01$, we would not disseminate our results because Ms. Teachwell's students would not be found to score significantly higher than statewide students. But perhaps the administrators would have been willing to recognize the students' accomplishments and implement Ms. Teachwell's successful teaching methods districtwide based on only a .05 α level.

The point is, predetermining what α level is necessary for the results to be useful is a bit presumptuous. Instead, we might better report the actual incurred error level—the *p* mentioned earlier in this module—and let the readers of the research decide for themselves whether or not the hypothesis has been supported at whatever error level they deem appropriate. This, in fact, is what most academic journals report. In the case of our study of Ms. Teachwell's class, the results would be reported as statistically significant at the .023 error level.

Of course, none of this "leaving it up to the reader" is to say that we don't have any prior beliefs about what error levels are desirable or acceptable. The traditional .05 and .01 values, while not "magical," have been around since before the time of computers and are, hence, entrenched in the tables. In point of fact, most academic journals print only studies with <.05 incurred error. Some journals print only studies with <.01 incurred error. These values are, however, only rules of thumb. Certainly, we would all agree that an incurred error of .24 or .37 is too high. Certainly, we would all agree that an incurred error of .0008 or .0026 is very low and, hence, desirable. In general, less incurred error is better.

CHECK YOURSELF!

What are some of the considerations in deciding on the maximum acceptable α level for a study?

CHECK YOURSELF!

Why is it preferable to report actual incurred *p* rather than only the α level at which a study was conducted?

Parameter Estimation: Point and Interval

In the case of Ms. Teachwell's class scores, we knew the population mean, and so we were able to estimate the probability that our sample mean came from such a population. That probability was only 2.3%. Thus, we concluded with 97.7% confidence that the arithmetic achievement of Ms. Teachwell's class was significantly higher than that of statewide students. In other words, we concluded that Ms. Teachwell's students were probably not a typical sample from the statewide population.

Well, if Ms. Teachwell's students are not typical of the statewide population, then what population *are* they typical of? We can't know for sure. What we can say, however, is that the population Ms. Teachwell's class is typical of is higher scoring than the statewide population. With this reasoning, we can determine the most likely mean of the population from which Ms. Teachwell's class probably came.

Note how the problem has changed. We are no longer asking about the probability that a particular sample came from a hypothesized population. Rather, we are asking what type of population a particular sample probably came from. This approach is called **parameter estimation**, because we are estimating a population mean—that is, a parameter.

Parameter estimation is fairly common in social science statistics. For example, suppose your college admissions officer wants to know the average IQ of incoming freshmen at your college. And suppose your college holds multiple freshman orientation sessions throughout the year and that 100 incoming freshmen are scheduled to attend one of these sessions the coming weekend. Now, suppose the admissions officer administers an IQ test to the 100 freshmen attending the weekend orientation session. And suppose their average IQ score is 112. Can the admissions officer then say that the average IQ of all incoming freshmen is 112? No, because he or she did not test all incoming freshmen. Only the 100 incoming freshmen who attended one orientation session were tested. However, if we assume that these 100 freshmen are a typical sample from the population of all incoming freshmen, then it makes sense to assume that the IQ of all incoming freshmen will be about 112. Thus, the admissions officer could use the sample mean as an estimate of the population mean. The estimated population mean is called a **point estimate.**

Alternatively, the admissions officer could use the sample mean of the 100 incoming freshmen to estimate a range of possible population means within which the mean of all incoming freshman probably falls. That is, rather than predicting an exact value for the population mean, it predicts a range of population means in which the true population mean probably falls. This range of population means is called an **interval estimate.**

> Art, like morality, consists of drawing the line somewhere.
>
> —Chesterton

How wide should that interval be? Well, the admissions officer can be 100% certain that the true mean population IQ of all incoming freshmen is between, say, 12 and 212, which is somewhere between profound retardation and extreme genius. However, such a range is not very informative, is it? It is so wide that it tells the admissions officer virtually nothing useful about the incoming freshman class. Well then, how about a range of 111.5 to 112.5? That's a very slim range—only 1 point. But now the range is so small that the admissions officer cannot be very certain that it will, in fact, include the true population mean of the incoming freshman class.

Fortunately, statisticians have provided guidelines regarding how wide the range should be. Typically, the limits of the range are set so that we have a specified level of confidence for the range to include the population parameter. Such a range is called a confidence interval. Do not confuse confidence in rejecting the null hypothesis, which I talked about earlier, with a confidence interval. To repeat, a **confidence interval**, symbolized **CI**, gives the range of scores within which a parameter probably falls, with a given degree of probability. The interval is said to be a 95% confidence interval if we can be 95% certain that the population parameter falls within the range. The interval is said to be a 99% confidence interval if we can be 99% certain that the population parameter falls within the range.

Did you hear about the man who took the Dale Carnegie course? He improved his confidence from 95% to 99%.

To review what I have said about parameter estimation,

- a point estimate predicts a specific value for the population parameter,
- an interval estimate predicts a range of values that probably includes the population parameter, and
- a confidence interval is an interval estimate having a specified probability of including the population parameter—usually 95% or 99%.

CHECK YOURSELF!

Contrast hypothesis testing and parameter estimation. How do the questions they answer differ?

Let's return to our study of Ms. Teachwell's class. Assume that for our current purposes, we don't know the statewide mean. All we know is the mean of Ms. Teachwell's class of 30 students. From this, we want to estimate the interval that probably includes the statewide mean, the parameter.

The formula for a confidence interval, symbolized CI, is

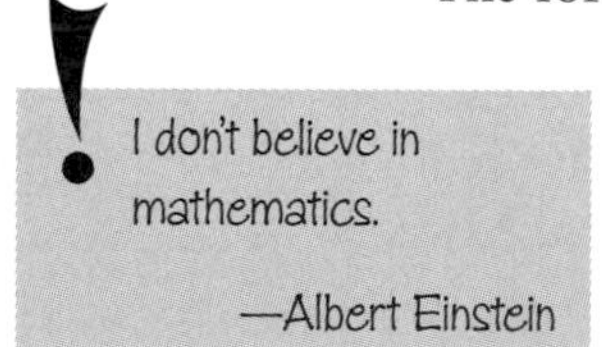

$$\text{CI} = M \pm (t_{\text{crit at } .5\alpha})(\sigma_M)$$

where

M = sample mean,

$t_{\text{crit at } .5\alpha}$ = tabled critical t at one-half the tabled α, and

σ_M = standard error of the mean.

Let's calculate the 95% confidence interval.

$$M \text{ for Ms. Teachwell's class} = 79.80$$

$$\text{Tabled } t_{\text{crit at directional } \alpha} = t_{\text{crit at } .025\ \alpha} = 2.045$$

$$\sigma_M = 1.34$$

$$\begin{aligned} \text{CI} &= 79.80 \pm (2.045)(1.34) \\ &= 79.80 \pm 2.74 \\ &= 79.80 + 2.74 \text{ and } 79.80 - 2.74 \\ &= 82.54 \text{ and } 77.06 \end{aligned}$$

Thus, we can be 95% confident that the statewide arithmetic mean falls between 77.06 and 82.54. Note that Ms. Teachwell's class mean of 79.80 falls in the middle of this range.

Now, let's calculate the 99% confidence interval.

$$M \text{ for Ms. Teachwell's class} = 79.80$$

$$\text{Tabled } t_{\text{crit at directional } \alpha} = t_{\text{crit at .005 } \alpha} = 2.756$$

$$\sigma_M = 1.34$$

$$\begin{aligned} CI &= 79.80 \pm (2.756)(1.34) \\ &= 79.80 \pm 3.693 \\ &= 79.80 + 3.693 \text{ and } 79.80 - 3.693 \\ &= 83.493 \text{ and } 76.107 \end{aligned}$$

Thus, we can be 99% confident that the statewide arithmetic mean falls between 76.107 and 83.493. Note that Ms. Teachwell's class mean of 79.80 falls in the middle of this range.

CHECK YOURSELF!

What is the difference between confidence and a confidence interval? What is the difference between a point estimate and an interval estimate? Why is there no formula for a point estimate?

PRACTICE

7. A college is expecting an entering freshman class of 700 students. The mean score on a test of writing skills for 100 freshmen attending one of several college orientation sessions is 86. The σ_M for the writing test is 1.5 points.
   a. What point estimate should the admissions officer make for the writing skills of the entire incoming freshman class?
   b. In constructing an interval estimate for the writing skills of the entire incoming freshman class, what score will fall in the center of the interval?
   c. The admissions officer can be 95% confident that the mean writing skills score for the entire incoming class falls between which two values?
   d. If the admissions officer were willing to be only 90% confident, would the confidence interval be larger or smaller? Why?
   e. If only 60 of the 100 students who attended the orientation session took the writing skills test, would the confidence interval be larger or smaller? Why?

8. A social psychologist measures the average time that 20 new transfer students at an orientation party mingle with other new transfer students before selecting a chair or corner in which to settle. The average time is 6.2 min. The σ_M for mingling is 0.33 min. Assume that these students are a representative sample of all new transfer students.
   a. What point estimate should the psychologist predict for mingle time at any other transfer orientation party?
   b. In constructing an interval estimate for mingle time, what time will fall in the center of the interval?
   c. The psychologist can be 90% confident that the mean mingle time for additional new transfer students will fall between which two values?

   d. If the psychologist wants to be 99% confident, would the confidence interval be larger or smaller? Why?
   e. If the psychologist had measured the mingle time of 28 new transfer students rather than 20, would the confidence interval be larger or smaller? Why?

9. A parole officer measures the average age of new felons. The average age of the 20 new felons in his district is 27.8 years, with a σ_M of 0.492 years.
   a. What point estimate should the parole officer predict for the average age of any future group of felons in his district?
   b. In constructing an interval estimate for new felon age, what age will fall in the center of the interval?
   c. The parole officer can be 95% confident that the mean felon age for a new felon will fall between which two values?
   d. If the parole officer had measured the age of 30 new felons rather than 20, would the confidence interval be wider or narrower? Why?

10. A personnel psychologist measures the average tenure of employees. The average number of years 41 employees work before resigning or retiring is 12.2 years, with a σ_M of 0.365 years.
    a. What point estimate should the personnel psychologist predict for a new employee's tenure?
    b. In constructing an interval estimate for new employee, what number of years will fall in the center of the interval?
    c. The personnel psychologist can be 99% confident that the mean number of years for new employees will fall between which two values?
    d. If the personnel psychologist had measured the age of 25 employees rather than 41, would the confidence interval be wider or narrower? Why?

11. On a reaction time task, 22 randomly selected participants take an average of 0.65 s to respond to a prompt. The σ_M is 0.03 s. What is the 95% confidence interval for reaction time for similar participants?

12. On an ability test, 15 randomly selected participants score an average of 74.8 points, with a σ_M of 1.263 points. What is the 95% confidence interval for future test takers' scores?

13. Exercise 5 in Module 17 addressed funds raised by a local chapter of Helps-A-Lot charity. Ms. DuGood's 632 contributors averaged $28.19. The σ_M was $0.31. Without adjusting for inflation, answer the following:
    a. What amount (point estimate) should Ms. DuGood expect the average contributor to donate to next year's local campaign?
    b. Ms. DuGood can be 99% confident that the average donation of next year's contributors will be between which two amounts?

14. Exercise 2 in Module 17 addressed the height of 10 children in Shining Sun Daycare Center. The children averaged 39.3 in. in height, and σ_M was .99.
    a. What is the best point estimate for the average height of the next class of 3-year-olds at Shining Sun Daycare Center?
    b. Betty can be 95% confident that children in the next class of 3-year-olds at Shining Sun Daycare Center will be between which two heights?

15. Exercise 7 in Module 17 addressed the life span of 30 Canadian geese. Using the mean and σ_M calculated in that exercise,
    a. What is the best point estimate of the next goose to be banded and radiomonitored?
    b. The wildlife biologist can be 95% certain that the life span of the next goose to be banded and radiomonitored will be between which two values?

16. Exercise 8 in Module 17 addressed the reading level of Merriweather's 46 second graders. Their mean reading level was 74.6%. Using the σ_M calculated in that exercise, Merriweather school administrators can be 99% certain that the reading level of the next second grader in the Merriweather schools will be between which two values?

17. Exercise 9 in Module 17 addressed race running times for 51 runners. Their mean running time was 12.1 min. Using the σ_M calculated in that exercise, the race director can be 99% certain that the running time of the next runner will be between which two values?

18. Exercise 10 in Module 17 addressed a manufacturing company's sales volume of the 10 franchise stores in one geographic area. Their mean sales volume was 33.6 million. Using the σ_M calculated in that exercise, the company can be 95% certain that the sales volume of the next franchise store in that geographic area will be between which two values?

SPSS Connection

Look again at the SPSS output that you produced in Module 17. Here it is again:

T-Test

One-Sample Statistics

	N	Mean	Std. Deviation	Std. Error Mean
Arithmetic Score	30	79.80	7.341	1.340

One-Sample Test

	Test Value = 77					
					95% Confidence Interval of the Difference	
	t	df	Sig. (2-tailed)	Mean Difference	Lower	Upper
Arithmetic Score	2.089	29	.046	2.800	.06	5.54

The two right-most cells in the bottom chart show values for a 95% confidence interval: 0.06 and 5.54. But those values do not look anything like what we calculated, do they? Our values for a 95% confidence interval were 77.06 and 82.54. Never fear. While a confidence interval is calculated as I presented it in this module, and its meaning also is as I presented it, SPSS instead gives values that one must use to obtain the values we calculated. In other words, the SPSS output stops short of the final calculations.

Also, while we calculated a single value that was added to and subtracted from our obtained sample mean of 79.80, the values SPSS prints out must be added to (not subtracted from) the test value—in other words, added to Ms. Teachwell's class mean of 77.0. Let's do that.

$$77.0 + .06 = 77.06$$

$$77.0 + 5.54 = 82.54$$

Thankfully, these were the two values we previously had calculated.

Visit the study site at www.sagepub.com/steinberg2e for practice quizzes and other study resources.

PART IX

The Two-Sample Test

"Ours Is Better Than Yours"

19

Standard Error of the Difference Between the Means

Terms: sampling distribution of the difference between the means, standard error of the difference between the means

Symbols: $\sigma_{M_1 - M_2}$, $t_{2\text{-samp}}$

Learning Objectives:

- Understand the difference between a sampling distribution of the mean and sampling distribution of the difference between the means
- Understand why the mean of the sampling distribution of the difference between the means is 0
- Understand the impact of sample size on the size of the standard error of the difference between the means
- Understand the impact of the size of the standard error of the difference between the means on the test statistic, t

One-Sample Versus Two-Sample Studies

So far, we have looked only at a single sample statistic being compared with one population parameter. However, the more common situation in research is that we have two samples—one that receives a treatment of interest and one that does not receive the treatment of interest. Consider the following examples:

Rat: a destructive rodent larger than a mouse

Ratiocinate: to reason

Ratiocinating about rats: what we are currently doing

Ratting on a ratiocinator: turning me in to the authorities

- We give one group of rats premium food and give another group of rats regular food. We want to determine whether the new rat food leads to faster maze running.
- We teach one group of children with a phonics-based reading curriculum, and we teach another group of similar children with a whole language-based reading curriculum. We want to determine which type of curriculum leads to higher reading achievement.
- We give one group of depressed clients counseling sessions and give another group of equally depressed clients an antidepressant medication. We want to determine whether the antidepressant medication alleviates depression better.

With two-sample research, we compare sample statistics of two samples with population parameters of two populations. Let's consider the depression example. We can think of the two samples as being drawn from two separate populations: the population of all depressed clients who receive the antidepressant medication and the population of all depressed clients who receive counseling. The question about the effectiveness of the treatments then boils down to the difference between two *populations* rather than the differences between the particular clients who happen to be in the two samples. In other words, is antidepressant medication more effective than counseling in alleviating depression (regardless of to whom it is given)?

It's a recession when your neighbor loses his job; it's a depression when you lose yours.

—former U.S. president Harry S. Truman

Our research hypothesis is that the antidepressant medication will lead to more relief in depression than will counseling. That is, clients who take the antidepressant medication will be less depressed. This is a directional hypothesis. Figure 19.1 is a diagram of the research hypothesis.

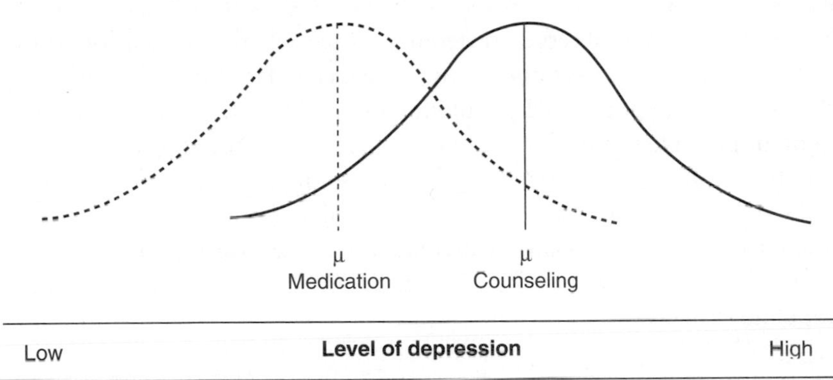

Figure 19.1 Diagram of the Research Hypothesis

NOTE: The solid line curve represents the depression level of clients who receive counseling, and the dashed line curve represents the depression level of clients who receive medication; "μ" is the population mean in each population. Notice the apparent difference in depression relief between the two treatments.

On the other hand, if the antidepressant medication offers no advantage in relieving depression, the means of the two populations would be the same. This is the null hypothesis. The null hypothesis states that there is no significant difference in depression relief between depressed clients given medication and those given counseling. Figure 19.2 is a diagram of the null hypothesis.

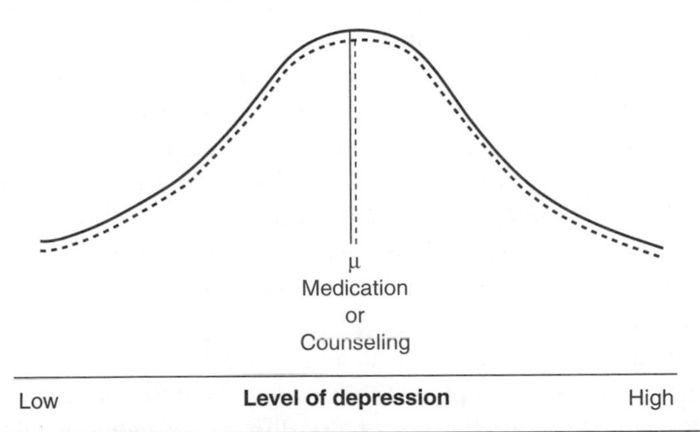

Figure 19.2 Diagram of the Null Hypothesis

NOTE: The solid line curve represents the depression level of clients who receive counseling, and the dashed line curve represents the depression level of clients who receive medication; "μ" is the population mean in each population. Notice that there is no apparent difference in depression relief between the two treatments.

Now, recall that it is always the null hypothesis that we test—with the goal being to reject it and thereby gain support for the research hypothesis. So let's assume for the moment that the null hypothesis is true. Even if the null hypothesis is true, we still do not expect to find exactly the same level of depression between the two samples of differently treated clients. Thus, the question is not whether the level of depression in our two samples of depressed clients is the same or different after treatment but, rather, *how much* different. Is the observed difference *a lot*, so that it is almost certainly due to the difference in the treatment? Or is the observed difference *only a little*, so that it is almost certainly due to sampling error and not due to the difference in treatment?

Sampling Distribution of the Difference Between the Means

How might we evaluate the observed difference to see whether it is a lot or only a little? Recall that in our previous one-sample example, we calculated how far our sample's mean was from the known or hypothesized population mean ($M - \mu$). Then, we created a sampling distribution of all possible sample means drawn from that population when the null hypothesis was true, found the average error unit for that sampling distribution (σ_M), and divided our observed ($M - \mu$) difference by those σ_M error units. That unit in the denominator was called the standard error of the mean. By dividing the numerator into denominator units, the resulting t statistic told us *how many* standardized error units our sample mean was away from the population mean:

$$t_{1\text{-samp}} = \frac{M - \mu}{\sigma_M}$$

Then, we used a table to determine whether the observed number of standardized error units (our t) was *a lot* (more than the tabled value and probably due to the treatment) or *only a little* (not more than the tabled value and probably due to mere sampling error).

For our current two-sample example, we need a similar method to evaluate the observed difference between our two sample means. However, because our two-sample example is based not on just one population but two populations (depressed clients who are given medication and depressed clients who are given counseling), we need a denominator unit that reflects the average error unit when the sampling distribution is between two different populations, not just within one population.

For the one-sample case, we created a sampling distribution of the mean for the t test denominator by hypothetically sampling and plotting all possible sample means from the population. For our two-sample case, we need to create a sampling distribution by hypothetically sampling and plotting all possible *pairs* of samples drawn from the two populations—one from the population of all depressed clients given medication and one from the population of all depressed clients given counseling. For each pair of samples, we will compute the difference between their mean depression levels. Then, we will plot all those differences between the pairs of means.

This distribution of differences between an infinite number of pairs of sample means drawn from two populations is called a **sampling distribution of the difference between the means.**

What will this sampling distribution of the differences between all possible pairs of means look like? Well, let's think it through:

- If, under the null hypothesis, there really is no difference in depression level between a client population treated with medication and a client population treated with counseling, then most samples drawn from those populations should show very little difference in mean depression levels.
- But we know that although most pairs of samples will show very little difference in their depression levels, some pairs will show a larger difference by mere chance (sampling error).
- We also know that if the null hypothesis is true, then the greater and greater the difference in depression level between a pair of samples, the less and less likely (less and less frequent) that difference will be.
- Finally, we know that if the null hypothesis is true, it is no more likely that clients treated with medication will, after treatment, be more depressed or less depressed than clients treated with counseling. There will be symmetry both above and below the average difference in means.

Given these facts, what will be the shape of the distribution of all possible pairs of samples? Yes, it will be the familiar normal curve.

Of course, the sampling distribution of the difference between the means is only theoretical because we cannot actually draw an infinite number of pairs of samples. Nevertheless, if we *could* draw an infinite number of pairs of samples of a given size from two populations and plot the difference between the samples' means, the sampling distribution of the difference between the means would look like Figure 19.3.

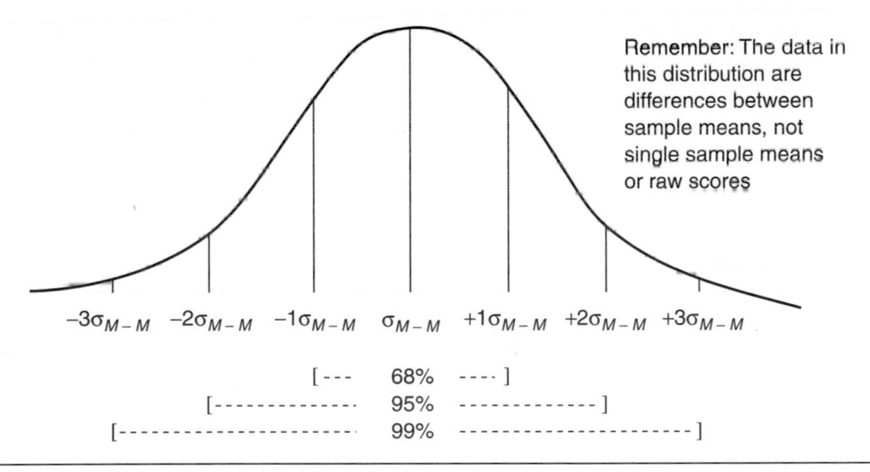

Figure 19.3 Sampling Distribution of the Difference Between the Means

What makes the sampling distribution of the difference between the means so useful is that we can create it even when we don't know what the population means really are! Let's see why.

- Suppose that the average level of depression for clients receiving the antidepressant medication is 44. Under the null hypothesis of no difference between the means, what would be the average level of depression for clients receiving counseling? Yes, of course, it would be 44.
- Suppose that the average level of depression for clients receiving the antidepressant medication is 16. Under the null hypothesis of no difference between the means, what

would be the average level of depression for clients receiving counseling? Yes, of course, it would be 16.

- Suppose that the average level of depression for clients receiving the antidepressant medication is 72. Under the null hypothesis of no difference between the means, what would be the average level of depression for clients receiving counseling? Yes, of course, it would be 72.

So regardless of the actual level of depression of the clients, whenever the null hypothesis is true, the average level of depression for clients given medication and clients given counseling will be the same.

Now, look again at Figure 19.3. What is the mean of this sampling distribution of the difference between the means?

The mean of this sampling distribution of differences between the means is 0 because, as we just saw, we expect the average level of depression for clients given medication and clients given counseling to be the same whenever the null hypothesis is true.

Now, look once more at Figure 19.3. This distribution, like any normal distribution, also has a standard deviation. But recall that the term *standard deviation* is reserved for distributions of raw scores. This distribution does not consist of raw scores. It consists of differences between all possible pairs of means—which are themselves summary statistics of samples. Recall that the standard deviation of a distribution of any summary statistic is called a *standard error*. Because the summary statistics in this distribution are differences between means, the standard deviation of this distribution is called the **standard error of the difference between the means**, symbolized $\sigma_{M_1 - M_2}$.

The standard error of the difference between the means is a standard deviation. Thus, it is the average amount of dispersion we can expect around the actual difference in sample means of clients in the two treatment conditions. In the example we are working with, it is the average dispersion around the difference in means for clients given medication and clients given counseling.

CHECK YOURSELF!

Why, under the null hypothesis, are most differences between sample means close to 0? Why does that cause the differences between sample means to be normally distributed?

CHECK YOURSELF!

Contrast a sampling distribution of the difference between the means with a sampling distribution of the mean: What are the data points in each distribution? Describe the theoretical process by which each distribution is created.

Calculating the Standard Error of the Difference Between the Means

Look once more at Figure 19.3. The standard deviation (now referred to as the standard error of the difference between the means) of this distribution is designated as $\sigma_{M_1 - M_2}$. The full formula for $\sigma_{M_1 - M_2}$ is somewhat intimidating. However, when both samples are the *same size*, and when the subjects in one sample are *independent* of (not the same subjects as and not linked in any way to) the subjects in the other sample, the formula reduces to a much more manageable one. In this module, we will stick to samples that are the same size and independent. For such samples, the formula for $\sigma_{M_1 - M}$ reduces to

$$\sigma_{M_1 - M_2} = \sqrt{\sigma^2_{M_1} + \sigma^2_{M_2}}$$

where

σ_{M_1} = the standard error of the mean of the first population and

σ_{M_2} = the standard error of the mean of the second population.

As you can see, we *pool* (add together) the two variances to get one overall variance. Then, we take the square root of that pooled variance to get a standard deviation. That is our standard error of the difference between the means.

When we calculate a two-sample *t* test in the next module, we will calculate $\sigma_{M_1 - M_2}$ from raw data. But calculating a $\sigma_{M_1 - M_2}$ first involves calculating the σ_M for each population. And as you recall from Module 17, calculating a σ_M for a population first involves calculating σ_{est} from the sample data. What a bother! In the next module we will, indeed, complete all those calculations. Rather than put you through that here as well, I will provide the two hypothetical values of σ_M for the next-to-last step of the process. Then, we need only plug the two values of σ_M into the $\sigma_{M_1 - M_2}$ formula.

Example 1: Assume that the value of σ_M for clients receiving medication is 1.92 and that for clients receiving counseling it is 2.76. What is the value of $\sigma_{M_1 - M_2}$?

$$\begin{aligned}
\sigma_{M_1 - M_2} &= \sqrt{\sigma^2_{M_1} + \sigma^2_{M_2}} \\
&= \sqrt{(1.92)^2 + (2.76)^2} \\
&= \sqrt{(3.686) + (7.618)^2} \\
&= \sqrt{11.304} \\
&= 3.36
\end{aligned}$$

Example 2: Assume that the value of σ_M for clients receiving medication is 2.53 and that for clients receiving counseling it is 1.21. What is the value of $\sigma_{M_1 - M_2}$?

$$\begin{aligned}
\sigma_{M_1 - M_2} &= \sqrt{\sigma^2_{M_1} + \sigma^2_{M_2}} \\
&= \sqrt{(2.53)^2 + (1.21)^2} \\
&= \sqrt{(6.401) + (1.464)^2} \\
&= \sqrt{7.865} \\
&= 2.80
\end{aligned}$$

Importance of the Size of the Standard Error of the Difference Between the Means

Look again at the formula for the standard error of the difference between the means:

$$\sigma_{M_1 - M_2} = \sqrt{\sigma^2_{M_1} + \sigma^2_{M_2}}$$

According to the formula, the size of the standard error of the difference between the means is influenced by only two factors. What are those factors?

Yes, it is influenced only by the size of the standard error of the mean (σ_M) of each population. Now, recall that each σ_M is a measure of its population's average dispersion. It then follows that

- the greater the dispersion in the sampling distributions of the two underlying populations, the greater the dispersion in the distribution of *pairs* of samples drawn from those two underlying populations and
- the less the dispersion in the sampling distributions of the two underlying populations, the less the dispersion in the distribution of *pairs* of samples drawn from those two underlying populations.

Because the size of $\sigma_{M_1 - M_2}$ is a function of the size of the two σ_M values, now is a good time for you to return to Module 15 to review the factors that influence the size of σ_M.

Is the size of $\sigma_{M_1 - M_2}$ really that important? Yes, it is. The $\sigma_{M_1 - M_2}$ will be the denominator in our two-sample t test. Its size will be critical in determining whether the treatment effect in the numerator (the level of depression with medication vs. counseling) is a lot or only a little. If we divide the treatment effect in the numerator by a big $\sigma_{M_1 - M_2}$ in the denominator, then the treatment effect won't seem very big. But if we divide the treatment effect in the numerator by a small $\sigma_{M_1 - M_2}$ in the denominator, then the treatment effect will seem very big.

Obviously, a small $\sigma_{M_1 - M_2}$ is more desirable for hypothesis testing. Note, however, that it is possible for the sample size to be *too* large, making even a tiny treatment effect in the numerator appear significant. A discussion of that important point will be left to a later module.

CHECK YOURSELF!

Explain why the size of $\sigma_{M_1 - M_2}$ is important in hypothesis testing.

PRACTICE

1. A study uses very small sample sizes. Considering the effect of sample size on σ_M (see Module 15), what effect will this have on the size of $\sigma_{M_1 - M_2}$?
2. As sample size increases in the treatment and control groups of a study, what effect will that have on the size of the standard error of the difference between the means?
3. The actual difference between the means of a treated group and an untreated group is 3 points. That amount might or might not be found to be significant because of sample size. Explain this in terms of the formula for the two-sample t-test statistic.
4. In two separate studies, the actual difference between the means of a treated group and an untreated group is 3 points. However, in one study, the $\sigma_{M_1 - M_2}$ is very large and so the 3 points is not found to be significant. In the other study, the $\sigma_{M_1 - M_2}$ is very small and so the 3 points is found to be significant. What might have caused this big difference in the $\sigma_{M_1 - M_2}$ for the two studies?

5. Calculate $\sigma_{M_1 - M_2}$ for a study of the effect of type of food on rats' running speed, in which the value of σ_M for rats getting premium food is 3.27, and for rats getting regular food, it is 2.54.
6. Calculate $\sigma_{M_1 - M_2}$ for a study of self-esteem between the sexes, in which the value of σ_M for women is 1.32 and that for men is 1.94. (*Review*: Is sex an *assigned* variable or a *found* variable? How might this complicate interpretation of the study's results?)
7. In a study of the effect of a training program on employee productivity, the σ_M for productivity in the group that received the training program was 2.16, and the σ_M for productivity in the group that did not receive the training program was 2.56. Calculate the $\sigma_{M_1 - M_2}$.
8. In a study of the effect of a new drug on the alleviation of asthma symptoms, the σ_M for symptom relief in the patient group that received the new drug is 1.45, and the σ_M for symptom relief in the group that did not receive the new drug is 1.22. Calculate the $\sigma_{M_1 - M_2}$.
9. In a study of the effect of a behavior modification program on a client's behavior, the σ_M for behavior incidents in the group that received the behavior modification program is 1.14, and the σ_M for behavior incidents in the group that did not receive the behavior modification program is 1.56. Calculate the $\sigma_{M_1 - M_2}$.
10. In a study of the effect of test humor in relieving students' test anxiety, the σ_M for anxiety level in the group that experienced test humor was 0.85, and the σ_M for anxiety level in the group that did not experience test humor was 0.96. Calculate the $\sigma_{M_1 - M_2}$.

Looking Ahead

Now, we have everything needed to compute a two-sample t test. We will give one sample of depressed clients antidepressant medication and another sample of clients counseling. Then, we will calculate the difference in their mean depression levels. Chances are that the mean depression level of the sample of clients given medication will not be exactly the same as the mean depression level of the sample of clients given counseling, even when the null hypothesis of no treatment effect is true. But how different are our two observed sample means? Are they only a little different? Or are they very different? We need a statistic to compute the probability of getting two sample means falling this far away from the expected zero difference. That statistic will scale our observed difference between our two sample means (numerator) against the average difference expected under the null hypothesis (denominator, which is the standard error of the difference between the means).

Visit the study site at www.sagepub.com/steinberg2e for practice quizzes and other study resources.

20

t Test With independent Samples and Equal Sample Sizes

Learning Objectives:

- Understand the similar logic underlying various test statistics
- Determine the degrees of freedom
- Calculate a two-sample *t* test for independent samples and equal sample sizes
- Use a table to interpret the calculated *t*
- Report results in APA format

A Two-Sample Study

Now that you know how to calculate the standard error of the difference between the means, you are ready to calculate a two-sample *t* test. This time, however, we will calculate $\sigma_{M_1 - M_2}$ from raw data. Recall that calculating a $\sigma_{M_1 - M_2}$ first involves calculating the σ_M for each population. And calculating a σ_M for a population first involves calculating the σ_{est} for each population based on sample data. Therefore, this module will include a lot of calculations.

For our example, assume that we randomly select 18 depressed clients. We then randomly assign the clients to one of two treatment groups (two samples), giving us 9 clients in each group. We treat one group with antidepressant medication and the other group with counseling. After a predetermined period of treatment, we measure the clients' depression levels, using a test where lower scores indicate lower depression and higher scores indicate higher depression. Here are the individual and the mean depression scores for each group:

Medication	*Counseling*
32	43
40	31
21	39
17	36
40	46
26	35
19	32

Medication	*Counseling*
44	44
20	37
$\Sigma = 259$	$\Sigma = 343$
$M_{\text{med}} = \frac{259}{9} = 28.778$	$M_{\text{med}} = \frac{343}{9} = 38.111$

Clearly, clients treated with medication were less depressed after treatment than clients treated with counseling: The mean depression levels of the two samples were 28.778 and 38.111 points, respectively. That's a difference of 9.333 points. But that difference does not necessarily indicate that medication is more effective than counseling. Recall that even if the null hypothesis is true and medication really is no more effective than counseling, we still do not expect any given sample of clients treated with medication to have exactly the same depression level as any given sample of clients treated with counseling. Although there will be no difference over an infinite number of pairs of samples, for any given pair of samples, the participants who were treated with medication could experience either more or less depression relief than those who were treated with counseling.

Thus, the question in rejecting the null hypothesis is not whether or not the mean depression levels for the two samples in our study are different but how different they are. Is the difference in depression level between the samples after treatment *only a little* and, therefore, probably due to mere sampling error? Or is the difference in depression level between the samples after treatment *a lot* and, therefore, probably due to something other than mere sampling error—in this case, due to the difference in treatment?

Inferential Logic and the Two-Sample *t* Test

We need a statistic to compute the probability of our observing two sample means as different as ours. As discussed in Module 19, the statistic we use will scale the observed difference between our two sample means (numerator) against the average difference expected under the null hypothesis (denominator, which is the standard error of the difference between the means). That statistic is a two-sample *t* test.

Q: What do they call a two-sample *t* test in England?

A: *t* for two.

The formula for a two-sample *t* test is

$$t_{\text{2-samp}} = \frac{(M_1 - M_2) - (\mu_1 - \mu_2)}{\sigma_{M_1 - M_2}}$$

where

M_1 = mean of the first sample,

M_2 = mean of the second sample,

μ_1 = mean of the first population,

μ_2 = mean of the second population, and

$\sigma_{M_1 - M_2}$ = standard error of the difference between the means.

But wait! The two-sample *t* test looks a lot like the *z* score, the normal deviate *Z* test, and the one-sample *t* test we previously encountered, doesn't it? Yes, it does. That's because each is an example of the prototype for any test of statistical significance. Recall that a *prototype* is a generic model. Here, again, is the prototype for any test of statistical significance:

$$\frac{\text{What did you get?} - \text{What did you expect?}}{\text{Standardized random error}}$$

When we studied z scores, the substitutions we made were as follows:

"What did you get?"	Raw score
"What did you expect?"	Sample mean
"Standardized random error"	Standard deviation

This gave us the formula

$$z\ \text{Score} = \frac{\text{Raw score} - \text{Sample mean}}{\text{Standard deviation}}$$

In symbols, this was

$$z\ \text{Score} = \frac{X - M}{s}$$

For the *normal deviate Z test* and for the *one-sample t test*, the substitutions we made were as follows:

"What did you get?"	Sample mean
"What did you expect?"	Population mean
"Standardized random error"	Standard error of the mean

This gave us the formula

$$Z_{\text{norm dev}} \text{ or } t_{\text{1-samp}} = \frac{\text{Sample mean} - \text{Population mean}}{\text{Standard error of the mean}}$$

In symbols, this was

$$Z_{\text{norm dev}} \text{ or } t_{\text{1-samp}} = \frac{M - \mu}{\sigma_M}$$

Well, now we have a *two-sample t test*. The substitutions we now make are as follows:

"What did you get?"	Difference between the two sample means
"What did you expect?"	Difference between the two population means
"Standardized random error"	Standard error of the difference between the means

This gives us the formula

$$t_{\text{2-samp}} = \frac{(\text{Difference between sample means}) - (\text{Difference between population means})}{\text{Standard error of the difference between the means}}$$

In symbols, this is

$$t_{2\text{-samp}} = \frac{(M_1 - M_2) - (\mu_1 - \mu_2)}{\sigma_{M_1 - M_2}}$$

Look again at the numerator of the formula. The second term, $\mu_1 - \mu_2$, is the expected difference between the population means. It is the expected difference in depression level between an infinite number of clients treated with medication and an infinite number of clients treated with counseling. Now recall that it is always the null hypothesis that we test. Under the null hypothesis, what do we expect that difference to be?

Yes, under the null hypothesis, the expected difference is 0. Therefore, the formula reduces to

$$t_{2\text{-samp}} = \frac{(M_1 - M_2) - 0}{\sigma_{M_1 - M_2}}$$

which further reduces to

$$t_{2\text{-samp}} = \frac{(M_1 - M_2)}{\sigma_{M_1 - M_2}}$$

Many sources show only the shortened formula above. However, to emphasize the similar logic underlying most inferential statistical tests, it is helpful to remember the fuller formula.

With the formula in hand, let's finally calculate our two-sample *t*. We want to determine whether the observed difference in depression level between the two samples was *a lot* (and therefore probably due to the difference in treatment between the samples) or *only a little* (and therefore probably due to mere sampling error).

CHECK YOURSELF!

Compare a two-sample *t* test with a one-sample *t* test. When would you use each one?

Calculating a Two-Sample *t* Test

In our study, Sample 1 is the medication group, and Sample 2 is the counseling group. Therefore, the formula is

$$\begin{aligned} t_{2\text{-samp}} &= \frac{(M_1 - M_2) - (\mu_1 - \mu_2)}{\sigma_{M_1 - M_2}} \\ &= \frac{(M_1 - M_2) - 0}{\sigma_{M_1 - M_2}} \\ &= \frac{M_1 - M_2}{\sigma_{M_1 - M_2}} \\ &= \frac{28.778 - 38.111}{\text{Whoops! First, we need to calculate } \sigma_{M_1 - M_2}} \end{aligned}$$

In this module, we are restricting ourselves to studies in which sample sizes are equal and the samples are independent. Recall that when samples are independent and sample sizes are equal, the formula for $\sigma_{M_1 - M_2}$ is

$$\sigma_{M_1 - M_2} = \sqrt{\sigma^2_{M_1} + \sigma^2_{M_2}}$$

where

σ_{M_1} = the standard error of the mean of the first population and

σ_{M_2} = the standard error of the mean of the second population.

But what *is* the σ_M for each population? Yes, unfortunately, we first have to compute each of those as well. Recall that the formula for σ_M is

$$\frac{\sigma}{\sqrt{n}} \quad \text{or} \quad \frac{\sigma_{\text{est}}}{\sqrt{n}}$$

> Life is good for only two things: discovering mathematics and teaching mathematics.
>
> —Simeon Poisson

where $\sigma_{\text{est}} = \sigma$ estimated from *s*.

Because we do not know the population standard deviation (σ) for either population, we have to use the second formula, which uses the *estimated* population standard deviation (σ_{est}) for each population.

But what *is* the estimated population standard deviation for each population? Yes, we estimate it from the sample data. That's where we must begin our calculations, so let's do that now.

Step 1: Find the two estimated population standard deviations.

Medication			*Counseling*		
X	X – M	(X – M)²	X	X – M	(X – M)²
32	3.222	10.381	43	4.889	23.902
40	11.222	125.933	31	–7.111	50.566
21	–7.778	60.497	39	0.889	0.790
17	–11.778	138.721	36	–2.111	4.456
40	11.222	125.933	46	7.889	62.236
26	–2.778	7.717	35	–3.111	9.678
19	–9.778	95.609	32	–6.111	37.344
44	15.222	231.709	44	5.889	34.680
20	–8.778	77.053	37	–1.111	1.234
Σ = 259		Σ = 873.553	Σ = 343		Σ = 224.886

$$M_{\text{med}} = \frac{259}{9} = 28.778 \qquad M_{\text{couns}} = \frac{343}{9} = 38.111$$

First, we need the estimated population variances from each sample variance. Recall from Module 17 that sample variances tend to underestimate the true population variance. And recall that we correct for this bias by dividing the sum of squared deviations by *n* – 1 rather than by *n*.

The estimated population variances for the medication and counseling populations, using $n - 1$ in the denominator, are as follows:

$$\sigma^2_{\text{med, est}} = \frac{\sum(X - M)^2}{n - 1} = \frac{873.553}{8} = 109.194 \qquad \sigma^2_{\text{couns, est}} = \frac{\sum(X - M)^2}{n - 1} = \frac{224.886}{8} = 28.111$$

Recall that the standard deviation is the square root of the variance. Therefore, the estimated population standard deviations for the medication and counseling populations are as follows:

$$\sigma_{\text{med, est}} = \sqrt{\sigma^2} = \sqrt{109.194} = 10.450 \qquad \sigma_{\text{couns, est}} = \sqrt{\sigma^2} = \sqrt{28.111} = 5.302$$

Step 2: Find the two standard errors of the means.

Now that we have the two estimated population standard deviations, we can compute the two standard errors of the means:

$$\sigma_{M_{\text{med}}} = \frac{\sigma_{\text{med, est}}}{\sqrt{n}} = \frac{10.450}{\sqrt{9}} = \frac{10.450}{3} = 3.483 \qquad \sigma_{M_{\text{couns}}} = \frac{\sigma_{\text{couns, est}}}{\sqrt{n}} = \frac{5.302}{\sqrt{9}} = \frac{5.302}{3} = 1.767$$

Step 3: Find the standard error of the difference between the means.

Now that we have our two standard errors of the means, we can calculate the standard error of the difference between the means, which we need for the denominator of the two-sample t:

$$\sigma_{M_1 - M_2} = \sqrt{\sigma^2_{M_1} + \sigma^2_{M_2}} = \sqrt{(3.483)^2 + (1.767)^2} = \sqrt{12.131 + 3.122} = \sqrt{15.253} = 3.91$$

So the *value* of our standard error of the difference between the means (which is the standard deviation of the sampling distribution of differences between an infinite number of pairs of sample means) is 3.91. We can add that value to our diagram of the sampling distribution of the differences between the means (Figure 20.1).

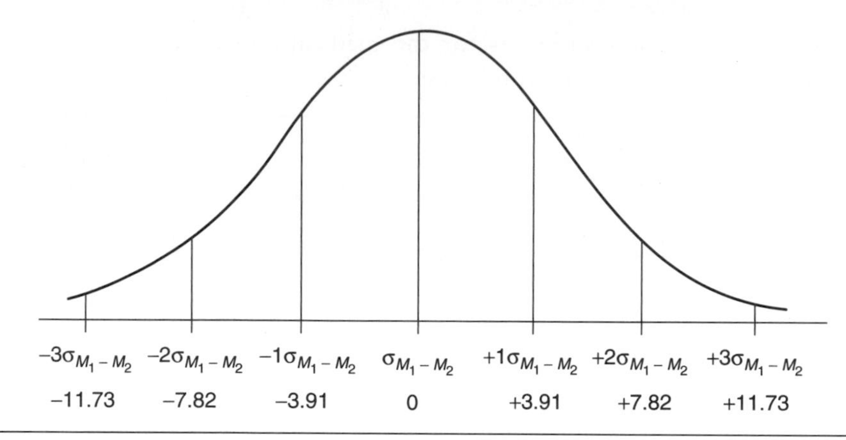

Figure 20.1 Sampling Distribution of the Difference Between the Means, Showing the Value of $\sigma_{M_1-M_2}$

Step 4: Find *t*.

Now that we know the standard error of the difference between the means, we can finally calculate the two-sample *t* test:

$$\begin{aligned} t_{\text{2-samp}} &= \frac{(M_1 - M_2) - (\mu_1 - \mu_2)}{\sigma_{M_1-M_2}} \\ &= \frac{(M_1 - M_2) - 0}{\sigma_{M_1-M_2}} \\ &= \frac{M_1 - M_2}{\sigma_{M_1-M_2}} \\ &= \frac{28.778 - 38.111}{3.91} \\ &= \frac{-9.333}{3.91} \\ &= -2.39 \end{aligned}$$

We are finally done. The observed −9.333 points difference in depression level between clients treated with medication and clients treated with counseling is −2.39 of the σ_{M-M} error units. But what does that mean?

CHECK YOURSELF!

List the steps for calculating a two-sample *t* test.

Interpreting a Two-Sample *t* Test

As usual, the question is whether the observed difference is *a lot*, and therefore probably due to the difference in treatment between the two samples, or whether it is *only a little*, and therefore probably due to mere sampling error. Let's see where our observed difference between the means falls within the sampling distribution of the differences between the means (Figure 20.2).

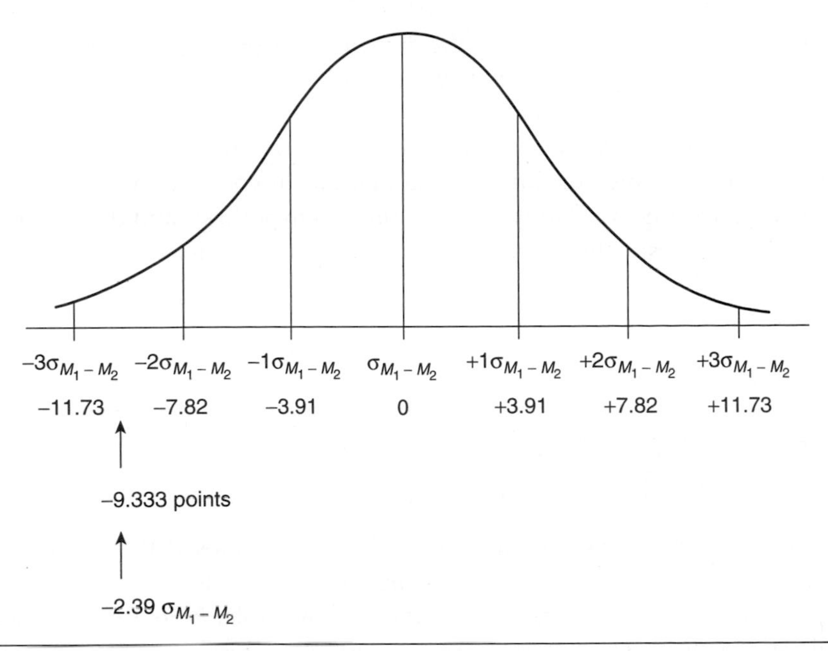

Figure 20.2 Sampling Distribution of the Difference Between the Means, Showing the Location of the Calculated *t*

Under the null hypothesis, the expected difference in depression level between clients given medication and those given counseling is 0—right in the middle of the distribution. From the diagram, we can see that the actual difference between the two samples is way down in the left tail of the distribution. It certainly does look like clients given medication were considerably less depressed than clients given counseling. But were they really less depressed, or is this just a random difference? What is the probability, if the null hypothesis is true and the two treatments really do not differ in effectiveness, that we would find this much difference in depression between the two differently treated samples?

As with the *z* score, the normal deviate *Z* test, and the one-sample *t* test, we answer that question by looking up the critical value in a table. That table is the *t* table, just as it was for the one-sample test. Appendix C contains the *t* table. A portion of that table is reproduced as Table 20.1.

We enter the *t* table at the correct degrees of freedom (*df*). Recall that the *df* for a *t* test is $n - 1$ for each sample. For the one-sample *t* test, that was $n - 1$. But this is a two-sample

Table 20.1 A Portion of the *t* Table

	Level of Significance for One-Tailed Test (%)			
	5	*2.5*	*1*	*.5*
	Level of Significance for Two-Tailed Test (%)			
df	*10*	*5*	*2*	*1*
16	1.746	2.120	2.584	2.921
17	1.740	2.110	2.567	2.898
18	1.734	2.101	2.552	2.878

t test; that is, we have two groups. Thus, the *df* is $(n - 1) + (n - 1)$, which is $N - 2$ (i.e., here $N = 2n$). Our total sample size *for both samples added together* is $9 + 9 = 18$. Thus, the correct *df* is $18 - 2 = 16$.

As before, the *t* table lists the minimum value that our calculated *t* must have for us to reject the null hypothesis and conclude instead that the difference in means is probably due to the difference in treatment. That is, to reject the null hypothesis and thereby gain support for the research hypothesis, the value of the calculated *t* must meet or exceed the value of the tabled critical *t*.

CHECK YOURSELF!

What do the entries in a *t* table tell you?

Recall that our hypothesis was directional: We had proposed that medication would result in *lower* depression levels than counseling would. Therefore, our hypothesis is *one-tailed*, and we must look in the one-tailed column of the table. Note that the figures in the table are absolute values. That is, they are the critical *t* values regardless of a positive or negative sign.

Assume that we were willing to make a Type 1 error 5% of the time. At $\alpha = .05$, can we reject the null hypothesis?

Yes, we can, because the calculated *t* of –2.39 meets or exceeds the one-tailed critical *t* of (–)1.746.

Now assume that we were willing to make a Type 1 error only 1% of the time. At $\alpha = .01$, can we reject the null hypothesis?

No, we cannot, because the calculated *t* of –2.39 does not meet or exceed the one-tailed critical *t* of (–)2.584.

The difference in participants' depression levels when given medication versus counseling was significant at the .05 error level. In a journal article, these results would be reported like this:

$$t(16) = -2.39, p < .05$$

This is read as follows: "*t* at 16 degrees of freedom is –2.39. There is less than a 5% chance that the difference in depression level is due to mere chance." Such a large observed difference in depression level is probably due to the difference in treatments.

PRACTICE

1. Assume that our research hypothesis for the above study is nondirectional. That is, while we believe that the type of treatment given will affect the level of depression, we have no idea which treatment—medication or counseling—will be more effective. Look up the critical *t* for this two-tailed hypothesis at both the .01 and .05 α levels. At which α level (if either) can we reject the null hypothesis?
2. A large furniture store stations salespeople near its entrance to greet customers and offer assistance in shopping. The salespeople, who work on a commission basis, tell the customers their name and hand them a business card. A psychologist thinks that the salespersons' intrusiveness might cause customers to buy less furniture rather than more furniture. She convinces the store's management to let her study the issue. Customers are

randomly selected to either receive or not receive a salesperson's offer of assistance immediately on entering the store. The amount of customers' purchases are then logged as they leave the store. Here are the data:

Amount of Purchase, in U.S. Dollars	
Immediate Assistance	*No Assistance Unless Requested*
0	761
2,274	0
0	2,592
0	0
0	1,037
362	0
855	84
0	0
0	672
1,273	0

a. What are the independent and dependent variables in this study?

b. State the null hypothesis and the directional (one-tailed) research hypothesis.

c. Calculate t and compare it with the tabled critical t at the .01 and .05 α levels. Can you reject the null hypothesis?

3. The Shine Company, which manufactures cleaning supplies, wants to determine whether or not adding a fragrance to a window cleaner leads people to believe that it cleans better than an unscented product. The company randomly selects 24 participants for a pilot study. The company gives 12 participants the scented cleaner and the other 12 participants the unscented cleaner. After using the cleaners for a month, the participants rate how well they thought the cleaner worked. Higher scores indicate more effective cleaning. Here are the ratings:

Unscented	*Scented*
6	8
5	8
7	7
5	9
6	7
8	8
4	9
7	9
5	6
6	5
6	7
7	6

a. Is the research hypothesis in this study directional or nondirectional?

b. State the research hypothesis.

c. Calculate t and compare it with the tabled critical t at the .05 α level. Can you reject the null hypothesis?

4. Elena Martin is campaigning for the city council. She has two types of lawn signs to distribute: large ones and small ones. She wonders if the size of the sign affects residents' willingness to display the signs on their property. Early in the campaign, her staff obtain a list of homeowners in each of the city's 10 voting districts who are registered in her political party and presumably not averse to advertising their support for her. The staff randomly selects homeowners in each of the 10 districts, to some of whom they send large signs and to others, small signs. Two weeks later, staff members drive by each home to which they sent the signs to record whether or not the sign is being displayed. Here are the percentages of homes displaying the signs in each district:

Large Sign	*Small Sign*
34	41
41	44
30	36
32	38
28	29
31	47
40	49
27	39
36	43
22	37

   a. Is the research hypothesis in this study directional or nondirectional?
   b. State the research hypothesis.
   c. Calculate t and compare it with the tabled critical t at the .01 α level. Can you reject the null hypothesis?

5. Carmine reads an article that says that male college students study less than female college students. Carmine wonders if this is really so. He asks 20 randomly selected students—10 males and 10 females—from his coed dorm to record their study times for a period of 4 weeks. Here are the students' average weekly study times, to the nearest half-hour:

Males	*Females*
13.5	15.5
6.5	10.0
8.0	10.5
14.5	7.0
12.5	13.0
16.0	12.5
12.0	11.0
9.5	8.5
7.0	8.0
11.5	13.0

   a. Is the research hypothesis in this study directional or nondirectional?
   b. State the research hypothesis.
   c. Calculate t and compare it with the tabled critical t at the .05 α level. Can you reject the null hypothesis?
   d. Do you think these samples are representative of all college males and females? To what populations can Carmine rightfully infer the results?

6. A cognitive psychologist is curious as to why women tend to physically turn a map's position to agree with the direction they are moving, while men tend to keep the map in its original position and mentally change orientations as they move. She wonders if there is a sex difference in ability to mentally rotate objects. She devises a computerized experiment in which she times correct responses to a mental visual rotation task at a given angle. Here are the times, in seconds, for 15 men and 15 women. Conduct a one-tailed t test at the .01 α level. Are men able to mentally rotate objects significantly faster than women?

Men	*Women*
0.25	0.29
0.27	0.30
0.23	0.25
0.24	0.23
0.22	0.22
0.28	0.26
0.23	0.33
0.22	0.27
0.24	0.29
0.28	0.26
0.22	0.29
0.21	0.27
0.23	0.22
0.23	0.29
0.20	0.25

7. A video arcade owner wonders if boys or girls spend more time playing video games once they enter the arcade. He records the playing time for 10 boys and 10 girls selected at random. No more than one person is selected from among a group of friends to minimize the effect of group behavior on individuals. Here are their times, in minutes. Conduct a two-tailed t test at the .01 α level. Do either boys or girls spend significantly more time playing the games, once they have entered the arcade?

Boys	*Girls*
124	64
36	22
39	105
48	19
40	32
79	43
63	24
143	33
36	98
22	30

8. An Internet provider wonders if its cable Internet customers are significantly more satisfied with one or the other of its two packages—free e-mail or free basic TV. The company surveys 12 customers currently receiving free e-mail with their cable Internet and 12 customers currently receiving free basic TV with their cable Internet. Satisfaction scores range from 1 to 10 based on answers to the survey questions. Here are the scores. Conduct a two-tailed t test at the .05 α level. Are customers significantly more satisfied with one package or the other?

Free E-mail	*Free Basic TV*
6	7
7	10

(Continued)

(Continued)

Free E-mail	*Free Basic TV*
5	8
7	5
8	9
6	10
10	7
9	6
6	5
7	6
5	8
7	6

9. An employee benefits administrator wonders if the amount of the insurance co-pay affects the number of times employees visit a doctor per year. The company offers two plans. One plan has a higher visit co-pay in return for a larger cap on annual coverage. The other plan has a lower visit co-pay but with a lower cap on annual coverage. Here are the number of visits in a given year for 14 members under each plan. Conduct a two-tailed t test at the .05 α level. Does either plan result in significantly fewer visits?

Higher Co-Pays	*Lower Co-Pays*
1	1
0	3
4	0
1	5
2	3
3	4
1	4
0	3
0	5
1	2
0	0
2	2
4	4
0	1

10. A police agency is interested in reducing the speed of cars on a particular stretch of highway. Lore has it that a visible police car with radar is more effective in reducing speed than a warning sign saying that the area is monitored by radar. A police sergeant wonders if the lore is correct, so he tests each method on the same stretch of highway at the same time of day and on the same day of the week (two consecutive weeks) to ensure similar driving volume and conditions. He then records the number of miles per hour (mph) each speeder drives over the speed limit. Here are the data of the excess speed of the eight drivers caught speeding under each condition. Conduct a one-tailed t test at the .05 α level. Does the police car result in lower speeds than the warning sign?

Excess mph With Police Car	*Excess mph With Warning Sign*
12	18
7	12
6	9
10	15
4	7
13	19
11	18
6	12

Looking Ahead

As we saw in Module 17, the ability to reject the null hypothesis depends not only on how different the observed group means are but also on what level of Type 1 error you are willing to accept. In Module 23, we will look at this concept of error in more detail, just as we did in Module 18. As it turns out, dichotomous decisions (reject/do not reject) are less meaningful than reports of actual Type 1 error.

SPSS Connection

Download the file **data_depression relief due to med couns.sav** from www.sagepub.com/steinberg2e. These data are used in the textbook example.

Alternatively, manually enter the 18 scores from the depression example in Module 20 into the SPSS **Data View** spreadsheet. Data entry for a *t* test with equal sample sizes is not intuitively obvious. In the textbook, the data are set up as two groups of 9 clients. In SPSS, all 18 scores (9 + 9) are entered in a single column. Then their group membership (medication vs. counseling) is entered in the second column. Thus, enter the data as follows:

32	m
40	m
21	m
17	m
40	m
26	m
19	m
44	m
20	m
43	c
31	c
39	c
36	c
46	c
35	c
32	c
44	c
37	c

Click on the **Variable View** tab to define the variables. Name the first variable **deprscor,** set the decimals at **0**, and label the variable as **Depression Score.** Name the second variable **typtreat,** and label the variable as **Type of Treatment.** Label the value as follows: **c = counseling** and **m = medication.**

If the file is not already in **Data View**, click that tab in the lower left of the screen.

In the toolbar at the top of the screen, click on **Analyze,** then **Compare Means,** then **Independent-Samples T Test.** Highlight the variable **Depression Score** in the left window, and then click on the **arrow** before the **Test Variable** window to send the variable into that window. This is the study's dependent variable. Click on the variable **Type of Treatment** in the left window, and then click on the **arrow** before the **Grouping Variable** window to send the variable into that window. This is the study's independent variable. Click on **Define Groups** beneath the Grouping Variable window. Enter **m** for Group 1 and enter **c** for Group 2. Click **Continue** and then **OK.** This is what you will see.

T-Test

Group Statistics

Type of Treatment		N	Mean	Std. Deviation	Std. Error Mean
Depression Score	Medication	9	28.78	10.450	3.483
	Counseling	9	38.11	5.302	1.767

Independent Samples Test

		Levene's Test for Equality of Variances		t-test for Equality of Means					95% Confidence Interval of the Difference	
		F	Sig.	t	df	Sig. (2-tailed)	Mean Difference	Std. Error Difference	Lower	Upper
Depression Score	Equal variances assumed	8.727	.009	-2.390	16	.030	-9.333	3.906	-17.614	-1.053
	Equal variances not assumed			-2.390	11.863	.034	-9.333	3.906	-17.855	-.812

Visit the study site at www.sagepub.com/steinberg2e for practice quizzes and other study resources.

21

t Test With Unequal Sample Sizes

Terms: special-case formula, generalized formula, weighting

Learning Objectives:

- Determine degrees of freedom
- Perform a two-sample *t* test for unequal sample sizes
- Use a table to interpret calculated *t*
- Report results in APA format

What Makes Sample Sizes Unequal?

I have a confession to make. The formula I gave in Module 20 for calculating a two-sample *t* test is not entirely correct. Or rather, it is correct, but only sometimes. The formula I gave is called a **special-case formula**—one that is true only under certain circumstances. Those circumstances are as follows:

1. Sample sizes are equal in both groups.
2. Samples are independent (i.e., not the same participants in both treatment groups and not different participants who have been matched or equated).

As long as those two conditions are met (and I made sure that they were in the examples I used), the formula I gave is correct.

Real research, however, doesn't always meet those conditions. Sometimes equal numbers of participants are not available for both treatment conditions. Other times, we start out with equal numbers of participants in both treatment conditions, but due to differential attrition, we end up with unequal numbers. In either of those situations, we do not have equal numbers of participants in each treatment group. Therefore, the special-case formula I gave earlier does not apply.

At still other times, the number of participants is equal in both treatment groups, but the participants in one group are not independent of the participants in the other treatment groups. Rather, the participants are related. Either the participants are matched between groups or the same participants are in both groups.

When sample sizes are not equal or when participants are not independent, we cannot use the special-case formula. The appropriate formula to use for analyzing data when sample sizes are not equal or when participants are related is a **generalized formula**. There are two types of generalized formulas: one for unequal sample sizes and one for nonindependent samples. In this module, I will present the formula for unequal sample sizes. In the next module, I will present the formula for nonindependent samples.

Comparison of Special-Case and Generalized Formulas

Here are both the special-case formula and the generalized formula for unequal sample sizes:

Special-case formula

$$t_{2\text{-samp}} = \frac{(M_1 - M_2) - (\mu_1 - \mu_2)}{\sigma_{M_1 - M_2}}$$

where

$$\sigma_{M_1 - M_2} = \sqrt{\sigma^2_{M_1} + \sigma^2_{M_2}}$$

Generalized formula for unequal sample sizes

$$t_{2\text{-samp}} = \frac{(M_1 - M_2) - (\mu_1 - \mu_2)}{\sigma_{M_1 - M_2}}$$

where

$$\sigma_{M_1 - M_2} = \sqrt{\left(\frac{SS_1 + SS_2}{n_1 + n_2 - 2}\right)\left(\frac{1}{n_1} + \frac{1}{n_2}\right)}$$

Before we delve further into the formulas, take a moment to compare the special-case and the generalized formulas. How are they the same? How are they different?

Yes, the numerators are the same, but the denominators are different. However, despite the different appearance of the denominators, the two formulas are equivalent when sample sizes are the same. To see why that is so, let me parse the denominator of the generalized formula so that it is clear what its terms do.

The denominator of the generalized formula includes the term

$$\frac{SS}{n - 1}$$

In words, that's the average (divided by $\approx n$) of the summed squared deviation scores (SS). It's been a while since you've seen that expression. Do you recall what that term is?

Yes, it's the estimated population variance! But note that the expression is under a square root sign. Do you remember what the square root of the estimated population variance is?

Yes, it's the estimated population standard deviation, or σ_{est}.

Now look at the term

$$\frac{1}{n}$$

The formula says to multiply by $1/n$. Going back to the math review in Module 1, dividing by a number is the same as multiplying by the reciprocal of that number. Reversing that rule, multiplying by the reciprocal of a number is the same as dividing by that number. Hence, by telling us to multiply the entire first term by $1/n$, it is really saying to *divide* the entire first term by n. Furthermore, the $1/n$ is under a square root sign. Thus, it says to divide the entire first term by the square root of the sample size.

Divide *what* by the square root of the sample size? Why, the first term that we just analyzed: $\sqrt{SS/(n-1)}$.

But the $\sqrt{SS/(n-1)}$ term was the estimated population standard deviation. Thus, this portion of the formula says to divide the estimated population standard deviation by the square root of the sample size.

Now, it has not been very long since you have seen *that* expression. Do you remember what we get when we divide the population standard deviation by the square root of the sample size? To remind you, in symbols that is

$$\frac{\sigma}{\sqrt{n}}$$

Yes, it is the standard error of the mean!

But all the above calculations were for the n_1 group only. The formula says to do all those same calculations again, this time for the n_2 group. After we do that, we will have two standard errors of the mean—one for the first sample and one for the second sample. The formula then says to add those two standard errors of the mean together and take the square root of that number. Now, you *do* remember what you get when you take the square root of the sum of two standard errors of the mean, don't you? (*Hint:* Look at the specialized formula.)

> Only professional mathematicians learn anything from proofs. Other people learn from explanations.
>
> —Ralph Boas

Of course. When you take the square root of the sum of the two standard errors of the mean, you get the standard error of the difference between the means!

The proof is complete: The numerators of the specialized and generalized formulas were identical to begin with, and now we see that the denominators, although they look very different, are also equivalent. Both denominators give the standard error of the difference between the means.

Thus, the specialized formula and the generalized formula lead to exactly the same t statistic—but only when the sample size in the two samples is the same. Note that, in computing the variances in the generalized formula, each SS is divided by its own n. This is called **weighting**. The process of weighting combines the variances of the two samples proportional to the number of subjects in each sample.

> A proof tells us where to concentrate our doubts.
>
> —Morris Kline

Think of weighting this way: Suppose two classes take the same test, but one class has 30 students and the other class has 10 students. And suppose the average for the first class is 90 and the average for the second class is 70. How can you figure the average for both classes combined? If you merely add the two averages together and divide by 2, you get 80. But that would be incorrect. There were three times as many students in the class that averaged 90 as there were in the class that averaged 70, so the 90 needs to be weighted by 3 against a weight of 1 for the 70. That would be $(90 + 90 + 90 + 70)/4 = 85$. Yes, the average for both classes combined is 85, not 80. Alternatively, if you knew each student's score in both classes, you could merely add up the scores for all students in both classes and then divide by 40, the total number of students. That gives each individual score an equal weight but ensures that 75% of the individual scores come from the class of 30 and 25% come from the class of 10. Thus, the two classes are effectively weighted.

The same principle applies when combining variances. Rather than just adding the two variances together and dividing by 2, the generalized formula for a t test weights each sample variance by the number of subjects in that sample before combining them.

What do you think happens when the two samples are the same size? Let's return to the two classroom tests. If both classes had 20 students in them, you could weight them both by 1 before adding them together, but that's the same as merely adding them together without weighting, isn't it? Thus, when samples sizes are equal, there is no need to weight the samples.

The same principle applies to variances. When sample sizes are equal, you can simply add the two variances together and divide by 2. But isn't that what the special-case formula did? Yes, it is. Thus, the generalized formula reduces algebraically to the special-case formula when sample sizes are equal.

"But this *is* the simplified version for the general public"

SOURCE: Reprinted with permission from Sydney Harris, ScienceCartoonsPlus.com.

CHECK YOURSELF!

Compare the special-case formula and the generalized formulas for a two-sample *t* test. How do they differ? What is the purpose of the added terms in the generalized formula?

More Clarification of the Underlying Logic

Now that you know that you must use the generalized formula except when the restricted conditions are met for the special-case formula, you may wonder why I didn't present the generalized formula in the first place. The answer is that I wanted you to see the underlying logic of the formula without getting bogged down in the calculations. Recall that several times already, I have presented an inferential formula and then said, "But wait! The (____) statistic looks a whole lot like the (____) statistic." And then I pointed out that the numerator always measures deviation from expectation and the denominator always scales that deviation against a standardized error term. In that way, the logic of each new inferential statistic has related to the logic of the ones you have previously learned.

Now suppose I had tried that with the generalized formula for the two-sample *t* test:

$$t_{\text{2-samp}} = \frac{(M_1 - M_2) - (\mu_1 - \mu_2)}{\sqrt{\left(\frac{SS_1 + SS_2}{n_1 + n_2 - 2}\right)\left(\frac{1}{n_1} + \frac{1}{n_2}\right)}}$$

Would you have been able to see the underlying logic in the generalized formula? Most people would not. Because of the complexity of the generalized formula, its logic is not intuitively obvious.

Instead, look at the special-case formula:

$$t_{2\text{-samp}} = \frac{(M_1 - M_2) - (\mu_1 - \mu_2)}{\sigma_{M_1 - M_2}}$$

Isn't it easier to see the underlying logic in the special-case formula? The numerator measures deviation from expectation, and the denominator scales that difference into a standardized error term.

Thus, while previously I was perhaps a bit evasive in not mentioning the conditions under which you are permitted to use the special-case *t*-test formula, the presentation was designed to clarify your understanding, not hide the facts. However, now that you know the conditions under which the special-case formula applies, you may use the special-case formula only when the special conditions are met: equal sample sizes and independent samples. When those conditions are not met, you will need to use the more complex generalized formula.

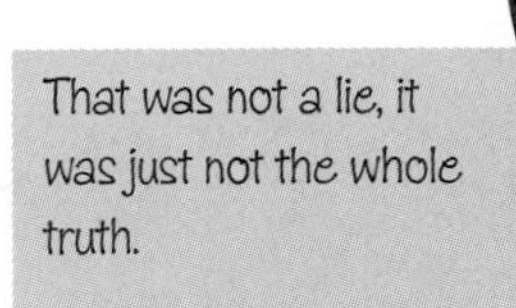

CHECK YOURSELF!

Under what conditions could you use the special-case formula for a two-sample *t* test? Under what conditions must you use the generalized formula?

Calculating a *t* Test With Unequal Sample Sizes

Let's calculate an example of a *t* test with unequal sample sizes. Sleep deprivation is an important health and performance issue for weary college students, especially as final exams loom and students pull all-nighters. So let's look at a study involving sleep deprivation.

Assume that an investigator hypothesizes that sleep deprivation increases recall errors for semantically related (with similar meanings) words. Twenty-four people volunteer to participate in the study. Half the participants are randomly assigned to remain awake throughout the night. The investigator stays with them in the sleep lab and gently prods them awake whenever they nod off. The other half of the participants are permitted to sleep on a comfortable air mattress throughout the night. In the morning, both groups are given a list of 50 words to memorize. After an hour of memorization, the participants are given a mixed list containing all the 50 words they had memorized along with 3 semantically related words for each of the stimulus words—for example, *infant* instead of *baby*, *ridiculous* instead of *ludicrous*, *compute* instead of *calculate*, and so on. Participants are told to indicate whether or not each of the 200 words appeared in the original set of 50 words. Participants' scores are the number of the original 50 memorized words they correctly identify.

> To sleep, perchance to dream . . .
>
> —Shakespeare

As an ethical condition of the study, participants are permitted to opt out of the study at any point throughout the night. In addition, the investigator is permitted to gently prod participants who fall asleep in the stay-awake condition no more than three separate times throughout the night. Participants who continue to fall asleep after three separate prods are no longer awakened and are simply dropped from the study. As a result, 3 of the 12 participants in the stay-awake condition are dropped from the study, and the final samples sizes are not equal.

> If we knew what we were doing, it wouldn't be called research, would it?
>
> —Albert Einstein

The following table gives the number of words (out of 50) correctly recalled for the 12 participants who were permitted to sleep and for the 9 remaining participants who were required to stay awake and actually did so:

Sleep	*Awake*
41	35
38	30
45	28
42	32
37	42
44	35
38	31
42	34
40	34
33	
39	
44	
$\Sigma = 483$	$\Sigma = 301$
$M_{\text{sleep}} = 483/12 = 40.250$	$M_{\text{awake}} = 301/9 = 33.444$

Just by inspecting the means, you can see that participants who were well rested made fewer errors than participants who were sleep deprived. The mean correct recall scores of the two groups were 40.250 words and 33.444 words, respectively. That's a difference of 6.806 words. However, that difference does not necessarily indicate that sleep deprivation increased recall errors. Even if the null hypothesis is true and sleep deprivation makes no difference in semantic confusion, we still do not expect any particular sample of participants deprived of sleep to recall exactly the same number of words as any particular sample of participants permitted to sleep. Thus, the question is not whether or not the two groups correctly recalled exactly the same number of words but how different their recall accuracy is. Is the difference in mean recall accuracy between the samples *only a little different* and, therefore, probably due to mere sampling error? Or is the difference in mean recall accuracy between the samples *very different* and, therefore, probably due to something other than mere sampling error—in this case, to the difference in amount of sleep?

We need a statistic to compute the probability of getting two sample means as different as those of the two observed samples. As before, that statistic is the two-sample t test. And as before, the statistic scales the difference between the observed and expected difference between the two sample means (numerator) against the standard error of the difference between the means (denominator). Because the sample sizes are unequal, we use the generalized formula:

$$t_{2\text{-samp}} = \frac{(M_1 - M_2) - (\mu_1 - \mu_2)}{\sqrt{\left(\frac{SS_1 + SS_2}{n_1 + n_2 - 2}\right)\left(\frac{1}{n_1} + \frac{1}{n_2}\right)}}$$

$$= \frac{M_1 - M_2}{\sqrt{\left(\frac{SS_1 + SS_2}{n_1 + n_2 - 2}\right)\left(\frac{1}{n_1} + \frac{1}{n_2}\right)}}$$

$$= \frac{40.250 - 33.444}{\text{Whoops! First we need to calculate the } SS \text{ to get the } \sigma_{M_1 - M_2}}$$

Sleep			*Awake*		
X	*X* – *M*	*(X* – *M)*²	*X*	*X* – *M*	*(X* – *M)*²
41	0.750	0.5625	35	1.556	2.4211
38	–2.250	5.0625	30	–3.444	11.8611
45	4.750	22.5625	28	–5.444	29.6371
42	1.750	3.0625	32	–1.444	2.0851
37	–3.250	10.5625	42	8.556	73.2051
44	3.750	14.0625	35	1.556	2.4211
38	–2.250	5.0625	31	–2.444	5.9731
42	1.750	3.0625	34	0.556	0.3091
40	–0.250	0.0625	34	0.556	0.3091
33	–7.250	52.5625			
39	–1.250	1.5625			
44	3.750	14.0625			
Σ = 483 *M* = 483/12 = 40.250		**Σ = 132.2500**	Σ = 301 *M* = 301/9 = 33.444		**Σ = 128.2219**

Now that we have the two sums of squared deviation scores—the two *SS* (which are highlighted in boldface)—we can fill in the remainder of the formula:

$$
\begin{aligned}
t_{2\text{-samp}} &= \frac{M_1 - M_2}{\sqrt{\left(\dfrac{SS_1 + SS_2}{n_1 + n_2 - 2}\right)\left(\dfrac{1}{n_1} + \dfrac{1}{n_2}\right)}} \\
&= \frac{40.250 - 33.444}{\sqrt{\left(\dfrac{132.2500 + 128.2219}{12 + 9 - 2}\right)\left(\dfrac{1}{12} + \dfrac{1}{9}\right)}} \\
&= \frac{6.806}{\sqrt{\left(\dfrac{260.4719}{19}\right)\left(\dfrac{1}{12} + \dfrac{1}{9}\right)}} \\
&= \frac{6.806}{\sqrt{(13.7090)(0.0833 + 0.1111)}} \\
&= \frac{6.806}{\sqrt{2.6650}} \\
&= \frac{6.806}{1.6325} = 4.170
\end{aligned}
$$

> The subject I most disliked was mathematics. I think the reason was that mathematics leaves no room for argument. If you made a mistake, that was all there was to it.
>
> —Malcolm X

We are finally done. Our observed 6.806 difference in correct word recall between participants who slept and participants who remained awake is 4.170 standard error units. But what does that mean?

Interpreting a *t* Test With Unequal Sample Sizes

As usual, the question is this: Is the observed difference *a lot*, and therefore probably due to the difference in sleep between the two samples? Or is it *only a little*, and therefore probably due to mere sampling error? Let's see where our observed difference between the means falls within the sampling distribution of the differences between the means (see Figure 21.1).

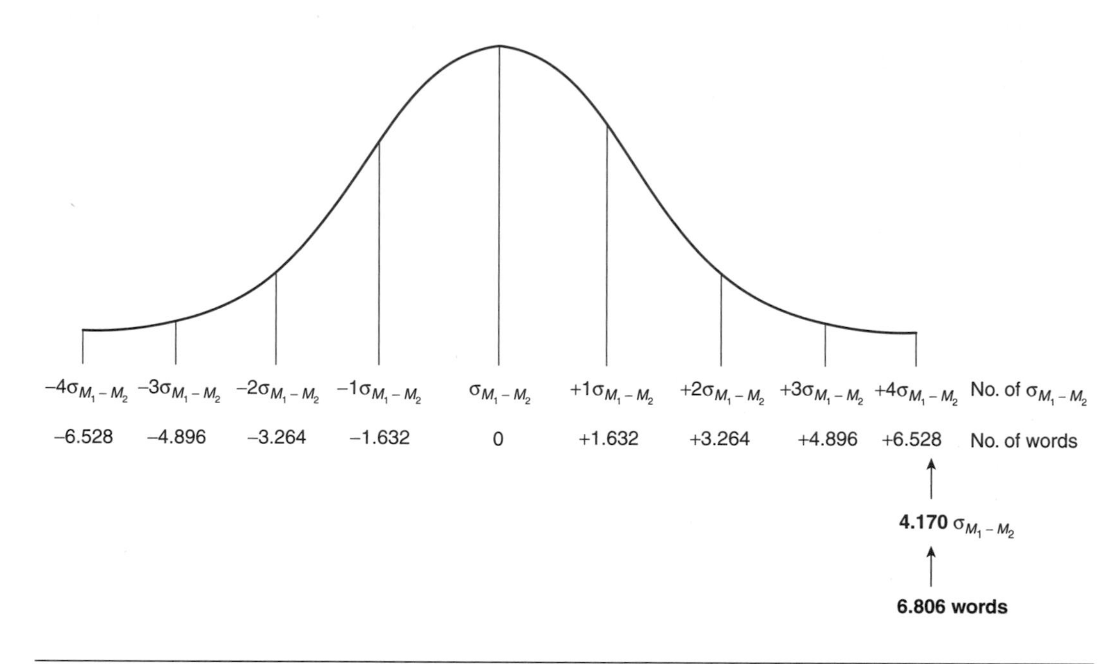

Figure 21.1 Location of Observed Difference Between the Sample Means

Under the null hypothesis, our result does seem quite unusual. But exactly how unusual is it? To determine that, we must look in the *t* table. Turn again to Appendix C. As before, we enter the table at the correct degrees of freedom (*df*). Recall that the *df* for a *t* test is total *N*, minus 1 for each group. This study has two samples, or two groups. Thus, the *df* is $N - 2$. The sample size *for both samples added together* is 12 + 9 = 21. Thus, the correct *df* is 21 – 2 = 19.

As before, the *t* table lists the minimum value that our observed *t* must be for us to reject the null hypothesis and conclude instead that the difference in means is probably due to the difference in treatment. Recall that our hypothesis was directional: We had proposed that sleep deprivation would decrease recall accuracy of semantically related words. Therefore, we must look in the one-tailed column of the table.

Assume that we are willing to make a Type 1 error 5% of the time. At $\alpha = .05$, can we reject the null hypothesis?

Yes, we can, because the calculated *t* of 4.170 meets or exceeds the one-tailed critical *t* of 1.729.

Now assume that we are willing to make a Type 1 error only 1% of the time. At $\alpha = .01$, can we reject the null hypothesis?

Yes, we still can, because the calculated *t* of 4.170 meets or exceeds the one-tailed critical *t* of 2.539.

CHECK YOURSELF!

Assume that you are willing to make a Type 1 error only one-half of one-tenth of 1% of the time. At $\alpha = .0005$, can you reject the null hypothesis?

Clearly, the number of words correctly recalled by well-rested versus sleep-deprived participants was very different. In a journal article, the results would be reported as follows:

$$t(19) = 4.170,\ p < .001.$$

This is read as "*t* at 19 degrees of freedom is 4.170. There is less than one chance in a thousand that the difference in recall is due to mere chance." We can be very confident that the difference in correct recall is due to the sleep deprivation and not due to mere sampling error.

PRACTICE

1. A researcher begins his study with 50 participants in one treatment and 50 participants in the other treatment. During the study, 4 participants in the first treatment drop out of the study, and 1 participant in the other treatment drops out of the study. Which *t* test should he use to analyze his result—equal sample size because that is what he started with? Or unequal sample size because that is what he ended up with?

2. The following data are from a study of reading achievement in children randomly assigned to be taught via a phonics curriculum ($N = 15$) versus a whole-language curriculum ($N = 10$).

Phonics	*Whole Language*
79	92
86	74
95	85
94	82
72	86
83	91
80	72
93	92
89	68
95	94
98	
72	
82	
93	
90	

Calculate *t* and compare it with the two-tailed critical *t* at the .05 α level. Is there a significant difference in reading achievement due to the curricula?

3. A local newspaper reports the results of the reading achievement study in Exercise 2 above with the following headline: "Children taught to read by phonics outscore students taught to read by whole language." In an editorial, the newspaper editor suggests that the study be replicated locally, giving the phonics curriculum to children who live on the east side of town and attend Eastside Elementary School and giving the whole-language curriculum to children who live on the west side of town and attend Westside Elementary School. Write a letter to the editor of the newspaper in which you
   a. clarify the interpretation of the results of the reported study and
   b. cite problems in the design of the local study suggested by the newspaper editor.

4. Mrs. Campbell is often awakened by the howling of dogs in the neighborhood. She wonders whether the dogs howl more often on rainy nights. During one particularly rainy month, she counts how many separate incidents of howling she hears per night and also records whether the night is rainy or clear. The data are as follows:

No. of Incidents on Rainy Nights	*No. of Incidents on Clear Nights*
9	8
14	9
17	15
13	7
10	6
21	3
11	19
14	11
7	4
19	7
16	5
14	10
	11
	6
	6
	4
	10
	13

   a. Calculate t and compare it with the one-tailed critical t at the .01 α level. Does rain significantly increase the number of howling incidents?
   b. *Think*: Why would a second rater/counter be advisable for this study?

5. Maizy Lewis runs a large day care center. She wonders whether the presence of other children affects any single child's willingness to perform simple motor tasks (assembling a puzzle, coloring a picture, etc.). She randomly selects 34 children. She isolates 18 of the children from their peers, one at a time, and asks them to perform seven simple motor tasks. She asks the remaining 16 children to do the same tasks, except that these children remain in the presence of their peers during the tasks. The numbers of tasks successfully completed by each child are given below:

No. of Completed Tasks for Isolated Children	*No. of Completed Tasks for Children With Peers*
4	3
5	2
6	1
6	4
5	7
7	5
4	3
3	1
6	2
5	2
7	7
3	6

No. of Completed Tasks for Isolated Children	*No. of Completed Tasks for Children With Peers*
4	3
4	3
5	2
5	5
7	4
6	
4	

Calculate t and compare it with the two-tailed critical t at the .05 α level. Does the presence or absence of other children significantly affect the number of tasks the children complete?

6. In Exercise 6 in Module 20, a cognitive psychologist was curious as to why women tend to physically turn a map's position to agree with the direction they are moving, while men tend to keep the map in its original position and mentally change orientations as they move. She wondered if there is a sex difference in ability to mentally rotate objects. She devised a computerized experiment in which she timed correct responses to a mental visual rotation task at a given angle. Now let's suppose that she did not obtain equal numbers of men and women for her study. For illustrative purposes, I have dropped the last five female cases from that previous example. Here are the times, in seconds, for 15 men and 10 women. Conduct a one-tailed t test at the .01 α level. Are men able to mentally rotate objects significantly faster than women?

Men	*Women*
0.25	0.29
0.27	0.30
0.23	0.25
0.24	0.23
0.22	0.22
0.28	0.26
0.23	0.33
0.22	0.27
0.24	0.29
0.28	0.26
0.22	
0.21	
0.23	
0.23	
0.20	

7. In Exercise 7 in Module 20, a video arcade owner wondered if boys or girls spend more time playing the games once they enter the arcade. He recorded the playing time for boys and girls selected at random. No more than one person was selected from among a group of friends to minimize the effect of group behavior on individuals. Now let's suppose that he did not record playing times for equal numbers of boys and girls. For illustrative purposes, I have added four female cases to the previous data. Here are the times, in minutes, for 10 boys and 14 girls. Conduct a two-tailed t test at the .01 α level. Do either boys or girls spend significantly more time playing the games once they have entered the arcade?

Boys	*Girls*
124	64
36	22
39	105
48	19
40	32
79	43
63	24
143	33
36	98
22	30
	34
	48
	22
	70

8. In Exercise 8 in Module 20, an Internet provider wondered if its cable Internet customers were significantly more satisfied with one or the other of its two packages—free e-mail or free basic TV. The company surveyed customers currently receiving free e-mail with their cable Internet and customers currently receiving free basic TV with their cable Internet. Satisfaction scores ranged from 1 to 10 based on answers to the survey questions. Now let's suppose that the company did not survey equal numbers of customers receiving free e-mail versus free basic TV. For illustrative purposes, I have dropped the last four free basic TV cases from that previous example, leaving 12 free e-mail customers and 8 free basic TV customers. Here are the scores. Conduct a two-tailed *t* test at the .05 α level. Are customers significantly more satisfied with one package or the other?

Free E-mail	*Free Basic TV*
6	7
7	10
5	8
7	5
8	9
6	10
10	7
9	6
6	
7	
5	
7	

9. In Exercise 9 in Module 20, an employee benefits administrator wondered if the amount of the insurance co-pay affected the number of times employees visit a doctor per year. The company offered two plans. One plan had a higher visit co-pay in return for a larger cap on annual coverage. The other plan had a lower visit co-pay but with a lower cap on annual coverage. Now let's suppose that he did not collect doctor visit data for equal numbers of employees in the two plans. For illustrative purposes, I have added four cases to the previous data for the lower co-pay plan. Here are the number of visits in a given year for 14 higher co-pay members and 18 lower co-pay members. Conduct a two-tailed *t* test at the .05 α level. Does either plan result in significantly fewer doctor visits?

Higher Co-Pays	*Lower Co-Pays*
1	1
0	3
4	0
1	5
2	3
3	4
1	4
0	3
0	5
1	2
0	0
2	2
4	4
0	1
	0
	3
	4
	2

10. In Exercise 10 in Module 20, a police agency was interested in reducing the speed on a particular stretch of highway. Lore had it that a visible police car with radar was more effective in reducing speeders than a warning sign saying that the area is monitored by radar. A police sergeant wondered if the lore was correct, so he tested each method on the same stretch of highway at the same time of day on the same day of the week recording how many miles per hour (mph) each speeder drive over the speed limit. Now let's suppose that he did not obtain equal numbers of speeders under each condition. For illustrative purposes, I have added 3 speeders to the police car condition giving 11 speeders with the police car and 8 speeders with the warning sign. Here are the excess miles per hour driven by the speeders. Conduct a one-tailed t test at the .05 α level. Does the police car result in lower speeds than the warning sign?

Excess mph With Police Car	*Excess mph With Warning Sign*
12	18
7	12
6	9
10	15
4	7
13	19
11	18
6	12
10	
7	
9	

SPSS Connection

Download the file **data_word memory due to sleep awake_ unequal N.sav** from www.sagepub.com/steinberg2e. These data are used in the textbook example.

Alternatively, manually enter the 21 scores from the depression example in Module 21 into the SPSS **Data View** spreadsheet. Data entry for a *t* test with unequal *N* is not intuitively obvious. In the textbook, the data are set up as two groups, one with 12 participants and one with 9 participants. In SPSS, all 21 scores (12 + 9) are entered in a single column. Then their group membership (slept vs. awake) is entered in the second column. Thus, enter the data as follows:

41	s
38	s
45	s
42	s
37	s
44	s
38	s
42	s
40	s
33	s
39	s
44	s
35	a
30	a
28	a
32	a
42	a
35	a
31	a
34	a
34	a

Click on the **Variable View** tab to define the variable. Name the first variable **numbword,** set the decimals at 0, and label the variable as **Number of Words Recalled.** Name the second variable **sleepcnd,** and label the variable as **Sleep Condition.** Label the value as follows: **s = slept** and **a = awake.**

If the file is not already in **Data View,** click that tab in the lower left of the screen.

In the toolbar at the top of the screen, click on **Analyze,** then **Compare Means,** then **Independent-Samples T Test.** Even though the samples are not equal in size, they are independent. SPSS will adjust the variances for the different sample sizes. Highlight the variable **Number of Words Recalled** in the left window, and then click on the **arrow** before the **Test Variable** window to send the variable into that window. This is the study's dependent variable. Click on the variable **Sleep Condition** in the left window, and then click on the **arrow** before the **Grouping Variable** window to send the variable into that window. This is the study's independent variable. Click on **Define Groups** beneath the Grouping Variable window. Enter s for Group 1, and enter **a** for Group 2. Click **Continue** and then **OK.** This is what you will see.

T-Test

Group Statistics

Sleep Condition		N	Mean	Std. Deviation	Std. Error Mean
Number of Words Recalled	Slept	12	40.25	3.467	1.001
	Awake	9	33.44	4.003	1.334

Independent Samples Test

		Levene's Test for Equality of Variances		t-test for Equality of Means						
									95% Confidence Interval of the Difference	
		F	Sig.	t	df	Sig. (2-tailed)	Mean Difference	Std. Error Difference	Lower	Upper
Number of Words Recalled	Equal variances assumed	.008	.929	4.168	19	.001	6.806	1.633	3.388	10.223
	Equal variances not assumed			4.080	15.878	.001	6.806	1.668	3.267	10.344

Visit the study site at www.sagepub.com/steinberg2e for practice quizzes and other study resources.

22

t Test With Related Samples

Terms: related samples, repeated measures, matched samples

Learning Objectives:

- Distinguish between independent and related samples
- Determine the degrees of freedom
- Calculate a two-sample *t* test for related samples
- Use a table to interpret calculated *t*
- Report results in APA format

What Makes Samples Related?

As mentioned briefly in Module 21, sometimes the number of participants is equal in both treatment groups, but the participants in one group are not independent of the participants in the other treatment group. We call this a **related-samples** study. Some statisticians call related samples *correlated samples* or *dependent samples*.

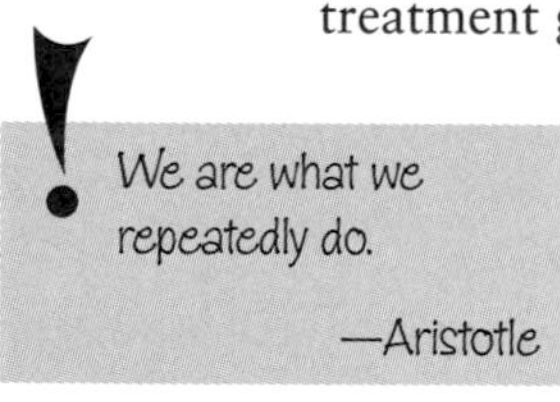

There are two types of related-samples designs. In one type, we use the same participants in both treatment conditions, giving the participants first one treatment and then the other treatment. This is called a **repeated measures** design. For example, in the memory and sleep deprivation study discussed in Module 21, some participants might first be put in the awake condition and later in the fully rested condition, while other participants would first be put in the fully rested condition and later in the awake condition.

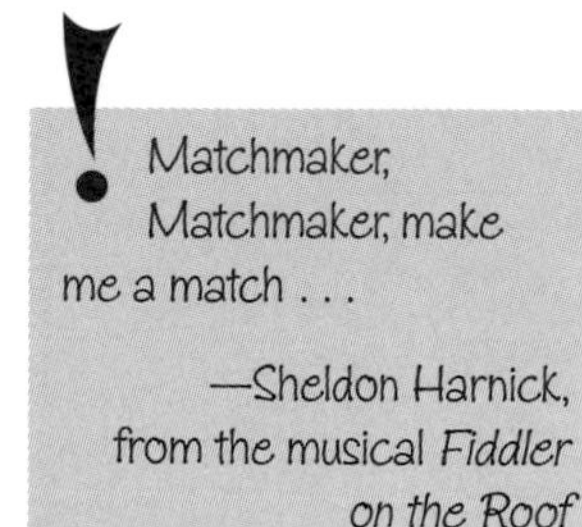

In the other type of related measures design, we use different participants in the two treatment groups, but first match the participants on some relevant extraneous variable before assigning them to treatment conditions. This is called a **matched-samples** design. Matching helps ensure that the participants in each treatment condition are equivalent on extraneous variables. This, in turn, allows the experimenter to conclude that the effect in the dependent variable is likely due to the independent variable and not to the extraneous variable. Common extraneous variables in psychological and educational research include intelligence, socioeconomic status, and amount of prior experience. In the case of the memory and sleep deprivation study, extraneous variables that might be related to recall error include participants' verbal IQ and the amount of nightly sleep participants typically need for their own optimal functioning.

In either a repeated measures or a matched subjects design, the participants in one treatment condition are no longer independent of the participants in the other treatment condition. For research purposes, this can be a very good thing. In independent-samples designs, it is always possible that the participants in one treatment condition, being different from the participants in the other treatment condition, are also different in ways directly related to the variable under study. For example, in the study of the effect of sleep deprivation on recall errors, the participants in one group may be more verbally intelligent and hence less prone to errors of recall than the participants in another group. Using the same participants in both treatment conditions or matching participants on verbal IQ eliminates this confounding variable. Moreover, for some types of research questions, repeated measures designs are not only preferred but also necessary. For example, if we want to see how subjects perform on a given task both before and after training, we must use the same subjects in both conditions. If we want to follow subjects' development over time—say children's reading achievement at the beginning of the year in comparison with their reading achievement at the end of the year—we must use the same subjects at both measurement times.

CHECK YOURSELF!

Is it possible to have a related-samples t test that also has unequal sample sizes? Why or why not?

CHECK YOURSELF!

Is it possible to have a repeated measures design if the independent variable is a preexisting variable such as age, gender, or socioeconomic status? Why or why not?

PRACTICE

1. State whether the investigator used independent samples, repeated measures, or matched samples:
   a. An investigator wants to know whether men or women select larger entree portions in a cafeteria food line. He watches men and women as they go through a food line and rates the relative size of their food portions.
   b. An investigator wants to know whether tutoring helps students' writing skills. He rates students' essays before the students receive tutoring and again after they receive tutoring.

2. State whether the investigator used independent samples, repeated measures, or matched samples:
   a. An investigator wants to know whether children from different socioeconomic levels differ in the age at which they learn to skip. She gathers information about children's socioeconomic levels and the ages at which they were first able to skip.
   b. An investigator wants to know whether students with asthma participate in more or fewer college sports than students who do not have asthma. She gathers information about the presence or absence of asthma in the studied students, as well as the number of college sports in which those students' parents participated while young (to control for family pressure to play sports).

3. State whether the investigator used independent samples, repeated measures, or matched samples:
   a. An investigator wants to know if parental marital status is related to children's reading achievement. She draws a sample of children from single-parent homes and a sample of children from married-parent homes. She compares the tested reading achievement level of both groups.
   b. An investigator wants to know if parental marital status is related to children's reading achievement. She also thinks that the parental marital status may determine socioeconomic level, which in turn may affect children's reading level. Therefore, she draws a sample of children from single-parent homes and a sample of children from married-parent homes, but being careful to draw equal numbers of children from among high, middle, and low socioeconomic strata within the married- and single-parent groups. She compares the tested reading achievement level of both groups.

4. State whether the investigator used independent samples, repeated measures, or matched samples:
   a. An investigator wants to know if elementary-age children who have experienced the death of a parent are helped by a counseling group consisting of other children who have experienced a death in the family. He randomly selects children for this special counseling group, comparing their emotional adjustment at the beginning of treatment with their emotional adjustment following a year of the specialized group counseling.
   b. An investigator wants to know if elementary-age children who have experienced the death of a parent are better helped by a counseling group consisting only of other children who have experienced a death in the family or by a counseling group consisting of children demonstrating a wide range of behavioral and emotional issues. He randomly assigns children to the two groups and compares each group's emotional adjustment following a year of group counseling.

5. In Exercise 7 in Module 20 and in Exercise 7 in Module 21, a video arcade owner wondered if boys or girls spend more time playing the games once they enter the arcade. He recorded the playing time for boys and girls selected at random. No more than one person was selected from among a group of friends to minimize the effect of group behavior on individuals. Why can this experiment not be conducted as a repeated measures design?

6. In Exercise 6 in Module 20 and in Exercise 6 in Module 21, a cognitive psychologist was curious as to why women tend to physically turn a map's position to agree with the direction they are moving, while men tend to keep the map in its original position and mentally change orientations as they move. She wondered if there is a sex difference in ability to mentally rotate objects. She devised a computerized experiment in which she timed correct responses to a mental visual rotation task at a given angle. Why can this experiment not be conducted as a repeated measures design?

Comparison of Special-Case and Related-Samples Formulas

Because the subjects in a related-samples design are either the same in both treatment conditions or matched to a similar subject in the other treatment condition, their responses will be more similar to each other than they would have been had the subjects in each treatment condition been independent of one another. This violates the assumption of independence in the special-case formula. Therefore, the special-case formula does not apply.

The appropriate formula to use for analyzing data when subjects are related is another *generalized formula*. There are two types of generalized formulas: one for unequal sample sizes and one for related samples. In Module 21, I discussed the formula for unequal sample sizes. In this module, I will present the formula for related samples. Here are both the special-case formula and the formula for related samples:

> If I am given a formula, and I am ignorant of its meaning, it cannot teach me anything, but if I already know it, what does the formula teach me?
>
> —St. Augustine

Special-case formula

$$t_{2\text{-samp}} = \frac{(M_1 - M_2) - (\mu_1 - \mu_2)}{\sigma_{M_1 - M_2}}$$

where

$$\sigma_{M_1 - M_2} = \sqrt{\sigma^2_{M_1} + \sigma^2_{M_2}}$$

Formula for related samples

$$t_{2\text{-samp}} = \frac{(M_1 - M_2) - (\mu_1 - \mu_2)}{\sigma_{M_1 - M_2}}$$

where

$$\sigma_{M_1 - M_2} = \sqrt{\sigma^2_{M_1} + \sigma^2_{M_2} - 2r\sigma_1\sigma_2}$$

Take a moment to compare the special-case and the related-samples formulas. How are they the same? How are they different?

Yes, the numerators are the same, but the denominators are different. However, despite the different appearance of the denominators, the two formulas are mathematically equivalent when samples are independent. To see why this is so, let me break down the denominator of the related-samples formula so that it is clear what its terms do.

The first two terms in the denominator are

$$\sigma^2_{M_1} + \sigma^2_{M_2}$$

Compare this with the denominator of the special-case formula. They are exactly the same, aren't they? Both say to add together the two squared standard errors of the means. However, the related-samples denominator has this additional term under the square root sign: $-2r\sigma_1\sigma_2$.

This is a *covariance* term. It measures the amount by which two sets of scores vary together. Let's look at it in some detail. Note that it has several parts:

$\sigma_1\sigma_2$: This says to multiply together the two population standard deviations, σ_1 and σ_2.

r: This is the correlation between the two sets of scores. If you have been reading the modules in roughly the order in which they have been presented in the textbook, you have not yet learned how to compute a correlation coefficient. For now, know that a correlation coefficient indicates the strength of the relationship between two sets of scores.

2: This says to double the result.

The part of this term I want you to focus on is the r. As already stated, r indicates the strength of the relationship between two sets of scores. In the example we have been discussing, r might measure the relationship in verbal IQ between participants in the sleep-deprived condition and participants in the well-rested condition. When samples are independent, the relationship between scores of participants in the two treatment conditions is zero. That is, there is no relationship. In the study of sleep deprivation and memory in Module 21, participants were independent of one another. Thus, there was no relationship between the verbal IQs of the participants in the sleep-deprived condition and the participants in the well-rested condition. Now recall that when we multiply anything by 0, the result is 0. Thus, when samples are independent, the entire r term becomes 0 and the denominator reduces to the special-case denominator:

> Proof is the idol before whom the pure Mathematician tortures himself.
>
> —Sir Arthur Eddington

$$2r\sigma_1\sigma_2 = (2)(\mathbf{0})(\sigma_1\sigma_2) = 0$$

Now see what this does to the denominator:

$$\begin{aligned}&\sqrt{\sigma^2_{M_1} + \sigma^2_{M_2} - 2r\sigma_1\sigma_2}\\ &= \sqrt{\sigma^2_{M_1} + \sigma^2_{M_2} - 0}\\ &= \sqrt{\sigma^2_{M_1} + \sigma^2_{M_2}}\end{aligned}$$

which is the denominator of the special-case formula!

So when samples are independent, the denominator of the related-samples t test is identical to the denominator of the independent-samples t test.

Now let's see what happens when the two samples are *not* independent. When two samples are related, the strength of the relationship between participants' scores in the two treatment conditions is not zero. On the contrary, some amount of relationship exists. Thus, the entire term is some nonzero number:

$$\sqrt{\sigma^2_{M_1} + \sigma^2_{M_2} - 2r\sigma_1\sigma_2} = \sqrt{\sigma^2_{M_1} + \sigma^2_{M_2} - \text{some nonzero number}}$$

Now, what does the formula say to do with that nonzero number? It says to subtract it from the rest of the denominator. Thus, the formula for the related-samples t test corrects for the correlation between participants in the two treatment conditions by removing it.

Finally, what happens to any number when you take something away from it—does that number become bigger or smaller? Of course, when you subtract something from a number, that number becomes smaller. Thus, when there is a relationship between participants, the formula for a related-samples t test decreases the size of the denominator.

Now recall that t is the numerator divided by the denominator. And we just saw that the denominator is made smaller in the related-samples formula. So will this adjustment make the final value of t bigger or smaller? (Try it yourself. Simply divide one number by any other number and then by any smaller number.)

Yes, when we make the denominator smaller, we increase the final value of t. Thus, by removing the effect of the correlation between the samples, the final value of t will be larger. Since the size of t is the indicator of statistical significance, this adjustment in the formula makes it more likely that we will find a statistically significant treatment effect between groups when samples are related.

CHECK YOURSELF!

Compare the special-case formula and related-samples formula for a two-sample t test. How do they differ? What is the purpose of the added term in the related-samples formula?

Advantage and Disadvantage of Related Samples

The major advantage of related-samples designs, as just discussed, is the increased likelihood of finding statistically significant treatment differences in the numerator. However, this increased likelihood of finding statistically significant treatment differences comes with a price. In the related-samples t test, the degrees of freedom for entering the t table is the number of *pairs* of subjects minus 1 rather than the total number of subjects minus 2. And we already know that smaller sample sizes (which yield smaller degrees of freedom) require a higher calculated t before the t is statistically significant. Thus, if your related subjects' design involves matching, you should match subjects only on variables expected to be substantially correlated with the dependent variable—for example, verbal IQ in the study of recall errors. Otherwise, degrees of freedom are lost unnecessarily, and the ability to detect differences between treatment groups is lost rather than gained.

CHECK YOURSELF!

For the study of recall errors under varying amounts of sleep, what might be another relevant matching variable? If you matched on this variable, what should be the effect on the t test statistic?

Computational Formula

Now that you understand the theoretical formula, we can calculate an example. However, if you have been working your way through the textbook modules in roughly the order they are presented, you do not yet know how to calculate a correlation coefficient (r). Moreover, using the theoretical formula for calculating a related-samples t test is quite tedious. There is an easier formula to use for hand calculation. It is called the *computational formula*. Some statisticians also call it the *direct-difference formula*. It looks like this:

$$t_{\text{matched samples}} = \frac{\overline{D} - \mu_D}{\sqrt{\dfrac{\sum D^2 - \dfrac{\left(\sum D\right)^2}{n}}{n(n-1)}}}$$

where D is the difference between paired scores.

In the computational formula, there is no correlation coefficient, and there are no standard errors of the means. We work directly from the raw scores and from the differences between those raw scores.

Note that this version of the statistic looks a lot more like a one-sample t test than a two-sample t test. Look at the numerator. It says to subtract a single population parameter (μ_D) from a single sample statistic (mean of D): $\overline{D} - \mu_D$.

Compare this with the numerator of a one-sample versus two-sample t test:

One Sample	*Two Sample*
$M - \mu$	$(M_1 - M_2) - (\mu_1 - \mu_2)$

Why does the related-samples numerator look much more like a one-sample t test than a two-sample t test? Weren't we conducting a two-sample t test? Well, yes and no. A study with related samples does have two treatment conditions—for example, sleep deprived or not sleep deprived. However, in a study with related samples, either the same subjects take both treatments or the subjects are matched between treatments on some relevant extraneous variable such as verbal intelligence or the amount of sleep subjects usually require for optimal functioning. Thus, in a related subjects design, there is only *one paired sample* across two treatment conditions, not two independent samples.

The computational formula indicates that, in addition to subjects' raw scores, we need to find

1. a difference score across the two treatment conditions for each subject pair (D);
2. the sum of these difference scores ($\sum D$);
3. the mean of these difference scores ($\overline{D}$);
4. the square of each difference score (D^2);
5. the mean difference score for the population (μ_D)—which, under the null hypothesis, will be 0; and
6. the sample size.

Each of these entries is simple to compute. In fact, the only difficult aspect of the computational formula is understanding how it leads to the same answer as the more complicated theoretical formula that we examined earlier. However, it is beyond the scope of this book to algebraically prove the equivalence of the two formulas.

Calculating a *t* Test With Related Samples

Research has shown that the color of food can dramatically affect people's judgment of how good that food tastes. Food that has an atypical color is usually rated as less tasty than food that has a typical color. Let's assume that we recruit 15 subjects to take part in a study. Participants are served award-winning mashed potatoes that are smooth, white, and lightly seasoned with salt, pepper, garlic, chives, and mild cheese. The same participants are also served mashed potatoes prepared identically except that the potatoes have been artificially colored blue. Participants are asked to rate the flavor of each serving, using a scale where 10 = *extremely delicious* and 1 = *repulsive tasting*. Because the same subjects are in both treatment conditions, the study is a repeated measures design. The data look like this:

Participant	*White Potatoes*	*Blue Potatoes*
1	8	5
2	9	4
3	9	8
4	6	8
5	10	7
6	8	7
7	7	4
8	8	6
9	5	6
10	10	7
11	10	8
12	7	6
13	8	8
14	10	6
15	6	3

Just by inspecting the data, it is obvious that participants found the white mashed potatoes tastier than the blue mashed potatoes. Although the computational formula does not require that we compute the two means, the two means are 8.067 and 6.200, respectively. That's a mean difference of 1.867 points. However, we will have to conduct a *t* test to determine whether or not this degree of difference is statistically significant. Here are the calculations:

			Difference	
Participant	*White Potatoes*	*Blue Potatoes*	*D*	D^2
1	8	5	3	9
2	9	4	5	25
3	9	8	1	1
4	6	8	–2	4
5	10	7	3	9
6	8	7	1	1
7	7	4	3	9
8	8	6	2	4
9	5	6	–1	1
10	10	7	3	9
11	10	8	2	4
12	7	6	1	1
13	8	8	0	0
14	10	6	4	16
15	6	3	3	9
			$\sum = 28$	$\sum = 102$

$$\overline{D} = \frac{\sum D}{n} = \frac{28}{15} = 1.8667$$

$$t_{\text{matched samples}} = \frac{\overline{D} - \mu_D}{\sqrt{\dfrac{\sum D^2 - \dfrac{\left(\sum D\right)^2}{n}}{n(n-1)}}}$$

$$= \frac{1.867 - 0}{\sqrt{\dfrac{102 - \dfrac{(28)^2}{15}}{15(14)}}}$$

$$= \frac{1.867}{\sqrt{\dfrac{102 - \dfrac{784}{15}}{210}}}$$

$$= \frac{1.867}{\sqrt{\dfrac{49.733}{210}}}$$

$$= \frac{1.867}{\sqrt{0.237}}$$

$$= \frac{1.867}{0.487}$$

$$= 3.834$$

For the things of this world cannot be made known without a knowledge of mathematics.

—Roger Bacon

We are finally done. The difference in participants' taste ratings between white mashed potatoes and blue mashed potatoes is 3.834 standardized error units. But what does that mean?

Interpreting a *t* Test With Related Samples

As usual, the question is this: Is the observed difference in taste ratings *a lot*, and therefore probably due to the color of the potatoes? Or is it *only a little*, and therefore probably due to mere sampling error? Let's see where our observed difference falls within the sampling distribution of the mean (remember, the computation algorithm was that of a one-sample *t* test) (Figure 22.1).

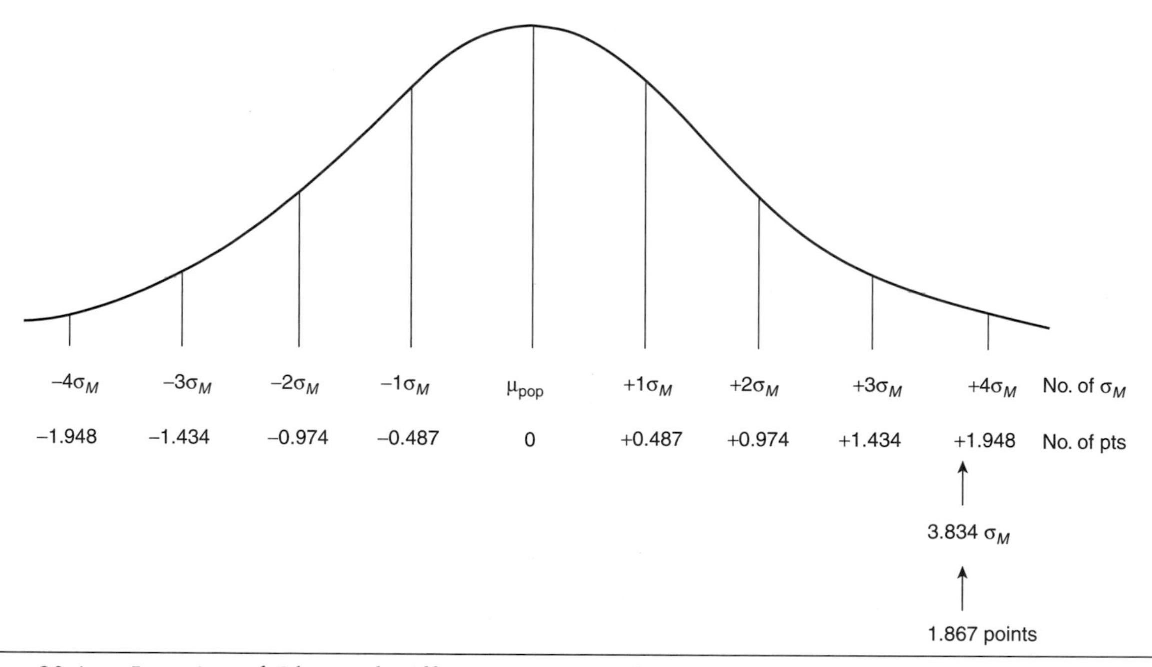

Figure 22.1 Location of Observed Difference Between the Sample Means

Yes, under the null hypothesis that result does seem very far from expected. But exactly how far is it? To determine that, we must look at the *t* table. Turn again to Appendix C. As before, we enter the *t* table at the correct degrees of freedom (*df*). Recall that the *df* for a *t* test is total *N*, minus 1 for each group, which is $N - 2$. But also recall that a related-samples *t* test has only *one paired sample*. Thus, the *df* for a related-samples *t* test is the same as for a one-sample *t* test: $N - 1$, where N = the total number of *pairs* of subjects. For this study, then, the correct *df* is $15 - 1 = 14$.

As before, the *t* table lists the minimum value that our observed *t* must be for us to reject the null hypothesis and conclude that the difference in mean ratings is probably due to the difference in treatment. Recall that our hypothesis was directional: We had proposed that participants would rate white mashed potatoes as significantly tastier than they would rate blue mashed potatoes. Therefore, we must look in the one-tailed column of the table.

Assume that we are willing to make a Type 1 error 5% of the time. At $\alpha = .05$, can we reject the null hypothesis?

Yes, we can, because the calculated *t* of 3.834 meets or exceeds the one-tailed critical *t* of 1.761.

Now assume that we are willing to make a Type 1 error only 1% of the time. At $\alpha = .01$, can we reject the null hypothesis?

Yes, we still can, because the calculated t of 3.834 meets or exceeds the one-tailed critical t of 2.624.

Let's keep going. Assume that we are willing to make a Type 1 error only one half of 1% of the time. At $\alpha = .005$, can we reject the null hypothesis?

Yes, we still can, because the calculated t of 3.834 meets or exceeds the one-tailed critical t of 2.977.

CHECK YOURSELF!

Assume that we are willing to make a Type 1 error only one half of one tenth of 1% of the time. At $\alpha = .0005$, can we reject the null hypothesis?

Clearly, the difference in participants' taste ratings of white mashed potatoes versus blue mashed potatoes was very large. In a journal article, the results would be reported like this:

$$t(14) = 3.834, p < .001.$$

This is read as "t at 14 degrees of freedom is 3.834. There is less than one chance in a thousand that the difference in taste ratings is due to mere chance." In other words, we can be very confident that the difference in taste ratings is due to the treatment variable, which was the color of the potatoes.

PRACTICE

7. The following data are from a study of aggression in 40 children (20 pairs) after viewing either a violent film or an educational film. Participants were first matched on gender and their typical aggression level. Here are participants' scores on an aggression test given after viewing the film. Higher scores indicate more aggression.

Pair	*Violent Film*	*Educational Film*
1	26	18
2	24	12
3	18	14
4	27	22
5	19	13
6	14	15
7	24	20
8	12	9
9	21	12
10	15	18
11	11	7
12	23	27
13	23	14
14	18	8
15	17	20
16	12	12
17	25	17
18	28	20
19	13	10
20	22	14

Calculate t and compare it with the one-tailed critical t at the .01 α level. Did the children who viewed the violent film show significantly more aggression?

8. Two professors teach different sections of the same course. One professor, Dr. Easy, has a reputation for being an easy grader. The other professor, Dr. Hard, has a reputation for being a hard grader. While it is true that the two professors' grades do tend to differ in the direction of their reputations, the division dean suspects that the reason for the difference may be that Dr. Easy's class typically includes more upper-class (UPC) students, while Dr. Hard's class typically includes more underclass (UNC) students, and underclass students are simply less well prepared for the course. To test her hypothesis, the division dean matches students between the two sections on class year. She leaves out students who don't have a match in the other section. Then, she compares students' final course grades. Here are the data:

Pair	*Dr. Easy*	*Dr. Hard*
1 (UPC)	87	85
2 (UNC)	75	70
3 (UNC)	82	86
4 (UPC)	94	84
5 (UPC)	77	83
6 (UNC)	76	75
7 (UNC)	68	71
8 (UPC)	90	90
9 (UPC)	76	84
10 (UNC)	69	67
11 (UNC)	86	80
12 (UPC)	93	89
13 (UNC)	94	90
14 (UPC)	83	89
15 (UPC)	90	94
16 (UNC)	86	85

a. State the division dean's research hypothesis. Note that this is one of those infrequent times when the research hypothesis will be supported by finding no difference rather than by finding a significant difference.

b. Calculate t and compare it with the one-tailed critical t at the .05 α level. Did Dr. Hard's students earn significantly lower grades than Dr. Easy's students?

9. Joe is taking a bowling course for his physical education elective in college. He wonders whether the course significantly improves students' bowling averages. He records his and his fellow students' bowling averages at the beginning of the course and again at the end of the course. Here are the scores:

Participants	*Bowling Average Before Course*	*Bowling Average After Course*
Amy	72	91
Brad	162	155
Carlos	145	152
Desi	183	190
Elena	123	134
Frank	167	175
Glenda	76	99
Hal	112	104

Participants	*Bowling Average Before Course*	*Bowling Average After Course*
Ingrid	124	134
Joe	137	156
Kessa	146	134
LaShawna	92	101
Missy	166	168
Nate	93	90

a. Calculate t and compare it with a one-tailed critical t at the .05 α level. Did the course result in significantly higher bowling averages?

b. The design of this study is not ideal because it is possible that students' bowling averages would have gone up over the semester even without instruction. *Think:* How could the design of this study be changed to eliminate this problem?

10. In Exercise 8 in Module 20 and in Exercise 8 in Module 21, an Internet provider wondered if its cable Internet customers were significantly more satisfied with one or the other of its two packages—free e-mail and free basic TV. The company surveyed customers currently receiving free e-mail with cable Internet and customers currently receiving free basic TV with cable Internet. Satisfaction scores ranged from 1 to 10, based on answers to the survey questions. Now let's suppose that the company did not survey separate customers receiving free e-mail versus free basic TV. Rather, it first provided its cable Internet customers with the free e-mail package for 6 months, then provided the same customers with free basic TV for 6 months. Because the same customers experienced both conditions, this is a repeated measures design. (For our purposes, we will ignore other problems with this design, such as possible order effect.) Here are the satisfaction scores for 12 customers for each condition. Conduct a two-tailed t test at the .05 α level. Are customers significantly more satisfied with one package or the other?

Customer No.	*Free E-mail*	*Free Basic TV*
1	6	7
2	7	10
3	5	8
4	7	5
5	8	9
6	6	10
7	10	7
8	9	6
9	6	5
10	7	6
11	5	8
12	7	6

11. In Exercise 9 in Module 20 and in Exercise 9 in Module 21, an employee benefits administrator wondered if the amount of the insurance co-pay affected the number of times employees visit a doctor per year. The company offered two plans. One plan had a higher visit co-pay in return for a larger cap on annual coverage. The other plan had a lower visit co-pay but with a lower cap on annual coverage. The employee benefits administrator compared doctor visit data for equal numbers of employees in the two plans. Now let's suppose that he did not merely collect doctor visit data for equal numbers of employees in the two plans. Rather, it first matched employees between the two plans on age, sex, preexisting health conditions, and number of doctor visits the employee had made in the previous year. Thus, this is a matched-samples design. Here are the

data for the 14 pairs of subjects. Conduct a two-tailed t test at the .05 α level. Does either plan result in significantly fewer doctor visits?

Pair No.	*Higher Co-Pays*	*Lower Co-Pays*
1	1	1
2	0	3
3	4	0
4	1	5
5	2	3
6	3	4
7	1	4
8	0	3
9	0	5
10	1	2
11	0	0
12	2	2
13	4	4
14	0	1

12. In Exercise 10 in Module 20 and in Exercise 10 in Module 21, a police agency was interested in reducing the speed on a particular stretch of highway. Lore had it that a visible police car with radar was more effective in reducing speed than a warning sign saying that the area is monitored by radar. A police sergeant wondered if the lore was correct, so he tested each method on the same stretch of highway for the same number of hours per day. Now let's suppose that, after catching speeders, the officers also recorded each speeder's prior speeding conviction record, sex, and number of years since obtaining a license. Subjects then were matched on these variables before including them in the study. Data from subjects without a match were discarded from the study (even though they received a speeding citation). Thus, this is a matched-samples design. Here are the excess miles per hour (mph) driven by the speeders from among the matched subjects. Conduct a one-tailed t test at the .05 α level. Does the police car result in lower speeds than the warning sign?

Driver Pair No.	*Excess mph With Police Car*	*Excess mph With Warning Sign*
1	12	18
2	7	12
3	6	9
4	10	15
5	4	7
6	13	19
7	11	18
8	6	12

SPSS Connection

Download the file **data_potato taste due to white blue.sav** from www.sagepub.com/steinberg2e. These data are used in the textbook example.

Alternatively, manually enter the 30 scores from the potato example in Module 21 into the SPSS **Data View** spreadsheet. Data entry for a related-samples t test differs in layout from that of an independent samples *t* test. The data are set up as in the textbook, as two groups of 15 clients. Thus, enter the data as follows:

8	5
9	4
9	8
6	8
10	7
8	7
7	4
8	6
5	6
10	7
10	8
7	6
8	8
10	6
6	3

Click on the **Variable View** tab to define the variable. Name the first variable **tastwhit,** set the decimals at 0, and label the variable as **Taste for White Potatoes.** Name the second variable **tastblue**, set the decimals at 0, and label the variable as **Taste for Blue Potatoes.**

If the file is not already in **Data View**, click that tab in the lower left of the screen.

In the toolbar at the top of the screen, click on **Analyze**, then **Compare Means**, then **Paired-Samples T Test**. Highlight the variable **Taste for White Potatoes** in the left window, and then click on the **arrow** before the **Paired Variables** window, to send the variable into that window. Click on the variable **Taste for Blue Potatoes** in the left window and then click on the **arrow** before the **Paired Variables** window to send the variable into that window. Click **OK**. This is what you will see.

T-Test

Paired Samples Statistics

		Mean	N	Std. Deviation	Std. Error Mean
Pair 1	Taste for White Potatoes	8.07	15	1.624	.419
	Taste for Blue Potatoes	6.20	15	1.612	.416

Paired Samples Correlations

		N	Correlation	Sig.
Pair 1	Taste for White Potatoes & Taste for Blue Potatoes	15	.322	.242

Paired Samples Test

		Paired Differences					t	df	Sig. (2-tailed)
					95% Confidence Interval of the Difference				
		Mean	Std. Deviation	Std. Error Mean	Lower	Upper			
Pair 1	Taste for White Potatoes Taste for Blue Potatoes	1.867	1.885	.487	.823	2.910	3.836	14	.002

Visit the study site at www.sagepub.com/steinberg2e for practice quizzes and other study resources.

23

Interpreting and Reporting Two-Sample *t*

Error, Confidence, and Parameter Estimates

Terms: confidence, parameter estimation, point estimate, interval estimate, confidence interval

Symbols: *p*, CI

Learning Objectives:

- Distinguish between tabled and incurred alpha
- Understand the relationship between error and confidence
- Estimate parameters—point and interval

What Is Confidence?

As we saw in Module 20 in the study of the effect of medication versus counseling for depression relief, our calculated t of -2.39 met the tabled value for statistical significance at $\alpha = .05$ but did not meet the tabled value for statistical significance at $\alpha = .01$. So if we reject the null hypothesis, how confident can we be that the difference in depression relief is due to the treatment and not due to mere sampling error? Said another way, if α is the probability that we wrongly reject the null hypothesis, then what is the probability that we do not wrongly reject the null hypothesis?

> Truth is arrived at by the painstaking process of eliminating the untrue.
>
> —Arthur Conan Doyle

Well, either we wrongly reject the null hypothesis or we don't. Thus, if α is the probability that we have made a Type 1 error, then the probability that we have *not* made a Type 1 error must be $1 - \alpha$. We call this probability that we have not made a Type 1 error our **confidence**. Said another way, confidence is the probability that the effect we found is real and not due to mere sampling error. Let's reexpress the tabled α levels as confidence levels:

At $\alpha = .10$, confidence $= 1 - \alpha = .90$.

At $\alpha = .05$, confidence $= 1 - \alpha = .95$.

At $\alpha = .025$, confidence $= 1 - \alpha = .975$.

At $\alpha = .01$, confidence $= 1 - \alpha = .99$.

CHECK YOURSELF!

At $\alpha = .005$, what is the confidence?

In our study of depression relief, we were able to reject the null hypothesis at $\alpha = .05$. Therefore, what is our confidence that we correctly rejected the null hypothesis and did not make a Type 1 error? Yes, it is $1 - \alpha$, or .95. That is, we can be 95% sure that the difference in performance between medication and counseling is due to the different treatments and not due to mere sampling error.

Refining Error and Confidence

The actual Type 1 error we incur in a study, and hence our actual confidence, is somewhat different from the values we just calculated. That's because a t table gives only a few α levels, not all possible α levels. Our calculated t met the tabled value under the .05 column, so we know we incurred .05 or less Type 1 error. We also know that our incurred error is not as low as .01 because we did not meet the tabled t value for .01. Thus, our incurred α is somewhere between .01 and .05.

The level of α that we actually incur in a study is reported as a probability, symbolized p. In real research, we typically use a statistical software program such as SPSS to find t. That same software provides actual incurred error, or p. Without the software, the best we can do is interpolate (estimate the location) of p between the tabled α columns. Table 23.1 shows a portion of the t table.

Table 23.1 A Portion of the t Table

	Level of Significance for One-Tailed Test (%)			
	5	*2.5*	*1*	*.5*
	Level of Significance for Two-Tailed Test (%)			
df	*10*	*5*	*2*	*1*
15	1.753	2.132	2.602	2.947
16	1.746	2.120	2.584	2.921
17	1.740	2.110	2.567	2.898

Recall that there were 18 people in the depression study. That's 16 df, so we look in the row labeled $df = 16$. At 16 df, here are the one-tailed critical t values for the three relevant Type 1 (α) error levels.

At $\alpha = .05$, critical $t = 1.746$.

At $\alpha = .025$, critical $t = 2.120$.

At $\alpha = .01$, critical $t = 2.584$.

Our calculated t was –2.39. (Recall that the table gives only positive values, but its entries apply to negatives values as well.) Thus, our calculated t met the tabled critical t at $\alpha = .05$ and at $\alpha = .025$, but it did not meet the tabled critical t at $\alpha = .01$. Therefore, the actual α for our study is somewhere between .025 and .01. Now, we interpolate the location of our calculated t between those two α columns.

- The tabled critical t for the .025 α column is 2.120. Our calculated t was 2.39. So the distance between the critical t and our calculated t is 2.120 – 2.39, or –0.27.
- The tabled critical t for the .01 α column is 2.584. Our calculated t was 2.39. So the distance between the critical t and our calculated t is 2.584 – 2.39, or 0.19.

Our calculated t falls slightly closer to the .01 α than to the .025 α. Thus, our incurred p is slightly closer to .01 than to .025. A p of about .015 is probably about right. That is, we are able to reject the null hypothesis with about a .015 Type 1 error level. This is the p that a software program such as SPSS would display.

CHECK YOURSELF!

What is the difference in meaning between α and p? Why would we use one rather than the other?

At the beginning of this module you learned that confidence = 1 – α. Now, we have refined our preselected error, α, to actual incurred error, p. So if confidence = 1 – error, and we refined error from the preselected α to incurred p, then what is our actual confidence in rejecting the null hypothesis in our study?

Yes, our actual confidence is $1 - p$, which is 1 – .015, or .985. We can be 98.5% confident that in rejecting the null hypothesis, we have not made a Type 1 error. Said another way, we can be 98.5% confident that the observed difference in depression relief between clients given medication and those given counseling is due to the treatment differences and not due to mere sampling error.

Here is the way the results would be reported in a journal article. The results are written in APA style:

$$t(16) = -2.39,\ p = .015 \text{ or } t(16) = -2.39,\ p < .025.$$

In words, this says that (1) the value of our calculated t at 16 *df* is –2.39 and (2) the probability that the difference between our two samples of clients was due to mere chance rather than to the difference in treatment is approximately 1.5% (or less than 2.5%).

PRACTICE

1. A researcher conducts a two-tailed two-sample t test on 28 subjects (14 subjects in each group) and finds that $t = 1.986$.
   a. How many degrees of freedom are there in this study?
   b. Estimate the actual Type 1 error for this study.
   c. If the researcher decides to reject the null hypothesis, how confident can the researcher be that the results are due to the treatment and not due to mere sampling error?
   d. Report the results as they would appear in an academic journal.

2. A researcher conducts a one-tailed two-sample *t* test on 32 subjects (16 subjects in each group) and finds that $t = 2.182$.
   a. How many degrees of freedom are there in this study?
   b. Estimate the actual Type 1 error for this study.
   c. If the researcher decides to reject the null hypothesis, how confident can the researcher be that the results are due to the treatment and not due to mere sampling error?
   d. Report the results as they would appear in an academic journal.

3. A researcher conducts a two-tailed independent samples *t* test on 30 subjects (13 subjects in one group and 17 subjects in the other group) and finds that $t = 2.25775$.
   a. How many degrees of freedom are in this study?
   b. About what is the actual Type 1 error for the study?
   c. How confident can the researcher be that the results are true and not due to mere sampling error?
   d. Report the results as they would appear in an academic journal.

4. A researcher conducts a one-tailed, matched sample *t* test on 15 pairs of subjects, and she finds that $t = 1.95305$.
   a. How many degrees of freedom are in this study?
   b. About what is the actual Type 1 error for the study?
   c. How confident can the researcher be that the results are true and not due to mere sampling error?
   d. Report the results as they would appear in an academic journal.

5. A researcher conducts a one-tailed, repeated measures *t* test on 15 subjects, and she finds that $t = 1.95305$.
   a. How many degrees of freedom are in this study?
   b. About what is the actual Type 1 error for the study?
   c. How confident can the researcher be that the results are true and not due to mere sampling error?
   d. Report the results as they would appear in an academic journal.

6. A researcher conducts a one-tailed independent samples *t* test on 20 subjects (10 subjects in each group) and finds that $t = 2.32665$
   a. How many degrees of freedom are in this study?
   b. About what is the actual Type 1 error for the study?
   c. How confident can the researcher be that the results are true and not due to mere sampling error?
   d. Report the results as they would appear in an academic journal.

7. You read a study that reports $p = .07$. What confidence can you have that the results are due to the treatment rather than to mere sampling error?

8. You read a study that reports $p = .04$. What confidence can you have that the results are due to the treatment rather than to mere sampling error?

9. You read a study that reports $p = .025$. What confidence can you have that the results are due to the treatment rather than to mere sampling error?

Decision Making With a Two-Sample *t* Test

To reject or not to reject, that is the question. Whether 'tis nobler in the eyes of man to take aim against . . . Wait, I already tried that line in Module 18, didn't I? So let's get back to interpreting our results. We now have two choices.

- We can say that the underlying populations of clients given the medication versus clients given counseling do not differ in depression relief, but we just happened to draw unusual samples in our study (Figure 23.1).

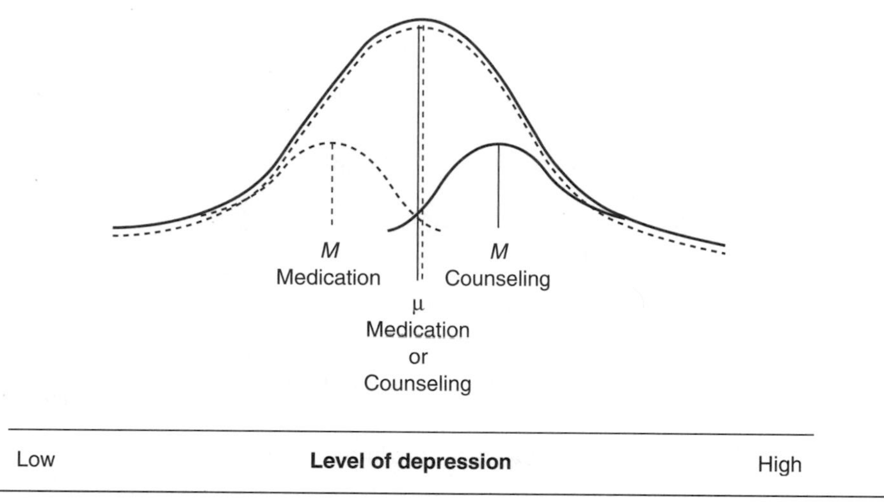

Figure 23.1 True Null Hypothesis, With Unusual Samples

In that case, we would retain (i.e., not reject) the null hypothesis that there is no significant difference between the populations due to type of treatment. The probability of this scenario is, as we just calculated, about .015.

- Alternatively, we can say that the observed difference in depression levels between the two samples of clients is too great for us to believe that the two underlying populations do not also differ in depression relief due to the treatment they receive (Figure 23.2).

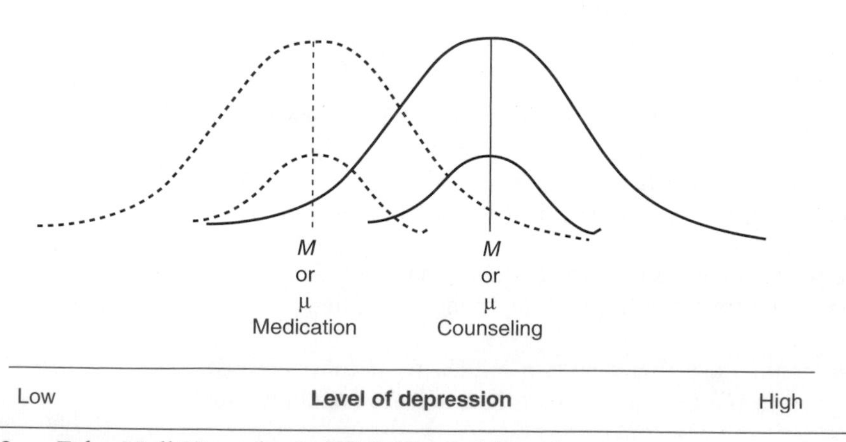

Figure 23.2 False Null Hypothesis, With Typical Samples

In that case, we would reject the null hypothesis that there is no significant difference between the underlying populations due to the type of treatment given. We would conclude, instead, that the reason we found the difference we did in the depression level of our differently treated samples is that the populations given those two treatments also differ in depression relief. In other words, the treatment effect is real. The probability of this scenario is, as we just calculated, about .985.

You may have noticed that the recommendation for whether or not we reject the null hypothesis, and hence whether or not the study's results are disseminated and put to use, depends wholly on the predetermined level of α. This is because, when we first learn how to calculate hypothesis tests such as *t*, either (a) our instructor tells us the value of α at which to conduct the test or (b) we preselect an appropriate value of α based on a subjective judgment of the risks involved, as discussed in Module 17. In the social sciences, that value is typically .05 for studies in which making a Type 1 error will not have serious consequences for people or programs. When the consequences will be more serious, .01 is adopted. After selecting an appropriate value of α, we then let the tables make the rejection decision for us. In the case of our study of depression relief, if we had conducted the study at $\alpha = .05$ or $\alpha = .025$, we would reject the null hypothesis. If we had conducted the study at $\alpha = .01$, we would not reject the null hypothesis.

Dichotomous Decisions Versus Reports of Actual *p*

If we were to share the results of our research with colleagues, our recommendation would differ depending on our predetermined level of α. At $\alpha = .05$, we would enthusiastically recommend the medication because depressed clients given the medication were significantly less depressed after treatment than clients given only counseling. However, at $\alpha = .01$, we would not recommend the medication because while depressed clients given medication were less depressed than clients given counseling, they were not significantly less depressed. We would assume that the additional relief in depression for clients given medication is due to mere sampling error.

As suggested previously, something about that situation doesn't seem quite right. Why should the treatments be judged either effective or not effective, with no middle ground? And why should a predetermined α level determine the either/or decision? Are there not degrees of support for the null and alternative hypotheses? Yes, there are. As we saw in Module 18, Rosnow and Rosenthal (1989) expressed the issue well when they wryly observed, "Surely, God loves the .06 as much as the .05!" (p. 1277). To reiterate, the authors are commenting on the inappropriateness of making all-or-nothing, reject or don't reject decisions.

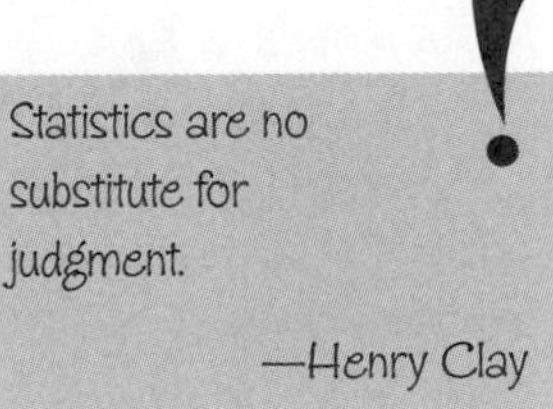

As discussed earlier, decisions on how much error is permissible are made on the basis of subjective variables such as danger to affected people if programs are changed or not changed as a result of the study, costs of changing or not changing programs as a result of the study, and so on. People—usually program managers—make those decisions. Statistics give us probabilities, but they can't make decisions.

Furthermore, who are we as researchers that we should decide on behalf of other users of our data what level of statistical significance is appropriate for their programs? Let's return to the study of depression relief. If we select a stringent α level for our study, say $\alpha = .01$, we would not disseminate our results because although those who received medication experienced more relief than those who received counseling, the relief was not *significantly* more. But perhaps physicians, pharmaceutical companies, or clients themselves would have been willing to settle for the modest depression relief detected by a 5% error rate rather than the more stringent 1% error rate.

The point is, predetermining what α level is necessary for the results to be useful is a bit presumptuous. Instead, we might better report the actual incurred error level—the p mentioned earlier in this module—and let the readers of the research decide for themselves whether or not the hypothesis has been supported at whatever error level they deem appropriate. This, in fact, is what most academic journals report. In the case of our study of the depression study, the results would be reported as statistically significant at the .015 error level.

Of course, none of this "leaving it up to the reader" is to say that we don't have any prior beliefs about what error levels are desirable or acceptable. The traditional .05 and .01 values, while not "magical," have been around since before computers and are, hence, entrenched in the tables. In point of fact, most academic journals print only studies with less than .05 Type 1 error. Other journals print only studies with less than .01 incurred error. Those values are, however, only rules of thumb. Certainly, we would all agree that an incurred error of .24 or .37 is too high. Certainly, we would all agree that an incurred error of .0008 or .0026 is very low and, hence, desirable. In general, less is better.

CHECK YOURSELF!

What are some of the considerations in deciding on the maximum acceptable α level for a study?

CHECK YOURSELF!

Why is it preferable to report actual incurred p rather than only the α level at which a study was conducted?

Parameter Estimation: Point and Interval

Under the null hypothesis, the expected difference in mean depression level between depressed clients given medication and those given counseling is 0. But we found a difference of 9.333 points on a test of depression, which translated into a difference of 2.39 standard error units. From that, we concluded that there is only a 1.5% probability that this difference was found by mere chance and a 98.5% probability that there is a real difference in depression levels between client groups given one treatment over the other.

> A habit of basing convictions upon evidence, and of giving to them only that degree or certainty which the evidence warrants, would, if it became general, cure most of the ills from which the world suffers.
>
> —Bertrand Russell

Although we are now 98.5% confident that the difference we found between our samples indicates the presence of a real treatment difference in the underlying populations, we do *not* know that the amount of treatment difference between the underlying populations is exactly 9.333 points. Although we found 9.333 points between our samples, the real difference in depression level between the populations due to treatment could be either more or less than 9.333 points.

So what *is* the real difference in depression level between depressed clients given medication and those given counseling? We cannot know for sure. We can only estimate the range of population differences from which our observed sample difference probably came.

Note how the problem has changed. We are no longer asking about the probability that the two samples came from differently treated underlying populations (medication vs. counseling). Rather, we are asking how different those two underlying populations (medication vs. counseling) really are. This approach, as we saw in Module 18, is called **parameter estimation**, because we are estimating the difference between population means, which are parameters.

For example, suppose your college admissions officer wants to know whether men and women in the incoming freshman class have equivalent high school GPAs (grade point averages). And suppose registration deposits have been received by only 50 men and 50 women so far. And suppose the admissions officer finds that the average high school GPA of the 50 incoming women is 0.15 higher on a 4.0 GPA scale than the average high school GPA of the 50 incoming men. Can the admissions officer then say that the GPA of women in next year's entering class will be higher than the men's GPA by 0.15? No, the admissions officer cannot because he or she does not know the high school GPA of all incoming men and women. He or she knows the GPAs only for 50 of the incoming men and 50 of the incoming women. However, if we assume that these 50 men and 50 women are typical samples from the populations of all incoming men and all incoming women, then it makes sense to assume that the difference in high school GPA between all incoming men and women will also be about 0.15, with the women scoring higher. Thus, the admissions officer could use the difference between the two sample means as a **point estimate** of the true difference between the two population means.

Alternatively, the admissions officer could use the difference in mean GPAs between the samples of incoming men and women to estimate a range within which the difference in GPA of all incoming men and women probably falls. This range of population parameters is called an **interval estimate**. That is, rather than predicting an exact value for the true population difference, it predicts a range in which the true population difference probably falls.

How wide should that interval be? Well, on a 4.0 GPA scale, the admissions officer can be nearly 100% certain that the average GPA of all incoming women will be somewhere between 2.85 points below and 3.15 points above the men's GPA. That's a range of three GPA points in either direction. However, such a range is not very informative. It is so wide that it tells the admissions officer virtually nothing about the true difference in high school GPA between all incoming men and women. Well then, how about a range of 0.145 to 0.155? Here, the admissions officer is saying that women will outscore men by an amount falling between those two numbers. That's a slim range—only 5/1,000 of a point in either direction. But now, the range is so small that the admissions officer cannot be very certain that it includes the true difference in high school GPA between all incoming men and women.

Fortunately, statisticians have provided guidelines regarding how wide the range should be. Typically, the limits of the range are set so that we have a specified level of confidence that the range includes the population parameter. Recall that such a range is called a **confidence interval**. I again remind you not to confuse confidence in rejecting the null hypothesis, which we talked about in earlier modules, with a confidence interval. To repeat, a confidence interval gives a range of scores within which the parameter probably falls. The interval is said to be a 95% confidence interval if we can be 95% certain that the population parameter falls within the range. The interval is said to be a 99% confidence interval if we can be 99% certain that the population parameter falls within the range.

Statistics means never having to say you're certain.

The following sum up what I have said about parameter estimation:

- A point estimate predicts a specific value for the population parameter.
- An interval estimate predicts a range of values that probably includes the population parameter.
- A confidence interval is an interval estimate having a specified probability of including the population parameter, usually 95% or 99%.

CHECK YOURSELF!

Contrast hypothesis testing and parameter estimation. How do the questions that they answer differ?

Let's return to our study of the difference in depression level between clients given medication and clients given counseling. The formula for a confidence interval is as follows:

$$CI = (M_1 - M_2) \pm t_{\text{crit at .5 } \alpha} (\sigma_{M_1 - M_2})$$

where

$M_1 - M_2$ = difference between sample means,

$t_{\text{crit at .5 } \alpha}$ = tabled critical t at one-half the tabled α, and

$\sigma_{M_1 - M_2}$ = standard error of the difference between the means.

Here is the 95% confidence interval for the true difference in depression level between depressed clients given medication and those given counseling:

$$\begin{aligned} CI &= (28.778 - 38.111) \pm (2.120)(3.91) \\ &= -9.333 \pm (2.120)(3.91) \\ &= -9.333 \pm 8.289 \\ &= -9.333 - 8.289 \text{ and } -9.333 - 8.289 \\ &= -1.044 \text{ and } -17.622 \end{aligned}$$

The first group in the formula (M_1) was the medication group. Thus, we can be 95% confident that depressed clients given medication will score between −1.044 points and −17.622 points lower on a depression scale than depressed clients given counseling. Note that our observed difference of −9.333 points falls in the middle of that range.

Here is the 99% confidence interval for the true difference in depression level between depressed clients given medication and those given counseling:

$$\begin{aligned} CI &= (28.778 - 38.111) \pm (2.921)(3.91) \\ &= -9.333 \pm (2.921)(3.91) \\ &= -9.333 \pm 11.421 \\ &= -9.333 + 11.421 \text{ and } -9.333 - 11.421 \\ &= +2.088 \text{ and } -20.754 \end{aligned}$$

Again, the first group in the formula (M_1) was the medication group. Thus, we can be 99% confident that depressed clients given medication will score between +2.088 points higher and −20.754 points lower than those given counseling. Note that our observed difference of −9.333 points falls in the middle of that range.

PRACTICE

10. A college is expecting an entering freshman class of 700. The sex ratio of the incoming class is exactly 350 women and 350 men. The mean score on a test of writing skills for 70 incoming female freshmen attending one of several college orientation sessions is 84. The mean score for 70 incoming male freshmen on the same writing test at the same orientation sessions is 78. The $\sigma_{M_1 - M_2}$ for the writing test is 2.15 points.

a. What point estimate should the admissions officer make for the difference in writing score between women and men in the entire incoming freshman class?

b. In constructing an interval estimate for the difference in writing score between men and women in the entire incoming freshman class, what difference score will fall in the center of the interval?

c. The admissions officer can be 95% confident that the difference in writing score for men and women in the entire incoming freshman class falls between which two values?

d. If the admissions officer were willing to be only 90% confident, would the confidence interval be larger or smaller? Why?

e. If only 50 of the 70 women and 50 of the 70 men who attended the orientation session took the writing skills test, would the confidence interval be larger or smaller? Why?

11. A social psychologist measures the average time that 20 new transfer students at an orientation party mingle with other new transfer students before selecting a chair or corner in which to settle. Ten of the new transfer students are from nearby colleges, and 10 are from distant colleges. The average mingle time for students transferring from nearby colleges is 6.2 min. The average mingle time for students transferring from distant colleges is 5.4 min. The $\sigma_{M_1-M_2}$ for mingle time is .47 min.

    a. What point estimate should the psychologist predict for the difference in mingle time between transfer students from nearby and distant colleges at another transfer orientation party?

    b. In constructing an interval estimate for the difference in mingle time between the nearby and distant transferees, what time difference will fall in the center of the interval?

    c. The psychologist can be 90% confident that the difference in mingle time for the additional nearby and distant transferees will fall between which two values?

    d. If the psychologist wants to be 99% confident, would the confidence interval be larger or smaller? Why?

    e. If the psychologist had measured the mingle time of 28 new transfer students rather than 20, would the confidence interval be larger or smaller? Why?

12. Exercise 6 in Module 20 addressed the speed at which men versus women mentally rotated an object. Fifteen men and 15 women mentally rotated the object. The men mentally rotated it in 0.0313 s less time than the women. The value of $\sigma_{M_1-M_2}$ was 0.01018.

    a. What is the best point estimate for the difference in mental rotation speed for the next pair of men and women?

    b. In constructing a confidence interval for the expected difference in mental rotation speed between men and women, what difference score will fall in the center of the interval?

    c. We can be 99% confident that the true difference in mental rotation speed between men and women is between which two values?

13. Exercise 7 in Module 20 addressed the time that boys and girls spent in a video arcade. Ten boys and 10 girls were observed. The boys spent 16 min longer than the girls. The value of the $\sigma_{M_1-M_2}$ was 16.29444.

    a. What is the best point estimate for the difference in times spent between the next pair of boys and girls?

    b. In constructing a confidence interval for the expected difference in time spent in video arcades between boys and girls, what difference score will fall in the center of the interval?

    c. We can be 99% confident that the true difference in time spent by boys over girls in video arcades is between which two values?

14. Exercise 8 in Module 20 addressed difference in satisfaction between two different Internet packages. Twelve customers receiving free e-mail and 12 customers receiving free basic TV were surveyed. Those with the free e-mail service were −0.3333 points less satisfied. The value of $\sigma_{M_1-M_2}$ was 0.66050.
    a. What is the best point estimate for the difference in satisfaction for the next customers surveyed from each package?
    b. In constructing a confidence interval for the expected difference in satisfaction, what difference score will fall in the center of the interval?
    c. We can be 95% confident that the true difference in satisfaction between the two packages is between which two values?

15. Exercise 9 in Module 20 addressed difference in number of annual doctor visits for employees having two different insurance co-pay plans. Annual visit numbers for 14 employees with a higher co-pay plan were compared with 14 employees with a lower co-pay plan. Those with the higher co-pay had −1.2858 fewer annual doctor visits. The value of the $\sigma_{M_1-M_2}$ was 0.59498.
    a. What is the best point estimate for the difference in annual doctor visits for the next customers surveyed from each co-pay plan?
    b. In constructing a confidence interval for the expected difference in number of annual doctor visits, what difference score will fall in the center of the interval?
    c. We can be 95% confident that the true difference in number of doctor visits between the two plans is between which two values?

16. Exercise 10 in Module 20 addressed the difference over the speed limit drivers went when speed was monitored by a visible police car versus a warning sign. A visible police car led to −5.125 fewer excess miles per hour than the warning sign for eight drivers in each condition. The value of the $\sigma_{M_1-M_2}$ was 1.96112.
    a. What is the best point estimate for the difference in excess speed for the next two speeders in each condition?
    b. In constructing a confidence interval for the expected difference in speed, what difference score will fall in the center of the interval?
    c. We can be 95% confident that the true difference in excess speed will be between which two values?

17. Exercise 3 in Module 20 addressed the rated effectiveness of two of the Shine Company's cleaning products: one with a fragrance added and one without a fragrance added. Twenty-four persons rated the scented cleaner as 1.72 points more effective than the unscented cleaner. The value of the $\sigma_{M_1-M_2}$ was 0.50.
    a. What is the best point estimate for the difference in effectiveness ratings of the next group of 24 participants who are asked to rate the effectiveness of the scented versus unscented cleaners?
    b. In constructing an interval estimate for the difference in effectiveness ratings for the scented versus unscented cleaners, what difference score will fall in the center of the interval?
    c. The Shine Company can be 95% confident that the true difference in effectiveness ratings between the scented and unscented cleaners falls between which two values?

18. Exercise 4 in Module 20 addressed the number of large versus small posters displayed by homeowners supporting the candidate Elena Martin. Posters of each type were sent to

homeowners supporting her in each of 10 districts. On average, supporters displayed 8.2 more small posters than large posters. The value of the $\sigma_{M_1 - M_2}$ was 2.62.

a. What is the best point estimate for the difference in number of large versus small posters displayed by the next group of Elena Martin's supporters in the 10 districts?

b. In constructing an interval estimate for the difference in number of large versus small posters displayed by the next group of Elena Martin's supporters, what score will fall in the center of the interval?

c. Elena Martin can be 99% confident that the true difference in the number of posters displayed by any 100 supportive homeowners sent each type of poster falls between which two values?

19. On a reaction time task, 11 randomly selected college students take an average of 0.60 s to respond to a prompt. On the same reaction time task, 11 randomly selected middle-aged adults take an average of 0.79 s to respond to the prompt. The value of $\sigma_{M_1 - M_2}$ is 0.048 s. What is the 95% confidence interval for the difference in reaction time for additional groups of college students versus middle-aged adults?

SPSS Connection

Look again at the SPSS output that you produced in Module 20. Here it is again:

T-Test

Group Statistics

Type of Treatment		N	Mean	Std. Deviation	Std. Error Mean
Depression Score	Medication	9	28.78	10.450	3.483
	Counseling	9	38.11	5.302	1.767

Independent Samples Test

		Levene's Test for Equality of Variances		t-test for Equality of Means					95% Confidence Interval of the Difference	
		F	Sig.	t	df	Sig. (2-tailed)	Mean Difference	Std. Error Difference	Lower	Upper
Depression Score	Equal variances assumed	8.727	.009	-2.390	16	.030	-9.333	3.906	-17.614	-1.053
	Equal variances not assumed			-2.390	11.863	.034	-9.333	3.906	-17.855	-.812

The two rightmost cells in the bottom chart show the following values for a 95% confidence interval: −17.614 and −1.053. Now recall that the values reported for confidence intervals for a one-sample *t* test differed significantly from those that we had calculated in the textbook because SPSS stopped short of the final calculations (see SPSS section in Module 18). We had to use information from the SPSS output to complete the calculations. Fortunately, this is not the case for two-sample *t* tests. For two-sample *t* tests, SPSS completes the calculations. Thus, the reported values agree with those we computed in the textbook (within rounding error), so no further calculation is needed.

Visit the study site at www.sagepub.com/steinberg2e for practice quizzes and other study resources.

PART X

The Multisample Test

"Ours Is Better Than Yours or Theirs"

24 ANOVA Logic

Sums of Squares, Partitioning, and Mean Squares

Terms: ANOVA, partitioning, total mean, grand mean, within group, between group, sum of squares, mean square

Symbols: F, MS_{bet}, MS_{with}

Learning Objectives:

- Know the assumptions underlying analysis of variance
- Distinguish between conditions under which it is appropriate to conduct a *t* test versus ANOVA
- Understand the relationship of within, between, and total variances
- Understand how the possible values of *F* follow from the distribution's shape
- Understand how changes in the within- or between-group variances influence the *F* statistic

When Do We Use ANOVA?

Until now, we have examined treatment differences between no more than two groups: a treatment group and a control group. For example, in our test of clients given medication versus counseling, the difference in depression levels between the two groups was 9.33 points in favor of medication. But what if we added a third treatment group—say, diet supplements? A *t* test can test the difference between the means of only two groups. With three or more groups, no single number represents the mean difference across all groups. We need to find another way to measure group treatment differences.

One thing we could do is test all possible pairs of means: medication versus counseling, counseling versus diet supplements, and medication versus diet supplements. That's 3 separate *t* tests. But if we did that, would the same 5% Type 1 error level that had been established for a single test hold truc across 3 separate tests? No, it would not. This is because the error probabilities associated with a *t* table assume a single *t* test between only two groups. If we did 3 separate *t* tests on the data, the error rate would be considerably higher than 5%. Think of it this way. With 20 *t* tests, the probability that at least one of the pairs will show a statistically significant difference by chance alone is nearly certain: After all, 5% is 1 test out of 20 tests. It follows that with 3 *t* tests (as in our example), the probability that any one of those tests will show a statistically significant difference by chance alone is less than it is with 20 t tests yet greater than it is with only 1 t test. Thus, a good reason for not using t tests in studies with more than two groups is that we quickly lose control of the Type 1 error rate.

Such calculations would also be extremely tedious. For example, suppose we tested not just three but five different treatment methods. That would involve doing separate t tests for the means of Group 5 against Group 4, Group 5 against Group 3, Group 5 against Group 2, Group 5 against Group 1, and also Group 4 against Group 3, Group 4 against Group 2, and so forth until finally reaching Group 2 against Group 1. Five groups would require 10 separate t tests. Not only would a Type 1 error be very likely, but who would want to compute all those t tests?

What we need is a procedure that can test the statistical significance of the difference between all the groups simultaneously while holding the Type 1 error level constant. As you may have guessed, there is such a test. It is called *analysis of variance*—or **ANOVA**, for short.

In ANOVA, we compute an F statistic instead of a t statistic, so we call it an F test instead of a t test. Also, the treatment effect in ANOVA is found in the differences between group *variances* rather than the differences between group *means*. This is because it is not possible for a single mean difference to represent treatment differences between more than two groups. Nevertheless, if the group means are similar, they will cluster close together. This makes the dispersion—the variance—between the means small. On the other hand, if the group means are very different from one another, they will be more dispersed. This makes the variance between the means large. Thus, ANOVA uses a between-groups variance measure to describe the mean differences between all groups.

CHECK YOURSELF!

You are studying the speed of runners after drinking different energy drinks. What will be the deciding factor in whether to conduct a t test or an F test on your data?

ANOVA Assumptions

Although ANOVA solves the multigroup problem, it is not always the appropriate procedure to use. Just as you learned in Module 5 that it is not appropriate to report a mean when a distribution is seriously skewed, so also it is not appropriate to report an F statistic unless certain assumptions about the data are met. Here are the assumptions for ANOVA:

- The populations from which the samples are drawn are normally distributed.
- The populations from which the samples are drawn have equal variances.

Of course, few populations are perfectly normal, and few populations have exactly equal variances. Fortunately, ANOVA is *robust* to moderate violations of its assumptions. This means that the results will hold even when the assumptions are not met. There are two conditions for accepting violations, however: (1) It is OK for populations to be nonnormal as long as the sample sizes are relatively large, and (2) it is OK for the variances to be unequal as long as the sample sizes are equal. This is why researchers planning to use an ANOVA technique try to have relatively large sample sizes and also equal sample sizes. Under these two conditions, moderate deviations from the assumptions will not affect the results.

CHECK YOURSELF!

A study has equal-sized small samples drawn from skewed distributions. Will ANOVA be appropriate? Why or why not?

CHECK YOURSELF!

A study has large samples drawn from normally distributed populations, but the sample sizes and variances differ. Will ANOVA be appropriate? Why or why not?

Partitioning of Deviation Scores

ANOVA analyzes variances. Because all members within any given group are given the same treatment, the variance within any group cannot be due to the treatment. Rather, this variance is due only to random error. Members in different treatment groups, on the other hand, receive different treatments. Therefore, the variance between different groups is due primarily to the different treatments, which is our independent variable. Some error variance creeps in between groups too. However, the bulk of between-group variance is due to the treatments.

> Q: Why couldn't the statistician lay floor tile?
>
> A: He didn't know how to partition some of the squares.
>
> —Gary Ramsayer

ANOVA analyzes these two variances. That is, it looks at the amount of variance between groups (treatment) relative to the amount of variance within groups (error). To do this, we must find the total amount of variance in the data and divide it into its between-group and within-group parts. This division of total variance into its within-group and between-group parts is called **partitioning**. Let me illustrate the process with an example.

Suppose we have three populations of depressed clients—those treated with medication, those treated with counseling, and those treated with diet supplements. If the three treatments bring about different amounts of depression relief, we could diagram the three treatment populations as shown in Figure 24.1.

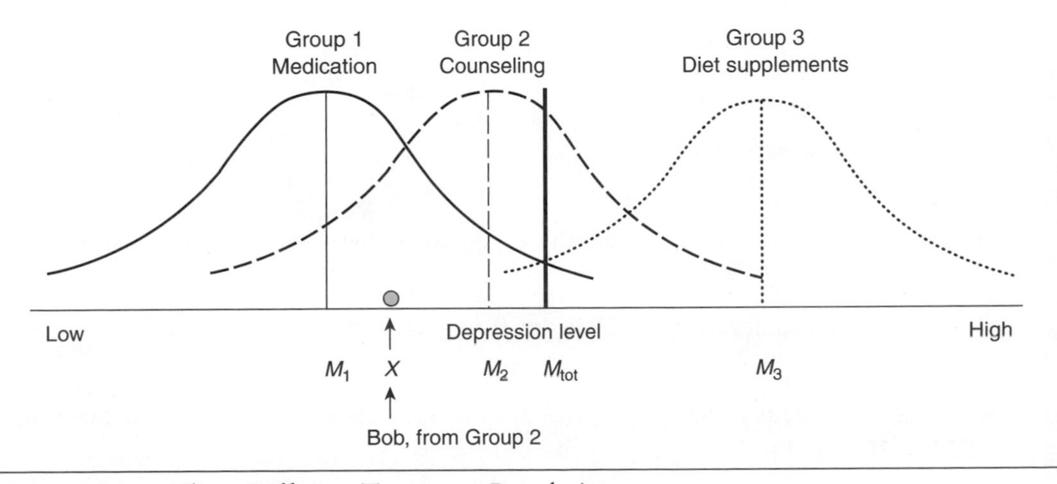

Figure 24.1 Three Different Treatment Populations

Each treated population has its own mean depression level. These are designated in Figure 24.1 by M_1, M_2, and M_3. We can also compute the mean of all three groups combined. This is designated by M_{tot}, which stands for the **total mean**, also sometimes called the **grand mean**.

Now suppose the depression level of a single client (we'll call him Bob) in Group 2 (counseling) falls at the point marked X on the graph. We can then say that Bob's score deviates from the total mean by the amount: $M_{tot} - X$. However, we can also break down Bob's total deviation score into two parts:

- First, the deviation of Bob's score from the mean of his own group: $M_2 - X$. The deviation of Bob's score from his own group's mean is called his **within-group** deviation because it is how far his score deviates from his own group's mean score.
- Second, the deviation of Bob's group's mean from the total mean: $M_{tot} - M_2$. The deviation of Bob's group's mean from the total mean is called Bob's **between-group** deviation because as one group differs more and more from the other groups, the overall grand mean (M_{tot}) is pulled in that same direction. Thus, the difference between any group mean and the total mean is an indirect measure of the difference between the group means.

CHECK YOURSELF!

Partitioning of deviations is at the heart of ANOVA. Therefore, understanding the partitioning process is important for your understanding of ANOVA. Stop now and locate the partitioned deviations in Figure 24.1. See if you can locate the between, within, and total deviations.

Now, if we can partition a single participant's score into within-group and between-group parts, we can do the same for the scores of every participant and then add those deviations. By doing that, we will have three summed deviations: (1) a summed within-group deviation for all participants, (2) a summed between-group deviation for all participants, and (3) a summed total deviation for all participants.

PRACTICE

1. Sketch overlapping normal curves for a three-group ANOVA (as in Figure 24.1). Locate and mark a score X at any point in Group 1. Then, using symbols, indicate that score's
   a. total deviation,
   b. within-group deviation, and
   c. between-group deviation.

2. If the total deviation is 683.78 and the within-group deviation is 457.52, what is the between-group deviation?

3. If the within-group deviation is 348.76 and the between-group deviation is 872.45, what is the total deviation?

4. If the between-group deviation is 245.22 and the within-group deviation is 106.33, what is the total deviation?

5. If the total deviation is 492.56 and the between-group deviation is 306.55, what is the within-group deviation?

From Deviation Scores to Variances

ANOVA does not partition deviations. Rather, it partitions variances. Recall that ANOVA stands for analysis of *variance*, not analysis of *deviation*. Thus, while part of the process is now in place, we're not quite there yet. We need to move from deviation scores to variances.

How do we move from deviation scores to variances? Recall from Module 6 that when we computed a variance, we (1) calculated the deviation of every score from the mean (i.e., we found the deviation scores), (2) squared each of those deviation scores, and (3) added all those squared deviation scores. That's the same process we are now following, except that we now refer to the sum of squared deviation scores as a **sum of squares**, symbolized *SS*. Finally, when we calculated a variance in Module 6, we (4) found the average of the squared deviation scores (the *SS*) by dividing by the number of scores. Thus, the variance we found in Module 6 was the *average of the squared deviation scores*.

Well, that's exactly what we do in ANOVA! Note that we have already done the first three steps. In fact, we have done it twice: First, we took the deviation of each participant's score from that participant's group mean (within-group deviation); second, we took the deviation of the mean of each participant's group from the overall total mean (between-group deviation). Then, for both the within-group and the between-group deviation scores, we

- squared those deviation scores for all participants and
- added the squared deviation scores for all participants to get the sum of the squared deviation scores, *SS*.

To get a variance, all we have to do is the fourth step:

- find the average of the squared deviation scores—that is, the average *SS*.

The only difference between calculating a variance via the method you learned in Module 6 and calculating a variance in ANOVA is that in ANOVA, instead of dividing by the number of subjects to find the average, we divide by the degrees of freedom. We do this because, as you learned in Module 15 and following, in inferential statistics we don't know the score of every person in the population. And whenever we don't know every score in a population, any parameter we compute for that population is merely an estimate of the population parameter rather than the population's true parameter. As you also learned in those modules, estimation formulas use degrees of freedom—a number smaller than the sample size—rather than sample size when computing an average. This corrects for the tendency of the sample statistic to underestimate the population parameter.

CHECK YOURSELF!

The variance used in ANOVA is an inferential statistic. How does its calculation differ from the variance computed as a descriptive statistic? Why is the change in the inferential formula necessary?

From Variances to Mean Squares

Once computed, these within-group and between-group variances are not called variances. Instead, they are called mean squares. Why aren't they called variances? I don't know. Fortunately for us, however, the definition of a mean square is clear regardless of the name change. Think of this: A *mean* is an average; *square* is shorthand for squared deviation score. Thus, a **mean square** is an average of the squared deviation scores. Compare the definition of a variance and a mean square below. As you can see, a mean square is simply a variance:

> *Who Moved My Cheese?*
>
> —title of book by Spencer Johnson, addressing unexpected change

- A variance is the average of the squared deviation scores.
- A mean square is the average of the squared deviation scores.

So we now have two variances—that is, two mean squares. The mean square between groups is symbolized as MS_{bet}. MS_{bet} consists of variance between (or among) the different groups. This variance is due primarily to the different treatments, which is our independent variable. However, it also includes some random error. In other words, MS_{bet} = Treatment + Error.

The mean square within groups is symbolized as MS_{with}. MS_{with} consists of variance within each group. Because all members within each group are given the same treatment, this variance cannot be due to the treatment. Rather, this variance is due only to random error. In other words, MS_{with} = Error.

From Mean Squares to F

ANOVA determines the amount of between-group variance relative to the amount of within-group variance. The test statistic is symbolized as F. Hence, ANOVA is sometimes called an F test. The formula for an F test is

$$F = \frac{MS_{bet}}{MS_{with}}$$

Recall that MS_{bet} is a measure of treatment effect along with some random error, while MS_{with} is a measure of only random error. In terms of treatment and error variance, then, the formula can be expressed as

$$F = \frac{\text{Treatment} + \text{Error}}{\text{Error}}$$

Note that the numerator contains both treatment effect and error, while the denominator contains only error. We can conclude thus:

- When there is *no* treatment effect, F is reduced to "error divided by error," which is +1.00. Thus, we expect a value close to +1.00 when there is no treatment effect.
- When there *is* a treatment effect, the numerator is greater than the denominator, so F is some value greater than +1.00.

Like the t distribution, the F distribution is not a single distribution but a family of distributions, one for each sample size. Unlike the t distribution, the F distribution is not symmetric. There are several reasons for this:

1. First, mean squares are squared values, and any value times itself is positive; therefore, F can never be negative.
2. Second, only a small proportion of F values fall below +1.00. Recall from the discussion above that the formula is Treatment + Error in the numerator, divided by Error in

the denominator. Thus, an F below +1.00 can occur only when there is no treatment effect and, by chance, the random error between groups in the numerator is less than the random error within groups in the denominator.

3. Third, cases begin to bunch up around +1.00. This happens when there is no treatment effect and the error variance is equal between the groups and within the groups, as expected.
4. Fourth, as the treatment effect gets stronger, the numerator gets larger, and hence, the value of F increases without limit.

With these four constraints, the F distribution is as shown in Figure 24.2.

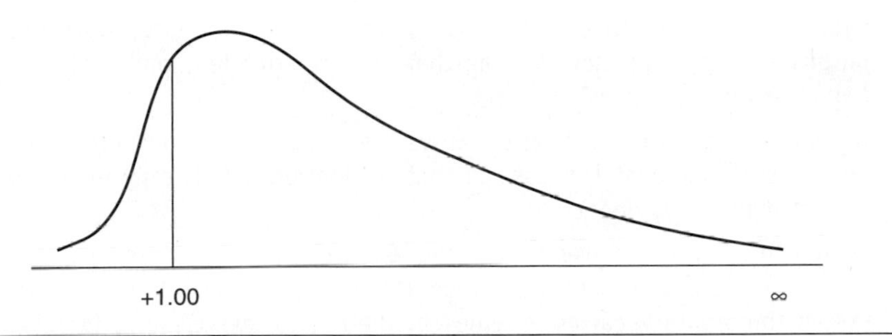

Figure 24.2 The *F* Distribution

Looking Ahead

In the next module, we will calculate an F test. Obviously, to get statistical significance, we will be looking for an F greater than +1.00. How much greater than +1.00 the F must be to be statistically significant is yet to be determined.

PRACTICE

6. You calculate F and get a value of +0.89. Is that value possible? Will it be statistically significant?
7. You calculate F and get a value of –0.23. Comment on this.
8. If $MS_{bet} = 100$ and $MS_{with} = 25$, what is F?
9. If $MS_{bet} = 800$ and $MS_{with} = 380$, what is F?
10. If $MS_{bet} = 300$ and F is 3.00, what is MS_{with}?
11. If $MS_{bet} = 500$ and F is 2.00, what is MS_{with}?
12. If $MS_{with} = 400$ and F is 2.00, what is MS_{bet}?
13. If $MS_{with} = 200$ and F is 3.00, what is MS_{bet}?

14. Use the values given in Exercise 8 for the following:
    a. What is *F* if the treatment effect is doubled but random error stays the same? For purposes of this question, assume that random error falls only in the denominator (which it primarily does).
    b. What is *F* if the treatment effect stays the same but random error is doubled? For purposes of this question, assume that random error falls only in the denominator (which it primarily does).
    c. Sketch three normal curves to represent Exercise 8. Below those curves, sketch three normal curves to represent the change described in Exercise 14a. Below those curves, sketch three normal curves to represent the change described in Exercise 14b.
15. Use the values given in Exercise 9 for the following:
    a. What is *F* if the treatment effect is doubled but random error stays the same? For purposes of this question, assume that random error falls only in the denominator (which it primarily does).
    b. What is *F* if the treatment effect stays the same but random error is doubled? For purposes of this question, assume that random error falls only in the denominator (which it primarily does).
    c. Sketch three normal curves to represent Exercise 9. Below those curves, sketch three normal curves to represent the change described in Exercise 15a. Below those curves, sketch three normal curves to represent the change described in Exercise 15b.
16. What will happen to the value of *F* if the treatment effect stays the same but the random error increases?
17. What will happen to the value of *F* if the treatment effect increases but the random error stays the same?

Visit the study site at www.sagepub.com/steinberg2e for practice quizzes and other study resources.

25

One-Way ANOVA

Independent Samples and Equal Sample Sizes

Terms: one-way ANOVA, ANOVA summary table

Learning Objectives:

- Understand the similar logic underlying various test statistics
- Determine the degrees of freedom
- Calculate a one-way ANOVA
- Use a table to interpret calculated *F*
- Display the results in an ANOVA summary table
- Report results in APA format

What Is a One-Way ANOVA?

Now that you understand the logic behind analysis of variance, we are ready to calculate an ANOVA. In this module, we will calculate a **one-way ANOVA**. One way means that the study has only one independent variable. You will learn about ANOVAs with more than one independent variable in Module 29. In addition, the steps and formulas in this module are for studies in which the sample sizes are equal and the samples are independent. As with *t* tests, samples of unequal size, as well as samples that are dependent (matched samples or repeated measures), require formula and procedural adjustments. However, unlike *t* tests, the formula adjustments for dependent ANOVA are beyond the scope of this textbook and will not be covered here.

Assume that we have the following data for the effect of three different treatments on depression (see Table 25.1). The dependent variable is the score on a clinical depression inventory, where lower scores indicate less depression.

Table 25.1 Depression Level After Treatment

Medication	*Counseling*	*Diet Supplement*
23	38	40
16	32	28
15	29	33
32	42	42
26	25	35
18	17	35
22	37	41
14	26	30
14	19	34
22	22	39
202	287	357
$M_{\text{med}} = \frac{202}{10} = 20.20$	$M_{\text{couns}} = \frac{287}{10} = 28.70$	$M_{\text{diet}} = \frac{357}{10} = 35.70$

Clearly, patients treated with medication had a lower level of depression than patients treated with counseling, who in turn had a lower level of depression than patients treated with diet supplements: The mean depression levels were 20.20, 28.70, and 35.70, respectively. But those mean differences do not necessarily indicate that one treatment was more effective in alleviating depression than another treatment. Recall that even if the null hypothesis is true and the type of treatment really does not alleviate depression, we still do not expect any sample of patients given any particular treatment to end up with exactly the same depression level as any other sample of patients given a different treatment. On average, over an infinite number of trials, there will be no difference. But for any given samples, patients given one type of treatment could show either more depression or less depression after treatment than those given another type of treatment, regardless of the truth of the null hypothesis.

Thus, the question is not whether or not the mean depression levels for the samples are exactly the same but how different they are. Is the mean depression level among the samples *only a little different*, and therefore probably due to mere sampling error? Or is the mean depression level among the samples *very different*, and therefore probably due to something other than mere sampling error—that is, due to the difference in type of treatment?

Inferential Logic and ANOVA

We need a statistic to compute the probability of getting differences in depression relief among the groups as great as that which we found in our observed samples, if indeed the null hypothesis is true and the treatments are equally effective. As discussed in Module 24, that statistic is an *F* test. The formula for an *F* test is as follows:

$$F = \frac{MS_{\text{bet obs}} - MS_{\text{bet exp}}}{MS_{\text{with}}}$$

where

$MS_{\text{bet obs}}$ = variance observed between groups,

$MS_{\text{bet exp}}$ = variance expected between groups, and

MS_{with} = variance within groups.

But wait! The logic of the F test looks a lot like the logic of the z score, normal deviate Z test, one-sample t test, and two-sample t test we previously encountered, doesn't it? Yes, it does. That's because each is an example of the prototype for any test of statistical significance. Recall the prototype:

$$\frac{\text{What did you get?} - \text{What did you expect?}}{\text{Standardized random error}}$$

When we studied z scores, the substitutions we made were as follows:

"What did you get?"	Observed raw score
"What did you expect?"	Sample mean
"Standardized random error"	Standard deviation

This gave us the formula

$$z\text{ Score} = \frac{X - M}{s}$$

For the *normal deviate Z test* and for the *one-sample t test*, the substitutions we made were as follows:

"What did you get?"	Observed sample mean
"What did you expect?"	Population mean
"Standardized random error"	Standard error of the mean

This gave us the formula

$$Z \text{ or } t_{1\text{-samp}} = \frac{M - \mu}{\sigma_M}$$

For the *two-sample t test*, the substitutions we made were as follows:

"What did you get?"	Observed difference between the two sample means
"What did you expect?"	Expected difference between the two population means
"Standardized random error"	Standard error of the difference between the means

This gave us the formula

$$t_{2\text{-samp}} = \frac{(M_1 - M_2) - (\mu_1 - \mu_2)}{\sigma_{M_1 - M_2}}$$

And now, we have the ANOVA F test:

$$F = \frac{MS_{\text{bet obs}} - MS_{\text{bet exp}}}{MS_{\text{with}}}$$

Substitutions are not as evident in ANOVA as in the previous test statistics because in ANOVA the numerator measures variance between groups rather than mean differences between groups. Also, the terms in both the numerator and the denominator are area measures (variances or mean squares) rather than linear measures (standard errors). Still, the logic is the same.

Let's take a look at what the substitutions are:

"What did you get?"	Observed difference between the sample variances
"What did you expect?"	Expected difference between the population variances
"Standardized random error"	Within-group variance

The first term in the numerator, $MS_{\text{bet obs}}$, is the observed difference between the sample variances. In other words, it's the observed difference in variance between clients given medication, clients given counseling, and clients given diet supplements.

The second term in the numerator, $MS_{\text{bet exp}}$, is the expected difference between the population variances. In other words, it's the expected difference in variance between an infinite number of clients given medication, an infinite number of clients given counseling, and an infinite number of clients given diet supplements. But recall that it is always the null hypothesis that we test. Under the null hypothesis, what do we expect that variance difference to be between groups?

Under the null hypothesis, we expect the difference in variance between groups to be 0. Therefore, the formula reduces to

$$F = \frac{MS_{\text{bet}} - 0}{MS_{\text{with}}}$$

This further reduces to

$$F = \frac{MS_{\text{bet}}}{MS_{\text{with}}}$$

Many sources show only this shortened formula. However, to understand the similar logic underlying most inferential statistical tests, it is helpful to remember the fuller formula.

CHECK YOURSELF!

Compare the numerator and denominator of the F statistic and the two-sample t statistic. How are the two formulas similar? How are they different?

Now, for the calculation. Recall from Module 24 that the sum of the squared deviation scores is called "sum of squares" for short and that it is indicated by the symbol SS. The worst part about calculating any ANOVA is calculating the three sums of squares: within group, between group, and total. It's not hard. It's just tedious.

We have a choice of formulas for calculating the three sums of squares. We can calculate them via deviation score formulas or via raw score formulas. The two sets of formulas (deviation score or raw score) are mathematically equivalent. Some instructors prefer the deviation score formulas; others prefer the raw score formulas. I will discuss both sets of formulas. Let your instructor be your guide for which set you should use. You might be directed to skip the section that covers formulas your instructor doesn't use.

Sums of Squares Formulas: Deviation Score Method

Here are the deviation score formulas for the three sums of squares:

$$SS_{bet} = n_g \sum_{1}^{k} (M_g - M_{tot})^2,$$ each group mean from the total mean, for each individual

$$SS_{with} = \sum_{1}^{k} \sum_{1}^{n} (X - M_g)^2,$$ each individual score from its group mean, for each group

$$SS_{tot} = \sum_{1}^{N} (X - M_{tot})^2,$$ each individual score from the total mean

If you see a formula in the *Physical Review* that extends over a quarter of a page, forget it. It's wrong. Nature isn't that complicated.

—Bernd T. Matthias

where

g = for that group,

k = number of groups, and

numbers below and above the Σ sign = the first and last cases, respectively.

SS_{bet} says to

1. take the difference between the mean of a group and the overall grand mean;
2. square that value;
3. do that for each group, beginning with the first group and ending with the kth group—that is, for all groups;
4. add those values; and
5. multiply by the number of cases in a group.

SS_{with} says to

1. take the difference between a score and the mean of that score's group;
2. square that value;
3. do that for each score within that group, beginning with the first score and ending with the nth score—that is, for all scores;
4. do that within each group, beginning with the first group and ending with the kth group—that is, for all groups; and
5. add those values.

SS_{tot} says to

1. take the difference between a score and the overall grand mean;
2. square that value;
3. do that for each score, beginning with the first score and ending with the Nth score—that is, all scores; and
4. add those values.

Note how the sources of variation in the three *SS* deviation formulas above were previously described and diagramed in Module 24. That diagram is repeated in Figure 25.1.

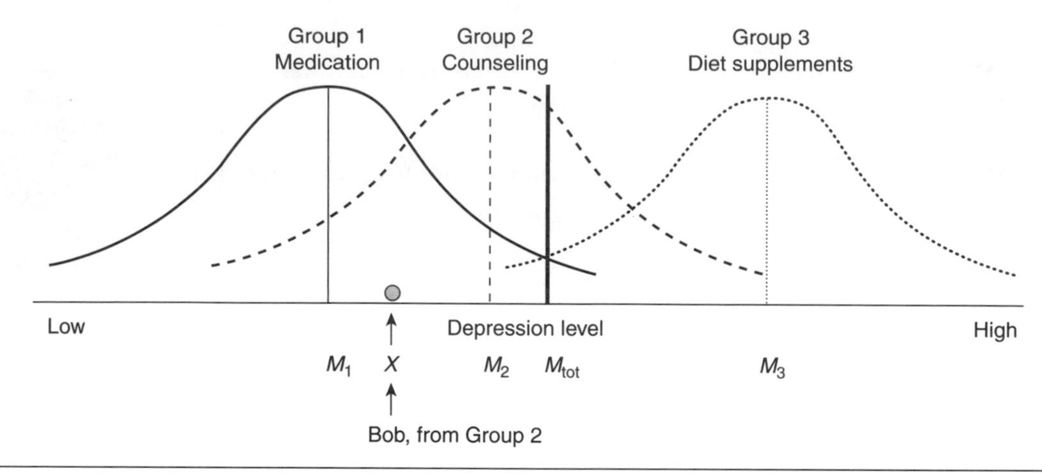

Figure 25.1 Three Different Treatment Populations

CHECK YOURSELF!

Module 24 asked you to use symbols (*X* and *M*) to indicate Bob's "between" deviation, his "within" deviation, and his "total" deviation? Did you do it? If not, do it now.

Again, note that each of these deviations is for Bob only. However, the three formulas tell you to do what you did for Bob for everyone, then square those deviations, and then add those squared deviations. When we do that, we get our three sums of squares.

Calculating Sums of Squares: Deviation Score Method

Here, once again, are the depression treatment data, but with columns added for the deviation formula entries (see Table 25.2).

Table 25.2 Depression Level After Treatment: Deviation Score Setup

Medication			*Counseling*			*Diet Supplement*		
X	*X* – *M*	$(X - M)^2$	*X*	*X* – *M*	$(X - M)^2$	*X*	*X* – *M*	$(X - M)^2$
23	2.80	7.84	38	9.30	86.49	40	4.30	18.49
16	–4.20	17.64	32	3.30	10.89	28	–7.70	59.29
15	–5.20	27.04	29	0.30	0.09	33	–2.70	7.29
32	11.80	139.24	42	13.30	176.89	42	6.30	39.69
26	5.80	33.64	25	3.70	13.69	35	–0.70	.49
18	–2.20	4.84	17	–11.70	136.89	35	–0.70	.49
22	1.80	3.24	37	8.30	68.89	41	5.30	28.09
14	–6.20	38.44	26	–2.70	7.29	30	–5.70	32.49
14	–6.20	38.44	19	–9.70	94.09	34	–1.70	2.89
22	1.80	3.24	22	–6.70	44.89	39	3.30	10.89
202		313.60	287		640.10	357		200.10

$M_{med} = \frac{202}{10} = 20.20$ $M_{couns} = \frac{287}{10} = 28.70$ $M_{diet} = \frac{357}{10} = 35.70$ $M_{total} = \frac{846}{30} = 28.30$

1. First, we compute SS_{bet}:

$$SS_{bet} = n_g \sum_{1}^{k} (M_g - M_{tot})^2$$
$$= 10[(20.20 - 28.20)^2 + (28.70 - 28.20)^2 + (35.70 - 28.20)^2]$$
$$= 10[(-8.00)^2 + (0.50)^2 + (7.50)^2]$$
$$= 10[(64.00 + 0.25 + 56.25)]$$
$$= (10)(120.50)$$
$$= 1205.00$$

2. Next, we compute SS_{with}:

$$SS_{with} = \sum_{1}^{k} \sum_{1}^{n} (X - M_g)^2$$
$$= 313.60 + 640.10 + 200.10$$
$$= 1153.80$$

3. Next, we compute SS_{tot}. Recall that SS_{tot} is not necessary for computation of the F statistic. However, recall also that SS_{bet} and SS_{with} should add to SS_{tot}. Thus, computing SS_{tot} allows us to check for calculation errors:

Old mathematicians never die. They just lose some of their functions.

$$SS_{tot} = \sum_{1}^{N} (X - M_{tot})^2$$

$$= (23 - 28.20)^2 + (16 - 28.20)^2 + (15 - 28.20)^2 + (32 - 28.20)^2 + (26 - 28.20)^2$$
$$+ (18 - 28.20)^2 + (22 - 28.20)^2 + (14 - 28.20)^2 + (14 - 28.20)^2 + (22 - 28.20)^2$$
$$+ (38 - 28.20)^2 + (32 - 28.20)^2 + (29 - 28.20)^2 + (42 - 28.20)^2 + (25 - 28.20)^2$$
$$+ (17 - 28.20)^2 + (37 - 28.20)^2 + (26 - 28.20)^2 + (19 - 28.20)^2 + (22 - 28.20)^2$$
$$+ (40 - 28.20)^2 + (28 - 28.20)^2 + (33 - 28.20)^2 + (42 - 28.20)^2 + (35 - 28.20)^2$$
$$+ (35 - 28.20)^2 + (41 - 28.20)^2 + (30 - 28.20)^2 + (34 - 28.20)^2 + (39 - 28.20)^2$$

$$= (-5.20)^2 + (-12.20)^2 + (-13.20)^2 + (3.80)^2 + (-2.20)^2 + (-10.20)^2 + (-6.20)^2$$
$$+ (-14.20)^2 + (-14.20)^2 + (-6.20)^2 + (9.80)^2 + (3.80)^2 + (0.80)^2 + (13.80)^2$$
$$+ (-3.20)^2 + (-11.20) + (8.80)^2 + (-2.20)^2 + (-9.20)^2 + (-6.20)^2 + (11.80)^2$$
$$+ (-0.20)^2 + (4.80)^2 + (13.80)^2 + (6.80)^2 + (6.80)^2 + (12.80)^2 + (1.80)^2 + (5.80)^2$$
$$+ (10.80)^2$$

$$= 27.04 + 148.84 + 174.24 + 14.44 + 4.84 + 104.04 + 38.44 + 201.64 + 201.64$$
$$+ 38.44 + 96.04 + 14.44 + 0.64 + 190.44 + 10.24 + 125.44 + 77.44 + 4.84$$
$$+ 84.64 + 38.44 + 139.24 + 0.04 + 23.04 + 190.44 + 46.24 + 46.24 + 163.84$$
$$+ 3.24 + 33.64 + 116.40$$

$$= 2358.80$$

4. *Check*: Do our SS_{bet} and our SS_{with} sum to our SS_{tot}? Yes, they do: 1205.00 + 1153.80 = 2358.80. Thus, we can proceed with the remaining steps.

Before proceeding with the remaining steps, let's go through the same calculations with the raw score formulas. As noted previously, the two sets of formulas (deviation score or raw score) are mathematically equivalent. The disadvantage of using raw score formulas is that their calculation algorithms do not map to the deviations depicted in Figure 25.1. That is, it is not obvious that you are computing between-group and within-group variances. The advantage of using raw score formulas is that they do not require calculation of deviation scores; therefore, they are much simpler to use than the deviation score formulas. Some instructors prefer one set of formulas, and some prefer the other. You should use whichever set your instructor prefers.

SOURCE: Peanuts: © United Feature Syndicate, Inc.

Sums of Squares Formulas: Raw Score Method

Here are the raw score formulas for the three sums of squares:

$$SS_{\text{bet}} = \sum_{1}^{k} \frac{\left(\sum X_{\text{g}}\right)^2}{n_{\text{g}}} - \frac{\left(\sum X_{\text{tot}}\right)^2}{N}$$

$$SS_{\text{with}} = \sum_{1}^{N} X^2 - \sum_{1}^{k} \frac{\left(\sum X_{\text{g}}\right)^2}{n_{\text{g}}}$$

$$SS_{\text{tot}} = \sum_{1}^{N} X^2 - \frac{\left(\sum X_{\text{tot}}\right)^2}{N}$$

where

g = for that group,

k = number of groups, and

numbers below and above the Σ sign = the first and last cases, respectively.

SS_{bet} says to

1. add all scores in a particular group,
2. square that sum, and
3. divide by the number of scores in that group; then,
4. repeat those steps for each group, beginning with the first group and ending with the last group, and

5. add up those values; then,
6. add all scores, starting with the first score and ending with the last score in all groups,
7. square that sum, and
8. divide that sum by the total number of scores in all groups; and finally,
9. subtract Step 8 from Step 5.

SS_{with} says to

1. square each score, starting with the first score and ending with the last score in all groups, and
2. add those squared scores; then,
3. add all scores in a particular group,
4. square that sum, and
5. divide by the number of scores in that group; then,
6. repeat those steps for each group, beginning with the first group and ending with the last group, and
7. add those values; and finally,
8. subtract Step 7 from Step 2.

SS_{tot} says to

1. square each score, starting with the first score and ending with the last score in all groups, and
2. add those squared scores; then,
3. add all scores, starting with the first score and ending with the last score in all groups,
4. square that sum, and
5. divide that sum by the total number of scores; and finally,
6. subtract Step 5 from Step 2.

Calculating Sums of Squares: Raw Score Method

According to the raw score formulas, we need the following values:

- The sum of the unsquared scores in each separate group
- The sum of all the unsquared scores
- The sum of the squared scores in each separate group
- The sum of all the squared scores
- The sample size in each separate group
- The overall sample size

Table 25.3 gives the depression treatment data but with columns added for the raw score formula entries.

Table 25.3 Depression Level After Treatment: Raw Score Setup

Medication		*Counseling*		*Diet Supplement*		
X	X^2	X	X^2	X	X^2	
23	529	38	1,444	40	1,600	
16	256	32	1,024	28	784	
15	225	29	841	33	1,089	
32	1,024	42	1,764	42	1,764	
26	676	25	625	35	1,225	
18	324	17	289	35	1,225	
22	484	37	1,369	41	1,681	
14	196	26	676	30	900	
14	196	19	361	34	1,156	
22	484	22	484	39	1,521	
202		287		357		$\rightarrow \sum X = 846$
	4,394		8,877		12,945	$\rightarrow \sum X^2 = 26,216$

1. First, we compute SS_{bet}:

> I admit that mathematics is a good thing. But excessive devotion to it is a bad thing.
>
> —Aldous Huxley

$$SS_{bet} = \sum_{1}^{k} \frac{\left(\sum X_g\right)^2}{n_g} - \frac{\left(\sum X_{tot}\right)^2}{N}$$

$$= \frac{(202)^2}{10} + \frac{(287)^2}{10} + \frac{(357)^2}{10} - \frac{(846)^2}{30}$$

$$= \frac{40804}{10} + \frac{82369}{10} + \frac{127449}{10} - \frac{715716}{30}$$

$$= (4080.40 + 8236.90 + 12744.90) - 23857.20$$

$$= 25062.20 - 23857.20$$

$$= 1205.00$$

2. Next, we compute SS_{with}:

$$SS_{with} = \sum_{1}^{N} X^2 - \sum_{1}^{k} \frac{\left(\sum X_g\right)^2}{n_g}$$

$$= 26216 - \left[\frac{(202)^2}{10} + \frac{(287)^2}{10} + \frac{(357)^2}{10}\right]$$

$$= 26216 - \left[\frac{40804}{10} + \frac{82369}{10} + \frac{127449}{10}\right]$$

$$= 26216 - (4080.40 + 8236.90 + 12744.90)$$

$$= 26216 - 25062.20$$

$$= 1153.80$$

3. Next, we compute SS_{tot}. Recall that SS_{tot} is not necessary for computation of the F statistic. However, recall also that SS_{bet} and SS_{with} should add to SS_{tot}. Thus, computing SS_{tot} allows us to check for calculation errors:

$$SS_{tot} = \sum_{1}^{N} X^2 - \frac{\left(\sum X_{tot}\right)^2}{N}$$
$$= 26216 - \frac{(846)^2}{30}$$
$$= 26216 - \frac{715716}{30}$$
$$= 26216 - 23857.20$$
$$= 2358.80$$

4. *Check*: Do our SS_{bet} and SS_{with} sum to our SS_{tot}? Yes, they do: 1205.00 + 1153.80 = 2358.80. Thus, we can proceed with the remaining calculations.

Remaining Steps: Mean Squares and *F*

5. Whether we have used deviation scores or raw scores, we do not yet have the mean squares (variances). We have only three sums of squared deviations—that is, three sums of squares. Recall from Module 6 and Module 24 that a variance is the average of the squared deviation scores. Thus, to get averages, we must divide each sum of squares by its appropriate degrees of freedom (*df*).

The *df* for each sum of squares is as follows:

df_{bet} = No. of groups – 1 = $k - 1$

df_{with} = No. of subjects – No. of groups = $N - k$

df_{tot} = No. of subjects – 1 = $N - 1$

For our study, these *df* values are as follows:

$df_{bet} = 3 - 1 = 2$

$df_{with} = 30 - 3 = 27$

$df_{tot} = 30 - 1 = 29$

The two mean squares needed for the *F* test are MS_{bet} and MS_{with}. Here are the calculations:

$$MS_{bet} = \frac{SS_{bet}}{df_{bet}} = \frac{1205.0}{2} = 602.50$$
$$MS_{with} = \frac{SS_{with}}{df_{with}} = \frac{1153.8}{27} = 42.733$$

6. Now that we have our two mean squares, we can finally calculate *F*. Recall that *F* is the ratio of between-group variance to within-group variance:

$$F = \frac{MS_{bet}}{MS_{with}} = \frac{602.50}{42.733} = 14.10$$

We are finally done. The ratio of between-group variance to within-group variance—that is, treatment effect to random error—is 14.10. But what does that mean?

CHECK YOURSELF!

In Module 24, you were directed to list the steps for calculating *F*. Did you do it? If not, do it now.

Interpreting a One-Way ANOVA

As usual, the question is whether the observed difference in observed depression level is *a lot*, and therefore probably due to the differing treatments among the samples, or *only a little*, and therefore probably due to mere sampling error.

Our previous test statistics (z and t) were symmetrically distributed. Thus, we first got a sense of the meaning of our z or t by placing it on the appropriate curve and seeing which tail it fell toward and how far into that tail it fell. We then looked up our z or t in a table to see if it met the tabled criterion for statistical significance. In contrast, the F statistic, as you learned in Module 24, is not symmetric. There are no negative values, and the positive values are positively skewed. Thus, we cannot get a sense of the meaning of our F by locating it on the curve, except that the farther it falls above +1.00, the more likely it is to be statistically significant.

We need to go directly to the table lookup. Appendix D contains the table for the F distribution. Table 25.4 is a portion of that table.

Table 25.4 A Portion of the *F* Table

		df_{bet}			
df_{with}	α (%)	*1*	*2*	*3*	*4*
26	5	4.22	3.37	2.98	2.74
	1	7.72	5.53	4.64	4.14
27	5	4.21	3.35	2.96	2.73
	1	7.68	5.49	4.60	4.11
28	5	4.20	3.34	2.95	2.71
	1	7.64	5.45	4.57	4.07

As with the t table, we must view the table at the correct degrees of freedom. For an F test, there are two *df*: one for between groups, df_{bet}, and one for within groups, df_{with}. In the F statistic, the mean square between groups is placed in the numerator, and the mean square within groups is placed in the denominator. Thus, many F tables refer to the *df* between groups as the "numerator" *df* and the *df* within groups as the "denominator" *df*. You should be able to read an F table with either type of notation: *df* between and *df* within, or numerator *df* and denominator *df*.

For our study, what are the two *df*s? When calculating the mean squares for our study, you learned that the *df* between groups was 2 and the *df* within groups was 27. (If you don't remember how to obtain these values, go back to the mean square calculations before continuing.) In the language of statisticians, the *df* between groups is always listed first, followed by the *df* within groups. Thus, we say that the *df*s for our study are 2 and 27.

The table lists only two levels of Type 1 error (α): 5% and 1%. These are the rows of the table. The row with the smaller-valued entries is for the 5% error. The row with the larger-valued entries is for the 1% error. As with other statistical tables, the *F* table lists the minimum value that our observed *F* must be for us to reject the null hypothesis and conclude instead that the differences among groups is probably due to the differences in treatment.

Our calculated *F* was 14.10. We want to determine whether or not our calculated value exceeds the tabled value at the specified Type 1 error level. Table 25.4 is the necessary portion of the *F* table.

Assume that we are willing to make a Type 1 error 5% of the time. At $\alpha = .05$, what is the critical *F*? Can we reject the null hypothesis?

The critical *F* for 2 and 27 *df* is 3.35. We can reject the null hypothesis because the calculated *F* of 14.10 meets or exceeds the critical *F* of 3.35.

Now, assume that we are willing to make a Type 1 error only 1% of the time. At $\alpha = .01$, what is the critical *F*? Can we reject the null hypothesis?

The critical *F* for 2 and 27 is 5.49. We can still reject the null hypothesis because the calculated *F* of 14.10 meets or exceeds the critical *F* of 5.49.

In a journal article, our result would be written as follows:

$$F(2, 27) = 14.10, p < .01.$$

This is read as "*F* at 2 and 27 degrees of freedom is 14.10. There is less than one chance in a hundred that the difference in depression level was due to mere chance." That is, we can be more than 99% certain that the difference in clients' depression level after treatment was due to the different treatments they received rather than due to mere chance.

Note that an *F* this large would be statistically significant at the .001 level as well. However, the table in Appendix D does not include that column. If it did, we would report that there is less than one chance in a thousand that the difference in depression was due to mere chance. Statistical software programs (such as SPSS) report error levels more refined than those possible through Appendix D.

The ANOVA Summary Table

Journals also summarize ANOVA calculations in an ANOVA summary table. An **ANOVA summary table** shows the partitioning of the between and within *SS*, the *df*, the *MS*, and the final *F*. It is also traditional to indicate statistical significance at the .05 level by a single asterisk, at the .01 level by a double asterisk, and at the .001 level by a triple asterisk. (Our *F* table did not show .001 values, so we cannot determine whether or not our obtained *F* was statistically significant at the .001 level.) Table 25.5 is the ANOVA summary table for our study.

Note how much information can be gleaned from an ANOVA summary table. Without even knowing what the study is about, the reader knows that there were three groups,

Table 25.5 ANOVA Summary Table for Effects of Three Different Treatments on Depression Level

Source	*SS*	*df*	*MS*	*F*
Between	1205.00	2	602.50	14.10**
Within	1153.80	27	42.733	
Total	2358.80	29		

**$p < .01$.

because $df_{bet} = k - 1 = 2$. The reader also knows that there were 30 subjects in the study, because $df_{tot} = N - 1 = 29$. Assuming equal numbers of subjects in each group (the case in our study), the reader also knows that there were 10 subjects in each group, because 30 subjects divided by three groups is 10 subjects per group. Finally, because of the two asterisks, the reader knows that the results were statistically significant at the .01 error level.

Again, if these data had been run through a statistical analysis software program rather than hand calculated and subjected to a table lookup, statistical significance would be reported at the .001 level, and the F would be followed by three asterisk symbols.

PRACTICE

1. The following data are from a study of the number of pounds participants lost after 3 months on a regimen of only diet and exercise versus diet and exercise with a personal trainer versus diet and exercise with a support group:

Only Diet and Exercise	*Diet, Exercise, and Personal Trainer*	*Diet, Exercise, and Support Group*
4	11	7
6	14	9
12	9	10
5	22	6
8	15	12
7	8	16
10	12	11
17	17	8
4	24	19
12	30	14
9	8	13
5	16	7

   a. What are the independent and dependent variables in this study?
   b. State the null hypothesis and the research hypothesis.
   c. Calculate F and compare it with the critical value of F at the .01 and .05 α levels. At what level of error can you reject the null hypothesis, if at all?
   d. Prepare an ANOVA summary table of the results.
   e. Write the results in APA journal format.

2. Twenty-four coworkers want to see whether various routes differ in the time it takes to get to work. Starting from points equidistant from work, eight coworkers take public transportation, eight drive on the highway, and eight take back roads. Below are their reported times in minutes for the commute:

Public	*Highway*	*Back Roads*
34	44	59
33	47	63
36	49	65
32	58	69
31	40	53
29	54	71
35	52	73
30	46	66

a. What are the independent and dependent variables in this study?

b. State the null hypothesis and the research hypothesis.

c. Calculate *F* and compare it with the critical value of *F* at the .01 and .05 α levels. At what level of error can you reject the null hypothesis, if at all?

d. Prepare an ANOVA summary table of the results.

e. Write the results in APA journal format.

3. Professor Testgiver wonders if the order in which questions are put on a test affects the test score. He administers different forms of the same test to subsets of his class. On one form, the more difficult questions are at the beginning and the easiest questions are at the end. In the next form, the easiest questions are at the beginning, and the more difficult questions are at the end. In the third form, items are randomly placed, not in order of their difficulty. Here are the students' test scores (out of 20):

Difficult First	*Easiest First*	*Random Order*
15	20	11
13	18	14
17	16	9
16	11	15
20	13	17
8	10	13
12	15	19
11	15	18
9	14	12
14	11	11

a. What are the independent and dependent variables in this study?

b. State the null hypothesis and the research hypothesis.

c. Calculate *F* and compare it with the critical value of *F* at the .01 and .05 α levels. At what level of error can you reject the null hypothesis, if at all?

d. Prepare an ANOVA summary table of the results.

e. Write the results in APA journal format.

4. A professor wonders if the number of hours students spend studying during the week before final exams is a function of their GPAs until that point. Ten students in four separate GPA brackets report the amount of time they spend studying during the final week of the semester. The GPA brackets are 2.0 to 2.5, 2.5 to 3.0, 3.0 to 3.5, and 3.5 to 4.0. Study time (in hours) during the week before final exams for the 10 students in each GPA bracket is listed below:

2.0–2.5	*2.5–3.0*	*3.0–3.5*	*3.5–4.0*
26	28	30	30
27	27	31	31
24	29	33	32
30	30	38	35
31	33	38	38
26	28	30	35
28	34	29	32
27	32	28	31
28	30	35	33
26	27	36	36

a. What are the independent and dependent variables in this study?
b. State the null hypothesis and the research hypothesis.
c. Calculate F and compare it with the critical value of F at the .01 and .05 α levels. At what level of error can you reject the null hypothesis, if at all?
d. Prepare an ANOVA summary table of the results.
e. Write the results in APA journal format.

5. More students than not gain weight during their first year of college. A nutritionist wonders if this is due to the type of food served in campus dining centers. She compares the number of pounds gained during the academic year by 10 randomly selected freshmen in each of three groups: lives and eats at home, lives on campus and eats in campus dining hall, lives in an apartment and prepares own meals. Here are the pounds gained by each student:

Eats at Home	*Eats in Campus Dining*	*Prepares Own Meals*
2	5	3
4	9	2
0	12	5
3	0	6
0	8	0
5	10	2
2	7	0
4	5	3
10	4	1
5	9	9

a. State the H_0.
b. Prepare an ANOVA summary table of the results.
c. Write the results in APA format.

6. Lore has it that small dogs bark more than larger dogs. To test the lore, a kennel owner logged the number of separate barking episodes per day for three size dogs: small, medium, and large. To represent each size fairly, four different breeds were included in each size categories. Here are the number of barking episodes for 12 dogs (three each of four breeds) for each size dog:

Small	*Medium*	*Large*
8	3	5
12	5	8
14	1	5
10	4	4
6	2	10
3	0	3
11	3	0
4	0	4
5	2	3
8	3	5
0	4	9
6	6	4

a. State the H_0.
b. Prepare an ANOVA summary table of the results.
c. Write the results in APA format

7. An energy researcher wonders which type of heating—electric, natural gas, or oil—is fastest in raising the temperature of a room. Eight different-sized rooms of identical construction are heated with each of the three types of energy source. Here are the times, in minutes, to raise the temperature by a given number of degrees:

Electric	*Natural Gas*	*Oil*
26	32	28
31	30	26
56	50	49
21	19	24
65	74	60
10	8	9
34	35	32
50	57	53

a. State the H_0.

b. Prepare an ANOVA summary table of the results.

c. Write the results in APA format

8. "Everyone knows" that certain college majors are associated with larger salaries for graduates. Is that really so? Here are the annual salaries (in thousands of dollars) for 8 randomly selected graduates from four different majors—English, accounting, psychology, and chemistry—5 years after graduation and with only a bachelor's degree:

English	*Accounting*	*Psychology*	*Chemistry*
32	78	42	56
46	55	36	45
28	40	70	50
70	92	43	78
36	102	37	60
42	80	57	77
55	76	45	66
34	67	40	76

a. State the H_0.

b. Prepare an ANOVA summary table of the results.

c. Write the results in APA format

9. Three groups of college students, 20 in each group, are given a list of eight nonsense syllables and required to learn the order in which the syllables appeared, by calling out the second syllable on the list when they were shown the first syllable, calling out the third syllable when they were shown the second, and so on. Group 1 is given 1 s in which to anticipate the next syllable; Group 2 is given 3 s; and Group 3 is given 6 s. Presentation of the lists of nonsense syllables continues until the person correctly anticipates all eight syllables. The dependent variable is the number of times (trials) the list of syllables has to be presented until the person correctly anticipates all eight syllables. Here are the results:

$$SS_{\text{tot}} = 257.06$$

$$SS_{\text{bet}} = 85.37$$

a. Complete the ANOVA summary table:

Source	*SS*	*df*	*MS*	*F*
Between	85.37			
Within				
Total	257.06			

b. From the *F* table, what is the critical value of *F* at the .05 error level?

c. What level of confidence can you have in rejecting the null hypothesis? Conversely, if you reject the null hypothesis, what is the probability that you have made a Type 1 error?

10. Examine the following ANOVA summary table:

Source	*SS*	*df*	*MS*	*F*
Between	495.34	4	123.84	8.16
Within	682.56	45	15.17	
Total	1,177.90	49		

a. How many groups were there in the study?

b. Assuming equal numbers of subjects per group, how many subjects were in each group?

c. What is the tabled critical value of *F* at the .01 level?

d. What level of confidence can you have in rejecting the null hypothesis? Conversely, if you reject the null hypothesis, what is the probability that you have made a Type 1 error?

11. Examine the following completed ANOVA table.

	SS	*df*	*MS*	*F*	*P*
Between groups	246.33	3	82.110	6.020	<.01
Within groups	927.47	68	13.639		
Total	1173.80	71			

a. How many groups were in the study?

b. Assuming equal number of subjects per group, how many subjects were in each group?

c. What is the tabled critical value of *F* at the .01 level?

12. A pharmaceutical company is running clinical trials on four experimental drugs—Drug A through Drug D—for treating a medical disorder. Four groups of 20 patients with identical diagnoses and symptoms are given one of the five drugs. Symptom improvement is measured on a scale with higher values indicating more improvement. SS_{bet} = 304.54 and SS_{with} = 1024.89. Complete the ANOVA table.

	SS	*df*	*MS*	*F*	*P*
Between drugs					
Within drugs					
Total					

SPSS Connection

Download the file **data_depression relief due to med couns dietsupp.sav** from www.sagepub.com/steinberg2e. These data are used in the textbook example.

Alternatively, manually enter the 30 scores from the depression example in Module 20 into the SPSS **Data View** spreadsheet. Data entry for a one-way ANOVA is not intuitively obvious. In the textbook, the data are set up as three groups of 10 clients. In SPSS, all 30 scores (10 + 10 + 10) are entered in a single column. Then their group membership (medication vs. counseling vs. diet supplements) is entered in the second column. In addition, and unlike *t* tests, the group membership variable must be coded as a numeric variable rather than a string variable (this has to do with the underlying regression calculation algorithm, despite the ANOVA design setup; regression will be discussed in the final modules of the textbook). Thus, enter the data as follows:

23	1
16	1
15	1
32	1
26	1
18	1
22	1
14	1
14	1
22	1
38	2
32	2
29	2
42	2
25	2
17	2
37	2
26	2
19	2
22	2
40	3
28	3
33	3
42	3
35	3
35	3
41	3
30	3
34	3
39	3

Click on the **Variable View** tab to define the variables. Name the first variable **deprscor,** set the decimals at 0, and label the variable as **Depression Score.** Name the second variable **typtreat,** set the variables at 0, and label the variable as **Type of Treatment.** Label the values as follows: **1 = Counseling, 2 = Medication,** and **3 = Diet Supplement.**

If the file is not already in **Data View,** click that tab in the lower left of the screen.

In the toolbar at the top of the screen, click on **Analyze,** then **Compare Means,** then **One-Way ANOVA.** Highlight the variable **Depression Score** in the left window and then click on the **arrow** before the **Dependent List** window to send the variable into that window. This is the study's dependent variable. Click on the variable **Type of Treatment** in the left window and then click on the **arrow** before the **Factor** window to send the variable into that window. This is the study's independent variable. (Do not be thrown by the change in terminology. Recall that the term *variance* also changed to *mean square* in ANOVA. A "factor" is simply ANOVA terminology for an independent variable.) Click **OK**. This is what you will see.

Oneway

ANOVA

Depression Score

	Sum of Squares	df	Mean Square	F	Sig.
Between Groups	1205.000	2	602.500	14.099	.000
Within Groups	1153.800	27	42.733		
Total	2358.800	29			

Visit the study site at www.sagepub.com/steinberg2e for practice quizzes and other study resources.

PART XI

Post Hoc Tests

"So Who's Responsible?"

Tukey HSD Test 26

Terms: post hoc, pairwise, honestly significant difference, studentized range statistic

Symbols: HSD, q

Learning Objectives:

- Distinguish between the information provided by an F test and a Tukey HSD
- Know when it is appropriate to calculate a Tukey HSD
- Calculate a Tukey HSD
- Construct a matrix of group mean differences
- Report results in APA format

Why Do We Need a Post Hoc Test?

When an F test is significant, we know that there is a significant difference among the means—somewhere. But we don't know exactly where that significant difference lies. That's because the F test compares all the group means simultaneously. Thus, the significant difference could lie between the first group and the second, the second group and the third, or the third group and the first. Or two of those pairs or all those pairs. We just don't know. To find out exactly where the significant difference lies, we must test each pair of means separately. That is what a post hoc test does. After the overall F has been found to be significant, a **post hoc** test locates the particular pairs of groups—that is, the particular independent variable treatments—responsible for the overall significant F. It follows that if the overall F is not statistically significant, then it is not necessary to compute post hoc tests because there is no significance to be found between pairs.

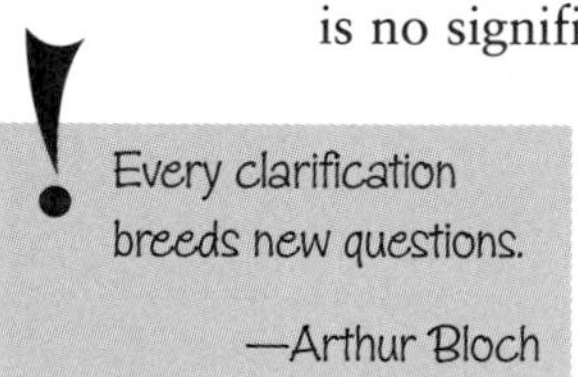

There are several different post hoc tests. All post hoc tests compare each pair of treatments in the study to determine which pair or pairs are significantly different from each other. We call this process **pairwise** comparison. Conceptually, the process can be likened to conducting a series of t tests or two-group ANOVAs.

Recall from Module 24 the discussion of why we conduct an overall ANOVA rather than multiple t tests on the data in the first place: Type 1 error escalates. So do post hoc tests allow error to escalate? No, they do not. This is because, although we actually compare only two means at a time, the sampling distribution and its table are based on the simultaneous comparison of multiple means—much like an ANOVA. Thus, the error rate is fixed.

In this module, we will look at the Tukey HSD, which stands for **honestly significant difference**. The HSD is the amount by which any two treatment group means must differ for those groups to be considered significantly different.

CHECK YOURSELF!

Why and when do we calculate a post hoc test? How does what is learned via a post hoc test differ from what is learned via an *F* test?

Calculating the Tukey HSD

For our study of the effect of various treatments on depression levels, there are three pairwise comparisons: medication versus counseling, medication versus diet supplements, and counseling versus diet supplements. Table 26.1 shows the data from that study.

Table 26.1 Depression Level After Treatment

Medication	*Counseling*	*Diet Supplement*
23	38	40
16	32	28
15	29	33
32	42	42
26	25	35
18	17	35
22	37	41
14	26	30
14	19	34
22	22	39
$\Sigma = 202$	$\Sigma = 287$	$\Sigma = 357$
$M = \frac{202}{10} = 20.20$	$M = \frac{287}{10} = 28.70$	$M = \frac{357}{10} = 35.70$

Table 26.2 shows the ANOVA summary from that study. Recall that the overall *F* test was statistically significant at the .01 level.

But is the significant difference in depression level due to the difference between medication and counseling, between medication and diet supplements, or between counseling and diet supplements? Or two of those pairs? Or all three? The HSD, which is the amount by which any two treatment group means must differ for those groups to be considered significantly different, will tell us that. Thus, we need to calculate each group's mean and determine how different each group's mean is from every other group's mean. Then, we will compare these differences with the HSD.

Table 26.2 ANOVA Summary Table for Effects of Three Different Treatments on Depression Level

Source	*SS*	*df*	*MS*	*F*
Between	1,205.00	2	602.50	14.10**
Within	1,153.80	27	42.733	
Total	2,358.80	29		

$^{**}p < .01$.

The formula for the HSD is

$$HSD = q\sqrt{\frac{MS_{with}}{n_g}}$$

where

q = studentized range statistic,

n_g = number of participants in each treatment group, and

MS_{with} = mean square within.

According to the formula, the first thing we have to do is find q, which stands for **studentized range statistic.** The studentized range statistic is based on a comparison of multiple means simultaneously. Therefore, it is the part of the HSD that controls escalating error. You can find q in Appendix E. A portion of that table is reproduced as Table 26.3.

Table 26.3 A Portion of the Studentized Range Table

		k = Number of Treatment Groups			
df_{with}	α	2	3	4	5
20	.05	2.95	3.58	3.96	4.23
	.01	4.02	4.64	5.02	5.29
24	.05	2.92	3.53	3.90	4.17
	.01	3.96	4.55	4.91	5.17
30	.05	2.89	3.49	3.85	4.10
	.01	3.89	4.45	4.80	5.05
40	.05	2.86	3.44	3.79	4.04
	.01	3.82	4.37	4.70	4.93

From Table 26.3, you can see that we first have to decide at what level of error to conduct the test. The table gives entries for two levels: .05 and .01. Let's pick the .05 error level.

The table also asks for the number of treatment groups in our study. How many groups do we have?

Yes, we have three groups: medication, counseling, and diet supplements. Note that, unlike the table for the overall F test, this table does *not* ask for the degrees of freedom between groups. Instead, it asks for the number of *groups*. This is one of the only statistical tables that asks for the number of groups rather than the degrees of freedom between groups, so be careful not to confuse the two.

The table also asks for the degrees of freedom within groups. We obtained the df_{with} when we computed the overall F test and reported it in Table 26.2. For this study, how many degrees of freedom are there within groups?

Yes, there are $N - k = 27$ df within groups.

Looking now at the table, what is the studentized range statistic, q, at the .05 α level, for three groups and 27 df_{with}?

Right, there is no row in the table for 27 df_{with}. There is a row for 24 df_{with} and a row for 30 df_{with}. Thus, we have to interpolate (estimate) between the two rows. For three groups, the tabled q for 24 df_{with} is 3.53; for 30 df_{with}, the tabled q is 3.49. Because 27 is halfway between 24 and 30, the interpolated q is halfway between 3.53 and 3.49. Thus, the interpolated q is 3.51.

We can now plug the q into the HSD formula. However, according to the formula, what two additional values do we need?

$$\text{HSD} = q\sqrt{\frac{MS_{\text{with}}}{n_g}}$$

Yes, we need the number of participants in each group, n_g, and the mean square within, MS_{with}. We calculated the MS_{with} when we computed the overall F test; it is listed in Table 26.2. There are 10 participants in each group, and the MS_{with} in Table 26.2 is 42.733. Plugging these values into the HSD formula, we get

$$\begin{aligned}\text{HSD} &= 3.51\sqrt{\frac{42.733}{10}}\\ &= 3.51\sqrt{4.273}\\ &= (3.51)(2.067)\\ &= 7.26\end{aligned}$$

So the Tukey HSD is 7.26. But what does that mean?

Interpreting the Tukey HSD

Recall that HSD is the smallest amount by which any two treatment group means must differ for those groups to be considered significantly different. Thus, we now need to calculate each group's mean and determine how different each group's mean is from every other group's mean. Only then can we compare each pair's difference in means with the HSD.

Here, again, is each group's mean:

Medication	*Counseling*	*Diet Supplements*
$M = \frac{202}{10} = 20.20$	$M = \frac{287}{10} = 28.70$	$M = \frac{357}{10} = 35.70$

The clearest way to view the difference between each pair of means is through a matrix. A matrix is a chart with the same row headings as column headings. For our study, the row and column headings are the three types of treatments—that is, the three groups. Here, we have each group's mean subtracted from the mean of each of the other groups. The dashed lines indicate each group compared with itself—a meaningless comparison. Note that the difference between the means in absolute value is the same above the dashed line as below the dashed line. Therefore, we need to make the subtractions only on one side of the dashed line, either above or below (Table 26.4).

Table 26.4 Tukey Matrix for the Effect of Three Treatments on Depression Level

	Medication	*Counseling*	*Diet Supplements*
Medication	—		
Counseling	28.70 – 20.20 = 8.50	—	
Diet supplements	35.70 – 20.20 = 15.50	35.70 – 28.70 = 7.00	—

Our calculated HSD was 7.26. That is the smallest amount by which any two means must differ to be the source of the significant overall *F*. Which of the pairwise comparisons meet(s) that criterion?

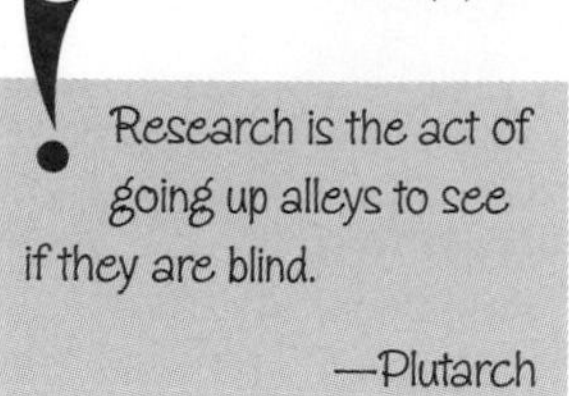

From the matrix, we see that clients given medication versus diet supplements (difference = 15.50) and clients given medication versus counseling (difference = 8.50) meet the critical 7.26 difference criterion. In contrast, clients given counseling versus diet supplements (difference = 7.00) do not meet the critical 7.26 difference criterion. Therefore, the significant overall *F* is due only to the difference in depression relief between medication and diet supplements and between medication and counseling. It is not due to the difference in depression relief between counseling and diet supplements.

In a journal article, a Tukey HSD result is typically part of the narrative. For example, our results might be written as follows:

HSD = 7.26, $\alpha = .05$.

Only two treatment pairs met the difference criterion: medication versus diet supplements ($d = 15.50$) and medication versus counseling ($d = 8.50$).

PRACTICE

1. Exercise 1 in Module 25 examined the relative effectiveness of three different weight loss regimens. Here is the ANOVA summary table:

Source	*SS*	*df*	*MS*	*F*
Between	321.50	2	160.75	6.145
Within	863.25	33	26.16	
Total	1,184.75	35		

Conduct a Tukey HSD on the ANOVA at $\alpha = .05$. Between which pair(s) does the statistical significance lie?

2. Exercise 2 in Module 25 examined the effect of difference in travel route on time taken to arrive at work. Here is the ANOVA summary table:

Source	*SS*	*df*	*MS*	*F*
Between	4,192.59	2	2,096.30	76.37
Within	576.38	21	27.45	
Total	4,768.97	23		

Conduct a Tukey HSD on the ANOVA at $\alpha = .01$. Between which pair(s) does the statistical significance lie?

3. Exercise 3 in Module 25 examined whether the order of question difficulty affects test score. Here is the ANOVA summary table:

Source	*SS*	*df*	*MS*	*F*
Between	3.2	2	1.6	0.138
Within	313.5	27	11.61	
Total	316.7	29		

Conduct a Tukey HSD on the ANOVA at $\alpha = .05$. Between which pair(s) does the statistical significance lie?

4. Exercise 4 in Module 25 examined the relationship between hours spent studying for final exams and overall GPA. Here is the ANOVA summary table:

Source	*SS*	*df*	*MS*	*F*
Between	235.0	3	78.33	10.09
Within	279.4	36	7.76	
Total	514.4	39		

Conduct a Tukey HSD on the ANOVA at $\alpha = .01$. Between which pair(s) does the statistical significance lie?

5. Exercise 5 in Module 25 examined freshmen weight gain due to where meals are eaten. Conduct a Tukey HSD at $\alpha = .05$. Between which pair(s) does the source of the statistical significance lie?

6. Exercise 6 in Module 25 examined whether the amount of barking is related to dog size. Conduct a Tukey HSD at $\alpha = .01$. Between which pair(s) does the source of the statistical significance lie?

7. Exercise 7 in Module 25 examined time to heat a room due to energy source. Why is it unnecessary to conduct a post hoc test to determine the pairs between which the source of the statistical significance lies?

8. Exercise 8 in Module 25 examined salary as a function of college major. Conduct a Tukey HSD at $\alpha = .01$. Between which pair(s) does the source of the statistical significance lie?

9. Assume the following data for a study:

Group 1	$M = 20$	$n = 10$
Group 2	$M = 30$	$n = 10$
Group 3	$M = 40$	$n = 10$

Source	*SS*	*df*	*MS*	*F*
Between	300	2	150	6.75
Within	600	27	22.22	
Total	500	29		

Conduct a Tukey HSD on the ANOVA at $\alpha = .05$. Between which pair(s) does the statistical significance lie?

10. Assume the following data for a study:

Group 1	$M = 146.98$	$n = 21$
Group 2	$M = 140.45$	$n = 21$
Group 3	$M = 147.04$	$n = 21$
Group 4	$M = 142.66$	$n = 21$

Source	*SS*	*df*	*MS*	*F*
Between	642.84	3	214.28	5.12
Within	3,346.52	80	41.83	
Total	3,989.36	83		

Conduct a Tukey HSD on the ANOVA at $\alpha = .01$. Between which pair(s) does the statistical significance lie?

SPSS Connection

Return to the SPSS file that you downloaded or manually data-entered in Module 25: **data_depression relief due to med couns dietsupp.sav** from www.sagepub.com/steinberg2e. These data are used in the textbook example.

If the file is not already in **Data View,** click that tab in the lower left of the screen. Follow the directions in the SPSS section of Module 15. However, *before* you click OK, click **Post-Hoc.** Mark the box for **Tukey.** Click **Continue.** Then click **OK.** This is what you will see.

Oneway

ANOVA

Depression Score

	Sum of Squares	df	Mean Square	F	Sig.
Between Groups	1205.000	2	602.500	14.099	.000
Within Groups	1153.800	27	42.733		
Total	2358.800	29			

Post Hoc Tests

Multiple Comparisons

Depression Score
Tukey HSD

(I) Type of Treatment	(J) Type of Treatment	Mean Difference (I-J)	Std. Error	Sig.	95% Confidence Interval	
					Lower Bound	Upper Bound
Medication	Counseling	-8.500*	2.923	.019	-15.75	-1.25
	Diet Supplements	-15.500*	2.923	.000	-22.75	-8.25
Counseling	Medication	8.500*	2.923	.019	1.25	15.75
	Diet Supplements	-7.000	2.923	.060	-14.25	.25
Diet Supplements	Medication	15.500*	2.923	.000	8.25	22.75
	Counseling	7.000	2.923	.060	-.25	14.25

* The mean difference is si gnificant at the 0.05 level.

Homogeneous Subsets

Depression Score

Tukey HSD [a]

Type of Treatment	N	Subset for alpha = 0.05	
		1	2
Medication	10	20.20	
Counseling	10		28.70
Diet Supplements	10		35.70
Sig.		1.000	.060

Means for groups in homogeneous subsets are displayed.
a. Uses Harmonic Mean Sample Size = 10.000.

The top table repeats the one-way ANOVA that you saw previously in Module 25. The Tukey post hoc test is shown in the next two tables. The bottom table gives the groups' means; the middle table gives the groups' difference in means. It is the middle table that most resembles the textbook calculations. SPSS prints each pair twice, subtracting first one group in the pair from the other and then reversing the order of the subtraction. Either set alone is sufficient.

Visit the study site at www.sagepub.com/steinberg2e for practice quizzes and other study resources.

27 Scheffé Test

Learning Objectives:

- Distinguish between the information provided by an *F* test and a Scheffé test
- Know when it is appropriate to calculate a Scheffé test
- Calculate a Scheffé test
- Report results in APA format

Why Do We Need a Post Hoc Test?

Recall that when an *F* test is significant, we know that there is a significant difference among the means—somewhere. But we don't know exactly where that significant difference lies. That's because the *F* test compares all the group means simultaneously. Thus, the significant difference could lie between the first group and the second group, the second group and the third group, or the third group and the first group. Or two of those pairs or all those pairs. Therefore, after the overall *F* has been found to be significant, we use a post hoc test to identify the particular treatments responsible for the overall significant *F*. Obviously, if the overall *F* test is not significant, there is no need to calculate a Scheffé to look for statistically significant pairs.

As mentioned in Module 26, there are several different post hoc tests. All post hoc tests compare each pair of treatments in the study to determine which pair or pairs are significantly different from each other. We call this process pairwise comparison. Conceptually, the process can be likened to conducting a series of *t* tests or two-group ANOVAs.

Also, recall from Module 24 the discussion of why we conduct an overall ANOVA rather than multiple *t* tests on the data in the first place: Type 1 error escalates. So do post hoc tests allow error to escalate? No, they do not. This is because, although we actually compare only two means at a time, the sampling distribution and its table are based on the simultaneous comparison of multiple means—much like an ANOVA. Thus, the error rate is fixed.

In Module 26, we examined one post hoc test, the Tukey HSD. In this module, we will examine another post hoc test, the Scheffé.

CHECK YOURSELF!

Why and when do we calculate a post hoc test? How does what is learned via a post hoc test differ from what is learned via an *F* test?

Calculating the Scheffé

For our study of the effect of various treatments on depression levels, there are three pairwise comparisons: medication versus counseling, medication versus diet supplements, and counseling versus diet supplements. The ANOVA summary table from that study is presented again in Table 27.1. Recall that the overall *F* test was statistically significant at the .01 error level.

Table 27.1 ANOVA Summary Table for Effects of Three Different Treatments on Depression Level

Source	*SS*	*df*	*MS*	*F*
Between	1,205.00	2	602.50	14.10**
Within	1,153.80	27	42.733	
Total	2,358.80	29		

**$p < .01$.

But is the significant difference in depression level due to the difference between medication and counseling, medication and diet supplements, or counseling and diet supplements? Or two of those pairs? Or all three? With the Scheffé test, we search for the source of the overall significance by forming a separate *F* ratio for each pair of treatments. That is, we do a series of two-group ANOVAs, each treatment against every other treatment. By dividing the treatments into all possible pairs, the Scheffé allows us to determine exactly where, among the many treatments, the significant difference lies.

Recall that *F* is the ratio of the between-groups *MS* over the within-groups *MS*. Thus, the three Scheffé *F* ratios are as follows:

$$F_{\text{med vs. diet}} = \frac{MS_{\text{bet med vs. diet}}}{MS_{\text{with}}}$$

$$F_{\text{med vs. couns}} = \frac{MS_{\text{bet med vs. couns}}}{MS_{\text{with}}}$$

$$F_{\text{couns vs. diet}} = \frac{MS_{\text{bet couns vs. diet}}}{MS_{\text{with}}}$$

SOURCE: Peanuts: © United Feature Syndicate, Inc.

The within-groups *MS* in the denominator of each Scheffé test is the same as that used in the overall ANOVA. From Table 27.1, we see that it is 42.733. However, the between-groups *MS* in the numerator of each of the three Scheffé tests differs from the overall *F* test. In a Scheffé test, the between-groups *MS* is between only the two treatments being compared rather than between all treatments. Therefore, we will need to calculate each between-groups *MS*.

Recall, also, that $MS_{bet} = SS_{bet}/df_{bet}$. Thus, before we can calculate each MS_{bet}, we first need to find each SS_{bet}. Fortunately for us, we calculated the full SS_{bet} when we conducted the overall *F* test. So we will simply use the necessary SS_{bet} portions from the overall ANOVA calculations. The overall ANOVA data are presented in Table 27.2.

Table 27.2 Depression Level After Treatment: Raw Score Calculation

Medication		*Counseling*		*Diet Supplement*		
X	X^2	X	X^2	X	X^2	
23	529	38	1,444	40	1,600	
16	256	32	1,024	28	784	
15	225	29	841	33	1089	
32	1,024	42	1,764	42	1,764	
26	676	25	625	35	1,225	
18	324	17	289	35	1,225	
22	484	37	1,369	41	1,681	
14	196	26	676	30	900	
14	196	19	361	34	1,156	
22	484	22	484	39	1,521	
202		287		357		$\rightarrow \sum X = 846$
	4,394		8,877		12,945	$\rightarrow \sum X^2 = 26{,}216$

Recall that to calculate SS_{bet} for the overall ANOVA, we used the following formula:

$$SS_{bet} = \sum_{1}^{k} \frac{\left(\sum X_g\right)^2}{n_g} - \frac{\left(\sum X_{tot}\right)^2}{N}$$

$$= (\text{Medication} + \text{Counseling} + \text{Diet Supplements}) - \text{Total}$$

$$= \left\{ \frac{(202)^2}{10} + \frac{(287)^2}{10} + \frac{(357)^2}{10} \right\} - \frac{(846)^2}{30}$$

$$= \ldots$$

Borrowing the parts we need for the Scheffé test,

$$SS \text{ for medication only} = \frac{(202)^2}{10}$$

$$SS \text{ for counseling only} = \frac{(287)^2}{10}$$

$$SS \text{ for diet supplements only} = \frac{(357)^2}{10}$$

> How difficult it often is for an experimenter to interpret his results without the aid of mathematics.
>
> —Lord Raleigh

The second term in the Scheffé formula, the total sums of squares, is the total *for the two treatments being compared* rather than for all treatments combined:

$$SS_{\text{bet}} = \sum_{1}^{k} \frac{\left(\sum X_g\right)^2}{n_g} - \frac{\left(\sum X_{\text{tot}}\right)^2}{N}$$

Let us now calculate SS_{bet} for all possible pairs of treatments.

$$SS_{\text{med vs. diet}} = \left\{ \overset{\text{Medication}}{\frac{(202)^2}{10}} + \overset{\text{Diet}}{\frac{(357)^2}{10}} \right\} - \overset{\text{Total for two treatments}}{\frac{(202+357)^2}{20}}$$

$$= \left\{ \frac{40804}{10} + \frac{127449}{10} \right\} - \frac{312481}{20}$$

$$= (4080.40 + 12744.90) - 15624.05$$

$$= 16825.30 - 15624.05$$

$$= 1201.25$$

$$SS_{\text{med vs. couns}} = \left\{ \overset{\text{Medication}}{\frac{(202)^2}{10}} + \overset{\text{Counseling}}{\frac{(287)^2}{10}} \right\} - \overset{\text{Total for two treatments}}{\frac{(202+287)^2}{20}}$$

$$= \left\{ \frac{40804}{10} + \frac{82369}{10} \right\} - \frac{239121}{10}$$

$$= (4080.40 + 8236.90) - 11956.05$$

$$= 12317.30 - 11956.05$$

$$= 361.25$$

$$SS_{\text{couns vs. diet}} = \left\{ \overset{\text{Counseling}}{\frac{(287)^2}{10}} + \overset{\text{Diet}}{\frac{(357)^2}{10}} \right\} - \overset{\text{Total for two treatments}}{\frac{(287+357)^2}{20}}$$

$$= \left\{ \frac{82369}{10} + \frac{127449}{10} \right\} - \frac{414736}{20}$$

$$= (8236.90 + 12744.90) - 20736.80$$

$$= 20981.80 - 20736.80$$

$$= 245$$

Now that we have each SS_{bet}, we can calculate each MS_{bet}. Recall that $MS_{\text{bet}} = SS_{\text{bet}}/df_{\text{bet}}$. In the Scheffé test, the df_{bet} for each comparison is whatever the *df* was for the overall ANOVA rather than between the two groups being compared. Our overall ANOVA had three groups. Thus, the $df_{\text{bet}} = K - 1 = 2$. Using 2 *df* in the denominator for each paired comparison rather than the 1 *df* that actually exists between the two treatments being compared contributes to the test's conservativeness in distributing Type 1 error. This is because the larger divisor results in a smaller *F* and hence requires a bigger treatment effect in the numerator for the final *F* to meet statistical significance. The increased stringency in declaring a result statistically significant means that we are also much less likely to make a Type 1 error.

$$MS_{\text{bet med vs. diet}} = \frac{SS_{\text{med vs. diet}}}{df_{\text{med vs. diet}}}$$
$$= \frac{1201.25}{2}$$
$$= 600.625$$

$$MS_{\text{bet med vs. couns}} = \frac{SS_{\text{med vs. couns}}}{df_{\text{med vs. couns}}}$$
$$= \frac{361.25}{2}$$
$$= 180.625$$

$$MS_{\text{bet couns vs. diet}} = \frac{SS_{\text{couns vs. diet}}}{df_{\text{couns vs. diet}}}$$
$$= \frac{245}{2}$$
$$= 122.50$$

Last, we calculate the three F tests:

$$F_{\text{med vs. diet}} = \frac{MS_{\text{bet med vs. diet}}}{MS_{\text{with}}}$$
$$= \frac{600.625}{42.733} \rightarrow \text{(From Table 27.1)}$$
$$= 14.055$$

$$F_{\text{med vs. couns}} = \frac{MS_{\text{bet med vs. couns}}}{MS_{\text{with}}}$$
$$= \frac{180.625}{42.733} \rightarrow \text{(From Table 27.1)}$$
$$= 4.227$$

$$F_{\text{couns vs. diet}} = \frac{MS_{\text{bet couns vs. diet}}}{MS_{\text{with}}}$$
$$= \frac{122.50}{42.733} \rightarrow \text{(From Table 27.1)}$$
$$= 2.867$$

We are finally done. We now have three Scheffé post hoc F statistics: 14.055, 4.227, and 2.867. But what do they mean?

Interpreting the Scheffé

We were trying to find which treatments account for the statistically significant overall *F*. Because the Scheffé test is calculated as an ANOVA, we use the ANOVA *F* table to interpret the Scheffé results. Thus, we must determine whether our observed Scheffé *F* tests meet or exceed the tabled critical *F*.

What do you think the *df* are for the Scheffé tests? Did you say 2 and 27? If you did, you are correct. The *df* are 2 and 27, just as they were for the overall ANOVA. This, too, contributes to the Scheffé's conservativeness.

The ANOVA table of *F* is found in Appendix D. Table 27.3 is a portion of that table.

Table 27.3 A Portion of the *F* Table for Use With the Scheffé Test

		df_{bet}			
df_{with}	α (%)	1	2	3	4
26	5%	4.22	3.37	2.98	2.74
	1%	7.72	5.53	4.64	4.14
27	5%	4.21	3.35	2.96	2.73
	1%	7.68	5.49	4.60	4.11
28	5%	4.20	3.34	2.95	2.71
	1%	7.64	5.45	4.57	4.07

Assume that we are willing to make a Type 1 error 5% of the time. At $\alpha = .05$ and $df = 2$ and 27, what is the critical *F*?

Yes, it is 3.35. This is the lowest value of *F* necessary to reject the null hypothesis for any pair of treatments. Now, we compare the tabled *F* with the three Scheffé post hoc *F*s that we calculated. Which pair(s) of treatments account for the statistically significant overall ANOVA? Here, again, are our three Scheffé *F*s:

$$F_{\text{med vs. diet}} = 14.055$$
$$F_{\text{med vs. couns}} = 4.227$$
$$F_{\text{couns vs. diet}} = 2.867$$

Yes, at $\alpha = .05$, there is a significant difference in depression level between clients given medication and those given diet supplements, and there is also a significant difference in depression level between clients given medication and those given counseling. In contrast, there is no significant difference in depression level between clients given counseling and those given diet supplements.

Now, assume that we are willing to make a Type 1 error only 1% of the time. At $\alpha = .01$ and $df = 2$ and 27, what is the critical *F*?

Yes, it is 5.49. Now, we compare the tabled critical *F* with the three Scheffé post hoc *F*s. Which pair(s) of treatments account for the statistically significant overall ANOVA? Here, again, are our three Scheffé *F*s:

$$F_{\text{med vs. diet}} = 14.055$$
$$F_{\text{med vs. couns}} = 4.227$$
$$F_{\text{couns vs. diet}} = 2.867$$

Yes, at $\alpha = .01$, there is a significant difference in depression level only between clients given medication and those given diet supplements. There is no significant difference in depression level between clients given medication and those given counseling, or between clients given counseling and those given diet supplements. Note that one of the pairs that was statistically significant at the .05 error level is not statistically significant at the .01 error level.

In a journal article, a Scheffé test result would typically become part of the narrative. For example, our results at the .05 error level might be written as follows:

> Scheffé $F = 3.35$, $\alpha = .05$. Only two treatment pairs met the difference criterion: medication versus diet supplements (15.50) and medication versus counseling (8.50).

CHECK YOURSELF!

We are less likely to make a Type 1 error with a Scheffé test than with a Tukey HSD. What aspect of the Scheffé test discussed in this module contributes to it constraining Type 1 error?

PRACTICE

1. Exercise 1 in Module 25 examined the relative effectiveness of three different weight loss regimens. Here is the ANOVA summary table:

Source	*SS*	*df*	*MS*	*F*
Between	321.50	2	160.75	6.145
Within	663.25	33	26.16	
Total	1,164.75	35		

Here is the raw score data table:

Only Diet and Exercise		*Diet, Exercise, and Personal Trainer*		*Diet, Exercise, and Support Group*	
X	*X*²	*X*	*X*²	*X*	*X*²
4	16	11	121	7	49
6	36	14	196	9	81
12	144	9	81	10	100
5	25	22	484	6	36
8	64	15	225	12	144
7	49	8	64	16	256
10	100	12	144	11	121
17	289	17	289	8	64
4	16	24	576	19	361
12	144	30	900	14	196
9	81	8	64	13	169
5	25	16	256	7	49
99	989	186	3,400	132	1,626

Conduct a Scheffé test on the ANOVA at $\alpha = .05$. Between what pair(s) does the statistical significance lie?

2. Exercise 2 in Module 25 examined the effect of difference in travel route on time taken to arrive at work. Here is the ANOVA summary table:

Source	*SS*	*df*	*MS*	*F*
Between	4,192.59	2	2,096.30	76.37
Within	576.38	21	27.45	
Total	4,768.97	23		

Here is the raw score data table:

Public		*Highway*		*Back Roads*	
X	X^2	*X*	X^2	*X*	X^2
34	1,156	44	1,936	59	3,481
33	1,089	47	2,209	63	3,969
36	1,296	49	2,401	65	4,225
32	1,024	58	3,364	69	4,761
31	961	40	1,600	53	2,809
29	841	54	2,916	71	5,041
35	1,225	52	2,704	73	5,329
30	900	46	2,116	66	4,356
260	8,492	390	19,246	519	33,971

Conduct a Scheffé test on the ANOVA at $\alpha = .01$. Between which pair(s) does the statistical significance lie?

3. Exercise 3 in Module 25 examined whether the order of question difficulty affects test score. Here is the ANOVA summary table:

Source	*SS*	*df*	*MS*	*F*
Between	3.2	2	1.6	0.138
Within	313.5	27	11.61	
Total	318.7	29		

Here is the raw score data table:

Difficult First		*Easy First*		*Random Order*	
X	X^2	*X*	X^2	*X*	X^2
15	225	20	400	11	121
13	169	18	324	14	196
17	289	16	256	9	81
16	256	11	121	15	225
20	400	13	169	17	289
8	64	10	100	13	169
12	144	15	225	19	361
11	121	15	225	18	324
9	81	14	196	12	144
14	196	11	121	11	121
135	1,945	143	2,137	139	2,031

Conduct a Scheffé test on the ANOVA at $\alpha = .05$. Between which pair(s) does the statistical significance lie?

4. Exercise 4 in Module 25 examined the relationship between hours spent studying for final exams and final GPA. Here is the ANOVA summary table:

Source	*SS*	*df*	*MS*	*F*
Between	235.0	3	78.33	10.09
Within	279.4	36	7.76	
Total	514.4	39		

Here is the raw score data table:

2.0–2.5		*2.5–3.0*		*3.0–3.5*		*3.5–4.0*	
X	*X*²	*X*	*X*²	*X*	*X*²	*X*	*X*²
26	676	28	784	30	900	30	900
27	729	27	729	31	961	31	961
24	576	29	841	33	1,089	32	1,024
30	900	30	900	38	1,444	35	1,225
31	961	33	1,089	38	1,444	38	1,444
26	676	28	784	30	900	35	1,225
28	784	34	1,156	29	841	32	1,024
27	729	32	1,024	28	784	31	961
28	784	30	900	35	1,225	33	1,089
26	676	27	729	36	1,296	36	1,296
273	7,491	298	8,936	328	10,884	333	11,149

Conduct a Scheffé test on the ANOVA at $\alpha = .01$. Between which pair(s) does the statistical significance lie?

SPSS Connection

SPSS output for a Scheffé post hoc test is obtained by the same method as the Tukey HSD discussed in the SPSS section of Module 26, except that you select the Scheffé box instead of the Tukey box. However, the output is not informative. While the calculation algorithm is a series of paired ANOVAs as shown in the textbook, the output produced by SPSS shows paired mean differences, much like the Tukey HSD. Because of the lack of concordance between the SPSS output and the textbook instruction, the Scheffé output is not shown here.

Visit the study site at www.sagepub.com/steinberg2e for practice quizzes and other study resources.

PART XII

More Than One Independent Variable

"Double Dutch Jump Rope"

28 Main Effects and Interaction Effects

Terms: factorial, main effect, interaction effect

Learning Objectives:

- Express a factorial ANOVA design in symbols
- Determine the number of possible main effects in an ANOVA
- Determine the number of possible interaction effects in an ANOVA
- From a graph of cell means, determine whether or not there are probable main effects or interaction effects
- From a table of cell means, determine whether or not there are probable main effects or interaction effects

What Is a Factorial ANOVA?

Recall that there are two situations for which we conduct an ANOVA rather than a *t* test. One situation is when the independent variable has more than two levels or conditions—say, three different treatments for depressed clients rather than only two. We looked at this situation in Modules 24 through 27.

We must also conduct an ANOVA rather than a *t* test when we have more than one independent variable in our study, regardless of the number of levels or conditions for each independent variable. Sticking with the depression example, not only might clients be given different treatments, but the treatments might be given either individually or as part of a group. Such a study is called **factorial** because it has more than one independent variable or "factor."

Factorial ANOVA Designs

For this module, let's change our working example. Let's consider the effect of two different independent variables on the number of words participants are able to remember. When we came across this example in Module 21, there was only one independent variable: sleep deprivation. Let's have three levels of sleep deprivation. Some participants are not permitted to sleep during the night before the recall test, some participants are allowed only 4 hr of sleep, and some participants are allowed 8 hr of sleep. Now suppose we add a second independent variable: memory training. Let's have two levels of memory training. Some get memory training and some do not get memory training. If we evenly divide the participants between the two independent variables (memory training and sleep deprivation) and then evenly divide them again between the treatment conditions within each independent variable, we have six unique groups across the two independent variables. The study's design can be seen in Table 28.1.

Table 28.1 Design Diagram for a Two-Way ANOVA

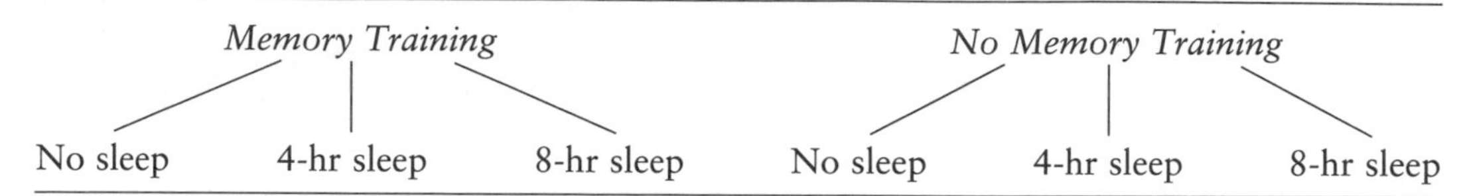

When a study has two independent variables, it is called a *two-way* ANOVA. For this study, the two independent variables are memory training and sleep deprivation. In symbols, this study is referred to as a 2 × 3 ANOVA, because there are two levels or conditions of the first independent variable (memory training, no memory training) and three levels or conditions of the second independent variable (no sleep, 4 hr of sleep, 8 hr of sleep).

Suppose we add a third independent variable to the study—say, time of day when the recall test is given. Assume that this new independent variable has three levels or conditions: morning, afternoon, or evening. What would such an ANOVA be called? How would it be symbolized?

It would be called a *three-way* ANOVA, because there are three independent variables: memory training, sleep deprivation, and time of day for the recall test. It would be symbolized as a 2 × 3 × 3 ANOVA because there are two conditions for the first independent variable, three conditions for the second independent variable, and three conditions for the third independent variable.

How would we diagram the three-way ANOVA? Working from our previous diagram for the two-way ANOVA, we can diagram the three-way ANOVA as shown in Table 28.2.

Table 28.2 Design Diagram for a Three-Way ANOVA

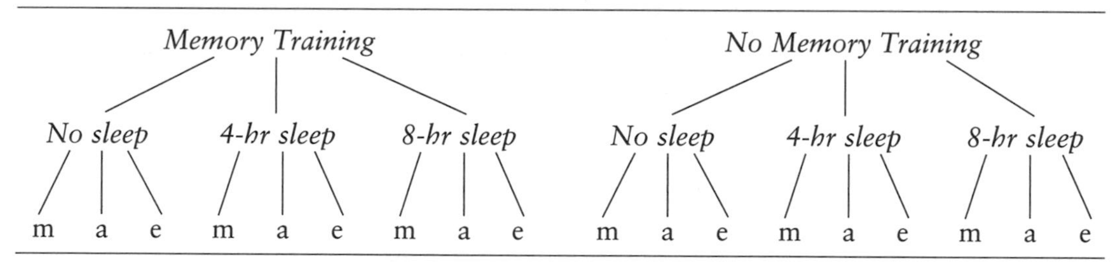

CHECK YOURSELF!

You are told that a study is a four-way ANOVA. From that statement, do you know how many independent variables are in the study? Do you know how many levels or conditions there are for each independent variable?

PRACTICE

1. You are told that a study is a 3 × 3 ANOVA. Make up variables and conditions and draw a diagram for the study.
2. You are told that a study is a 3 × 2 × 2 ANOVA. Make up variables and conditions and draw a diagram for the study.

Number and Type of Hypotheses

A factorial ANOVA combines several different hypotheses in a single analysis. One hypothesis is that the independent variable has an effect on the dependent variable. This is called the main effect. There is one of these hypotheses for each independent variable in the study. Main effects are similar to the one-way ANOVAs that we conducted in Module 25.

> It is a capital mistake to theorize before one has the data.
>
> —Sir Arthur Conan Doyle

A second type of hypothesis in a factorial ANOVA is that one independent variable has an effect on the dependent variable, but only as a function of the level or condition of the second independent variable. This is called an interaction effect. Interaction effects are unique to factorial ANOVAs. We cannot have an interaction effect in a one-way ANOVA because there is only one independent variable.

Let's look more closely at main and interaction effects. For simplicity's sake, let's return to a two-way ANOVA for the remainder of this module. Our study will have two independent variables: memory training and sleep deprivation.

Main Effects

When there is a **main effect**, one independent variable has a significant effect on the dependent variable, regardless of the treatment level or condition of the other independent variable(s).

There can be as many main effects as there are independent variables. Thus, in our two-way ANOVA example, we could have a main effect for memory training on the number of words remembered, regardless of sleep status. We could also have a main effect for sleep deprivation on the number of words remembered, regardless of memory training status.

To look for main effects, let's plot the average number of words remembered for each of the six groups. Because only two axes fit on a flat piece of paper, it is customary to plot the dependent variable on the vertical Y-axis, plot one of the independent variables on the horizontal X-axis, and plot the other independent variable via separate lines or dots. Figure 28.1 shows such a graph.

Figure 28.1 Unconnected Group Means for a Two-Way ANOVA With Two Main Effects and No Interaction Effect

From the graph in Figure 28.1, there appear to be two main effects. There is a main effect for memory training because, regardless of degree of sleep deprivation, participants who are given memory training remember more words than those who are not given memory training. Stop now and examine Figure 28.1 to confirm that you can find the main effect for memory training. To do this, look at the relative positions for mean words remembered in the memory-training and no-memory-training conditions. Notice that participants remember more words in the memory-training condition than in the no-memory-training condition, regardless of whether they have 8 hr of sleep, 4 hr of sleep, or no sleep.

In addition, there appears to be a main effect for sleep deprivation because, regardless of whether or not participants receive memory training, those who get more sleep remember more words than those who get less sleep. Again, stop now and examine Figure 28.1 to confirm that you can find the main effect for sleep. To do this, look at the relative positions for mean words remembered for the different amounts of sleep. Notice that participants who have 8 hr of sleep remember more words than those who have 4 hr of sleep, who in turn remember more words than those who have no sleep, regardless of whether or not they receive memory training.

Now let's connect the group means via straight lines. The graph then looks like Figure 28.2.

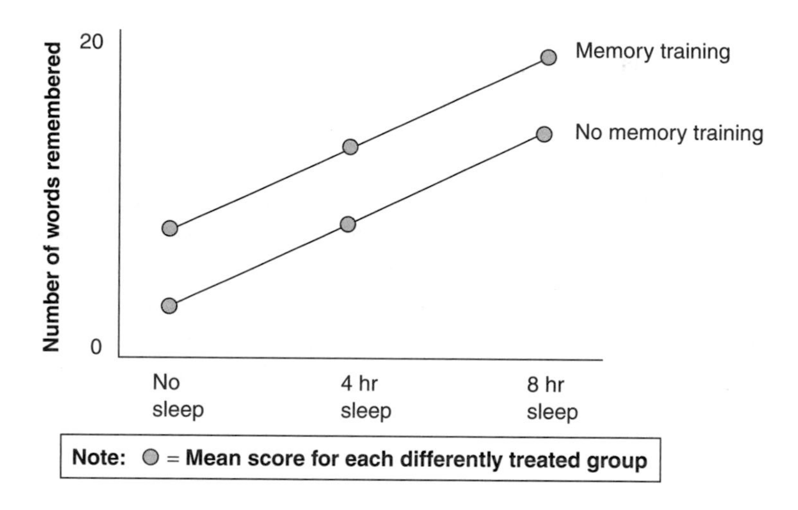

Figure 28.2 Connected Group Means for a Two-Way ANOVA With Two Main Effects and No Interaction Effect

Notice that the lines are parallel. When there are one or more main effects and no interaction effect, the lines connecting the group means will be parallel.

SOURCE: By permission of John L. Hart FLP and Creators Syndicate, Inc. (B.C.)

Some statisticians prefer to look for main effects via a table rather than a graph. Table 28.3 shows the same data.

Table 28.3 Group Means for a Two-Way ANOVA With Two Main Effects and No Interaction Effect

	No Sleep	*4-hr Sleep*	*8-hr Sleep*	*Memory Condition Row Means*
Memory training	10	15	20	15
No memory training	5	10	15	10
Sleep Condition Column Means	7.5	12.5	17.5	

The table cells give the group means. Row means to the far right give the mean across all groups for a given level or condition of one independent variable (in this case, memory-training conditions). Column means along the bottom give the mean across all groups for a given level or condition of the other independent variable (in this case, sleep deprivation conditions).

From the row and column means, we can see that participants who receive memory training (first row) remember more words than those who do not receive memory training (second row), regardless of the amount of sleep they have. Similarly, participants who have 8 hr of sleep (third column) remember more words than those who have only 4 hr of sleep (middle column), who in turn remember more words than those who have no sleep (left column), regardless of whether or not they receive memory training. Thus, there appear to be main effects for both memory training and amount of sleep.

SOURCE: By permission of Dave Coverly and Creators Syndicate, Inc. (Speed Bump)

PRACTICE

3. A psychologist conducts a 2 × 3 × 2 ANOVA. How many main effects are possible?
4. A psychologist conducts a 3 × 5 ANOVA. How many main effects are possible?
5. A researcher conducts a 3 × 6 ANOVA. How many main effects are possible?
6. A researcher conducts a 2 × 2 × 4 ANOVA. How many main effects are possible?

Interaction Effects

When there is an **interaction effect**, one independent variable has a significant effect on the dependent variable, but only under certain levels or conditions of the other independent variable.

There can be as many interaction effects as there are *combinations* of the independent variables. In a two-way ANOVA, there can be only one interaction effect. For our memory study, the one interaction would be memory-training condition by sleep-deprivation condition. The interaction is symbolized as M × S (using the first letter of each independent variable). In a three-way ANOVA, there would be four possible interactions. For the three-way

ANOVA discussed earlier, the four interaction effects would be: (1) memory-training condition by sleep condition, symbolized M × S; (2) sleep condition by time of the recall test, symbolized S × T; (3) memory-training condition by time of the recall test, symbolized M × T; and (4) a three-way interaction of memory-training condition by sleep condition by time of the recall test, symbolized M × S × T.

Let's return to our two-way ANOVA and graph a hypothetical interaction effect (Figure 28.3).

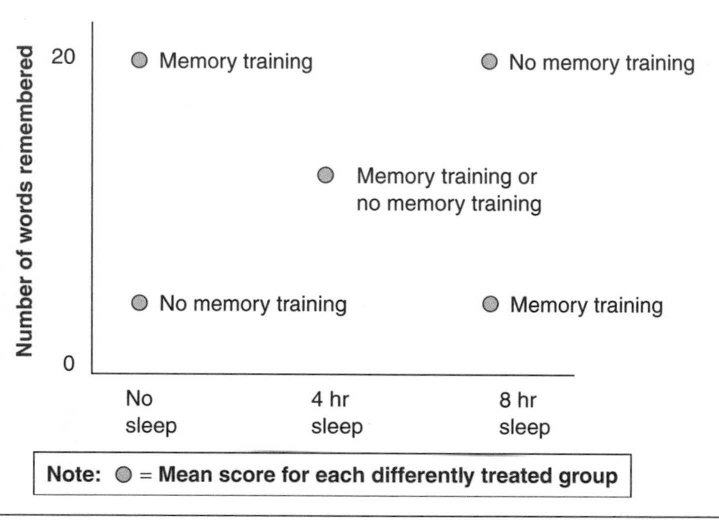

Figure 28.3 Unconnected Group Means for a Two-Way ANOVA With an Interaction Effect and No Main Effects

> To understand is to perceive patterns.
>
> —Isaiah Berlin, in *Historical Inevitability*

As depicted in Figure 28.3, participants who receive memory training remember more words than those who do not receive memory training, but only if those participants receive no sleep. When the participants receive 8 hr of sleep, memory training not only doesn't help participants' recall but also decreases it. Stop now and look at Figure 28.3 to confirm that you can find the interaction effect via the position of the group means.

Now let's connect the group means via straight lines. The graph then looks like Figure 28.4.

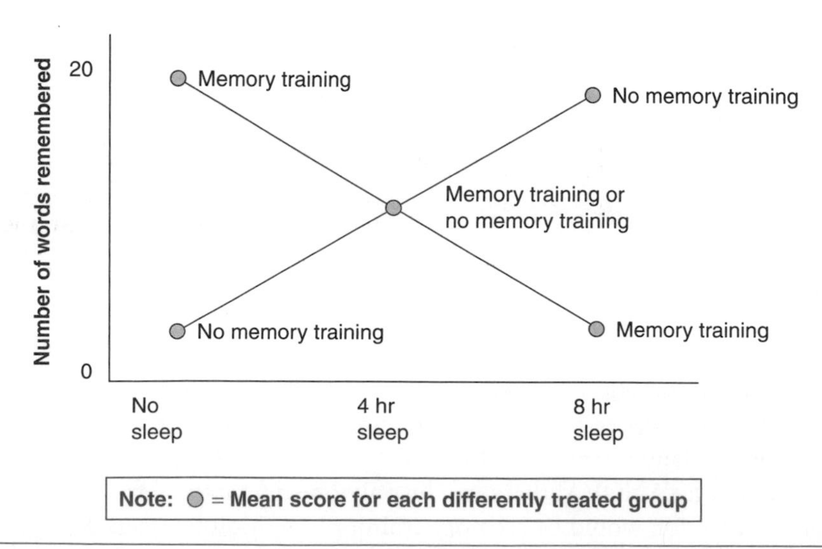

Figure 28.4 Connected Group Means for a Two-Way ANOVA With an Interaction Effect and No Main Effects

Notice that the lines are not parallel. When there is an interaction effect, the lines connecting the group means are not parallel. The lines cross at some point, either within the graph or, if the lines were to be extended, someplace off the graph.

Moreover, although there is a clear interaction effect, there are no main effects. Look at the average (middle) point across the three sleep conditions on both memory condition lines. The number of words remembered is identical. Similarly, look at the average (middle) point between the two memory condition lines at each of the three sleep conditions. The number of words remembered is identical.

Again, some statisticians prefer to look for interaction effects via a table rather than a graph. Table 28.4 shows the same data.

Table 28.4 Group Means for a Two-Way ANOVA With an Interaction Effect and No Main Effects

	No Sleep	*4-hr Sleep*	*8-hr Sleep*	*Memory Condition Row Means*
Memory training	20	12.5	5	12.5
No memory training	5	12.5	20	12.5
Sleep Condition Column Means	12.5	12.5	12.5	

From the group means within the cells, we can see that the relationship between number of words remembered and memory training is inconsistent. Participants given memory training (first data row) remember more words than participants who are not given memory training (second data row), but only if they get no sleep (top left cell). Similarly, the relationship between number of words remembered and sleep deprivation is inconsistent. Participants who get 8 hr of sleep (right data column) remember more words than participants who have either 4 hr of sleep or no sleep (middle and left columns), but only if they are given no memory training (bottom right cell). Hence, the two independent variables interact in their effect on the dependent variable.

Moreover, although there is a clear interaction effect, the lack of any main effect can be seen in the row and column means (in this case, 12.5). The group means within the cells arising from different combinations of conditions on the two independent variables cancel each other, making the row and column means across each condition equivalent.

PRACTICE

7. A psychologist conducts a 2 × 3 × 2 ANOVA. How many interaction effects can there be?
8. A psychologist conducts a 3 × 5 ANOVA. How many interaction effects can there be?
9. A researcher conducts a 3 × 6 ANOVA. How many interaction effects are possible?
10. A researcher conducts a 2 × 2 × 4 ANOVA. How many interaction effects are possible?

Looking Ahead

In this module, we looked for main effects and interaction effects graphically and in tables. It is good practice to either graph or table your data before calculating any statistics because graphs and tables give a sense of what the data show and, hence, what the statistics will probably be. However, the true test of whether or not there is a main effect or an interaction effect lies not with a graph or table but with the calculated statistics. For example, suppose the results look like Figure 28.5 (which, by the way, is more likely than the exaggerated results shown in Figure 28.4).

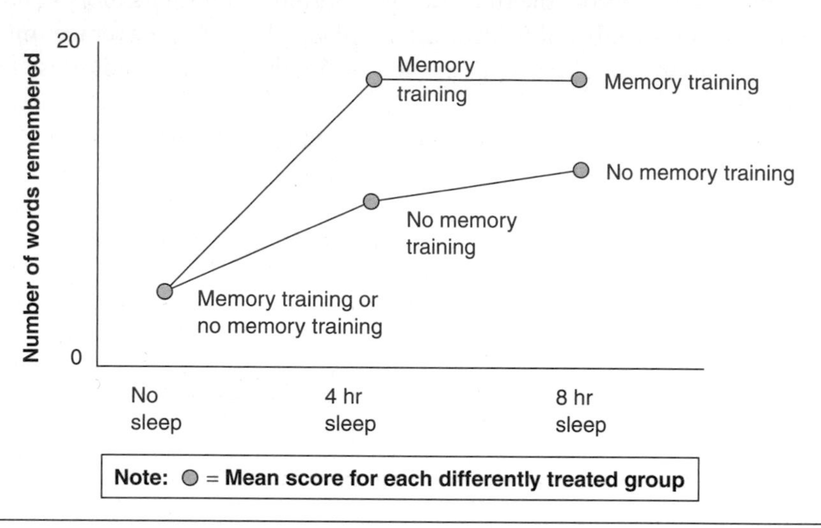

Figure 28.5 Graph of Connected Group Means for a Two-Way ANOVA With a Less Obvious Interaction Effect

> Although the pattern prevailed, the breaks were everywhere.
>
> —Gwendolyn Brooks, U.S. poet

In Figure 28.5, the lines are neither completely parallel nor crossed within the graph. The lines are somewhat parallel, but are they parallel enough for there to be a main effect? They would cross if the lines were extended beyond the left vertical axis at the same angles at which they approach the axis, but is the divergence enough for there to be an interaction effect?

The situation is just as ambiguous when the data are presented in table form (Table 28.5).

Table 28.5 Group Means for a Two-Way ANOVA With a Less Obvious Interaction Effect

	No Sleep	*4-hr Sleep*	*8-hr Sleep*	*Memory Condition Row Means*
Memory Training	5	17	17	13
No Memory Training	5	10	12	9
Sleep Condition Column means	5	13.5	14.5	

> Certitude is not the test of certainty. We have been cocksure of many things that are not so.
>
> —Oliver Wendell Holmes

From the pattern of group means in the table's cells and in its row and column means, memory training appears to result in more correct words than no memory training, but not in the no-sleep condition. In the no-sleep condition, memory training has no effect on the number of words correctly remembered (5). Similarly, more sleep appears to increase the number of words correctly remembered when the move is from no sleep to 4 hr of sleep, but when the move is from 4 hr of sleep to 8 hr of sleep, the number of additional words correctly remembered is negligible (13.5 vs. 14.5).

Determining whether or not there are main effects or interaction effects merely by looking at graphs or tables of group means is akin to trying to determine whether or not there is a treatment effect in a one-way ANOVA merely by comparing group means. Just as the means in a one-way ANOVA are rarely identical between the treated and untreated groups even when the null hypothesis is true, so also the lines in a factorial ANOVA are rarely exactly parallel when there is a main effect or perfectly crossed when there is an interaction effect. The real question is not whether the results are exactly parallel or perfectly crossed, but whether the divergence is *a lot* or *only a little.*

For a one-way ANOVA, we calculated an *F* statistic to test the deviation from expectation for statistical significance. We must now do the same for a factorial ANOVA. That is, whereas in this module we made visual judgments as to whether or not there were main effects or an interaction effect, in the next module, we will calculate *F* statistics to determine whether the main effects and interaction effect are statistically significant.

PRACTICE

11. A psychologist is studying the effect of the number of hours of television children view and the primary type of program children view on the children's level of aggression. Here are the mean aggression levels for each group (higher scores indicate more aggression):

Hours per Week	*Type of Show*	*Aggression Level*
<3	Educational	16
3–10	Educational	11
11–20	Educational	8
>20	Educational	3
<3	Action-adventure	52
3–10	Action-adventure	58
11–20	Action-adventure	69
>20	Action-adventure	71
<3	Situation comedy	22
3–10	Situation comedy	30
11–20	Situation comedy	38
>20	Situation comedy	43

a. What are the independent variables? What is the dependent variable?
b. How many main effects can there be in this study? How many interaction effects can there be?
c. Express the design in words.
d. Express the design in symbols.
e. Graph the group means.
f. Create a table for the group means.
g. Does there appear to be a main effect? If so, for which independent variable(s)? Explain your decision.
h. Does there appear to be an interaction effect? Explain your decision.

12. A college registrar is studying the effect of length of commute (in miles) and gender on the number of absences per semester in the students' first registered M-W-F class of the day. Here are the mean number of absences per semester in the first registered M-W-F class of the day for each group:

Commute Miles	*Gender*	*No. of Absences*
<10	Male	1.9
11–20	Male	2.2
>20	Male	2.8
<10	Female	1.3
11–20	Female	1.6
>20	Female	2.1

   a. What are the independent variables? What is the dependent variable?
   b. How many main effects can there be in this study? How many interaction effects can there be?
   c. Express the design in words.
   d. Express the design in symbols.
   e. Graph the group means.
   f. Create a table for the group means.
   g. Does there appear to be a main effect? If so, for which independent variable(s)? Explain your decision.
   h. Does there appear to be an interaction effect? Explain your decision.

13. A criminologist is studying the effect of parental marital status and highest parental educational level and on the age at which a male child commits his first crime. Here are the mean ages for first crime for each group:

Parents' Marital Status	*Parents' Highest Education*	*Age at Male Child's First Crime*
M	<HS	21
M	HS but <AA	21
M	AA but <BA	24
M	BA or greater	29
NM	<HS	17
NM	HS but <AA	18
NM	AA but <BA	23
NM	BA or greater	25

   a. What are the independent variables? What is the dependent variable?
   b. How many main effects are possible? How many interaction effects are possible?
   c. Is this design a ___ way ANOVA or is it a ___ × ___ way ANOVA?
   d. Draw the design as a tree diagram.
   e. Graph the group means.
   f. Create a table for the group means.
   g. Do there appear to be any main effects? If so, for which independent variable(s)? Explain your decisions.
   h. Do there appear to be any interaction effects? Explain your decision.

14. A sociologist is studying the age at which girls have their first child as a function of the parents' socioeconomic status (SES) and mother's religiosity. Here are the mean ages for girl's first child for each group:

Parents' SES	*Mother's Religiosity*	*Age at Girl's First Child*
Below average	Low	16
Below average	Medium	19
Below average	High	19
Average	Low	20
Average	Medium	20
Average	High	23
Above average	Low	24
Above average	Medium	24
Above average	High	27

a. What are the independent variables? What is the dependent variable?

b. How many main effects are possible? How many interaction effects are possible?

c. Is this design a ___ way ANOVA or is it a ___ × ___ way ANOVA?

d. Draw the design as a tree diagram.

e. Graph the group means.

f. Create a table for the group means.

g. Do there appear to be any main effects? If so, for which independent variable(s)? Explain your decisions.

h. Do there appear to be any interaction effects? Explain your decision.

15. For the following 2 × 2 ANOVA tables, decide whether or not there appears to be (a) a main effect for IV1, (b) a main effect for IV2, and/or (c) an interaction effect. Explain your decisions.

15.1.

	IV1-A	*IV1-B*	*Row Mean*
IV2-A	10	10	10
IV2-B	5	5	5
Column mean	7.5	7.5	

15.2.

	IV1-A	*IV1-B*	*Row Mean*
IV2-A	10	5	7.5
IV2-B	5	10	7.5
Column mean	7.5	7.5	

15.3.

	IV1-A	*IV1-B*	*Row Mean*
IV2-A	10	5	7.5
IV2-B	10	5	7.5
Column mean	10	5	

15.4.

	IV1-A	*IV1-B*	*Row Mean*
IV2-A	15	5	10
IV2-B	10	10	10
Column mean	12.5	7.5	

16. For the following 2 × 2 ANOVA tables, decide whether or not there appears to be (a) a main effect for IV1, (b) a main effect for IV2, and/or (c) an interaction effect.

16.1.

	IV1-A	*IV1-B*	*Row Mean*
IV2-A	10	10	10
IV2-B	10	10	10
Column mean	10	10	

16.2.

	IV1-A	*IV1-B*	*Row Mean*
IV2-A	5	10	7.5
IV2-B	10	5	7.5
Column mean	7.5	7.5	

16.3.

	IV1-A	*IV1-B*	*Row Mean*
IV2-A	5	10	7.5
IV2-B	15	20	17.5
Column mean	10	15	

16.4.

	IV1-A	*IV1-B*	*Row Mean*
IV2-A	5	5	5
IV2-B	5	10	7.5
Column mean	5	7.5	

17. For the following 3 × 2 ANOVA tables, decide whether or not there appears to be (a) a main effect for IV1, (b) a main effect for IV2, and/or (c) an interaction effect. Explain your decisions.

17.1.

	IV1-A	*IV1-B*	*Row Mean*
IV2-A	15	5	10
IV2-B	10	10	10
IV2-C	5	15	10
Column mean	10	10	

17.2.

	IV1-A	*IV1-B*	*Row Mean*
IV2-A	15	10	12.5
IV2-B	10	10	10
IV2-C	5	15	7.5
Column mean	10	10	

17.3.

	IV1-A	*IV1-B*	*Row Mean*
IV2-A	15	5	10
IV2-B	15	5	10
IV2-C	15	5	10
Column mean	15	5	

17.4.

	IV1-A	*IV1-B*	*Row Mean*
IV2-A	15	15	15
IV2-B	10	10	10
IV2-C	5	5	5
Column mean	10	10	

18. For the following graphs of 2 × 2 ANOVAs, decide whether or not there appears to be (a) a main effect for IV1 (graphed as dots along the horizontal *X*-axis), (b) a main effect for IV2 (graphed as two separate lines), and/or (c) an interaction effect. Explain your answers.

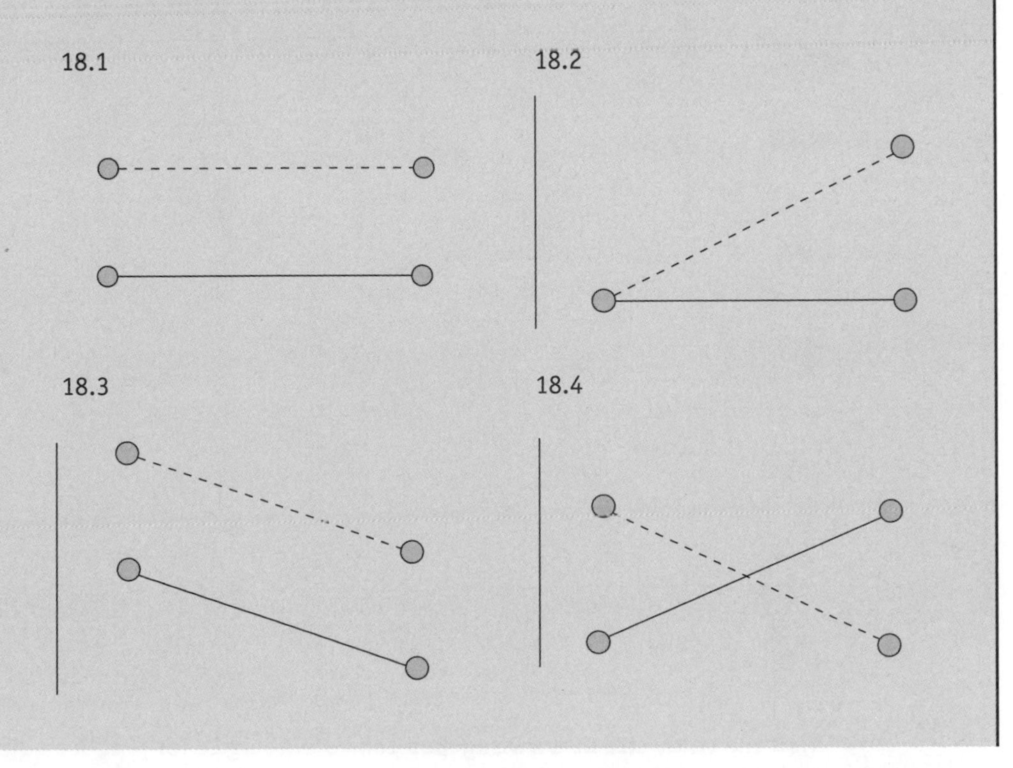

19. For the following graphs of 2 × 2 ANOVAs, the DV is on the vertical axis; the IV1 is on the horizontal axis; and the IV2 is the two separate lines. Decide whether or not there appears to be (a) a main effect for IV1, (b) a main effect for IV2, and/or (c) an interaction effect.

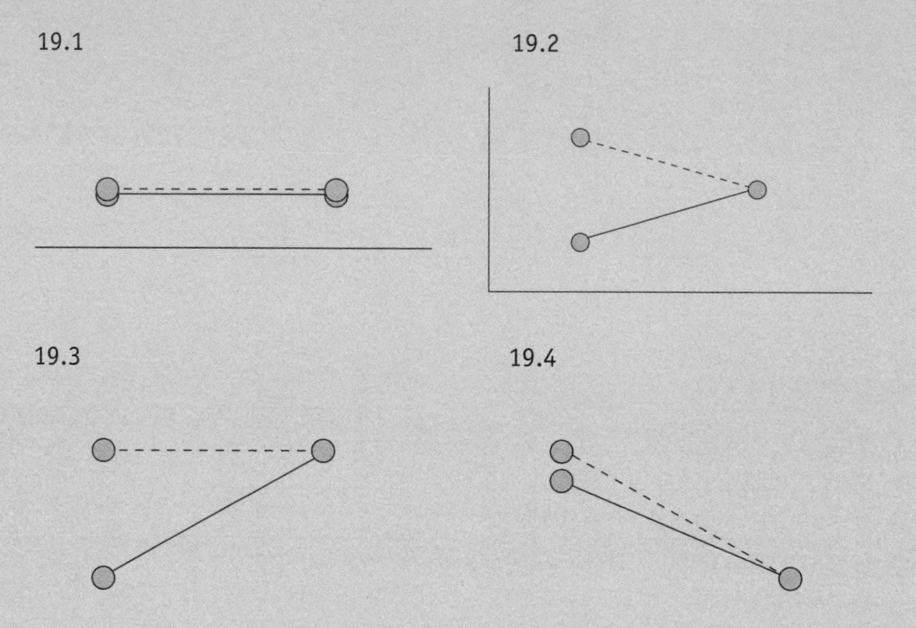

20. For the following graphs of 3 × 3 ANOVAs, decide whether or not there appears to be (a) a main effect for IV1 (graphed as dots along the horizontal X-axis), (b) a main effect for IV2 (graphed as three separate lines), and/or (c) an interaction effect. Explain your answers.

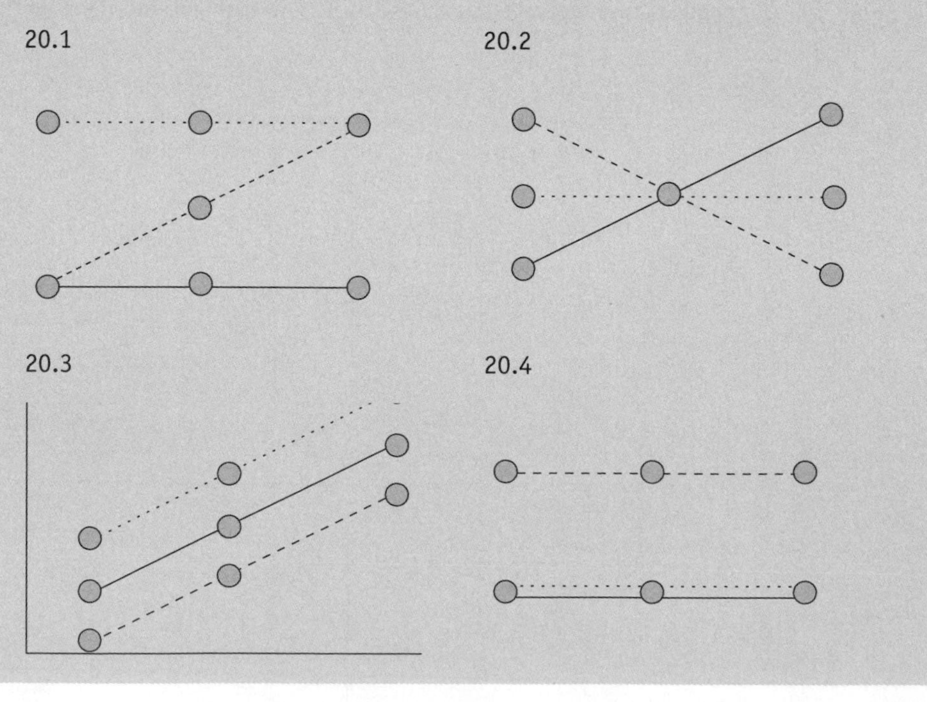

Visit the study site at www.sagepub.com/steinberg2e for practice quizzes and other study resources.

29 Factorial ANOVA

Terms: interaction effect, between conditions

Learning Objectives:

- Understand the similar logic underlying various test statistics
- Determine the degrees of freedom
- Calculate a factorial ANOVA
- Use a table to interpret F
- Present results in an ANOVA table
- Report results in APA format

Review of Factorial ANOVA Designs

Now that you understand the logic behind factorial ANOVA, you are ready to calculate an example. As you learned in Module 28, factorial ANOVAs can have any number of independent variables. However, the minimal design is a two-way ANOVA. Among two-way ANOVAs, the minimal design is a 2 × 2 ANOVA—that is, only two levels of the first independent variable and only two levels of the second independent variable.

Let's return to the memory example we were working with in Module 28. For calculation simplicity (an oxymoron for factorial ANOVA!), we will limit the study to two memory conditions and two sleep conditions—in other words, a 2 × 2 ANOVA. The design for this study is depicted in Table 29.1.

Q: What do we call a statistical superstar?

A: A nova.

Table 29.1 Design Diagram for a Two-Way ANOVA

Memory Training		*No Memory Training*	
Full sleep	No sleep	Full sleep	No sleep

For those who prefer tables, the study is depicted in Table 29.2.

Table 29.2 Design Table for a Two-Way ANOVA

	Memory Training	*No Memory Training*
Full sleep		
No sleep		

Notice that there are four different groups, or cells, in the study. The four cells consist of participants with

1. memory training, full sleep;
2. no memory training, full sleep;
3. memory training, no sleep; and
4. no memory training, no sleep.

We will examine data from these four groups or cells to determine if there are main effects, interaction effects, or both. Recall that there can be as many main effects as there are independent variables. Thus, in our example we could have a main effect for memory training, regardless of sleep condition. We could also have a main effect for sleep deprivation, regardless of memory-training condition. Recall also that there can be as many interaction effects as there are combinations of the independent variables. For our two-way ANOVA, there can be only one interaction effect: memory training by sleep deprivation.

CHECK YOURSELF!

You are studying the effectiveness of various treatments and need to decide whether to analyze the data via a one-way ANOVA or a factorial ANOVA. What do you need to know in order to make that decision?

Data Setup and Preliminary Expectations

For ease of calculation, our study will have only 7 participants in each of the four groups, for a total of 28 participants. The number of words remembered for all 28 participants is given in Table 29.3.

Cell means are shown in boldface within the cells. A cell mean is the average score for participants within that given group. For example, the mean for participants who had full sleep and were given memory training is 10.43. Row means, column means, and the overall mean are shown in boldface outside the table. Row means outside the table refer to the effect of sleep deprivation regardless of memory training; column means outside the table refer to the effect of memory training regardless of sleep deprivation. Because row means refer to one independent variable and column means refer to the other independent variable, their relative values are indicators of main effects.

Examine the means. What are the means for each independent variable condition? This will tell us if we can expect to find main effects.

From inspection of the means, it appears that getting a full night's rest aids recall regardless of memory training (8.79 vs. 5.36 words). Therefore, we can expect our test statistic to indicate a main effect for sleep deprivation.

From inspection of the means, it also appears that memory-training aids recall regardless of amount of sleep (7.79 vs. 6.36 words). However, the effect of memory training on recall is not as great as the effect of getting a full night's sleep on recall. Therefore, our test statistic might or might not indicate a main effect for memory training.

Analyzing data for main effects is equivalent to doing a series of one-way ANOVAs—one for each independent variable. However, the advantage of a factorial ANOVA is that we can look for interaction effects as well. Examine the means again, this time looking to see if the trend in one independent variable (memory training) is consistent or inconsistent across the other independent variable (amount of sleep).

Table 29.3 Data Table for a Two-Way ANOVA of Sleep Deprivation and Memory Training on Word Recall

	Memory Training	*No Memory Training*	
Full sleep			
	9	9	
	11	6	
	9	7	
	12	8	
	12	7	
	9	8	
	11	5	
	$\Sigma = 73$	$\Sigma = 50$	
	$M = 10.43$	$M = 7.14$	$M_{\text{full sleep}} = 8.79$
No sleep			
	4	6	
	7	4	
	6	6	
	4	6	
	5	4	
	4	6	
	6	7	
	$\Sigma = 36$	$\Sigma = 39$	
	$M = 5.14$	$M = 5.57$	$M_{\text{no sleep}} = 5.36$
	$M_{\text{memory training}} = 7.79$	$M_{\text{no memory training}} = 6.36$	$M_{\text{tot}} = 7.07$

From inspection of the means, it appears that memory training is much more effective on recall when participants have had a good night's rest than when they have not had any sleep. Participants recalled 10.43 words after memory training when they were well rested, but only 5.14 words after memory training when they were sleep deprived. Without memory training, participants recalled only 5.57 words when sleep deprived, and the number of words recalled improved only to 7.14 words with full sleep. Thus, it seems that the effectiveness of memory training depended on the amount of sleep the participants had. Therefore, we can expect our test statistic to indicate an **interaction effect** between memory training and amount of sleep.

Sums of Squares Formulas

Formulas for a factorial ANOVA follow the same logic as in a one-way ANOVA. There are simply more of them. Recall that in a one-way ANOVA, we partitioned the total sums of squares (SS_{tot}) into two parts: the sums of squares between treatments (SS_{bet}) and the sums of squares within treatments (SS_{with}). This was the diagram for our one-way ANOVA in Module 25 (see Figure 29.1).

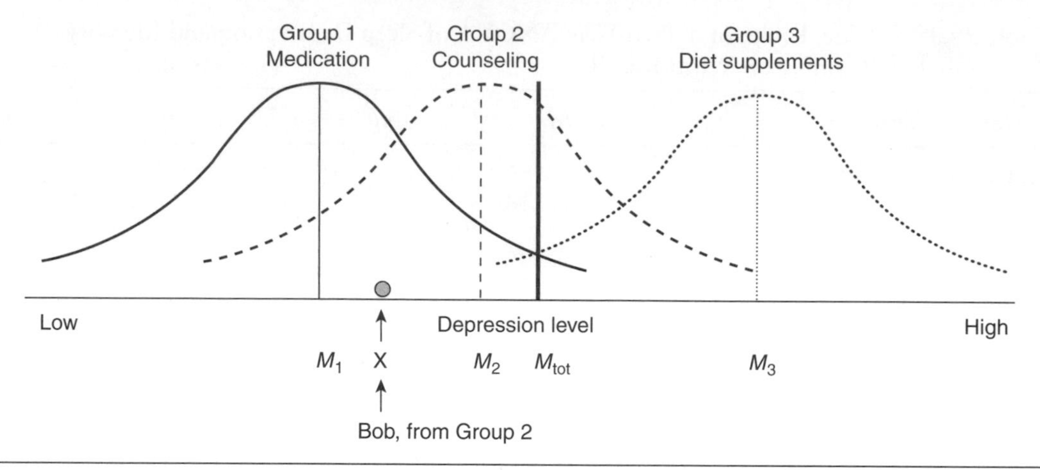

Figure 29.1 Three Different Treatment Populations

Unfortunately, Figure 29.1 displays levels or conditions for only a single independent variable: type of treatment. In contrast, a factorial ANOVA has not just one independent variable but two or more independent variables. The two independent variables in our 2 × 2 study are sleep deprivation and memory training. How can we diagram that? We can't. We could display each independent variable in a separate diagram, as shown in Figure 29.2.

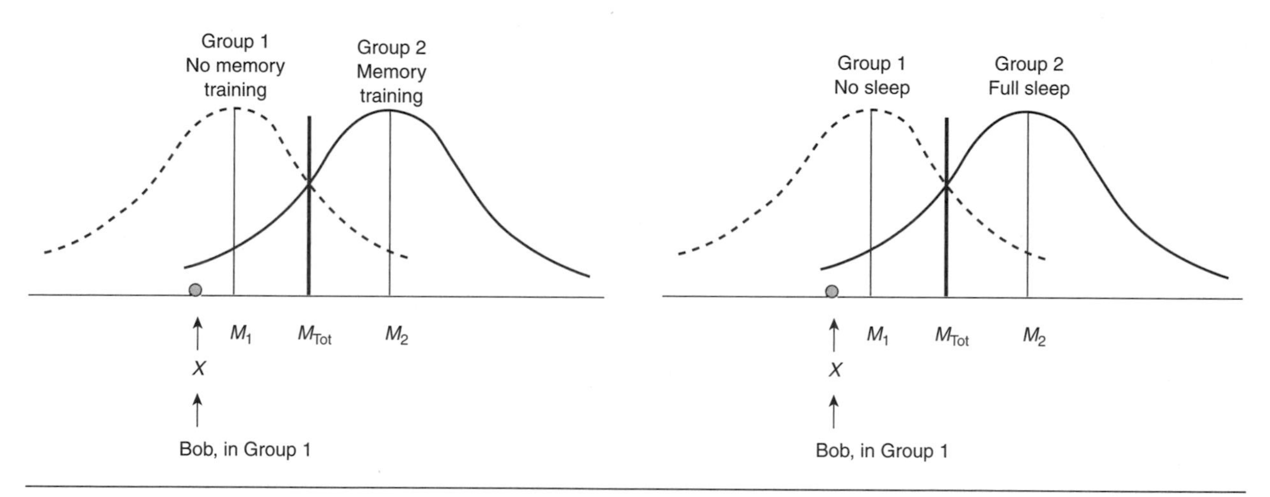

Figure 29.2 Two Separate Independent Variables

Unfortunately, Figure 29.2 does not show that some of the participants in the full-sleep condition are in the memory-training condition while others are in the no-memory-training condition. It also does not show that some of the participants in the no-sleep condition are in the memory-training condition while others are in the no-memory-training condition. In short, unless we could lay part of the top diagram over part of the bottom diagram, we cannot account for both independent variables operating at the same time. For that reason, Figure 29.2 shows only two one-way ANOVAs, not a factorial ANOVA. For a factorial ANOVA, we will have to forgo such a diagram.

Nevertheless, partitioning of the sums of squares follows the same logic as in a one-way ANOVA. Recall that in a one-way ANOVA, the deviation score formulas explicitly indicate the source of each variance:

$$SS_{\text{bet}} = n_g \sum_1^k (M_g - M_{\text{tot}})^2 \rightarrow \text{each group mean from the total mean, for each individual}$$

$$SS_{\text{with}} = \sum_1^k \sum_1^n (X - M_g)^2 \rightarrow \text{each individual score from its group mean, for each group}$$

$$SS_{\text{tot}} = \sum_1^k (X - M_{\text{tot}})^2 \rightarrow \text{each individual score from the total mean}$$

For a factorial ANOVA, we make two formula changes. First, we further partition the sums of squares between treatments into the effects for each separate independent variable. We refer to this as **between conditions.** In our study, the SS_{bet} is partitioned into the sums of squares between sleep conditions ($SS_{\text{bet sleep}}$) and the sums of squares between memory-training conditions ($SS_{\text{bet memory}}$). One of the independent variables appears in the rows of the table, and the other independent variable appears in the columns of the table. Hence, the two independent variables are sometimes referred to as SS_{row} and SS_{column}. However, I will call them $SS_{\text{bet sleep}}$ and $SS_{\text{bet memory}}$ so that there is no confusion over which independent variable is under consideration. Table 29.4 shows the row and column factors in a two-way ANOVA.

Table 29.4 Row and Column Factors in a Two-Way ANOVA

Columns → *↓ Rows*	*Memory Training*	*No Memory Training*
Full sleep		
No sleep		

In addition to the additional *between* sums of squares, a factorial ANOVA must also compute the *interaction* sums of squares—in this case, the interaction of sleep and memory training ($SS_{\text{sleep} \times \text{memory}}$). Recall that the partitioned sums of squares add to the total sums of squares. Therefore, the sums of squares for the interaction is whatever part of the total sums of squares not already accounted for by the sums of squares within or the various sums of squares between.

As in the one-way ANOVA, calculation of these sums of squares is not difficult, just tedious. As before, we can use either deviation score formulas or raw score formulas. Here are the deviation score formulas for a factorial ANOVA:

$$SS_{\text{tot}} = \sum_1^N (X - M_{\text{tot}})^2 \rightarrow \text{each individual score from the total mean}$$

$$SS_{\text{with}} = \sum_1^k \sum_1^n (X - M_{\text{cell}})^2 \rightarrow \text{each individual score from the group mean}$$

$$SS_{\text{bet sleep}} = \sum_1^k (M_{\text{sleep cells}} - M_{\text{tot}})^2 \rightarrow \text{each group mean for the first factor from the total mean}$$

$$SS_{\text{bet memory}} = \sum_1^k (M_{\text{memory cells}} - M_{\text{tot}})^2 \rightarrow \text{each group mean for the second factor from the total mean}$$

$$SS_{\text{sleep} \times \text{memory}} = SS_{\text{tot}} - (SS_{\text{with}} + SS_{\text{bet sleep}} + SS_{\text{bet memory}})$$

Recall that deviation-score formulas are computationally explicit about the sources of variance. However, they are tedious to use. This is even more the case for factorial ANOVA.

With raw score formulas, there is no need to calculate means or deviations from means. Thus, they are simpler to use. The drawback of raw score formulas is that the sources of variance are no longer computationally explicit. Nevertheless, for ease of calculation we will use the raw score formulas. I will not show the deviation score calculations.

Here are the raw score formulas for a factorial ANOVA:

$$SS_{\text{tot}} = n_{\text{g}} \sum_{1}^{k} X^2 - \frac{\left(\sum X_{\text{tot}}\right)^2}{N}$$

$$SS_{\text{with}} = \sum_{1}^{k} X^2 - \sum_{1}^{k} \frac{\left(\sum X_{\text{cell}}\right)^2}{n_{\text{cell}}}$$

$$SS_{\text{bet sleep}} = \sum_{1}^{k} \frac{\left(\sum X_{\text{row}}\right)^2}{n_{\text{row}}} - \frac{\left(\sum X_{\text{tot}}\right)^2}{N}$$

$$SS_{\text{bet memory}} = \sum_{1}^{k} \frac{\left(\sum X_{\text{column}}\right)^2}{n_{\text{column}}} - \frac{\left(\sum X_{\text{tot}}\right)^2}{N}$$

$$SS_{\text{sleep} \times \text{memory}} = SS_{\text{tot}} - (SS_{\text{with}} + SS_{\text{bet sleep}} + SS_{\text{bet memory}})$$

SOURCE: Reprinted with permission of the Carolina Biological Supply Co.

Calculating Factorial ANOVA Sums of Squares: Raw Score Method

Table 29.5 gives the data for our 28 participants. Also included inside and outside each cell are the various sums, squares, sums of squares, and squared sums that are required by the raw score formulas.

Table 29.5 Data Table for Use With Raw Score Formulas

	Memory Training	*No Memory Training*	
Full sleep			
	9	9	
	11	6	
	9	7	
	12	8	
	12	7	
	9	8	
	11	5	
	$n_{\text{cell}} = 7$	$n_{\text{cell}} = 7$	$n_{\text{full sleep}} = 14$
	$\Sigma X_{\text{cell}} = 73$	$\Sigma X_{\text{cell}} = 50$	$\Sigma X_{\text{full sleep}} = 123$
	$\Sigma X^2_{\text{cell}} = 773$	$\Sigma X^2_{\text{cell}} = 368$	$\Sigma X^2_{\text{full cell}} = 1,141$
No sleep			
	4	6	
	7	4	
	6	6	
	4	6	
	5	4	
	4	6	
	6	7	
	$n_{\text{cell}} = 7$	$n_{\text{cell}} = 7$	$n_{\text{no sleep}} = 14$
	$\Sigma X_{\text{cell}} = 36$	$\Sigma X_{\text{cell}} = 39$	$\Sigma X_{\text{no sleep}} = 75$
	$\Sigma X^2_{\text{cell}} = 194$	$\Sigma X^2_{\text{cell}} = 225$	$\Sigma X^2_{\text{no cell}} = 419$
	$n_{\text{mem}} = 14$	$n_{\text{no memory}} = 14$	$N_{\text{tot}} = 28$
	$\Sigma X_{\text{memory}} = 109$	$\Sigma X_{\text{no memory}} = 89$	$\Sigma X_{\text{tot}} = 198$
	$\Sigma X^2_{\text{memory}} = 967$	$\Sigma X^2_{\text{no memory}} = 593$	$\Sigma X^2_{\text{tot}} = 1,560$

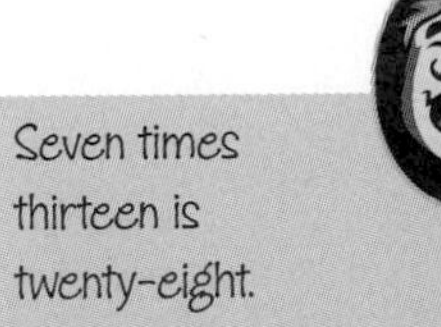

Seven times thirteen is twenty-eight.

Proof:

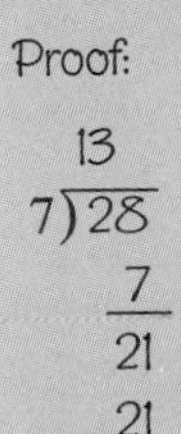

—from the movie *In the Navy*, starring Abbott and Costello

Here are the steps for calculating a factorial ANOVA:

1. First, compute the total sum of squares:

$$SS_{\text{tot}} = \sum_{1}^{N} X^2 - \frac{\left(\sum X_{\text{tot}}\right)^2}{N}$$

$$= 1560 - \frac{(198)^2}{28}$$

$$= 1560 - \frac{39204}{28}$$

$$= 1560 - 1400.143$$

$$= 159.857$$

2. Next, compute the sums of squares within:

$$
\begin{aligned}
SS_{\text{with}} &= \sum_{1}^{N} X^2 - \sum_{1}^{k} \frac{\left(\sum X_{\text{cell}}\right)^2}{n_{\text{cell}}} \\
&= 1560 - \left[\frac{(73)^2}{7} + \frac{(50)^2}{7} + \frac{(36)^2}{7} + \frac{(39)^2}{7}\right] \\
&= 1560 - \left[\frac{5329}{7} + \frac{2500}{7} + \frac{1296}{7} + \frac{1521}{7}\right] \\
&= 1560 - (761.286 + 357.143 + 185.143 + 217.286) \\
&= 1560 - 1520.858 \\
&= 39.142
\end{aligned}
$$

3. Next, compute each of the sums of squares between—that is, between sleep conditions and between memory conditions:

$$
\begin{aligned}
SS_{\text{bet sleep}} &= \sum_{1}^{k} \frac{(\sum X_{\text{row}})^2}{n_{\text{row}}} - \frac{(\sum X_{\text{tot}})^2}{N} \\
&= \left[\frac{(123)^2}{14} + \frac{(75)^2}{14}\right] - \frac{(198)^2}{28} \\
&= \left[\frac{15129}{14} + \frac{5625}{14}\right] - \frac{39204}{28} \\
&= (1080.643 + 401.786) - 1400.143 \\
&= 1482.43 - 1400.143 \\
&= 82.286
\end{aligned}
$$

$$
\begin{aligned}
SS_{\text{bet sleep}} &= \sum_{1}^{k} \frac{(\sum X_{\text{column}})^2}{n_{\text{column}}} - \frac{(\sum X_{\text{tot}})^2}{N} \\
&= \left[\frac{(109)^2}{14} + \frac{(89)^2}{14}\right] - \frac{(198)^2}{28} \\
&= \left[\frac{11881}{14} + \frac{7921}{14}\right] - \frac{39204}{28} \\
&= (848.643 + 565.786) - 1400.143 \\
&= 1414.429 - 1400.143 \\
&= 14.286
\end{aligned}
$$

4. Finally, determine the sums of squares interaction (by subtraction):

$$
\begin{aligned}
SS_{\text{sleep} \times \text{memory}} &= SS_{\text{tot}} - (SS_{\text{with}} + SS_{\text{bet sleep}} + SS_{\text{bet memory}}) \\
&= 159.857 - (39.142 + 82.286 + 14.286) \\
&= 159.857 - 135.714 \\
&= 24.143
\end{aligned}
$$

Are we done? No, we are not yet done. We now have five sums of squared deviations, but we are supposed to be analyzing variances, not sums of squares. To calculate an F test, we need variances.

SOURCE: Reprinted with permission from Sydney Harris, ScienceCartoonsPlus.com.

Factorial Mean Squares and *F*s

Recall that in ANOVA, a variance is called a *mean square.* And recall that the formula for a mean square is:

$$MS = \frac{SS}{df}$$

Also recall that a variance, or mean square, is the average of the squared deviation scores. That is why the formula divides the sums of squares by the degrees of freedom. But we have several different sums of squares: within, between, and interaction. Therefore, we have to divide each of our sums of squares by its appropriate degrees of freedom.

And what are those appropriate degrees of freedom? Here are the formulas:

$$df_{\text{tot}} = N - 1$$
$$df_{\text{with}} = (\text{No. of rows})(\text{No. of columns})(n_{\text{cell}} - 1)$$
$$df_{\text{bet sleep}} = \text{No. of rows} - 1$$
$$df_{\text{bet memory}} = \text{No. of columns} - 1$$
$$df_{\text{interaction}} = df_{\text{sleep} \times \text{memory}} = (\text{No. of rows} - 1)(\text{No. of columns} - 1)$$

1. First, we calculate the degrees of freedom:

$$df_{\text{tot}} = 28 - 1 = 27$$
$$df_{\text{with}} = (2)(2)(7 - 1) = 24$$
$$df_{\text{bet sleep}} = 2 - 1 = 1$$
$$df_{\text{bet memory}} = 2 - 1 = 1$$
$$df_{\text{sleep} \times \text{memory}} = (2 - 1)(2 - 1) = 1$$

2. Then, we calculate the within, between, and interaction mean squares (variances):

$$MS_{\text{with}} = \frac{SS_{\text{with}}}{df_{\text{with}}} = \frac{39.142}{24} = 1.631$$
$$MS_{\text{bet sleep}} = \frac{SS_{\text{bet sleep}}}{df_{\text{bet sleep}}} = \frac{82.286}{1} = 82.286$$
$$MS_{\text{bet memory}} = \frac{SS_{\text{bet memory}}}{df_{\text{bet memory}}} = \frac{14.286}{1} = 14.286$$
$$MS_{\text{sleep} \times \text{memory}} = \frac{SS_{\text{sleep} \times \text{memory}}}{df_{\text{sleep} \times \text{memory}}} = \frac{24.143}{1} = 24.143$$

3. Finally, we calculate F.

Our study has two possible main effects and one possible interaction effect. Thus, we must calculate three separate F statistics, one for each possible effect:

$$F_{\text{sleep}} = \frac{MS_{\text{bet sleep}}}{MS_{\text{with}}} = \frac{82.286}{1.631} = 50.451$$
$$F_{\text{memory}} = \frac{MS_{\text{bet memory}}}{MS_{\text{with}}} = \frac{14.286}{1.631} = 8.76$$
$$F_{\text{sleep} \times \text{memory}} = \frac{MS_{\text{sleep} \times \text{memory}}}{MS_{\text{with}}} = \frac{24.143}{1.631} = 14.803$$

We are finally done. Our F statistics for the three possible effects are 50.45, 8.76, and 14.80. But what do they mean?

Interpreting a Factorial *F* Test

As usual, the key question is whether the observed difference from expectation is *a lot*, and therefore probably due to the sleep, to the memory training, or to the interaction between sleep and memory training, or whether the difference from expectation is *only a little*, and therefore probably due to mere sampling error. To answer that question, we look in the F table to see whether our calculated F meets the tabled critical value of F. If our calculated F meets the tabled critical value of F, the difference from expectation is deemed to be a lot, and the null hypothesis is rejected. If the calculated F does not meet the tabled critical value of F, the difference from expectation is deemed to be only a little and due to mere sampling error, and so the null hypothesis is not rejected.

> ! I know not anything more pleasant, or more instructive, than to compare experience with expectation.
>
> —Samuel Johnson, 1758

As before, we must view the table at the correct degrees of freedom (df). Recall that for an F test, there are two dfs: one for the number of groups in the study (df_{bet} or df numerator) and one for the sample size (df_{with} or df denominator).

The df_{with} in the denominator is the same for each of our three F tests. In our study, the df_{with} is 24. However, the df_{bet} in the numerator differs depending on which effect we are evaluating: the effect of sleep, the effect of memory, or the effect of the interaction between sleep and memory. In our 2×2 study, each of our independent variables had only two levels or conditions; therefore, the dfs were $2 - 1$, $2 - 1$, and $2 - 1$, respectively—or 1, 1, and 1, respectively. Obviously, if our study had been other than 2×2, the three values for df_{bet} would not each have been only 1.

Appendix D contains the table for the family of *F* distributions. A portion of the table is reproduced in Table 29.6.

Table 29.6 A Portion of the *F* Table

		df_{bet}		
df_{with}	α (%)	*1*	*2*	*3*
23	5	4.28	3.42	3.03
	1	7.88	5.66	4.76
24	5	4.26	3.40	3.01
	1	7.82	5.61	4.72
25	5	4.24	3.39	2.99
	1	7.77	5.57	4.99

We just saw that the degrees of freedom for each of the three *F* tests are 1_{bet} and 24_{with}. What is the tabled value of *F* at the .05 error level for 1 and 24 *df*?

Yes, the tabled critical value of *F* for all three effects in our study is 4.26. Can we reject the null hypotheses?

Yes, we can reject all three null hypotheses. There is a main effect for sleep deprivation because our calculated *F* of 50.49 exceeds the tabled critical value of 4.26. There is a main effect for memory training because our calculated *F* of 8.76 exceeds the tabled critical value of 4.26. And there is an interaction effect for sleep deprivation × memory training because our calculated *F* of 14.76 exceeds the tabled critical value of 4.26.

Now assume that we are willing to make a Type 1 error only 1% of the time. Are all three effects still statistically significant? Which effects are statistically significant and which are not?

At $\alpha = .01$, the tabled critical value of *F* is 7.82. The calculated *F* for sleep deprivation, 50.45, still exceeds the tabled critical value of 7.82. The main effect for memory training, 8.76, also still exceeds the tabled critical value of 7.82. The calculated *F* for the interaction of sleep deprivation and memory training, 14.80, also still exceeds the tabled critical value of 7.82. Thus, at 99% confidence, we still can reject all three null hypotheses.

In a journal article, these results would be written as

$$F_{\text{sleep}}(1, 24) = 50.45, \quad \alpha < .01$$
$$F_{\text{memory}}(1, 24) = 8.76, \quad \alpha < .01$$
$$F_{\text{sleep} \times \text{memory}}(1, 24) = 14.80, \quad \alpha < .01$$

This is read as, "*F* for sleep at 1 and 24 *df* is 50.45. There is less than one chance in a hundred that the increase in recall was due to mere chance" (and so forth, for the remaining two effects).

The Factorial ANOVA Summary Table

Researchers display factorial ANOVA results in an ANOVA table similar to the one you saw in Module 25. However, a factorial ANOVA table includes rows for the additional main effects and for the interaction effect. Table 29.7 gives the ANOVA results for our study.

Table 29.7 Factorial ANOVA Summary Table

Source	*SS*	*df*	*MS*	F
Sleep	82.286	1	82.286	50.45**
Memory	14.286	1	14.286	8.76**
Sleep × memory	24.143	1	24.143	14.80**
Within	39.142	24	1.631	
Total	159.857	27		

$^{**}p < .01$.

Note how much information can be gleaned from the ANOVA table. Without even knowing what the study is about, the reader knows that there were two independent variables because only one interaction effect is displayed. The reader also knows that each of the two independent variables had only two conditions because each df_{bet} is either $C - 1$ or $R - 1$, both of which are equal to 1. The reader also knows that there were 28 participants in the study, because $df_{tot} = N - 1 = 27$. Assuming equal numbers of participants in each group (the case in our study), the reader also knows that there were 7 participants in each group, because 28 participants divided by four groups across the two variables is 7 participants per group. Finally, because of the two asterisks, the reader knows that each effect was statistically significant at the .01 error level.

PRACTICE

1. The following data are the time to complete a task (in minutes) for males and females when subjected to noise or no noise. $N = 24$, with each participant appearing in only one experimental condition (cell):

	Very Noisy	*No Noise*
Male	3.72	2.94
Male	3.56	2.76
Male	3.11	3.10
Male	3.62	2.52
Male	3.24	2.77
Male	3.89	2.65
Female	3.45	2.33
Female	3.80	2.67
Female	3.21	2.91
Female	3.77	2.38
Female	3.02	2.51
Female	3.10	2.34

   a. What is the dependent variable? What are the independent variables?
   b. Express the design in symbols (e.g., a 3 × 4 × 3 ANOVA).
   c. How many main effects are possible? How many interaction effects are possible?
   d. Calculate and graph the cell means.
   e. Interpret the graph: What do you expect the statistics to show for main effects and/or interaction effects?

f. Calculate the complete ANOVA.
g. Display the results in an ANOVA summary table.
h. At the .05 error level, is there a main effect? If so, for which independent variable(s)?
i. At the .05 error level, is there an interaction effect?
j. Write the results in APA journal format.

2. The following data are the number of cans of beverage fraternity members versus new plebes drink during an evening at an on-campus versus off-campus party. $N = 32$, with each participant appearing in only one experimental condition (cell):

Frat or Plebe Status	*On Campus*	*Off Campus*
Fraternity member	2	4
Fraternity member	3	3
Fraternity member	1	4
Fraternity member	2	4
Fraternity member	4	3
Fraternity member	2	5
Fraternity member	2	6
Fraternity member	4	5
New plebe	3	5
New plebe	2	2
New plebe	4	4
New plebe	1	6
New plebe	3	3
New plebe	2	4
New plebe	3	5
New plebe	3	4

a. What is the dependent variable? What are the independent variables?
b. Express the design in symbols (e.g., a 3 × 4 × 3 ANOVA).
c. How many main effects are possible? How many interaction effects are possible?
d. Calculate and graph the cell means.
e. Interpret the graph: What do you expect the statistics to show for main effects and/or interaction effects?
f. Calculate the complete ANOVA.
g. Display the results in an ANOVA summary table.
h. At the .05 error level, is there a main effect? If so, for which independent variable(s)?
i. At the .05 error level, is there an interaction effect?
j. Write the results in APA journal format.

3. The following data are the number of milliseconds it takes for children who are 9, 13, or 17 years of age to decide if a stimulus on a target card appeared previously on a memory card, when the memory card contained either 1, 3, or 5 stimuli. $N = 45$, with each participant appearing in only one experimental condition (cell):

Age (Years)	*One Stimulus*	*Three Stimuli*	*Five Stimuli*
9	468	579	872
9	442	624	856

(Continued)

(Continued)

Age (Years)	*One Stimulus*	*Three Stimuli*	*Five Stimuli*
9	532	686	931
9	477	653	789
9	502	635	793
13	367	512	659
13	404	579	731
13	398	612	765
13	427	634	683
13	345	548	612
17	249	478	563
17	314	460	572
17	242	379	512
17	279	382	542
17	238	356	501

a. What is the dependent variable? What are the independent variables?
b. Express the design in symbols (e.g., a 3 × 4 × 3 ANOVA).
c. How many main effects are possible? How many interaction effects are possible?
d. Calculate and graph the cell means.
e. Interpret the graph: What do you expect the statistics to show for main effects and/or interaction effects?
f. Calculate the complete ANOVA.
g. Display the results in an ANOVA summary table.
h. At the .01 error level, is there a main effect? If so, for which independent variable(s)?
i. At the .01 error level, is there an interaction effect?
j. Write the results in APA journal format.

4. The following data are the number of milliseconds it takes for children who are 9, 13, or 17 years of age to decide if a stimulus on a target card appeared previously on a memory card, when the memory card contained either 1, 3, or 5 stimuli and when participants were either well rested or sleep deprived. $N = 90$, with each participant appearing in only one experimental condition (cell):

	Well Rested			*Sleep Deprived*		
Age (Years)	*One Stimulus*	*Three Stimuli*	*Five Stimuli*	*One Stimulus*	*Three Stimuli*	*Five Stimuli*
9	468	579	872	612	812	1,076
9	442	624	856	654	865	1,123
9	532	686	931	612	785	1,280
9	477	653	789	587	812	1,034
9	502	635	793	643	885	1,256
13	367	512	659	567	756	980
13	404	579	731	503	742	956
13	398	612	765	512	678	879
13	427	634	683	492	660	934
13	345	548	612	546	774	893
17	249	478	563	348	678	784
17	314	460	572	452	656	698
17	242	379	512	438	657	602
17	279	382	542	372	568	782
17	238	356	501	411	570	739

a. What is the dependent variable? What are the independent variables?
b. Express the design in symbols (e.g., a 3 × 4 × 3 ANOVA).
c. How many main effects are possible? How many interaction effects are possible?
d. Calculate the complete ANOVA.
e. Display the results in an ANOVA summary table.
f. At the .01 error level, is there a main effect? If so, for which independent variable(s)?
g. At the .01 error level, is there an interaction effect? If so, for which interaction?
h. Write the results in APA journal format.

5. Exercise 5 in Module 25 addressed freshmen weight gain as a function of where they ate meals. Let's assume that the data in that exercise was for males only. Now the nutritionist collects data for females as well. She wants to see if meal site or gender or both affects weight gain. Here are the data.

	Eats at Home	*Eats in Campus Dining*	*Prepares Own Meals*
Male	2	5	3
Male	4	9	2
Male	0	12	5
Male	3	0	6
Male	0	8	0
Male	5	10	2
Male	2	7	0
Male	4	5	3
Male	10	4	1
Male	5	9	9
Female	4	4	6
Female	0	14	2
Female	3	5	5
Female	6	9	0
Female	3	2	3
Female	2	7	2
Female	9	4	7
Female	2	11	2
Female	0	3	4
Female	1	0	1

a. Calculate the ANOVA and present the results in an ANOVA summary table.
b. Write the results in APA format.

6. Exercise 6 in Module 25 addressed the number of times that dogs bark as a function of dog size. Now let's assume that the data in that exercise were for spayed/neutered (s/n) dogs only. The kennel owner now wants to determine if dog size or neuter/spay status or both affect the number of barking episodes. Here are the data.

	Small	*Medium*	*Large*
s/n	8	3	5
s/n	12	5	8
s/n	14	1	5
s/n	10	4	4
s/n	6	2	10

(Continued)

(Continued)

	Small	Medium	Large
s/n	3	0	3
s/n	11	3	0
s/n	4	0	4
s/n	5	2	3
s/n	8	3	5
s/n	0	4	9
s/n	6	6	4
Not s/n	10	1	8
Not s/n	6	3	4
Not s/n	13	6	3
Not s/n	2	0	2
Not s/n	6	2	9
Not s/n	0	0	0
Not s/n	4	4	3
Not s/n	11	3	4
Not s/n	3	1	6
Not s/n	7	4	5
Not s/n	6	2	11
Not s/n	2	3	7

a. Calculate the ANOVA and present the results in an ANOVA summary table.

b. Write the results in APA format.

7. Exercise 7 in Module 25 addressed the time to heat a room as a function of energy source. Now let's assume that the data in that exercise were only for energy-efficient rooms (thermopane windows, insulated joints, and solar-panel ceilings). Now the energy researcher wonders if energy source or energy-efficient construction (Energy Eff) or both affect time to heat the room. Here are the data.

	Electric	Natural Gas	Oil
Energy Eff	26	32	28
Energy Eff	31	30	26
Energy Eff	56	50	49
Energy Eff	21	19	24
Energy Eff	65	74	60
Energy Eff	10	8	9
Energy Eff	34	35	32
Energy Eff	50	57	53
Not Energy Eff	72	68	70
Not Energy Eff	45	42	51
Not Energy Eff	58	65	61
Not Energy Eff	47	43	40
Not Energy Eff	39	42	37
Not Energy Eff	18	14	18
Not Energy Eff	79	83	87
Not Energy Eff	62	56	59

a. Calculate the ANOVA and present the results in an ANOVA summary table.
b. Write the results in APA format.

8. Exercise 8 in Module 25 addressed salaries as a function of specific college majors. Now let's assume that the data were for those with only a bachelor's degree (Bach) in their field. Now the researcher wants to know if having a master's degree (Master) or specific college major or both determine salary. Here are the data.

	English	*Accounting*	*Psychology*	*Chemistry*
Bach	32	78	42	56
Bach	46	55	36	45
Bach	28	40	70	50
Bach	70	92	43	78
Bach	36	102	37	60
Bach	42	80	57	77
Bach	55	76	45	66
Bach	34	67	40	76
Master	47	85	70	87
Master	71	112	45	62
Master	56	72	44	45
Master	42	48	61	53
Master	39	57	38	55
Master	51	49	47	82
Master	56	87	48	76
Master	60	92	55	59

a. Calculate the ANOVA and present the results in an ANOVA summary table.
b. Write the results in APA format.

SPSS Connection

Download the file **data_word memory due to sleep awake mem nomem.sav** from www .sagepub.com/steinberg2e. These data are used in the textbook example.

Alternatively, manually enter the 28 scores from the example in Module 29 into the SPSS **Data View** spreadsheet. Data entry for a two-way ANOVA is not intuitively obvious. In the textbook, the data are set up as four groups of seven clients, forming a 2 × 2 table. In SPSS, all 28 scores (7 + 7 + 7 + 7) are entered in a single column. Then their group membership on the first independent variable (sleep vs. awake) is entered in the second column. Then their group membership on the second independent variable (memory training vs. no memory training) is entered in the third column. In addition, and unlike *t* tests, the group membership variable must be coded as a numeric variable rather than a string variable (this has to do with the underlying regression calculation algorithm, despite the ANOVA design setup. Regression will be discussed in the final modules of the textbook). Thus, enter the data as follows:

9	1	1
11	1	1
9	1	1
12	1	1
12	1	1
9	1	1
11	1	1
9	1	0
6	1	0
7	1	0
8	1	0
7	1	0
8	1	0
5	1	0
4	0	1
7	0	1
6	0	1
4	0	1
5	0	1
4	0	1
6	0	1
6	0	0
4	0	0
6	0	0
6	0	0
4	0	0
6	0	0
7	0	0

Click on the **Variable View** tab to define the variables. Name the first variable **numbword**, set the decimals at 0, and label the variable as **Number of Words**. Name the second variable **sleepsts**, set the decimals at 0, and label the variable as **Sleep Status**. Label the value as follows: **1** = **Slept** and 0 = **Awake**. Name the third variable **memtrsts**, set the decimals at 0, and label the variable **Memory Training Status**. Label the values as follows: **1** = **Memory Training** and 0 = **No Memory Training.**

If the file is not already in **Data View**, click that tab in the lower left of the screen.

In the toolbar at the top of the screen, click on **Analyze**, then **General Linear Model**, then **Univariate.** Highlight the variable **Number of Words** in the left window, and then click on the **arrow** before the **Dependent Variable** window, to send the variable into that window. This is the study's dependent variable. Click on the variable **Sleep Status** in the left window, and then click on the **arrow** before the **Fixed Factor** window, to send the variable into that window. This is the study's first independent variable. Click on the variable **Memory Training Status** in the left window, and then click on the **arrow** before the **Fixed Factor** window, to send the variable into that window. This is the study's second independent variable. (Do not be thrown by the change in terminology. Recall that the term *variance* also changed to *mean square* in ANOVA. A *factor* is simply ANOVA terminology for an independent variable.) Click **OK**. This is what you will see.

Univariate Analysis of Variance

Between-Subjects Factors

		N
Sleep Status	0	14
	1	14
Memory Training Status	0	14
	1	14

Tests of Between-Subjects Effects

Dependent Variable:Number of Words

Source	Type III Sum of Squares	df	Mean Square	F	Sig.
Corrected Model	120.714[a]	3	40.238	24.672	.000
Intercept	1400.143	1	1400.143	858.482	.000
Sleepsts	82.286	1	82.286	50.453	.000
Memtrsts	14.286	1	14.286	8.759	.007
Sleepsts * Memtrsts	24.143	1	24.143	14.803	.001
Error	39.143	24	1.631		
Total	1560.000	28			
Corrected Total	159.857	27			

a. R Squared = .755 (Adjusted R Squared = .725)

The bottom chart is the ANOVA summary table. Because SPSS runs the ANOVA via a multiple regression calculation algorithm (you have not yet learned about multiple regression), it includes a few extra rows in the table that we can ignore for now—"Corrected Model", "Intercept", and "Total". The ANOVA summary table to match the one demonstrated in this module consists of the remainder of the table—**sleepsts, memtrsts, sleepsts*memtrsts, error, and corrected total.** *Error* also is a regression algorithm term. It is simply ANOVA's "within-group."

Visit the study site at www.sagepub.com/steinberg2e for practice quizzes and other study resources.

PART XIII

Nonparametric Statistics

"Without Form or Void"

30

One-Variable Chi-Square

Goodness of Fit

Terms: parametric, nonparametric, goodness of fit

Symbol: χ^2

Learning Objectives:

- Understand the similar logic underlying various test statistics
- Determine the degrees of freedom
- Calculate a one-variable chi-square
- Use a table to interpret χ^2
- Report results in APA format

What Is a Nonparametric Test?

Recall from Module 1 that a parameter is any summary population value. A population mean is a parameter. So is a population standard deviation. Each of the test statistics we have calculated thus far—normal deviate Z, t, and F—have been based on a probability distribution for a population parameter. Thus, each of the tests has been **parametric**.

There are situations, however, when parametric tests are not appropriate. Some of those situations include the following:

- *Badly skewed data*: Suppose the data are not normally distributed but are instead seriously skewed. For example, income is positively skewed, with most people falling in the low and middle ranges and relatively few falling in the high ranges. Certainly, it would not be appropriate to interpret such a seriously skewed variable against a symmetric distribution such as a normal curve or a t curve.
- *Extremely small sample size*: With small samples—say, 20 or 25—we use a t table because it reflects the greater error and a slightly different distribution shape between small samples and large samples. But suppose we have an extremely small sample—say, only 5 in each group. With only 5 subjects in each group, the samples cannot possibly assume symmetry. Even a distribution that is inherently normal will not acquire a symmetric shape until there are a sufficient number of scores. Thus, even a t table would not be appropriate for interpreting data from extremely small samples.
- *Data not on an interval scale*: Suppose our data are not in score form but, rather, in frequencies. For example, we may count the number of people having various hair colors, the number of people married or not married, or the number of people who have committed various types of crimes. In such studies, there are no dependent variable

"scores." There are only categorical classifications (nominal data) and frequency counts for each category. Without scores, there can be no means. And without means, there can be no deviation scores, variances, or standard deviations. There are no population parameters.

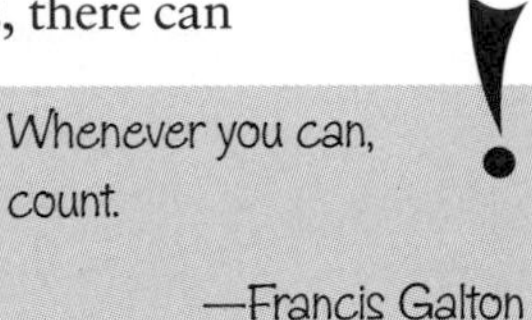

Fortunately, statisticians have created statistics to use that are not based on population parameters. These statistics are **nonparametric**, reflecting the fact that population parameters such as means and standard deviations are either missing or irrelevant.

> Not everything that can be counted counts, and not everything that counts can be counted.
>
> —Albert Einstein

There are dozens of nonparametric statistical tests. Each one reflects specifics regarding the data scale (nominal, ordinal) or the nature of the research question. For example, are the frequencies significantly different from expectation? Are the shapes of the distributions for the two groups significantly different? Is the pattern of responses in the two groups the same or different? Frequencies, shapes, patterns—there are no population parameters involved.

In this module and the next, we will look at one nonparametric test: chi-square. **Chi-square**, symbolized χ^2, tests the statistical significance of the difference in *frequencies* in two or more different nominal categories. Compare this with the parametric *t* test. A *t* test compares means, whereas a χ^2 compares frequencies.

Chi-Square as a Goodness-of-Fit Test

In this module, we will calculate a one-variable χ^2, also called a **goodness-of-fit** test. The goal of a χ^2 goodness-of-fit test is to determine whether a set of frequencies or proportions is similar to—and therefore "fits" with—a hypothesized set of frequencies or proportions. Those other frequencies or proportions might be from a larger group of which the smaller group is a part. In that case, a χ^2 goodness-of-fit test is similar to a one-sample *t* test: It determines if a given sample is similar to, and therefore representative of, a given population. Or the hypothesized set of frequencies or proportions might be that of equality—that they are the same across all categories.

Chi-square goodness-of-fit tests are fairly common. For example, we might compare the proportion of jelly beans of each color in a given bag of jelly beans to the proportion of jelly beans of each color that the manufacturer claims to produce. Or we might compare the sex of live births at an area hospital to determine if the number of boy babies and girl babies is equal. Or we might compare the political party registrations in a particular voting district with the political party registrations of the city in which the voting district falls. For such studies, there is only one variable—jelly bean color, sex, or political party.

That variable may be divided into as many categories as desired: Red, Orange, Yellow, Green, Blue, Purple, Pink, Black; or Male, Female; or Democrat, Republican, Conservative, Libertarian, Socialist, Communist, Green Party. However, there is still only one variable. The χ^2 goodness-of-fit test will determine whether or not the relative frequencies in the observed categories are similar to, or statistically different from, the hypothesized relative frequencies within those same categories.

Formula for Chi-Square

The formula for χ^2 is

$$\chi^2 = \sum \left[\frac{(f_o - f_e)^2}{f_e} \right]$$

where f_o = frequency observed and f_e = frequency expected.

Inferential Logic and Chi-Square

But wait! The logic of the χ^2 looks a lot like the logic of the z score, normal deviate Z test, one-sample t test, two-sample t test, and F test that we previously encountered, doesn't it? Yes, it does. That's because each is an example of the prototype for any test of statistical significance. Recall the prototype:

$$\frac{\text{What did you get?} - \text{What did you expect?}}{\text{Standardized random error}}$$

When we studied z scores, we made the following substitutions:

"What did you get?"	Raw score
"What did you expect?"	Sample mean
"Standardized random error"	Standard deviation

This gave us the formula

$$z\text{ Score} = \frac{\text{Raw score} - \text{Sample mean}}{\text{Standard deviation}}$$

In symbols, this was

$$z\text{ Score} = \frac{X - M}{s}$$

For the *normal deviate Z test* and for the *one-sample t test*, the substitutions we made were as follows:

"What did you get?"	Sample mean
"What did you expect?"	Population mean
"Standardized random error"	Standard error of the mean

This gave us the formula

$$Z_{\text{norm dev}} \text{ or } t_{\text{1-samp}} = \frac{\text{Sample mean} - \text{Population mean}}{\text{Standard error of the mean}}$$

In symbols, this was

$$Z_{\text{norm dev}} \text{ or } t_{\text{1-samp}} = \frac{M - \mu}{\sigma_M}$$

For the *two-sample t test*, the substitutions we made were as follows:

"What did you get?"	Difference between the two sample means
"What did you expect?"	Difference between the two population means
"Standardized random error"	Standard error of the difference between the means

This gave us the formula

$$t_{2\text{-samp}} = \frac{\text{(Difference between sample means)} - \text{(Difference between population means)}}{\text{Standard error of the difference between the means}}$$

In symbols, this is

$$t_{2\text{-samp}} = \frac{(M_1 - M_2) - (\mu_1 - \mu_2)}{\sigma_{M_1 - M_2}}$$

For the ANOVA F test, the substitutions we made were as follows:

"What did you get?"	Observed difference between the sample variances
"What did you expect?"	Expected difference between the population variances
"Standardized random error"	Within-group variance

This gave us the formula

$$F = \frac{MS_{\text{bet obs}} - MS_{\text{bet exp}}}{MS_{\text{with}}}$$

And now we have the χ^2. The substitutions we now make are as follows:

"What did you get?"	Observed frequency (there is no parametric mean)
"What did you expect?"	Expected frequency (there is no parametric mean)
"Standardized random error"	Expected frequency (there is no standardized parametric error term because there is no parameter distribution)

This gives us the formula

$$\chi^2 = \sum \left[\frac{(f_o - f_e)^2}{f_e} \right]$$

CHECK YOURSELF!

Compare a parametric test with a nonparametric test. When would you use each one?

Calculating a χ^2 Goodness of Fit

Let's start with an example. Assume that a city has 11,000 registered voters. Of those, 614 are registered as Libertarians, 840 as Green Party, 3,622 as Democrats, and 5,924 as Republicans. Now, assume that a given voting district within that city has 768 registered voters. Of those, 33 are registered as Libertarians, 57 as Green Party, 322 as Democrats, and 356 as Republicans. We want to know whether or not the voter registrations in the smaller district are significantly different from those in the larger city.

> If the Republicans will stop telling lies about the Democrats, we will stop telling the truth about them.
>
> —Adlai Stevenson

What do we need to answer that question? According to the formula, we need the observed voter registrations and the expected voter registrations. For the 768 voters in the district, we already know the observed number of voters registered for each political party. What we need to know is the expected numbers. How might we determine the expected numbers?

The toughest part of any χ^2 is determining the expected frequencies. Information within the hypothesis provides the expected values. The research hypothesis in this example is that political party registrations in the district are different from those of the city in which the district lies. The null hypothesis is that they are the same. Because the null hypothesis states that the percentages of registrations in the district should be the same as in the city, the percentages in the city determine the expected percentages in the district.

> ! Sensible and responsible women do not want to vote.
>
> —Grover Cleveland, U.S. president, 1905

Of the 11,000 voters in the city, the percentages registered in each political party are as follows:

Libertarian =	614/11,000 =	0.056 =	5.6%
Green Party =	840/11,000 =	0.076 =	7.6%
Democrat =	3,622/11,000 =	0.329 =	32.9%
Republican =	5,924/11,000 =	0.539 =	53.9%

If we expect the percentages in the district to be the same as those in the city, we simply apply those percentages to the 768 voters in the district. For the district, the expected numbers of registrants in each party become

Libertarian =	(0.056)(768) =	43
Green Party =	(0.076)(768) =	58
Democrat =	(0.329)(768) =	253
Republican =	(0.539)(768) =	414

We already had the observed number of registrants in each party in the district, and now we know the expected number of registrants in each party in the district. That's all the data we need to calculate a χ^2 goodness of fit. Before we plug the observed and expected registrations into the χ^2 formula, however, it is helpful to create a table of the data (see Table 30.1).

Table 30.1 Observed and Expected Registrations in Each Political Party in the District

Libertarian		*Green Party*		*Democrat*		*Republican*	
Observed	*Expected*	*Observed*	*Expected*	*Observed*	*Expected*	*Observed*	*Expected*
33	43	57	58	322	253	356	414

With the table completed, we can quickly find the numbers to plug into the χ^2 formula:

$$\begin{aligned}
\chi^2 &= \sum\left[\frac{(f_o - f_e)^2}{f_e}\right] \\
&= \frac{(33-43)^2}{43} + \frac{(57-58)^2}{58} + \frac{(322-253)^2}{253} + \frac{(356-414)^2}{414} \\
&= \frac{(-10)^2}{43} + \frac{(-1)^2}{58} + \frac{(69)^2}{253} + \frac{(-58)^2}{414} \\
&= \frac{100}{43} + \frac{1}{58} + \frac{4761}{253} + \frac{3364}{414} \\
&= 2.33 + .02 + 18.82 + 8.13 \\
&= 29.30
\end{aligned}$$

So our χ^2 is 29.30. But what does that mean?

CHECK YOURSELF!

List the steps for calculating a χ^2 goodness-of-fit test.

Interpreting a χ^2 Goodness of Fit

By now, you know that the key question is whether the observed difference from expectation is *a lot*, and therefore probably reflects a real difference in political party affiliation between the district and the city, or whether it is *only a little*, and therefore due merely to chance variation. By now, you also know that we look in a table to see if the value we calculated meets the tabled criterion. The table we use is called the χ^2 table.

The χ^2 table is located in Appendix F. Table 30.2 is a portion of the table.

Table 30.2 A Portion of the χ^2 Table

df	*.10*	*.05*	*.025*	*.01*	*.005*
1	2.71	3.84	5.02	6.63	7.88
2	4.61	5.99	7.38	9.21	10.60
3	6.25	7.81	9.35	11.34	12.84

Notice that there are no one-tailed or two-tailed columns in the χ^2 table. That is, there is no direction to the test statistic. Why? Look again at the formula. The numerator is a squared value. Because any number squared is positive, χ^2 cannot be negative. More to the point, the lack of directionality follows from the research question. The question is not the disparity in *total* political party registration proportions in the district versus the city. After all, as the proportion in one political party goes up, the proportion in another political party must go down. Thus, the proportions will sum to 100% for both the district and the city. Rather, the question is the disparity of the political party registration proportions in *each* political party. That is, a large χ^2 may be due to significantly fewer registrants or to significantly more registrants in any political party than expected. Discrepancies in either direction will increase the χ^2.

> FDR will be a one-term president
>
> —Mark Sullivan, columnist for the *New York Herald Tribune*, 1935

As usual, we must view the table at the correct degrees of freedom (*df*). For a one-variable χ^2, *df* is the number of categories minus 1. This is typically indicated as $c - 1$. How many categories are there in our study? What is the *df*?

Yes, there are four categories (Libertarian, Green Party, Democrat, and Republican). Thus, there are $4 - 1 = 3$ *df*. We enter the χ^2 table at 3 *df*.

The columns in the table are for the typical Type 1 error levels: .01, .02, .05, and .10. The entries in the body of the table give the minimum value that the calculated χ^2 must be to reject the null hypothesis. That is, they give the minimum value that the calculated χ^2 must be to conclude that the difference from expectation is probably real and not due to mere chance.

Our calculated χ^2 was 29.30. Assume that we are willing to make a Type 1 error 5% of the time. At $\alpha = .05$, what is the tabled critical χ^2? Can we reject the null hypothesis?

The tabled critical χ^2 is 7.81. Yes, we can reject the null hypothesis because the calculated χ^2 of 29.30 meets or exceeds the tabled critical χ^2 of 7.81.

Now, assume that we are willing to make a Type 1 error only 1% of the time. At $\alpha = .01$, what is the tabled critical χ^2? Can we reject the null hypothesis?

The tabled critical χ^2 is 11.34. Yes, we can still reject the null hypothesis because the calculated χ^2 of 29.30 meets or exceeds the tabled critical χ^2 of 11.34. We can be more than 99% certain that the difference in political party registration between the district and the city is "real" rather than due to mere chance.

In a journal article, our result would be written as

$$\chi^2(3, n = 768) = 29.30, p < .01.$$

This is read as "χ^2 at 3 degrees of freedom and sample size of 768 is 29.30. The probability is less than one in a hundred that this outcome occurred by mere chance." It appears that political party affiliation in this district is significantly different from the city.

Notice that the sample size is given in a report of χ^2, although it was not given in reports of t or F. This is because df in reports of t or F are based on the sample size; hence, the reader can figure out the sample size. In a χ^2, in contrast, df is based on the number of categories. Therefore, the reader needs to be told the sample size directly.

CHECK YOURSELF!

What do the entries in a χ^2 table tell you?

Looking Ahead

This module has discussed one-variable χ^2 tests. You may have guessed from the title of the module that not all χ^2 tests have only one variable. For example, suppose we wanted to know if the number of people who actually cast a vote on election day differs by political party registration. Our study would then have two variables: political party registration and voting status. Chi-square studies with two variables are fairly common. Do freshmen, sophomores, juniors, and seniors differ in the number of times they visit the college health center? Do boys and girls differ in the frequency with which they get out of their seats during class? Do children, teenagers, and adults differ in the types of snacks they prefer? In the next module, we will look at χ^2 studies when two variables are operating at once.

PRACTICE

1. You read a study that says 36% of college students nationwide change their intended major between their freshman and senior years. You get permission to do an exit poll of your college's graduating seniors. Of the 647 respondents, 268 report having changed majors at some time during their college years.
   a. State the research hypothesis.
   b. What is the df for this study?
   c. Complete the following table:

Changed Major		*Didn't Change Major*	
No. Observed	*No. Expected*	*No. Observed*	*No. Expected*

   d. Calculate χ^2.
   e. Interpret the result at the .01 error level. Is the research hypothesis supported?

2. A dorm has 86 men and 132 women living in it. The number of men and women at the college is 753 and 1,063, respectively. The dean of students wonders if this dorm's sex ratio reflects the sex ratio of the college.
   a. State the research hypothesis.
   b. What is the *df* for this study?
   c. Complete the following table:

Men		*Women*	
No. Observed	*No. Expected*	*No. Observed*	*No. Expected*

   d. Calculate χ^2.
   e. Interpret the result at the .05 error level. Is the research hypothesis supported?

3. LaTawna runs a used textbook business near City University. One semester, she sells 882 used textbooks to City University students. Of these, 387 are for 100-level, 147 for 200-level, 180 for 300-level, 135 for 400-level, and 33 for 500-level courses. During the same time period, City University's on-campus bookstore sells the following percentage of used textbooks: 37% for 100-level, 18% for 200-level, 10% for 300-level, 19% for 400-level, and 16% for 500-level courses.
   a. State the research hypothesis.
   b. What is the *df* for this study?
   c. Complete the following table:

100 Level		*200 Level*		*300 Level*		*400 Level*		*500 Level*	
No. Obs.	*No. Exp.*	*No. Obs.*	*No. Exp.*	*No. Obs.*	*No. Exp.*	*No. Obs.*	*No. Exp.*	*No. Obs.*	*No. Exp.*

   d. Calculate χ^2.
   e. Interpret the result at the .01 error level. Is the research hypothesis supported?

4. A published study says that 11% of university instructors are alumni of that college, 24% are from other colleges within the state, and 65% are from colleges in other states or countries. At one university, 63 instructors are alumni, 113 are from other colleges within the state, and 206 are from other states and countries.
   a. State the research hypothesis.
   b. What is the *df* for this study?
   c. Complete the following table:

Alumni		*Within State*		*Other State/Country*	
No. Observed	*No. Expected*	*No. Observed*	*No. Expected*	*No. Observed*	*No. Expected*

   d. Calculate χ^2.
   e. Interpret the result at the .05 error level. Is the research hypothesis supported?

5. A recent poll of young college women between the ages of 18 and 21 found that 44% want to marry before they are 25, 32% want to marry between the ages of 25 and 30, 15% want to marry between the ages of 30 and 35, and 9% want to marry after the age of 35 or not at all. You poll 100 women aged 18 to 21 at your college and find that 28 want to marry before they are 25, 40 want to marry between the ages of 25 and 30, 19

want to marry between the ages of 30 and 35, and 13 want to marry after the age of 35 or not at all.

a. Create a table of observed and expected frequencies.

b. Calculate χ^2.

c. Interpret the result at the .05 and .01 error levels. Is the research hypothesis supported? If so, with what level of confidence?

6. Amy is the new resident director at her university. In the summer before school begins, she orders "for him" and "for her" welcome packages for the male and female residents of her dorm. Not yet knowing the genders of the incoming students in her dorm, she orders packages based on the university's collegewide percentages of the previous year: 63% male and 37% female. When the school year begins, 362 students move into her dorm. 194 are male and 168 are female.

   a. Create a table of observed and expected frequencies.

   b. Calculate χ^2.

   c. Interpret the result at the .05 and .01 error levels. Is the research hypothesis supported? If so, with what level of confidence?

7. The tourism director for an ocean resort town claims that 57% of the adults who visit the area stay in hotels, 26% rent condos, 10% stay in campgrounds, and 7% stay with relatives who live in the area. Of the 800 people vacationing in the area this week, 504 are staying in a hotel, 190 are renting a condo, 82 are camping, and 24 are staying with relatives who live in the area. Are this week's accommodation choices significantly different from what the tourism director claims?

   a. Create a table of observed and expected frequencies.

   b. Calculate χ^2.

   c. At what level of confidence can you reject the null hypothesis?

8. A study of college student cell phone use found usage to be broken down as text messaging 82%, phone calls 10%, and Internet surfing 8%. A dorm director asks students to log their use for one day. On average, the students used their phones 65 times that day, broken down as 43 text messages, 14 phone calls, and 8 Internet surfs. Is this dorm's residents' usage significantly different from what the prior study had indicated?

   a. Create a table of observed and expected frequencies.

   b. Calculate χ^2.

   c. At what level of confidence can you reject the null hypothesis?

9. A lakeside food vendor knows from past experience that his sales ratios for lunches are 43% hamburgers, 37% hot dogs, 12% sausage rolls, and 8% veggie burgers. Today he sells 189 lunches. Of those, 72 are hamburgers, 71 are hot dogs, 27 are sausage rolls, and 19 are veggie burgers. Was today's food sale pattern significantly different from the past food sale pattern?

   a. Create a table of observed and expected frequencies.

   b. Calculate χ^2.

   c. At what level of confidence can you reject the null hypothesis?

10. High schoolers at a given high school show the following preferences for the one sport in which they are most interested: 30% prefer football, 25% prefer basketball, 11% prefer baseball, 5% prefer soccer, 3% prefer track, and 26% have no interest in any sport. A total of 450 students attend this high school. The athletic director's survey of the current entering class shows the following preferences: 105 football, 117 basketball, 60 baseball, 36 soccer, 25 track, and 107 who are not interested in any sport. Do the sport preferences of this year's entering class differ significantly from past students' preferences?
    a. Create a table of observed and expected frequencies.
    b. Calculate χ^2.
    c. At what level of confidence can you reject the null hypothesis?

SPSS Connection

Download the file **data_political party registrations.sav** from www.sagepub.com/steinberg2e. These data are used in the textbook example.

Alternatively, manually enter the categories frequencies from the political party example in Module 30 into the SPSS **Data View** spreadsheet. Data entry for a chi-square goodness-of-fit test is not intuitively obvious. In the textbook, the data are set up as four categories. But for a chi-square goodness-of-fit test in SPSS, the categories form one variable placed in one column, the observed frequencies are a second variable placed in another column, and the expected frequencies are input at the time of analysis. The category designations can be coded as numeric or string (nominal). However, SPSS will print the categories as if they had a natural order—which, for a string variable, will be alphabetical. If that is not the order in which you want your categories to appear, it is best to code your category values numerically. Then you can create a narrative label for the numeric codes, using the Label Values function. Thus, set up the data like this:

1	33
2	57
3	322
4	356

Click on the **Variable View** tab to define the variables. Name the first variable **polparty**, set the decimals at 0, and label the variable as **Political Party**. Label the values as follows: **1 = Libertarian, 2 = Green, 3 = Democrat**, and **4 = Republican**. Name the second variable **obsregis**, set the decimals at 0, and label the variable as **Observed Registrations.**

If the file is not already in **Data View**, click that tab in the lower left of the screen.

Data analysis in SPSS for a chi-square goodness-of-fit test, like its data entry, also is not intuitively obvious. SPSS does not ask you to transfer the observed frequencies variable (the "obsregis", in our example) into the analysis window, as it does for most other types of analysis. Instead, prior to requesting the chi-square analysis, it asks you to "weight" the category variables (the "polparty", in our example) by the observed frequencies. On the toolbar at the top of the screen, click on **Data**, then **Weight Cases**. Select the button for **Weight cases by**. Then highlight the variable **Observed Registrations** in the left window and click on the **arrow** before the **Frequency Variable** window to send the variable into that window. This is the set of frequencies we will test. Click **OK**.

On the toolbar at the top of the screen, click on **Analyze**, then **Nonparametric Tests**, then **Legacy Dialogs**, then **Chi Square**. Highlight the variable **Political Party** in the left window, and then click on the **arrow** before the **Test Variable** window to send the variable into that window. This is the variable we will test. (Recall that the observed frequencies for this variable were selected in the prior Data-Weight Cases step.)

So far, we have told the system what variable we are testing and also what the *observed* frequencies are for that variable. Before doing the analysis, we must tell the system what the *expected* frequencies are for that variable. Note that the expected frequencies are not part of the actual data spreadsheet. Instead, we manually enter the expected frequencies to which the observed frequencies will be compared. In the **Expected Values** box, SPSS gives us two choices. "All Categories Equal" tests the hypothesis that we expect equal numbers of registrants across the four political parties. This is not the case in our study. **Values** allows us to input hypothesized values. This is the case in our study. Therefore, click on **Values.** In the blank box to the right, enter the expected frequencies of the first category (in this case, for the Libertarian party). That expected value is **43.** Click **Add.** Repeat the expected values for the remaining categories: 43 (already entered, so do not enter this one again), **58, 253,** and **414.** It is important that the expected values be entered *in the same order* as those categories were listed in the **obsregis** variable in the **Data View** spreadsheet, because SPSS will pair the obsregis data with inputted expected frequencies row-for-row. Now click **OK.** This is what you will see.

NPar Tests

Chi-Square Test

Frequencies

Political Party

	Observed N	Expected N	Residual
Libertarian	33	43.0	-10.0
Green	57	58.0	-1.0
Democrat	322	253.0	69.0
Republican	356	414.0	-58.0
Total	768		

Test Statistics

	Political Party
Chi-square	29.287[a]
df	3
Asymp. Sig.	.000

a. 0 cells (.0%) have expected frequencies less than 5. The minimum expected cell frequency is 43.0.

Visit the study site at www.sagepub.com/steinberg2e for practice quizzes and other study resources.

31 Two-Variable Chi-Square

Test of Independence

Term: test of independence

Symbol: χ^2

Learning Objectives:

- Understand the similar logic underlying various test statistics
- Know the assumptions underlying a χ^2 test of independence
- Find expected frequencies from row and column totals
- Determine degrees of freedom
- Calculate a two-variable χ^2
- Use a table to interpret χ^2
- Report results in APA format

Chi-Square as a Test of Independence

In Module 30, we calculated a χ^2 goodness of fit, which is a one-variable test. Although there were multiple categories within that one variable (Libertarian, Green Party, Democratic, and Republican), there was still only one variable—political party affiliation. In this module, we will calculate a two-variable χ^2, also called a **test of independence**. As with any χ^2, there are no scores and, hence, no means or standard deviations. The data, as in Module 30, are frequencies.

The goal of a two-variable χ^2 is to determine whether or not the first variable is related to—or independent of—the second variable. In that way, it is similar to the test for an interaction effect in ANOVA: Is the outcome in one variable related to the outcome in some other variable?

Chi-square tests of independence are fairly common. To continue with our example from Module 30, we might investigate whether or not persons registered with various political parties are differentially likely to actually go to the polls and cast a vote on election day. Note that there are two variables in that example: (1) political party affiliation and (2) voting status. In another study, we might determine whether boys or girls are more likely to be placed in a gifted education program. There are two variables in that study: (1) the child's gender and (2) whether or not the child is placed in a gifted education program.

CHECK YOURSELF!

Assume that you want to determine whether the type of crime committed by juveniles is different from the type of crime committed by adults. What are the two variables in this study?

PRACTICE

1. You read a report saying that women and men differ in their preferred car colors. What are the two variables in this study?
2. You read a report saying that people who buy antiques online tend to be more highly educated than people who buy antiques at live auctions. What are the two variables in this study?
3. You read a report saying that the hobbies of same-sex friends are more similar than the hobbies of same-sex classmates. What are the two variables in this study?
4. You read a report saying that students who take at least a year off from school between high school and college are less apt to drop out than those who go to college directly from high school. What are the two variables in this study?
5. You read a report saying that those who listen to music before they drift off to sleep tend to be extraverts, while those who read books before they drift off to sleep tend to be introverts. What are the two variables in this study?

Prerequisites for a Chi-Square Test of Independence

To calculate any χ^2, the data must be in frequencies. However, a χ^2 test of independence requires that two additional criteria be met. First, the expected frequency in every cell must be at least 5. The actual frequency may be <5, but not the expected frequency. The difficulty posed by tiny expected frequencies is similar to that posed by small sample sizes in parametric tests: Error quickly overshadows the effect of the independent variable(s). There is a correction formula to use when expected frequency in any cell is <5. However, when any expected frequency is that small, it is probably best not to calculate χ^2 at all. Alternatively, collapse two or more cells until the expected frequency in the new category is at least 5. For example, suppose a florist is testing whether or not men and women differ in their choice of corsage flower. She uses the following flower categories: rose, carnation, orchid, and lily. Now, suppose the expected frequency for lily is <5. She could change the flower categories to rose, carnation, and "all other" and then calculate χ^2.

Second, the two variables must be able to at least theoretically operate independently. To meet this criterion, the subjects in the various categories must differ, and at the same time they must be comparable in relevant characteristics. For example, if we want to know if men and women differ in political party affiliation, we cannot sample husbands and wives. This is because husbands and wives tend to agree in their political party affiliations more than men and women chosen at random would. Similarly, if we want to know if freshmen or seniors are more likely to hold a job while attending college, we cannot sample freshmen only from among students who live on campus and seniors only from among students who live off campus. The different living arrangements, on campus and off campus, are probably related to having or not having an automobile. This additional variable, having a car, probably relates to the student holding a job more than the student's class year does. The problem posed by nonindependent variables in χ^2 is similar to what it was for parametric tests: It introduces a relationship between the variables. Recall that for the parametric *t* test and *F* test, there was a formula adjustment to account for the dependency. In the nonparametric χ^2, there is no comparable formula adjustment for dependency. Hence, only variables that are theoretically independent of one another may be used.

CHECK YOURSELF!

You want to test whether the number of auto accidents people have is related to their age. How might you sample subjects so that the two variables are independent of one another? How might you sample subjects so that the two variables are not independent of each other? Why are the variables in your latter example not independent?

Formula for a Chi-Square

The formula for a two-variable χ^2 is the same as for a one-variable χ^2. Recall the formula

$$\chi^2 = \sum \left[\frac{(f_o - f_e)^2}{f_e}\right]$$

where

f_o = frequency observed and

f_e = frequency expected.

The only calculation difference between a one-variable χ^2 and a two-variable χ^2 lies in the number of categories across which we sum—that is, the number of "$f_o - f_e$" cells. In all other ways, the calculation is identical.

Finding Expected Frequencies

Let's return to the question of whether or not there is a relationship between gender and placement in a gifted education program. The null hypothesis is that there is no significant relationship between gender and program placement. For our investigation, assume that an elementary school has 640 students: 346 boys and 294 girls. The school's gifted education program has 50 slots. The 50 slots are currently filled with 22 boys and 28 girls. Are these frequencies significantly different from what we would expect if there really were no relationship between gender and program placement?

As before, the first step in calculating any χ^2 is determining the expected frequencies. Expected frequencies always follow from the null hypothesis and are usually calculated from information within the data set. The null hypothesis is that there is no relationship between students' genders and their program placement. Given that hypothesis, how many boys and how many girls would you expect to be placed in the 50 slots in the gifted program?

Did you say "25 and 25"? If you did, that would not be correct. There are not equal numbers of boys and girls in the elementary school population. Thus, there is no reason to expect that equal numbers of girls and boys would be selected for the gifted program. Rather, the expected frequencies should be allotted proportional to the number of boys and girls in the school population. These proportions are as follows:

$$346 \text{ boys} + 294 \text{ girls} = 640 \text{ total students}$$

$$\text{Boys} = \frac{346}{640} = 0.54 \quad \text{Girls} = \frac{294}{640} = 0.46$$

Now we apply these proportions to the 50 slots in the gifted program and calculate the expected frequencies as

$$\text{Boys' } f_e = (0.54)\ (50 \text{ slots}) = 27 \text{ boys expected}$$

$$\text{Girls' } f_e = (0.46)\ (50 \text{ slots}) = 23 \text{ girls expected}$$

Many statisticians find it more straightforward to find expected frequencies via a formula that uses column and row totals to determine and apply proportions. Figure 31.1 is a blank table showing gifted program placement status as the table's rows and gender as the table's columns. Row and column totals are listed outside the table.

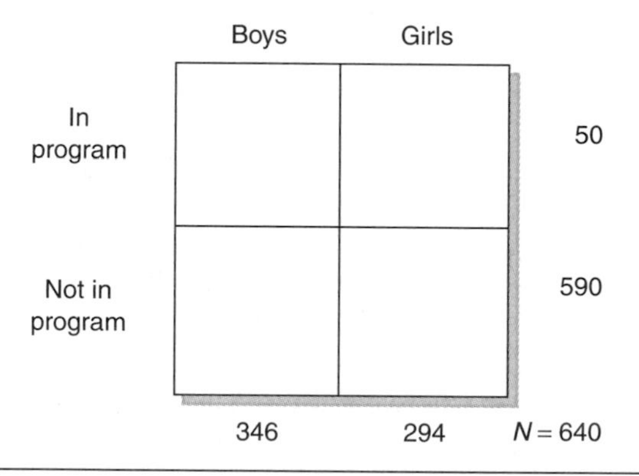

Figure 31.1 Blank Table of Observed and Expected Program Placement for Boys and Girls

To find the expected frequency (E) in any cell, we use the formula

$$E = \frac{\text{RT} \times \text{CT}}{N}$$

where

RT = row total,

CT = column total, and

N = total in population.

According to the formula, cell entries are calculated by cross-multiplying the row and column totals and then dividing by the total sample size. Here are the calculations:

Top left cell: (50)(346)/640 = 17300/640 = 27.03

Top right cell: (50)(294)/640 = 14700/640 = 22.97

Bottom left cell: (590)(346)/640 = 204140/640 = 318.97

Bottom right cell: (590)(294)/640 = 173460/640 = 271.03

Figure 31.2 is again a table, but this time showing the observed and expected frequencies entered. Expected frequencies are rounded to the nearest whole person and are shown in parentheses.

	Boys	Girls	
In program	22 (27)	28 (23)	50
Not in program	324 (319)	266 (271)	590
	346	294	$N = 640$

Figure 31.2 Completed Table of Observed and Expected Program Placement for Boys and Girls

Calculating a Chi-Square Test of Independence

With the table completed, we can quickly find the numbers to plug into the χ^2 formula. To ensure that we don't miss any cells, it is helpful to list one variable (program placement) along the top and the other variable (gender) below it, as shown below:

$$\chi^2 = \sum \left[\frac{(f_o - f_e)^2}{f_e} \right]$$

In gifted program: Boys, Girls; Not in gifted program: Boys, Girls

$$= \left[\frac{(22-27)^2}{27} + \frac{(28-23)^2}{23} + \frac{(324-319)^2}{319} + \frac{(266-271)^2}{271} \right]$$

$$= \frac{(-5)^2}{27} + \frac{(+5)^2}{23} + \frac{(+5)^2}{319} + \frac{(-5)^2}{271}$$

$$= \frac{25}{27} + \frac{25}{23} + \frac{25}{319} + \frac{25}{271}$$

$$= 0.93 + 1.09 + 0.08 + 0.09$$

$$= 2.19$$

So our χ^2 test of independence is 2.19. But what does that mean?

Interpreting a Chi-Square Test of Independence

We must look in a χ^2 table to see whether or not our calculated value meets the tabled critical value for statistical significance. As usual, we enter the table at the correct degrees of freedom (*df*). However, degrees of freedom are calculated differently for a two-variable χ^2 than for a one-variable χ^2. For a two-variable χ^2, we multiply the number of categories on the first variable minus 1 by the number of categories on the second variable minus 1. This is usually indicated as (No. of columns minus 1) (No. of rows minus 1), or more simply as $(C - 1)(R - 1)$.

Recall that the first variable is program placement and the second variable is gender. Therefore, what is the *df* for this study?

There are two classifications of program placement (in, out) and two classifications of gender (male, female). Therefore, $df = (2 - 1)(2 - 1) = 1$.

The χ^2 table is located in Appendix F. A portion of the table is reproduced as Table 31.1.

Table 31.1 A Portion of the χ^2 Table

df	*.10*	*.05*	*.025*	*.01*	*.005*
1	2.71	3.84	5.02	6.63	7.88
2	4.61	5.99	7.38	9.21	10.60
3	6.25	7.81	9.35	11.34	12.84

Our calculated χ^2 was 2.19. Assume that we are willing to make a Type 1 error 5% of the time. At $\alpha = .05$, what is the tabled critical χ^2? Can we reject the null hypothesis?

The tabled critical χ^2 is 3.84. No, we cannot reject the null hypothesis because the calculated χ^2 of 2.19 does not meet the tabled critical χ^2 of 3.84. The observed difference in the number of boys versus girls selected for the gifted program is not big enough for us to conclude that there is a significant relationship between students' gender and selection for the gifted program.

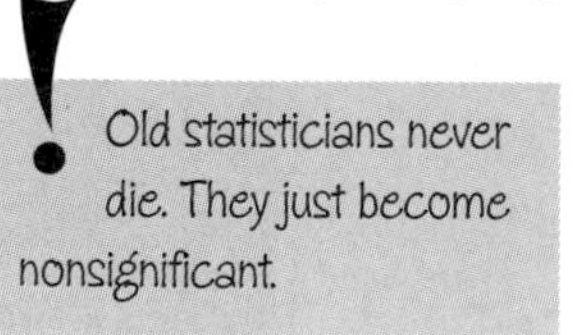

We didn't meet the .05 criterion. Assume that we are willing to make a Type 1 error 10% of the time (that's 10%, not 1%). At $\alpha = .10$, what is the tabled critical χ^2? Can we reject the null hypothesis?

The tabled critical χ^2 is 2.71. No, we still cannot reject the null hypothesis because the calculated χ^2 of 2.19 still does not meet the tabled critical χ^2 of 2.71.

In general, journals publish only results that meet some reasonable level of statistical significance. A χ^2 having a Type 1 error of more than 10% would not qualify as a statistically significant result by anyone's standards. However, occasionally *non*significance is an important finding. For example, if the current belief is that one gender is systematically excluded from gifted programs and yet our study finds that belief to be unfounded, a journal might publish our study to dispel the current belief.

In a journal, results would be reported as

$$\chi^2(1, n = 640) = 2.19, p > .10.$$

This is read as "Chi-square at 1 degree of freedom and sample size of 640 is 2.19. There is more than a 10% chance that this outcome occurred by chance alone." Placement in the gifted program is not related to gender.

Notice that the sample size is given in a report of χ^2, although it was not given in reports of *t* or *F*. This is because *df* in reports of *t* or *F* are based on the sample size; hence, the reader can figure out the sample size. In a χ^2, in contrast, *df* is based on the number of categories. Therefore, the reader needs to be told the sample size directly.

CHECK YOURSELF!

Compare a χ^2 goodness-of-fit test with a χ^2 test of independence.

	Goodness of Fit	*Test of Independence*
Measurement scale		
Number of variables		
Finding expected frequencies		
Determining *df*		

PRACTICE

6. A college career officer wonders if the employment fields of graduates are more directly related to field of study for some majors than for others. She collects data on majors and jobs for 400 of the most recently employed graduates. Here is what she finds. Of the 240 who were employed in a field related to their majors, 48 were business majors, 49 were humanities majors, 96 were social science majors, and 47 were science majors. Of the 160 who were employed in a field not related to their major, 32 were business majors, 51 were humanities majors, 64 were social science majors, and 13 were science majors.
   a. What are the variables in this study?
   b. State the null hypothesis.
   c. What is the *df* for this study?
   d. Complete the following table. Use row and columns totals to compute expected frequencies. Put the expected frequencies in parentheses.

	Business	Humanities	Social Science	Science
Related				
Unrelated				

   e. Calculate χ^2.
   f. Interpret the result at the .01 and the .05 error levels. Can you reject the null hypothesis?

7. A psychology student wonders if under-class students (freshmen and sophomores) differ in their vegetable preference from upper-class students (juniors and seniors). She randomly samples students and asks them their preferred vegetable among four choices: carrots, green beans, corn, and spinach. Of the 123 under-class students sampled, 40 prefer carrots, 32 prefer green beans, 43 prefer corn, and 8 prefer spinach. Of the 77 upper-class students sampled, 38 prefer carrots, 10 prefer green beans, 17 prefer corn, and 12 prefer spinach.
   a. What are the two variables in this study?
   b. State the null hypothesis.

c. What is the *df* for this study?

d. Complete the following table. Fill in observed frequencies and row and column totals. Use row and column totals to compute expected frequencies. Put the expected frequencies in parentheses.

	Carrots	Beans	Corn	Spinach
Underclass				
Upperclass				

e. Calculate χ^2.

f. Interpret the result at the .01 error level: Can you reject the null hypothesis?

8. You want to know whether men or women are more likely to take a set of stairs or an escalator. You position yourself in a shopping mall where a set of stairs and an escalator are near each other and equally accessible. During 1 hr, you record the gender and choice of each lone adult who enters either the stairs or the escalator. Of 34 lone men, 21 choose the stairs and 13 choose the escalator. Of 69 lone women, 31 choose the stairs and 38 choose the escalator.

a. What are the two variables in this study?

b. State the research hypothesis.

c. What is the *df* for this study?

d. Complete the following table. Fill in observed frequencies and row and column totals. Use row and column totals to compute expected frequencies. Put the expected frequencies in parentheses.

	Stairs	Escalator
Men		
Women		

e. Calculate χ^2.

f. Interpret the result at the .01 and .05 error levels: Is the research hypothesis supported? If so, with what level of confidence?

9. To minimize spoilage due to shelf life, a vendor wants to determine if snacks placed in vending machines in male versus female dorms sell differently. During the past month, sales in the male dorms were 222 bags of chips and 94 candy bars. During that same time period, sales in the female dorms were 268 bags of chips and 124 candy bars.

a. What are the variables in this study?

b. State the null hypothesis.

c. What is the *df* for this study?

d. Complete the following table. Use row and column totals to compute expected frequencies. Put the expected frequencies in parentheses.

	Chips	Candy
Male Dorm		
Female Dorm		

e. Calculate χ^2.

f. Interpret the result at the .01 an the .05 error levels. Can you reject the null hypothesis?

10. There are 25 children in a class: 15 boys and 10 girls. When report cards are issued, 10 "A" grades are given in mathematics. Seven of the "A" grades go to boys and 3 go to girls. The parents of the girls complain that boys receive significantly more "A" grades than girls do.

    a. What are the two variables in this study?

    b. State the null hypothesis.

    c. What is the *df* for this study?

    d. Create a table of observed and expected frequencies. Use row and column totals to compute expected frequencies. Put the expected frequencies in parentheses.

    e. Calculate χ^2.

    f. Interpret the result at the .01 and .05 error levels: Can you reject the null hypothesis? If so, with what level of confidence?

    g. Even if the complaint were statistically justified, that would not necessarily imply that there is gender bias in the instruction or grading. What aspect of the design of this study makes it difficult to assign cause to the independent variable, gender?

11. Exercise 7 in Module 30 addressed the type of accommodations chosen by visitors to a resort area. Now assume that the tourism director wants to know if the type of accommodation is related to age. She gathers average daily accommodations data as a function of the visitor's age. Here are the observed adult frequencies for an average week. Place the expected frequencies in the parentheses. Then calculate χ^2. Is there a significant relationship between type of accommodation and age?

	Hotel Room	*Rented Condo*	*Campground*	*Relative's House*
<30 years	58 ( )	93 ( )	60 ( )	39 ( )
30–50 years	87 ( )	131 ( )	62 ( )	70 ( )
>50 years	94 ( )	70 ( )	18 ( )	18 ( )

12. Exercise 8 in Module 30 addressed the college students' cell phone usage. Now assume that the dorm director wants to know if the usage is related to gender. She gathers cell phone usage data for several days as a function of the user's sex. Here are the observed

average frequencies. Place the expected frequencies in the parentheses. Then calculate χ^2. Is there a significant relationship between cell phone use category and gender?

	Text Messaging	*Phone Calls*	*Internet Surfing*
Male	256 ( )	31 ( )	19 ( )
Female	367 ( )	43 ( )	22 ( )

13. Exercise 9 in Module 30 addressed a lakeside food vendor's lunch sales. Now assume that the vendor wants to know if food choice is related to day of the week. He wonders if weekend customers have a different eating preference than weekday customers. He gathers food sales data for a midweek day and for a weekend day. Here are the observed frequencies. Place the expected frequencies in the parentheses. Then calculate χ^2. Is there a significant relationship between food choice and day of the week?

	Hamburger	*Hot Dog*	*Sausage Roll*	*Veggie Burger*
Midweek	70 ( )	63 ( )	32 ( )	17 ( )
Weekend	96 ( )	88 ( )	24 ( )	28 ( )

14. Exercise 10 in Module 30 addressed sport fan preferences of high schoolers. Now assume that the athletic director wants to know if preferences differ between high schools. There are three high schools in the district. Based on a survey of all three schools, here are the preference data for each school. Place the expected frequencies in the parentheses. Then calculate χ^2. Is there a significant relationship between sport preference and school?

	Football	*Basketball*	*Baseball*	*Soccer*	*Track*	*No Interest*
East High	105 ( )	117 ( )	60 ( )	36 ( )	25 ( )	107 ( )
Central High	97 ( )	85 ( )	40 ( )	56 ( )	19 ( )	59 ( )
West High	114 ( )	134 ( )	46 ( )	38 ( )	33 ( )	86 ( )

SPSS Connection

Download the file **data_gender and gifted program placement.sav** from www.sagepub.com/steinberg2e. These data are used in the textbook example.

Alternatively, manually enter the categories frequencies from the political party example in Module 31 into the SPSS **Data View** spreadsheet. Data entry for a chi-square test of independence is not intuitively obvious. In the textbook, the data are set up as four categories. But for a chi-square test of independence in SPSS, the categories of the first variable are placed in one column, the categories of the second variable are placed in a second column, and the observed frequencies form a third variable that is placed in a third column. Unlike the one-variable chi-square goodness-of-fit test, the expected frequencies for a two-variable chi-square test of independence are calculated by SPSS from the row and column totals and so do not have to be input.

The category designations can be coded as numeric or string (nominal). However, SPSS will print the categories as if they had a natural order—which, for a string variable, will be alphabetical. If that is not the order in which you want your categories to appear, it is best to code your category values numerically. Then you can create a narrative label for the numeric codes, using the Label Values function. Set up the data like this:

i	1	22
i	2	28
n	1	324
n	2	266

Click on the **Variable View** tab to define the variables. Name the first variable **progplac,** set the Type as **String,** and label the variable as **Program Placement.** Label the values as follows: **i = In Program** and **n = Not In Program.** Name the second variable **gender,** set the Type as **String,** and label the variable as **Gender.** Label the values as follows: **1 = Male** and **2 = Female.** Name the "final" variable **obscount,** set the decimals at 0, and label the variable as **Observed Count.**

If the file is not already in **Data View,** click that tab in the lower left of the screen.

Data analysis in SPSS for a chi-square test of independence, like its data entry, is not intuitively obvious. SPSS does not ask you to transfer the observed frequencies variable (the "obscount", in our example) into the analysis window, as it does for most other types of analysis. Instead, prior to requesting the chi-square analysis, it asks you "weight" the cells (the intersection of "progplac" and "gender", in our example) by the observed frequencies. On the toolbar at the top of the screen, click on **Data,** then **Weight Cases.** Select the button **Weight cases by.** Highlight the variable **Observed Count** in the left window and click on the **arrow** before the **Frequency Variable** window to send the variable into that window. This is the set of frequencies we will test.

The placement of this statistical test also is not intuitively obvious. You might expect it to be found under nonparametric tests, then chi-square—just as the chi-square goodness-of-fit test was. However, that is not the location of the chi-square test of independence. Instead, it is a *statistical option* under a different analytic procedure. On the toolbar at the top of the screen, click on **Analyze,** then **Descriptives,** then **Crosstabs.** Highlight the variable **Program Placement** in the left window and then click on the **arrow** before the **Rows** window to send the variable into that window. Click on the variable **Gender** in the left window and then click on the **arrow** before the Columns window to send the variable into that window. Click on the **Statistics** button in the upper right and check the box for **Chi Square.** Click **Continue.** If you want to see expected counts as well as observed counts within the cells (recommended if you hand-calculated chi-square and are checking your calculations in SPSS), click the **Cells** button in the upper right and check the box for **Expected.** Click **Continue** and then **OK.** This is what you will see.

Crosstabs

Case Processing Summary

	Cases					
	Valid		Missing		Total	
	N	Percent	N	Percent	N	Percent
Program Placement * Gender	640	100.0%	0	.0%	640	100.0%

Program Placement * Gender Crosstabulation

			Gender		Total
			Male	Female	
Program Placement	In Program	Count	22	28	50
		Expected Count	27.0	23.0	50.0
	Not In Program	Count	324	266	590
		Expected Count	319.0	271.0	590.0
Total		Count	346	294	640
		Expected Count	346.0	294.0	640.0

Chi-Square Tests

	Value	df	Asymp. Sig. (2-sided)	Exact Sig. (2-sided)	Exact Sig. (1-sided)
Pearson Chi-Square	2.211[a]	1	.137		
Continuity Correction[b]	1.794	1	.180		
Likelihood Ratio	2.203	1	.138		
Fisher's Exact Test				.142	.090
N of Valid Cases	640				

a. 0 cells (.0%) have expected count less than 5. The minimum expected count is 22.97.

b. Computed only for a 2 x 2 table

Visit the study site at www.sagepub.com/steinberg2e for practice quizzes and other study resources.

PART XIV

Effect Size and Power

"How Much Is Enough?"

32

Measures of Effect Size

Terms: effect size, Cohen's *d*, eta, effect size *r*, phi, Cramer's *V*

Symbols: d, r, η, ϕ, V

Learning Objectives:

- Distinguish between statistical significance and practical or clinical importance
- Understand the influence of sample size on statistical significance
- Know which measure of effect size is appropriate for each test of statistical significance
- Calculate various measures of effect size
- Know the guidelines for interpreting effect size

What Is Effect Size?

Each of the inferential statistics you have learned thus far has been designed to decide between retaining and rejecting the null hypothesis. Each answered the question "Is the observed difference only a little, or is it a lot?" And each defined "a lot" in terms of statistical significance. In other words, it asked whether or not the observed difference was greater than what would have been expected by mere chance. Using that criterion, if the observed difference was beyond the critical value (say, at the .01 or .05 α level), then it was judged to be a lot, and the null hypothesis was rejected. But if the observed difference was not beyond the critical value, then it was judged not to be a lot, and the null hypothesis was not rejected.

> ! Truth in science can be defined as the working hypothesis best suited to open the way to the next better one.
>
> —Konrad Lorenz, ethologist

However, if the observed difference is beyond the critical value and, hence, statistically significant, does that necessarily mean that the treatment is also effective? In other words, is statistical significance the final determinant of an effective treatment?

No, statistical significance is not the final determinant of an effective treatment. Although it is true that if a treatment does not bring about a statistically significant difference then it is probably also not effective, the reverse is not true. A treatment can result in a statistically significant outcome and yet still not be clinically effective.

The reason for this state of affairs lies with the influence of sample size on statistical significance. An increase in sample size increases the likelihood of finding statistical significance. Why? Well, let's take the *t* test as an example. As we saw in the number-drawing exercise in Module 15, the larger the sample size, the smaller the standard error of the mean. Furthermore, because the standard error of the mean is the denominator in a *t* test, the smaller the standard error of the mean, the greater the value of the calculated *t*. And large values of *t* are statistically significant. It follows that if we want to find statistical significance, all we have to do is use a big enough sample!

Yes, it's true. If sample size is big enough, even the smallest difference between sample means will be statistically significant. Yet the difference may be too small in practical terms to make a real-world difference.

Suppose, for example, that a new rat food causes rats to run 1/100 seconds faster than those given the old food. Or suppose that a new medication causes depressed patients to score 1/10 of a point less depressed on a test of depression than those given some other medication. Or suppose that children taught by a new curriculum score 1 point higher on a reading achievement test than those given the old curriculum. Each of these results might be statistically significant if the sample size had been 10,000 subjects or even 1,000 subjects. But is such a tiny difference in scores practically useful? In other words, is the difference in scores worth changing foods, therapies, or curricula? Probably not. This is because statistical significance does not guarantee a meaningful difference.

The question before us is "What constitutes a meaningful difference, as opposed to a merely statistically significant difference?" To answer that question, we need a measure of the practical or clinical importance of an effect. Such measures have come to be called **effect size.**

The criteria for judging effect size are necessarily subjective. That is, what may be judged a meaningful treatment effect by one researcher may be judged as too small to bother with by another researcher. For example, one school district may be willing to fund a new curriculum and retrain all teachers in return for a 2-point gain in reading achievement scores, but another district might require promise of a 10-point gain before spending the money and effort to change the curricula. Who is right? They both are—in their own situations. There are no tables of critical values for effect size as there are in hypothesis testing.

There are, however, guidelines. Guidelines are values statisticians have suggested for small, medium, and large effect sizes. Nevertheless, it is still up to the user to determine whether a small, medium, or large effect size will be useful for his or her own purposes.

> Whoever undertakes to set himself up as a judge of Truth and Knowledge is shipwrecked by the laughter of the gods.
>
> —Albert Einstein

CHECK YOURSELF!

What is the difference in meaning between statistical significance and effect size?

For Two-Sample *t* Tests

One method for measuring effect size scales the numerator over the standard deviation of the dependent variable scores rather than over the standard error. This statistic is called **Cohen's *d*.** For a two-sample *t* test, the formula for Cohen's *d* is

$$d = \frac{M_1 - M_2}{S_{DV}}$$

where

$M_1 - M_2$ = difference between the sample means and

S_{DV} = standard deviation of the dependent variable for all subjects.

Recall from Module 8 what happens when we divide by any value: It *rescales* the numerator into the denominator units. Thus, Cohen's d converts the difference between the means into standard deviation units. That is, it tells how many standard deviation units the observed difference between the means is. Look at the two formulas below. How does Cohen's d differ from a t test?

$$d = \frac{M_1 - M_2}{S_{DV}} \text{ and } t = \frac{M_1 - M_2}{\sigma_{M_1 - M_2}}$$

Yes, the t test rescales the difference between the means into standard error units, while Cohen's d rescales the same difference between the means into standard deviation units.

Scaling the numerator in terms of standard deviation units rather than standard error units expresses it against an internal criterion for that particular study. The advantage of scaling over standard deviation in Cohen's d is that the standard deviation is not materially affected by sample size. In contrast, the standard error term in the t test is heavily dependent on sample size.

Let's calculate Cohen's d for the two-sample t test of the effect of medication versus counseling on depression level. Recall from Module 20 that the means for that study were

$$M_{\text{med}} = 28.778 \text{ and } M_{\text{couns}} = 38.11.$$

The Cohen's d formula also requires the standard deviation of the dependent variable (S_{DV}) for all subjects in the study. That value is 9.10.

With this information, we can calculate Cohen's d:

$$\begin{aligned} d &= \frac{28.78 - 38.11}{9.10} \\ &= \frac{-9.333}{9.10} \\ &= -1.026 \end{aligned}$$

How big should Cohen's d be for the effect to be considered meaningful? Meaningful effect size is a matter of judgment. Is 0.25 standard deviations enough? How about 0.5 standard deviations? Or should we require 2 or 3 standard deviations? Experts differ in their judgments of how many standard deviations is a meaningful effect size. However, Jacob Cohen (1988), author of the statistic, suggested the following guidelines, and most statisticians follow his suggestions:

Small effect = 0.2 standard deviation

Medium effect = 0.5 standard deviation

Large effect = 0.8 standard deviation

Thus, in our study of treatment and depression level, not only were the results statistically significant (recall that α was <.02), but the effect size was also large (−1.026).

Other measures of effect size, applicable to a wide array of research designs, are based on correlation theory. If you are progressing through these modules in the order in which they are presented, you have not yet learned about correlation. For now, it is enough to know that a correlation coefficient measures the magnitude of the association between two variables and that the coefficient can range from −1.00 to +1.00, with higher values (either + or −) indicating a greater association between the two variables. A correlation of 0 indicates no relationship.

In the case of effect size, the two variables being correlated are the independent variable and the dependent variable. Taking our depression study, the independent variable was the type of treatment (medication vs. counseling), and the dependent variable was the depression level after treatment. Thus, a measure of effect size calculates the proportion of depression relief that is due to the medication versus counseling treatments. The higher the correlation, the greater the effect of the independent variable on the dependent variable.

One such statistic for a two-sample *t* test is called effect size *r*. It is measured by the formula

$$\text{Effect size } r = \sqrt{\frac{t^2}{t^2 + df}}$$

where

t = calculated value of the t test and

df = degrees of freedom for the t test.

Let's calculate effect size *r* for the study of medication versus counseling on depression level. Recall from Module 20 that the statistics for that study were

$N = 18$ (9 in each sample),

$df = 16$ (N – the number of samples = 18 – 2),

$t = 2.39$, and

$\alpha < .02$.

From this information, we calculate effect size *r*:

$$\begin{aligned}\text{Effect size } r &= \sqrt{\frac{(2.39)^2}{(2.39)^2 + 16}}\\ &= \sqrt{\frac{5.712}{5.712 + 16}}\\ &= \sqrt{\frac{5.712}{21.712}}\\ &= \sqrt{.263}\\ &= .513\end{aligned}$$

How big should effect size *r* be for the effect to be considered useful? That, again, is a matter of judgment. Is .10 enough? How about .40? Or should we require .80 or .90? Experts differ in their judgments. However, most statisticians (Cohen, 1988) have settled on the following guidelines for correlational measures of effect size:

Small effect = less than .25

Medium effect = .25 to .40

Large effect = .40 or more

Thus, in our study of medication versus counseling for depression, not only were the results statistically significant (α was <.02), but the effect size was large (.513).

For ANOVA *F* Tests

With ANOVA, effect size is measured by another correlational measure called **eta**, symbolized η. Like other correlations, its range is from 0.00 to +1.00. The formula is

$$\text{Effect size } \eta = \sqrt{\frac{SS_{\text{bet}}}{SS_{\text{tot}}}}$$

This formula, more clearly than the others, asks what proportion of the total variation (denominator) is due to the difference in treatments (numerator). Because the two sums of squares in the formula are variance measures (area measures), the formula takes the square root of the proportion to return to a standard deviation (linear measure).

Let's calculate η for our three-group study of the impact of medication, counseling, or diet supplement on depression level. Recall from Module 25 that the sums of squares were as follows:

$SS_{\text{bet}} = 1205.00$

$SS_{\text{with}} = 1153.80$

$SS_{\text{tot}} = 2358.80$

From this information, we can now calculate η:

$$\begin{aligned}\text{Effect size } \eta &= \sqrt{\frac{SS_{\text{bet}}}{SS_{\text{tot}}}}\\ &= \sqrt{\frac{1205.00}{2358.80}}\\ &= \sqrt{.511}\\ &= .715\end{aligned}$$

How big should η be to be considered useful? Again, it is partly a matter of judgment. However, because its possible range is the same as for effect size *r*, most statisticians have settled on similar guidelines:

Small effect = less than .25

Medium effect = .25 to .40

Large effect = .40 or more

Thus, in our study of the effect of three treatments on depression relief, not only were the results statistically significant (recall that α was <.01), but the effect size was large (.715).

For Chi-Square Tests

Chi-square (χ^2) tests the statistical significance of the difference in observed frequencies from expected frequencies. As with *t* and *F*, it is possible to obtain a χ^2 that is statistically significant but meaningless in practical terms. Thus, we need a measure of effect size for χ^2 as well.

Effect size statistics for χ^2 vary according to whether the χ^2 is a goodness-of-fit test (one variable) or a test of independence (two variables). For the two-variable test, the effect size statistic also differs according to whether the number of categories in each variable is 2×2 or some other size. We will take these situations one at a time.

For a Goodness-of-Fit Test

When we calculated a one-variable χ^2 goodness of fit in Module 30, we tested the hypothesis that the proportions of voters in a particular voting district were significantly different in their political party affiliations from voters in the town in which that voting district lay.

To calculate effect size for a one-variable goodness of fit, we use the formula

$$\text{Effect size } r = \sqrt{\frac{\chi^2}{(N)(c-1)}}$$

where

N = total sample size and

c = number of categories.

The statistics for the voting study were as follows:

$N = 768$

$\chi^2 = 29.30$

$c = 4$ (Green Party, Libertarian, Democrat, Republican)

$\alpha < .01$

Plugging the values into the formula, we get

$$\begin{aligned}\text{Effect size } r &= \sqrt{\frac{29.30}{(768)(4-1)}}\\ &= \sqrt{\frac{29.30}{2304}}\\ &= \sqrt{.013}\\ &= .114\end{aligned}$$

God forbid that Truth should be confined to mathematical demonstration!

—William Blake, *Notes on Reynold's Discourses*, 1808

How big should the effect size be to be considered meaningful? You certainly expect me to say that it is partly a matter of judgment, and you are right. However, unlike t or F, chi-square statistics are nonparametric. The possible range for the correlational effect is restricted due to the nominal scale of the data. This makes the effect size correlations appear greater than they would in a parametric study. This is presumably why Cohen (1988, pp. 226–227) suggests slightly more stringent guidelines for interpreting effect size for a χ^2 study. The guidelines for a χ^2 effect size are as follows:

Small effect = less than .30

Medium effect = .30 to .50

Large effect = more than .50

Thus, our .114 effect size is small.

Now, isn't this outcome interesting? Although the initial χ^2 had been statistically significant at the .01 error level, the effect size of .114 is small. In other words, the statistically significant difference in political party patterns between this district and the overall town is not practically meaningful. What accounts for this result? Why might the difference be statistically significant and yet be so small in practical terms?

Yes, it's the sample size problem that I discussed at the beginning of this module. Recall that if sample size is large enough, then practically any nonzero difference from expectation will be statistically significant. That's why we calculate measures of effect size: to see whether the statistical significance translates into an effect that is large enough to be useful or whether the statistical significance is due merely to the large sample size. In the study of political party affiliations, the effect size is too small to be of any practical importance. The statistical significance was due merely to the large sample size.

For a Test of Independence

There are two different effect size statistics for a two-variable χ^2 test of independence. The appropriate measure of effect size when there are exactly two categories on each variable (i.e., 2×2) is called **phi**, symbolized ϕ. Phi is another type of correlation. Its formula is

$$\phi = \sqrt{\frac{\chi^2}{N}}$$

where N = total sample size.

In Module 31, we calculated the χ^2 test of independence for program placement (gifted or nongifted) as a function of gender (male or female). The statistics for that study were as follows:

$\chi^2 = 2.19$

$N = 50$

$\alpha > .10$

Plugging the values into the formula, we get

$$\begin{aligned}\phi &= \sqrt{\frac{2.19}{50}} \\ &= \sqrt{.0438} \\ &= .21\end{aligned}$$

Using the more stringent guidelines (less than .30 = small, .30 to .50 = medium, and more than .50 = large), our effect size of .21 is small. This should not surprise us. Remember that the χ^2 was not statistically significant ($\alpha > .10$). If the overall test statistic is not statistically significant, then the effect size cannot be very big (although the reverse is not true, as we saw with the voting example).

And what about a χ^2 test of independence with category numbers other than 2×2? The appropriate measure of effect size for such studies is **Cramer's *V***, symbolized V. Its formula is

$$V = \sqrt{\frac{\phi^2}{(\text{the smaller of } R \text{ or } C) - 1}}$$

where R = number of rows and C = number of columns.

Note that to use this formula, we must first compute ϕ. For studies having only two categories on either of the two variables (rows or columns), Cramer's V and ϕ will be identical.

For studies with larger numbers of categories in both rows and columns, the two statistics will not be identical.

Exercise 5 in Module 31 examined the relationship between college year (lower class or upper class) and vegetable preference (carrots, beans, corn, spinach). That study was a 2×4 χ^2 test of independence. Here are the statistics from that study:

$\chi^2 = 12.43$

$N = 200$

$\alpha < .01$

$R = 2$ (lower class, upper class)

$C = 4$ (carrots, beans, corn, spinach)

Plugging the values into the phi formula, we get

$$\phi = \sqrt{\frac{12.43}{200}}$$
$$= \sqrt{.0622}$$
$$= .249$$

Plugging the ϕ value into the Cramer's V formula, we get

$$V = \sqrt{\frac{(.249)^2}{2-1}}$$
$$= \sqrt{\frac{.062}{1}}$$
$$= \sqrt{.062}$$
$$= .249$$

Cramer's V is interpreted by the same guidelines as ϕ: less than .30 = small, .30 to .50 = medium, and more than .50 = large. Our effect size of .249 is small. Recall that the original χ^2 was statistically significant at <.01. This is a study whose results are highly significant but whose effect size is small. The reason for this is, again, sample size. The observed difference in vegetable preferences by class years is not as meaningful for our sample of 200 people as the same difference in vegetable preferences would have been for a smaller number of people.

CHECK YOURSELF!

List the appropriate effect size statistic(s) for each test of statistical significance.

Test Statistic	*Effect Size Statistic(s)*
t test	
F test	
One-variable χ^2	
Two-variable χ^2	

PRACTICE

1. Exercise 2 in Module 20 examined whether having salespeople offer shopping assistance increased or decreased the amount of furniture sold. There were 10 subjects in each group: assisted or unassisted. The mean amount of furniture sold (in dollars) for the 10 customers in the assisted group was $476.40. The mean amount of furniture sold for the 10 customers in the unassisted group was $514.60. The standard deviation for furniture sold across both groups combined is 760.75. A t test showed no significant difference in the amount of furniture sold between the two conditions ($t = -.107$, $\alpha > .10$).
   a. Calculate the effect size for this study, using both Cohen's d and effect size r.
   b. Interpret the result: Was there a practical difference in amount of furniture sold between the assisted and unassisted conditions?

2. Exercise 3 in Module 20 examined whether adding a fragrance to a window cleaner led people to believe that it cleans better than an unscented product. There were 12 subjects in each group: scented or unscented. The mean cleaning effectiveness rating for the group getting the scented cleaner was 7.42. The mean cleaning effectiveness rating for the group getting the unscented cleaner was 6.00. The standard deviation for cleaning effectiveness ratings across all subjects is 1.369. A t test showed that the scented cleaner was rated as a significantly more effective cleaner ($t = 2.84$, $\alpha < .01$).
   a. Calculate the effect size for this study, using both Cohen's d and effect size r.
   b. Interpret the result: Was there a practical difference in cleaning effectiveness between scented and unscented cleaners?

3. Exercise 5 in Module 20 examined whether male or female college students studied more. There were 10 males and 10 females in the study. The mean study time for males was 11.1 hr/week. The mean study time for females was 10.9 hr/week. The standard deviation for study time across both males and females is 2.81. A t test showed no significant difference in study time between males and females ($t = 0.15$, $\alpha > .10$).
   a. Calculate the effect size for this study, using both Cohen's d and effect size r.
   b. Interpret the result: Was there a practical difference in study times between males and females?

4. Exercise 2 in Module 25 examined whether the route that workers take to work affects commute time. Here are the results:

Source	*SS*	*df*	*MS*	*F*
Between	4,192.585	2	2,096.29	76.38**
Within	576.375	21	27.45	
Total	4,768.96	23		

   a. Compute effect size η for this study.
   b. Interpret the result: Is there a practical difference in commute time between the three routes?

5. Exercise 3 in Module 25 examined whether the order of questions on an exam affects test score. Here are the results:

Source	*SS*	*df*	*MS*	*F*
Between	3.20	2	1.60	0.138
Within	313.50	27	11.61	
Total	316.70	29		

a. Compute effect size η for this study.

b. Interpret the result: Is there a practical difference in test score based on the question order?

6. Exercise 4 in Module 25 examined whether the number of hours students spend studying during the week before final exams is a function of their GPAs until that point. Here are the results:

Source	*SS*	*df*	*MS*	*F*
Between	235.0	3	78.33	10.09**
Within	279.4	36	7.76	
Total	514.4	39		

a. Compute effect size η for this study.

b. Interpret the result: Is there a practical difference in study time between students with different GPAs?

7. Exercise 3 in Module 30 examined whether the proportions of textbooks an off-campus bookseller sells to a university's students were different across five grade levels from the proportion of textbooks sold by that university's on-campus bookstore. The off-campus bookseller sold 882 textbooks. The χ^2 goodness of fit was 197.70, $\alpha < .01$.

a. Compute the effect size for this study.

b. Interpret the result: Is there a practical difference in the proportions of books sold across grade levels between the off-campus and on-campus sellers?

8. Exercise 4 in Module 30 examined whether the faculty in a particular college differ from nationwide faculty in three locations from which they earned their undergraduate degrees. There were 382 faculty at this college. The χ^2 goodness of fit was 22.40, $\alpha < .01$.

a. Compute the effect size for this study.

b. Interpret the result: Is there a practical difference in the location from which faculty at this college versus nationwide faculty earned their undergraduate degrees?

9. Exercise 5 in Module 30 examined whether the age at which young women in a particular college say they want to marry differed from the age at which young women nationwide say they want to marry. One hundred female college students from this college were polled, and their responses were compared with the nationwide data in four marriage age preference categories. The χ^2 goodness of fit was 10.67, $\alpha < .025$.

a. Compute the effect size for this study.

b. Interpret the result: Is there a practical difference between the age at which women at this college versus women nationwide want to marry?

10. Exercise 6 in Module 31 examined whether gender was related to taking a set of stairs versus an escalator. One hundred and three people were observed. The χ^2 test of independence was 2.81, which was not statistically significant.

a. Compute effect size ϕ for this study.

b. Interpret the result: Is there a practical difference in the proportion of men and women taking the stairs versus the escalator?

11. Exercise 7 in Module 31 examined whether undergraduate or graduate student status was related to the students' city of origin (Baltimore or New York City). One hundred and three men and women were observed. The χ^2 test of independence was 36.61, $\alpha < .01$.
    a. Compute effect size ϕ for this study.
    b. Interpret the result: Is there a practical difference in the proportion of undergraduate versus graduate students coming from Baltimore versus New York City?

12. Exercise 4 in Module 31 examined whether drinking frequency (frequent, infrequent) was related to class year (freshmen, sophomore, junior, senior). A total of 8,644 students were polled. The χ^2 test of independence was 155.16, $\alpha < .01$.
    a. Compute Cramer's V effect size for this study.
    b. Interpret the result: Is there a practical difference in the drinking frequency between students of different college years?

Visit the study site at www.sagepub.com/steinberg2e for practice quizzes and other study resources.

33 Power and the Factors Affecting It

Term: power

Learning Objectives:

- Know the mathematical relationship between Type 2 error and power
- Understand the effect that various factors have on power
- Understand the interrelationship of statistical significance, effect size, and power
- Use graphs or tables to determine power, given known values of other variables
- Use graphs or tables to determine effect size, given known values of other variables
- Use graphs or tables to determine necessary sample size for a desired power, given known values of other variables

What Is Power?

In previous modules, we discussed statistics for testing null hypotheses. Each of the statistics assumed the null hypothesis to be true. To reject the null hypothesis, we looked for a big enough difference between treated and untreated groups. In Module 13, you learned that Type 1 error, α, occurs when there is really no treatment effect in the populations, but we nevertheless find one in our samples. Thus, with Type 1 error, we incorrectly reject the null hypothesis. This situation is shown in Figure 33.1.

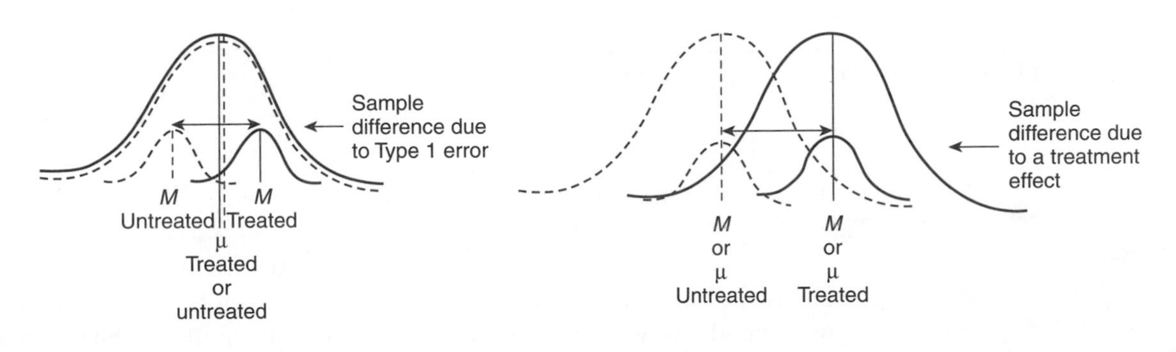

Figure 33.1 Rejecting the Null Hypothesis: Type 1 Error Versus a Treatment Effect

So is that enough? No, that's not enough. Experimental studies need to be able to correctly detect not only when there is *not* a real difference in treatments but also when there *is* a real difference in treatments. We are moving to a new topic—Type 2 error and its inverse, power.

The alternative hypothesis states that the null hypothesis is false. In other words, there really is a treatment effect in the underlying populations. In Module 13, you learned that Type 2 error, β, occurs when there really is a treatment effect in the populations, but we do not find it in our samples. With a Type 2 error, we incorrectly retain the null hypothesis. This situation is shown in Figure 33.2.

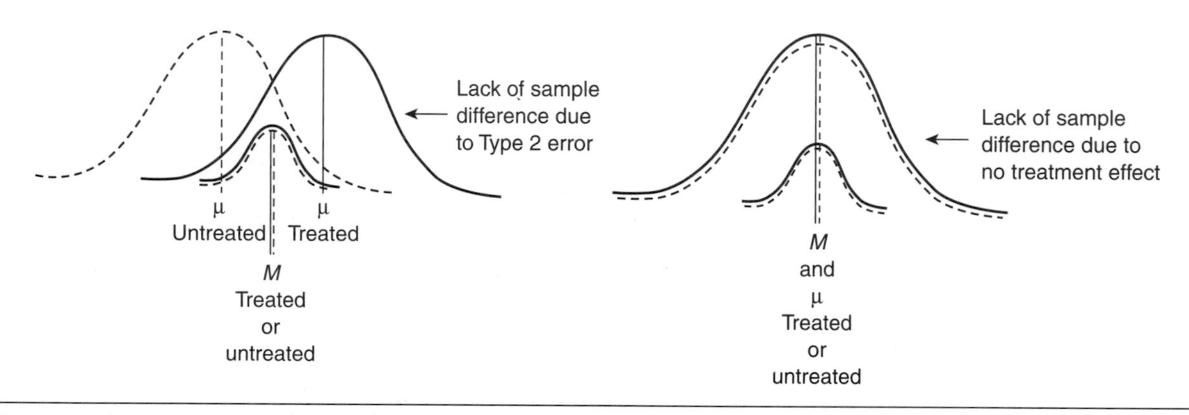

Figure 33.2 Retaining the Null Hypothesis: Type 2 Error Versus No Treatment Effect

Real truth of the H_0: in the population(s)

Your conclusion about the H_0: based on sample data	True	False	
Retain	Hit	Miss (Type 2) (β)	
Reject	Miss (Type 1) (α)	Hit	← Power cell

Figure 33.3 Decision Table With Power Cell

> When you have eliminated the impossible, whatever remains, however improbable, must be the truth.
>
> —Sir Arthur Conan Doyle

Recall, also, that Module 13 depicted both types of errors in a decision table (Figure 33.3).

The left column of the decision table refers to the null hypothesis. When the difference between treated and untreated groups is not big enough to reject the null hypothesis, we either correctly retain the null hypothesis (hit) or make a Type 1 error (α). These are the only two possibilities with a true null hypothesis. Because we have previously tested only the null hypothesis, α is the only error we have measured.

Now we will work within the right column of the decision table. Thus, we will be concerned about β. Note the new notation in the decision table—the power cell. **Power** occurs when the null hypothesis is really false, and we correctly reject the null hypothesis. To put it another way, power is when there really is a difference between groups due to the treatment, and we do find it.

Now look again at the decision table. When the null hypothesis is really false (right column), either we correctly find it to be false (power—which is a hit) or we fail to find it to be false (β). These are the only two possibilities with a false null hypothesis. Because either one or the other must occur, the two possibilities taken together add to 1. It follows that we can calculate power by knowing the value of Type 2 error. That is, power = $1 - \beta$. Thus, if β is .36, then power is .64. Conversely, we can calculate Type 2 error by knowing the power of our study. That is, $\beta = 1 -$ power. Thus, if power is .82, then β is .18.

CHECK YOURSELF!

State the relationship between power and β.

Until fairly recently, many studies had very little power. That is, the researcher found no statistically significant result and, hence, retained the null hypothesis, but the researcher did not know if the lack of observed difference between the means was because there really was no treatment effect or because of a Type 2 error. That is, it is quite possible that the treatment really was effective but the study simply lacked power to detect it.

Because of this unfortunate past situation, researchers now design studies that can detect not only when the null hypothesis is really true but also when it is really false, and the results are misleading the researcher to think that it is true. All journals published by the American Psychological Association (APA), as well as many other journals of rigor, now require that authors submit power estimates for their studies. In this module, we will look at factors that increase the power of a study as well as methods for estimating a study's actual power.

CHECK YOURSELF!

The results from a rigorous study must meet three requirements. One is statistical significance. Based on this module and the previous module, what are the other two requirements?

PRACTICE

1. Calculate the following.
   a. When power is 80%, what is the Type 2 error?
   b. When Type 2 error is 12%, what is the power?
2. Calculate the following.
   a. When power is 84%, what is the Type 2 error?
   b. When Type 2 error is 25%, what is the power?

Factors Affecting Power

Because power occurs when there really is a treatment effect and we detect it, power is a very good thing. We want lots of power in our studies. So what factors affect power? The factors include size of the Type 1 error, directionality of the alternative (research) hypothesis, size of the actual difference between the means of treatment groups (effect size), amount of error variance, and sample size. We will look at each of these factors in a moment. But first, let me explain the diagrams that accompany each factor.

In each of the diagrams to follow, Type 1 error is indicated by the small area beyond the dotted line in the tail of the H_0 distribution. This is where Type 1 error has fallen all along. What's new is the addition of a second overlapping distribution. The overlapping distribution represents scores in a treatment group when the treatment does have an effect—that is, when the alternative hypothesis is true (the null hypothesis is false). Therefore, the overlapping distribution is labeled H_A. Power is indicated by the portion of the overlapping H_A distribution falling to the right of the dotted line in the H_0 distribution. For clarity, the area of power is shaded. Now let's look at the factors that influence power.

> If you want to test a man's character, give him power.
>
> —Abraham Lincoln

Size of Type 1 Error

As Type 1 error increases, power increases. Figure 33.4 can help us understand this principle.

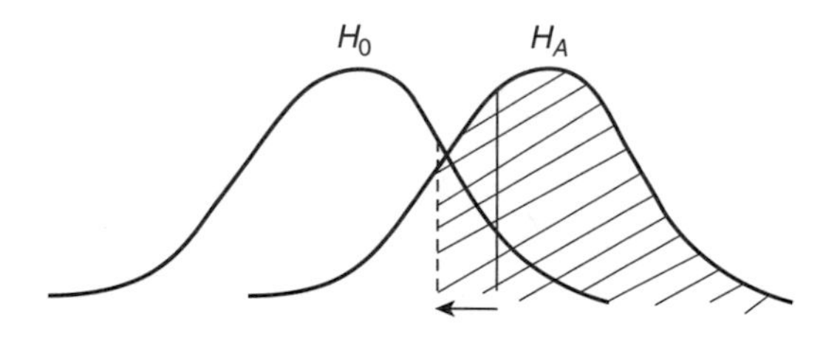

Figure 33.4 Power and the Amount of Type 1 Error

Increasing Type 1 error means accepting a greater chance of making a Type 1 error. For example, we could accept a 10% chance of Type 1 error rather than a 5% chance. Figure 33.4 shows that, as Type 1 error increases under the H_0 (indicated by the direction of the arrow), power (the shaded area under H_A) also increases.

Directionality of the Alternative Hypothesis

For any given Type 1 error level, directionality increases power. Figure 33.5 can help us understand this principle.

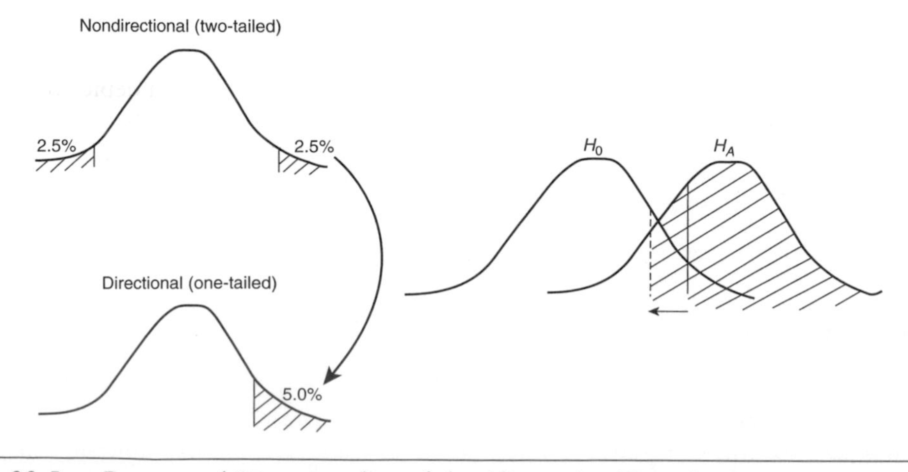

Figure 33.5 Power and Directionality of the Alternative Hypothesis

Recall that the alternative hypotheses can be either directional (one-tailed) or nondirectional (two-tailed). Let's pick a Type 1 error level—say, 5%. In a nondirectional hypothesis, that 5% is split so that 2.5% is in each tail. A nondirectional hypothesis tests whether the experimental group scores higher than the control group or the control group scores higher than the experimental group—either outcome. In contrast, in a directional hypothesis, the whole 5% is placed in one tail. It tests only whether the experimental group does better than the control group or only whether the control group does better than the experimental group—one or the other but not both.

The two diagrams on the left show that Type 1 error in a single tail increases (indicated by the arrow) as the hypothesis switches from nondirectional to directional. In Figure 33.5, it goes from 2.5% to 5%. Because that is the tail of the H_0 distribution that overlaps with the H_A distribution, power (shaded area in the H_A) increases. This is shown by the arrow in Figure 33.5 on the right.

Size of the Actual Difference Between the Means

As the actual difference between the means (you might recognize this as the effect size) increases, power increases. Figure 33.6 helps us understand this principle.

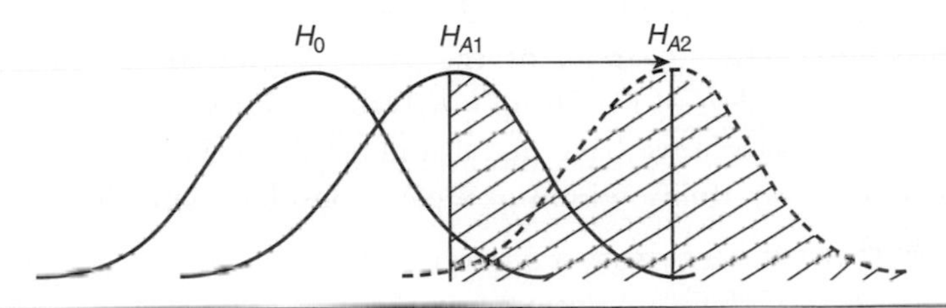

Figure 33.6 Power and the Actual Difference Between the Means

If the experimental and control groups differ in their response to treatment by only a little (from H_0 to H_{A1} in Figure 33.6), we might or might not detect the difference in our samples. But as the difference between the experimental and control groups increases (from H_0 through H_{A1} and on to H_{A2} in Figure 33.6), we are more likely to detect the difference between our H_0 and H_A samples. In Figure 33.6, power (shaded area under H_A) increases as the difference between the means increases. In the extreme case, we could depict yet another alternative hypothesis, H_{A3}, so far away from H_0 that the two distributions do not overlap at all. In that case, we would have 100% power: No matter which cases ended up in our samples, we would definitely find the difference, because even the lowest members of H_A would outscore the highest members of H_0.

The principle we have just examined deals with the size of the difference between the means. Now recall the formula for a two-sample t test:

$$t_{2\text{-samp}} = \frac{M_1 - M_2}{\sigma_{M_1 - M_2}}$$

Where does this difference between the means lie? Does it lie in the numerator or in the denominator?

It resides in the numerator. Thus, anything that increases the difference between the means increases our ability to find treatment differences. Anything that decreases the difference between the means decreases our ability to find treatment differences.

Amount of Error Variance

The greater the error variance (or the standard deviation), the less the power. Figure 33.7 helps us understand this principle.

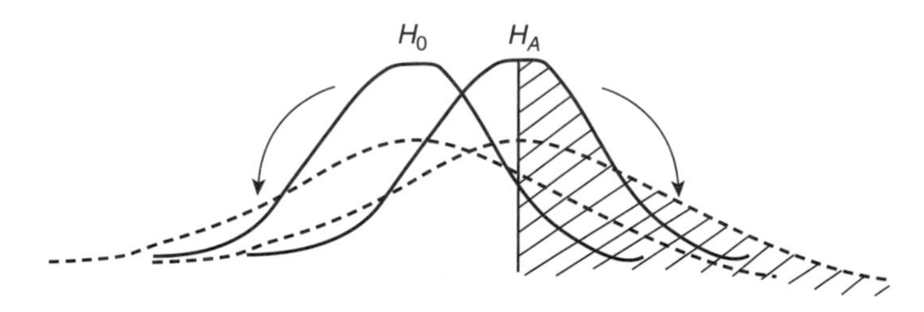

Figure 33.7 Power and the Amount of Error Variance

> A man is like a fraction whose numerator is who he is and whose denominator is what he thinks of himself. The larger the denominator, the smaller the fraction.
>
> — Leo Tolstoy

In Figure 33.7, as the distributions become more variable within their own groups (indicated by the arrow), there is less overlap between the H_0 and H_A groups. Hence, power (shaded area in H_A) decreases. H_A, H_0, or both can increase in variability. The rule remains the same: Within-group variability decreases power, regardless of the group in which it occurs.

The principle we have just examined deals with the size of the standard deviation (variability within groups). Consider again the formula for a two-sample *t* test:

$$t_{2\text{-samp}} = \frac{M_1 - M_2}{\sigma_{M_1 - M_2}}$$

Where does this within-group variability reside in the *t* statistic? Does it lie in the numerator or in the denominator? To answer that, recall that $\sigma_{M_1 - M_2}$ is calculated from the σ_M for each group and that one of the variables in the σ_M is the size of the standard deviation.

So it resides in the denominator. Thus, anything that increases the variability within groups will decrease our ability to find any treatment difference that does exist. That is, too much within-group error variance masks the treatment effect between groups in the numerator. Conversely, anything that decreases the variability within groups will increase our ability to find any treatment differences that do exist. That is, less within-group variance will allow the treatment effect between groups in the numerator to be more apparent.

Sample Size

The bigger the sample size, the greater the power. This principle is shown in Figure 33.8.

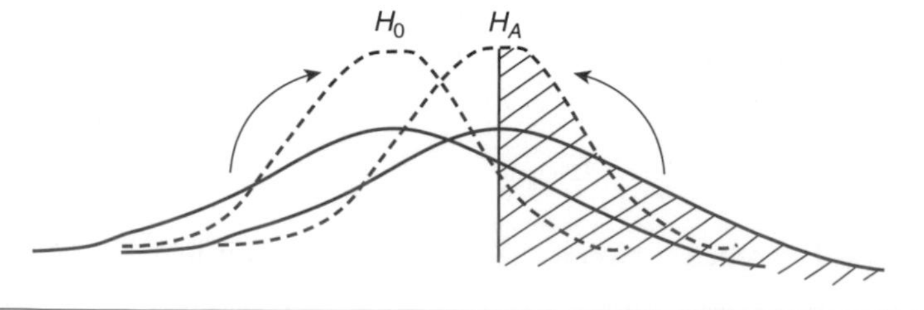

Figure 33.8 Power and Sample Size

Figure 33.8 shows that, as sample size increases, within-group variability (error variance) decreases. Thus, power (the shaded portion of H_A) increases.

The principle we have just examined deals with sample size. Let's again consider the formula for a two-sample t test:

$$t_{2\text{-samp}} = \frac{M_1 - M_2}{\sigma_{M_1 - M_2}}$$

Where does sample size reside in the t statistic? Does it lie in the numerator or in the denominator? To answer that, recall that $\sigma_{M_1 - M_2}$ is calculated from the σ_M for each group and that one of the variables in the σ_M is the sample size.

It resides in the denominator. Now recall the number-drawing exercise (Exercise 7) you performed in Module 15 when we studied the standard error of the mean. From that exercise, you learned that the larger the sample size, the smaller the σ_M (error variance). Thus, sample size is one variable that affects the value of the σ_M and, hence, the $\sigma_{M_1 - M_2}$.

We also learned in the preceding section that anything that increases the value of the denominator will decrease the ability to find any treatment difference that does exist in the numerator. And conversely, anything that decreases the value of the denominator will increase the ability to find any treatment difference that does exist in the numerator. Because larger sample sizes decrease the value of the denominator of our test statistic (t, F), larger sample sizes increase our ability to find treatment differences in the numerator.

CHECK YOURSELF!

List five factors affecting power. For each factor, tell the direction of the effect.

Factor	*Direction of Effect*
1.	
2.	
3.	
4.	
5.	

PRACTICE

Directions for this set of practice exercises: A researcher is conducting studies under the following conditions. For each study,

a. State whether the change in condition will tend to increase or decrease the study's power.
b. If the change in condition affects primarily the numerator or the denominator of the test statistic, state which part is primarily affected.

3. Fifty subjects are randomly selected and agree to participate in an experiment. However, on the day of the experiment, only 32 subjects show up.
4. A researcher had originally intended to select subjects for placement into treatment groups (i.e., the independent variable) based on a median split of their scores on a

placement test for the independent variable. However, now he has decided to put in one group only people scoring at least 1 standard deviation above the mean on the placement test and put in the other group only people scoring at least 1 standard deviation below the mean on the placement test. Sample size remains the same.

5. A research assistant is helping to collect observational data. However, he is careless in scoring the observations. As a result, dependent variable scores are inaccurate and prone to error.

6. A researcher had originally planned to conduct a study while allowing only 1% Type 1 error. However, now he has decided to permit 5% Type 1 error.

7. Just before conducting an experiment, the researcher switches his hypothesis from nondirectional to directional. He maintains the same total Type 1 error.

Putting It Together: Alpha, Power, Effect Size, and Sample Size

In my undergraduate course in economics, the professor graphed the effect of a change in an economic variable on an economic outcome. As I recall, he graphed the effect of a rise in loan interest rates on the purchasing power of consumers. However, he always qualified the effect with a Latin phrase: "ceteris paribus." If you have taken a course in either economics or Latin perhaps you've heard this phrase. It means "all other things remaining the same." In other words, what is the effect of a change in this variable when this variable is the *only* one to change? Unfortunately, a single variable is rarely the only one to change. For example, an increase in loan interest rates slows consumer spending in some areas while spurring compensatory spending in other areas. This, in turn, brings about changes in yet other areas and . . . well, suddenly the relationship is no longer simple.

We see this same complexity in behavioral research. Recall that one reason for conducting a factorial ANOVA rather than a series of t tests is that independent variables rarely affect the dependent variable in a simple fashion. Rather, the independent variables tend to *interact* to bring about a more complex outcome.

The same principle applies to power analysis. If all other things remained constant, 100% power would be the ideal. However, as we gain power to detect a false null hypothesis we also increase the risk of falsely rejecting a true null hypothesis. That is, as power goes up, Type 1 error also goes up. Certainly, that's not a good outcome. In addition, some methods of increasing power—for example, increasing sample size—also increase the chance of finding statistical significance when the meaningful effect size is actually quite small. That, too, is not a good outcome. So what is the best amount of power? Although no single value is best, the best trade-off among competing goals hovers somewhere around 80% power.

Now let's look at how we can estimate the actual power of a study. Power can be estimated from formulas originally found in a book by Jacob Cohen (1988). However, the formulas are cumbersome to use. In the same book, Cohen presents a series of tables derived from the formulas. Power tables show the interaction between six variables: power, alpha, effect size, directionality of the hypothesis, independence of the samples, and sample size. The tables aid in determining a study's power, given a known alpha, effect size, directionality, independence, and sample size. They also aid in determining ahead of time the necessary sample size for a study with a given alpha, desired power, projected effect size, given directionality, and given independence.

Because a power table must portray interaction among six variables, which is more than can be effectively shown in a single table, we use a series of tables to depict the interaction of only three of the variables: power, effect size, and sample size. The remaining three variables—alpha, directionality, and independence—are depicted via separate tables for each condition.

The entire set of tables is found in Cohen's book. For illustrative purposes, we will look at just three tables. Table 33.1 is for a two-sample nondirectional independent t test. Table 33.2 is for a one-way ANOVA. Table 33.3 is for a chi-square test of independence. Each table assumes .05 α. For the two parametric tests (t test and ANOVA), independent samples and equal sample sizes are also assumed.

Table 33.1 Sample Size Needed per Group to Obtain .80 or .90 Power, Given Various Cohen's *d*s or Effect Size *r*s, for a Two-Sample Independent *t* Test With Equal Sample Sizes and .05 Nondirectional α

Cohen's d	*Effect Size r*	*.80 Power*	*.90 Power*
.10	.063	1,571	2,102
.20 (small)	.100 (small)	393	526
.30	.148	175	234
.40	.196	99	132
.50 (medium)	.243 (medium)	64	85
.60	.287	45	59
.70	.330	33	44
.80 (large)	.371 (large)	26	34
.90	.447	17	22
1.00	.514	12	16

SOURCE: Adapted from Cohen (1988).

Table 33.2 Sample Size Needed per Group to Obtain .80 or .90 Power, Given Various Effect Size ηs, for an Independent Sample's One-Way ANOVA With Equal Sample Sizes and .05 α

	.80 Power			*.90 Power*		
	Number of Groups			*Number of Groups*		
Effect Size η	*3*	*4*	*5*	*3*	*4*	*5*
.100 (small)	322	271	240	417	350	310
.196	80	68	60	106	88	78
.243 (medium)	52	44	39	68	58	50
.287	36	31	27	48	40	35
.371 (large)	21	18	16	27	23	20
.447	14	12	11	18	15	13
.573	8	7	6	10	9	8

SOURCE: Adapted from Cohen (1988).

Table 33.3 Total Sample Size Needed to Obtain .80 or .90 Power, Given Various Effect Size ϕs or *V*s, for a Two-Variable Chi-Square Test of Independence at .05 α and 1 to 5 *df*

	.80 Power					*.90 Power*				
Effect Size V	*1 df*	*2 df*	*3 df*	*4 df*	*5 df*	*1 df*	*2 df*	*3 df*	*4 df*	*5 df*
.100 (small)	783	960	—	—	—	—	—	—	—	—
.200	196	240	275	300	322	262	320	350	388	417
.300 (medium)	88	108	120	134	143	117	140	160	170	185
.400	49	60	68	75	80	66	80	90	97	104
.500 (large)	32	38	44	48	51	43	50	57	62	66

SOURCE: Adapted from Cohen (1988).

Examine the three tables. It is apparent from each table that, as effect size (left column) increases, necessary sample size (table entries) decreases. This makes sense because, as you learned in Module 32, the bigger the actual treatment effect (i.e., the bigger the difference between the means), the greater the probability of finding that effect. When the actual difference between means is very large, we will find it even with a small number of subjects.

It is also apparent from each table that, as desired power (across the top) increases, necessary sample size (table entries) also increases. This, too, makes sense. Recall from the discussion earlier in this module that smaller samples contain greater error variance and that error variance lies in the denominator of the significance test statistic. Too much error variance in the denominator masks the treatment effect in the numerator. Thus, the more certain we want to be of actually finding an existing treatment effect, the more subjects we need.

Now, let's try reading one of the tables. Reading down the .80 power column in Table 33.1, when Cohen's *d* is .20 or when effect size *r* is .100, a study with .80 power requires 393 subjects per group. However, when Cohen's *d* is .50 or when effect size *r* is .243, only 64 subjects per group are needed. With a Cohen's *d* of .80 or with effect size *r* of .371, only 26 subjects per group are needed.

The other way of reading the table is to estimate actual power once a study is complete. From the same table, if we conduct a study with 26 subjects per group and find Cohen's *d* to be .80 or find effect size *r* to be .371, the power of our study is .80. But if we find Cohen's *d* to be .50 or find effect size *r* to be .243 and have only those same 26 subjects, we have fallen short of 80% power. Complex formulas and more complete tables would tell us exactly how short our power fell.

Each of the tables reads similarly. Note that sample size is per group for the parametric studies (*t* test, ANOVA), but sample size is total for chi-square. This is because chi-square has no "groups," only categories across variables.

PRACTICE

Directions for this set of practice exercises: Use Power Tables 33.1, 33.2, and 33.3 to answer the questions.

8. You are designing a study to be evaluated with a two-sample nondirectional independent *t* test at $\alpha = .05$. You want .90 power. You expect Cohen's *d* to be about .40. How many subjects will you need per group?

9. You are designing a study to be evaluated with a two-sample nondirectional independent *t* test at $\alpha = .05$. You want .80 power. You expect Cohen's *d* to be about .70. How many subjects will you need per group?

10. You are designing a study to be evaluated with a two-sample nondirectional independent t test at $\alpha = .05$. You are able to obtain 25 subjects per group. Approximately what will effect size r have to be for you to obtain .80 power?
11. You are designing a study to be evaluated with a two-sample nondirectional independent t test at $\alpha = .05$. You are able to obtain 65 subjects per group. Approximately what will Cohen's d have to be for you to obtain .80 power?
12. You calculate a two-sample nondirectional independent t test at $\alpha = .05$. Your study has 130 subjects per group and effect size r is .20. To the nearest 10%, what is the power of your study?
13. You calculate a two-sample nondirectional independent t test at $\alpha = .05$. Your study has 45 subjects per group and Cohen's d is .60. To the nearest 10%, what is the power of your study?
14. You are designing a three-group study to be evaluated with a one-way ANOVA at $\alpha = .05$. You expect effect size η to be about .29. How many subjects will you need per group for .80 power?
15. You are designing a four-group study to be evaluated with an ANOVA at $\alpha = .05$. You are able to obtain 40 subjects per group. Approximately what will effect size η have to be for .90 power?
16. You calculate a one-way ANOVA at $\alpha = .05$. Your study has four groups with 30 subjects per group. Effect size η is .30. To the nearest 10%, what is the power of your study?
17. You are designing a study to be evaluated with a chi-square test of independence at 3 df and $\alpha = .05$. You expect Cramer's V effect size to be about .30. How many total subjects will you need for .80 power?
18. You are designing a study to be evaluated with a chi-square test of independence at 4 df and $\alpha = .05$. You are able to obtain 60 total subjects. Approximately what will Cramer's V effect size need to be for .90 power?
19. You calculate a chi-square test of independence at $\alpha = .05$. Your study has 1 df, 200 total subjects, and effect size ϕ is .19. To the nearest 10%, what is the power of your study?

Looking Ahead

This module completes the instruction on hypothesis testing (Modules 10 to 33). You have learned how to word hypotheses, sample subjects, conceptualize sampling distributions, calculate tests of hypotheses, determine statistical significance, construct confidence intervals, determine effect size, and plan for and estimate power. In the remaining modules, we will turn our attention to correlation and prediction. The design and interpretation of correlational studies are quite different from the experimental studies with which we have worked thus far. With correlational studies, we are looking to establish only relationships, not causality. Instruction concludes with an introduction to multiple regression. This technique uses correlational (relational) statistics to analyze data from either correlational or experimental designs. When used for the latter, causation can again be inferred.

Visit the study site at www.sagepub.com/steinberg2e for practice quizzes and other study resources.

PART XV

Correlation

"Whither Thou Goest, I Will Go"

34 Relationship Strength and Direction

Terms: reliability, prediction, scatterplot, bivariate, strength, positive direction, negative direction, linear, curvilinear, outliers

Learning Objectives:

- Distinguish between correlational and experimental studies
- Create a scatterplot
- Estimate the strength and direction of a set of data based on its scatterplot
- Understand the effect of outliers on the strength and direction of a correlation

Experimental Versus Correlational Studies

In Modules 12 to 33, we measured the effect of an independent variable on a dependent variable. Then, we tested the result for statistical significance, effect size, and power. In each of the studies, there were at least two groups that differed on the independent variable. We then examined whether the independent variable (the defining difference between the groups) caused the effect in the dependent variable.

In this module, we begin to look at data from a new perspective. We will look at correlational studies. In a correlational study, we have only a single group of subjects rather than two or more groups. In addition, each of the subjects has a score on two different variables. Also, in a correlational study, we do not seek cause-and-effect relationships between independent and dependent variables. Rather, we simply want to know whether or not the scores on two variables are related.

Sometimes correlational studies are used to establish the properties of the tests themselves. The SAT, for example, is given on multiple test dates throughout the year. Students taking the test on one date do not answer the same questions as students taking the test on another date. Rather, there are parallel forms of the test—a different form for each date. Scores have the same meaning regardless of which form students take because the test forms are comparable. But how do the test developers know that the scores are comparable? During the test development process, they gave the same students (note the single group of subjects) two different forms of the test (note the two variables). Then, they compared the students' scores on both tests (note the correlation). They found that the scores were similar for the same students on both forms of the test. This type of correlation is called test **reliability.**

Most of the time, correlational studies are used for **prediction** rather than for establishing the reliability of the tests themselves. That is, we seek to establish relationships so that the score of a person on one variable can be used to predict that person's probable score on a second variable. For example, once a relationship is established between the number of hours children watch television and children's academic performance in school, we can predict any given child's probable academic performance in school just by knowing the number of hours of television he or she watches. Similarly, a researcher interested in prediction may want to know the relationship between

- the amount of time students study and their grade on a test,
- the amount of antidepressant medication clients take and their reported mood level,
- air temperature and crime rate,
- income and years of education,
- height and weight, and
- IQ and shoe size (do you think there is any relationship?).

CHECK YOURSELF!

What are two common uses of correlations?

CHECK YOURSELF!

Complete the following table to compare and contrast studies analyzed by three different statistics.

	No. of Groups	*No. of Variables*	*Purpose of Study or Type of Conclusion to Be Drawn*
t test			
F test			
Correlation			

PRACTICE

1. Which of these studies might be analyzed with a correlation?
   a. The relationship between birth order and academic achievement
   b. The number of inches babies grow when breast-fed versus bottle-fed
   c. Whether the amount of time runners practice is related to their running times during a tournament
   d. Which of three treatments best helps hyperactive children stay on task

2. Which of these studies might be analyzed with a correlation?
   a. The cost of various cars when new and the amount of time before those cars need repair
   b. The number of illnesses reported per year by vegetarians versus nonvegetarians

   c. The duration of fever for viral versus bacterial infections
   d. The number of miles dieters walked and the amount of weight they lost

3. Which of these studies might be analyzed with a correlation?
   a. The level of depression in various communities and suicide rate in those communities
   b. The physical fitness level and number of miles run
   c. The effectiveness of Drug A versus Drug B in reducing fever
   d. The difference in age at marriage between Guyanese youth and Somalian youth

4. Which of these studies might be analyzed with a correlation?
   a. The which group has a higher reading comprehension level—dyslexic children versus ADHD (attention deficit hyperactivity disorder) children
   b. The amount of religious commitment and frequency of attendance at religious services
   c. The number of doctor visits and income level
   d. The soft drink preferences of college students

Plotting Correlation Data

A **scatterplot** displays the scores of individual cases on two variables. It visually displays the degree to which scores on one variable are related to scores on another variable. Because it depicts two variables, it is said to be **bivariate** (bi = two, variate = variable). Unlike the graphs you created in Module 4, in a scatterplot the *Y*-axis does not indicate frequency. Rather, both axes indicate scores. One axis indicates the score on one variable; the other axis indicates the score on the other variable. Individual cases are represented by a dot or some other symbol.

To create a scatterplot, locate a subject's score for Variable 1 on the *X*-axis and the same subject's score for Variable 2 on the *Y*-axis. Mentally draw perpendicular lines from the axes to the interior of the graph. The point at which the two lines intersect is the location of that student's scores on both variables (Figure 34.1).

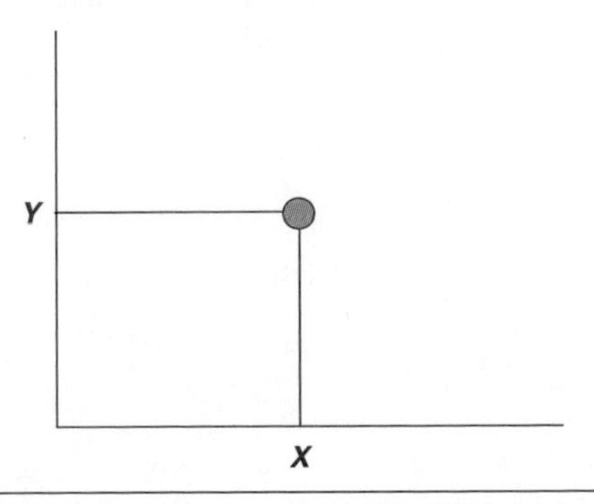

Figure 34.1 Drawing a Scatterplot

Assume that 10 students receive the following scores on Quiz 1 and Quiz 2. The scatterplot for this set of data is displayed in Figure 34.2, with the location of the first student, Abby, indicated.

Student	*Quiz 1*	*Quiz 2*
Abby	9	10
Babs	7	8
Clyde	10	7
DeShawn	9	8
Emily	5	5
Fred	6	4
Gino	10	8
Hortense	9	6
Ingrid	6	6
Jorge	10	9

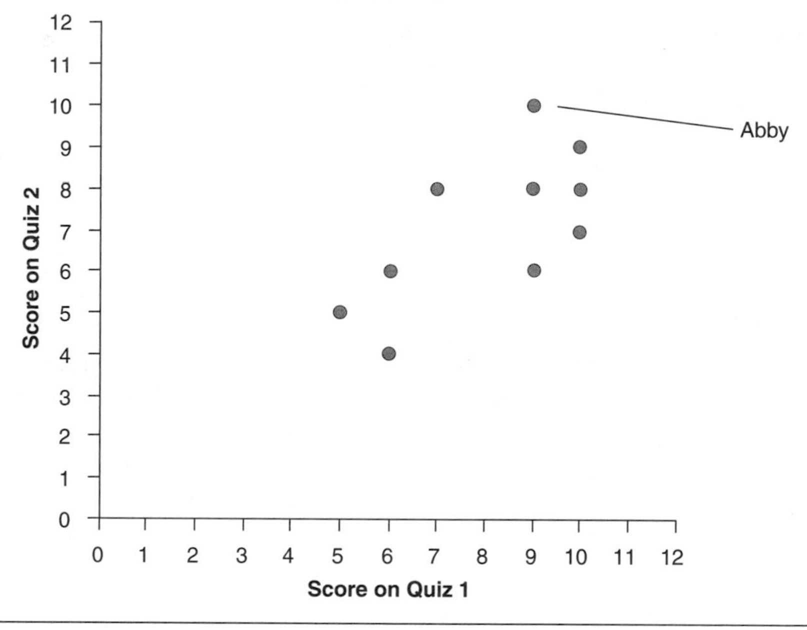

Figure 34.2 Scores on Quiz 1 and Quiz 2

PRACTICE

5. Create a scatterplot of the following heights and weights of ninth-grade girls:

Subject	*Height*	*Weight*
Quashonna	66	132
Ronette	68	121
Sherelle	62	105
Telise	60	107
Ulinda	67	130
Vonnie	60	110
Winona	68	136
Xelei	70	148
Yvette	63	96
Zoe	69	154

6. Create a scatterplot of the following data on the average annual household income and the average cost of a 3-bedroom, 1½-bath home (both to the nearest thousand dollars) in 10 geographic areas:

Area	*Income*	*House*
1	32	104
2	46	232
3	29	95
4	67	291
5	42	165
6	78	482
7	35	128
8	52	347
9	39	119
10	43	143

7. Create a scatterplot of the following number of times 15 teenage girls look at themselves in a mirror per day and their score on a test of self-esteem (scored 1–10, where higher scores = more self-esteem).

Girl #	*Level of Self-Esteem*	*Number of Mirror Looks*
1	5	3
2	7	4
3	3	1
4	6	1
5	7	4
6	9	6
7	5	4
8	6	8
9	3	2
10	5	3
11	6	2
12	5	1
13	7	4
14	4	2
15	9	5

8. Create a scatterplot of the following number of hours 12 students spend playing computer games per day and their academic average (in percentage).

Student #	*Hours Playing Computer Games*	*Academic Average*
1	1.5	84
2	3.8	76
3	0.5	85
4	2.3	79
5	1.2	88
6	2.9	82
7	1.0	95
8	0.4	93
9	3.0	78
10	4.2	80
11	1.1	90
12	0.3	91

Relationship Strength

The relationship between two sets of scores has two characteristics: strength and direction. Let's look at both strength and direction in more detail.

The **strength** of a relationship tells the degree to which scores on one variable are related to scores on the other variable. Strength is expressed from .00 to 1.00. The higher the numerical value (regardless of sign), the stronger the relationship. A correlation of .82 is strong, while a correlation of .13 is weak. In a perfect relationship, all data points fall along a straight line. For example, by knowing the temperature on the Celsius scale, we can exactly predict the temperature on the Fahrenheit scale. Thus, the correlation between Celsius and Fahrenheit is 1.00.

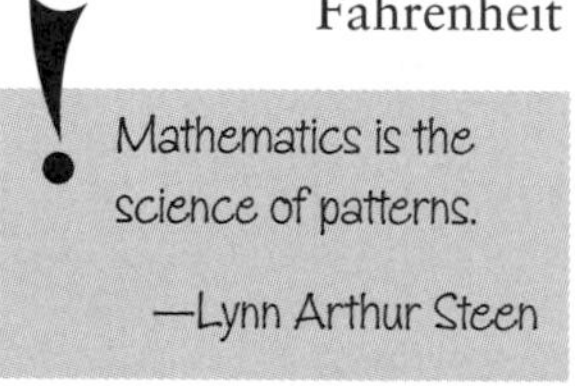

A correlation of .00, at the other extreme, indicates no relationship. For example, there is no relationship between adult IQ and shoe size. Adults with high, medium, or low IQs are equally likely to have small, medium, or large shoe sizes. Thus, the data points fall in a circular "blob." Here is how these two scatterplots would look (Figure 34.3).

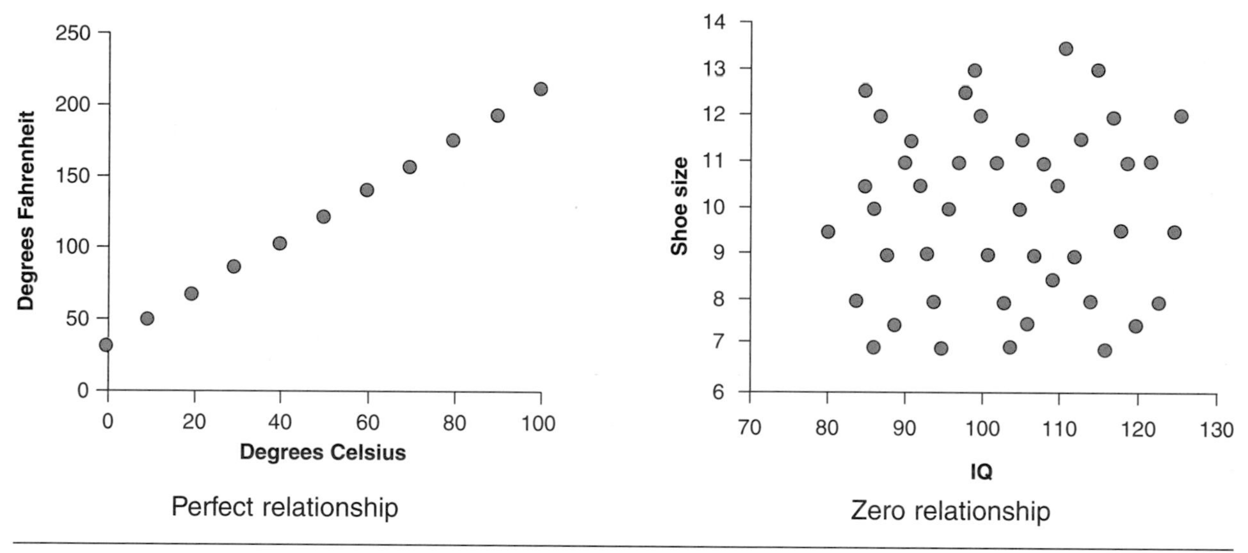

Figure 34.3 Scatterplots of Perfect and Zero Relationships

Although we sometimes find perfect relationships in the physical sciences, relationships are rarely perfect in the social sciences. With human beings, most variables are only moderately related. SAT score predicts freshman GPA, for example, but only moderately. Other variables such as motivation or test anxiety come into play. Thus, some people who score very high on the SAT do poorly in college, and some people who score low on the SAT do very well in college. Similarly, the number of absences during a semester predicts grade on a final exam but only moderately. Some students who never miss a class do poorly on the final exam, and some students who are frequently absent do well on the final exam. Other variables such as prior knowledge, diligence in reading the textbook, and obtaining missed notes come into play.

The data for most social science relationships fall neither on a straight line nor in a circular blob. Rather, they fall in an ellipse. The thinner the ellipse, the closer the points fall to a straight line and, hence, the stronger (closer to 1.00) the relationship. The wider the ellipse, the farther the points fall from a straight line and, hence, the weaker (closer to .00) the relationship. Figure 34.4 shows two typical moderate relationships in the social sciences.

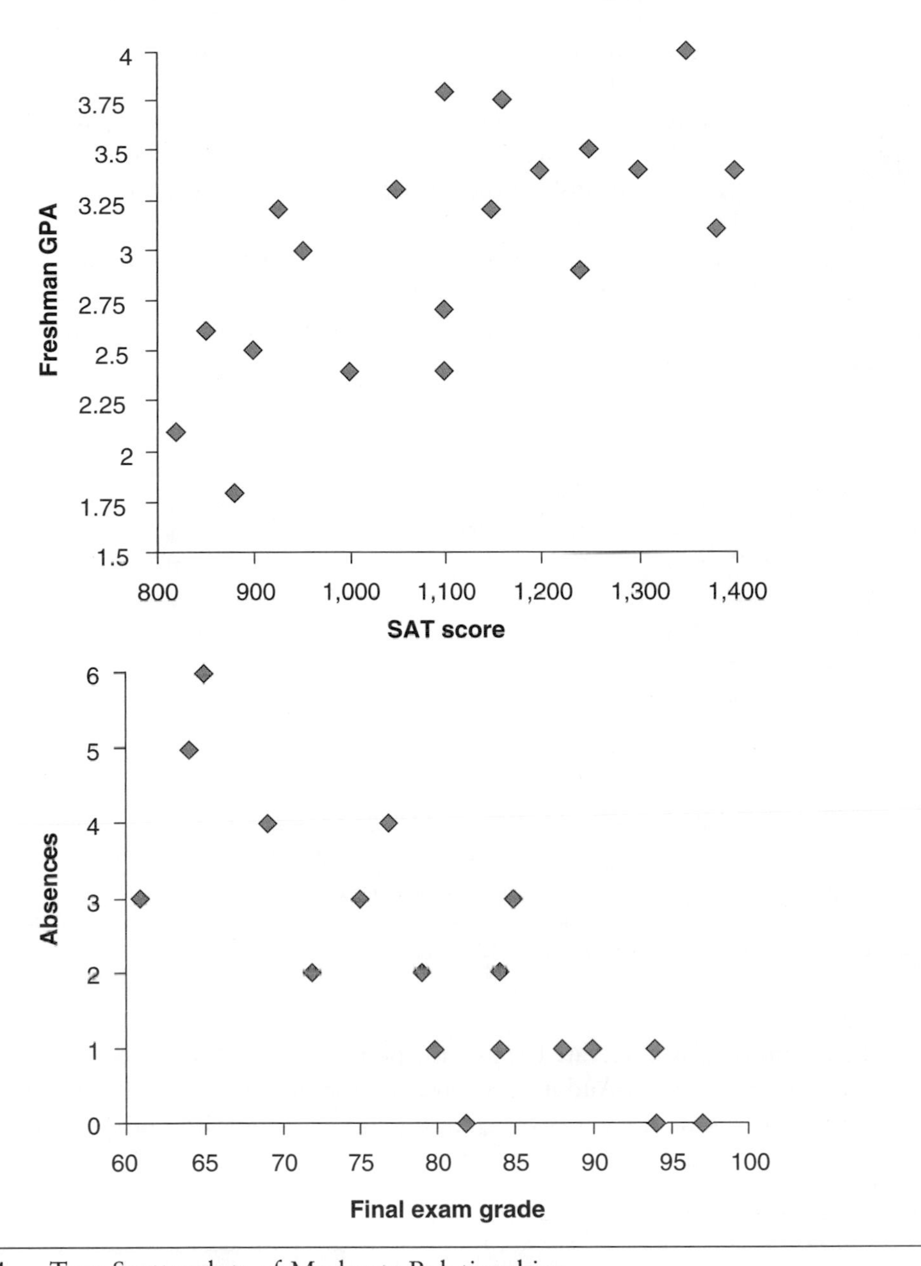

Figure 34.4 Two Scatterplots of Moderate Relationships

Relationship Direction

The **direction** of a relationship tells whether or not the values on two variables go up and down together. Direction is indicated by a positive or a negative sign. If two variables are **positively** correlated, then as the values on one variable go up, so do the values on the other variable. For example, the relationship between SAT score and freshman college GPA is positive. Thus, students who receive higher scores on the SAT are more likely to receive higher grades during their freshman year of college. You can see this pattern in the upper scatterplot in Figure 34.4. Note the direction of the data points. With a positive relationship, the data points go from the bottom left to the upper right.

If two variables are **negatively** correlated, then as the values of one variable go up, the values of the other variable go down. For example, the relationship between the number

> He seemed to have very strong intuitions but unfortunately of negative sign.
>
> —the biologist Francis Crick, referring to René Thom, in *What Mad Pursuit*

of absences in a course and score on the final exam in that course is negative. In other words, the more often students are absent from class, the lower their grades tend to be on the final exam. You can see this pattern in the lower scatterplot in Figure 34.4. Note the direction of the data points. With a negative relationship, the data points go from the upper left to the lower right.

Let's return to the set of quiz scores from the beginning of this module. Judging from the scatterplot shown in Figure 34.5, what is the approximate strength of the relationship? What is the direction of the relationship?

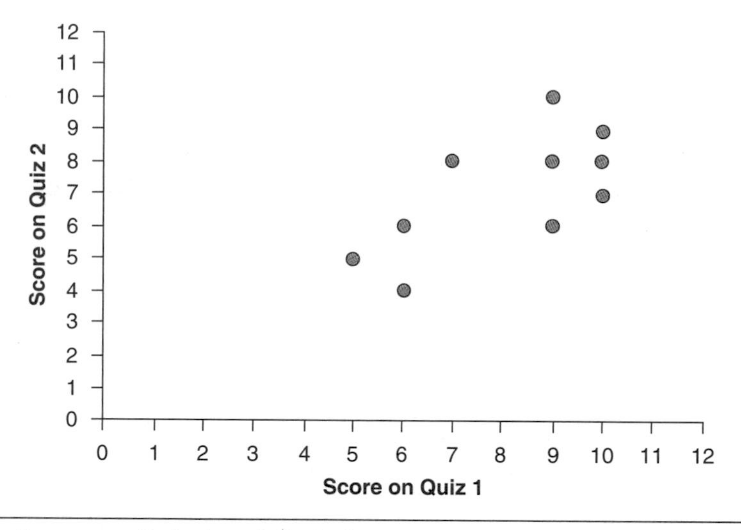

Figure 34.5 Scores on Quiz 1 and Quiz 2

The relationship is moderate because the points form an ellipse about halfway between a straight line and a circle. And it is positive because the general trend is from lower left to upper right. That is, as the scores on Quiz 1 go up, the scores on Quiz 2 also go up.

CHECK YOURSELF!

What two terms are used to describe a correlational relationship? What does each term indicate?

PRACTICE

9. For each of the following, indicate whether the expected relationship between the two variables will be positive (+), negative (–), or zero (0):
   - ____ a. The average number of calories eaten per day and body weight
   - ____ b. Running speed and general physical condition
   - ____ c. Length of hair and introversion
   - ____ d. The amount of education and the time spent on welfare
   - ____ e. Per capita consumption of alcohol and suicide rate

10. For each of the following, indicate whether the expected relationship between the two variables will be positive (+), negative (–), or zero (0):
    _____ a. Air temperature and the amount of snow on the ground
    _____ b. The number of minutes of exercise per day and score on a physical fitness test
    _____ c. The number of years having a driver's license and age
    _____ d. The number of pages in a textbook and cost of that textbook
    _____ e. Age at which a child takes its first step and educational level of the parents

11. For each of the following, indicate whether the expected relationship between the two variables most likely will be positive (+), negative (–), or zero (0):
    _____ a. number of months dating current romantic partner and academic GPA
    _____ b. amount of rainfall and how quickly the grass grows
    _____ c. how much sleep one gets and how tired one feels the next morning
    _____ d. number of cars in a family and number of pets in a family
    _____ e. age at first marriage and education level

12. For each of the following, indicate whether the expected relationship between the two variables will be positive (+), negative (–), or zero (0):
    _____ a. size of house and size of electric bill
    _____ b. height of parents and height of children
    _____ c. air humidity level and people's energy level
    _____ d. number of books read per year and age at which got first eyeglasses
    _____ e. hours spent at the beach and depth of tan

Linear and Nonlinear Relationships

The relationships we have examined thus far have been linear. In **linear** relationships, the trend in the data is best described by a straight line. That is, we could fit a straight line in the center of the scatterplot to indicate the trend in the data. Figure 34.6 shows examples of two straight-line trends.

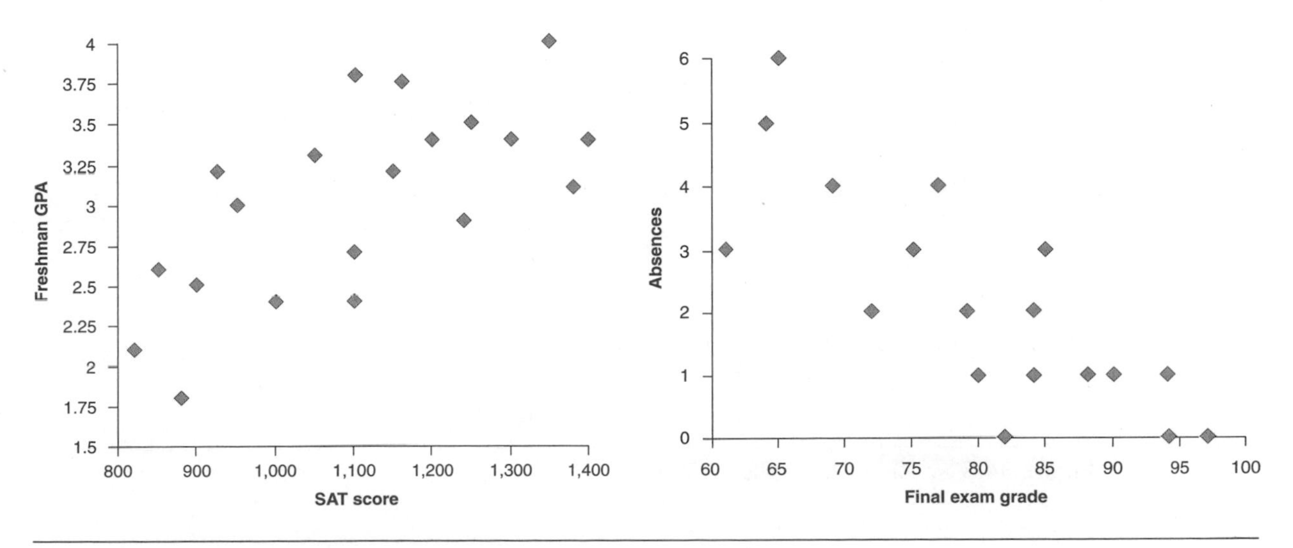

Figure 34.6 Two Linear Relationships

Not all relationships are linear. Some are curvilinear. In a **curvilinear** relationship, the trend in the data changes direction. For example, the relationship between test score and test anxiety is curvilinear. High levels of test anxiety impair test performance, but so do low levels of test anxiety. The best test performance occurs among test takers having moderate anxiety. Thus, the relationship takes on an inverted U shape, as you can see in Figure 34.7.

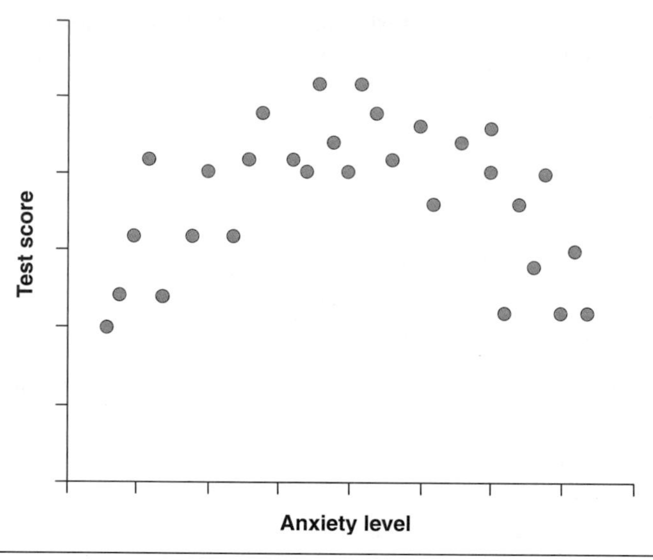

Figure 34.7 An Inverted U Relationship

The relationship mentioned at the beginning of this module between the amount of antidepressant medication taken and reported mood level is also curvilinear. You can see this pattern in Figure 34.8.

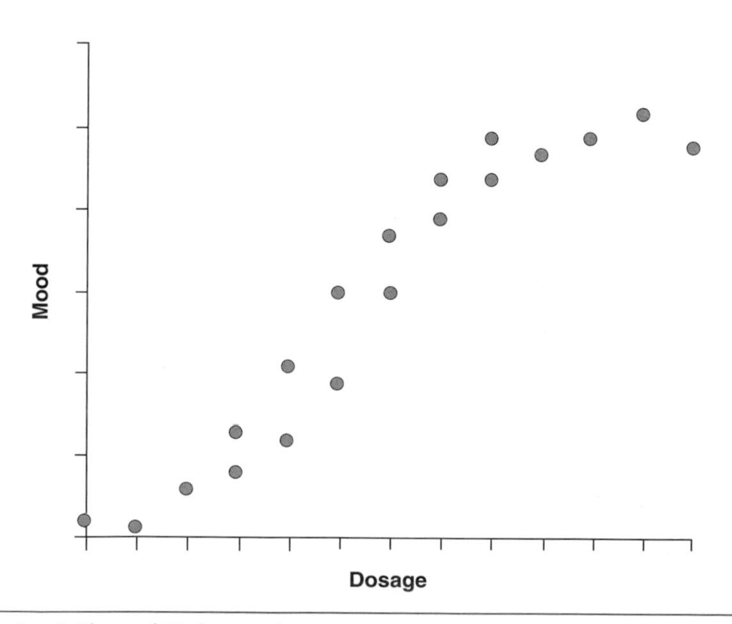

Figure 34.8 An S-Shaped Relationship

Note from the graph that mood elevates sharply with initial dosage, then elevates more slowly with a further increased dosage, and finally levels off at a still higher dosage. Thus, the relationship takes on a leaning S shape.

If we try to fit a straight line to curvilinear data, many data points will fall off the line. This is especially true for the relationship between test score and test anxiety. However, if we fit a curved line to the data, the data points hug the line quite closely. Figure 34.9 shows how a curved line fits the data better than a straight line does.

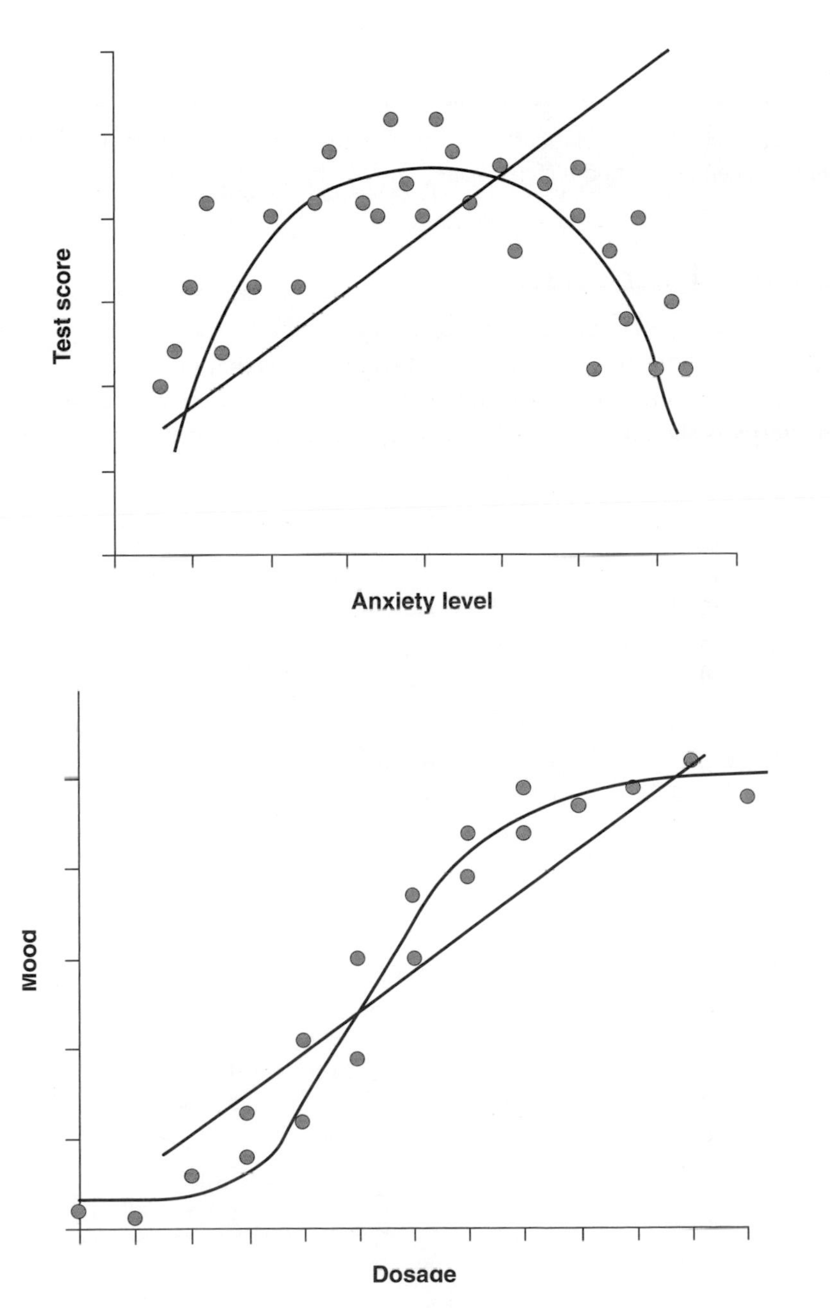

Figure 34.9 Curvilinear Data Fitted With Linear and Curvilinear Lines

Recall that the closer data fall to the line, the stronger the relationship. Because the data in Figure 34.9 closely hug a curved line, fitting a curved line to the data will yield a relatively high correlation—something close to 1.00. Conversely, because the same data fall quite far from a straight line, fitting a straight line to the data will yield a relatively low correlation—something close to .00. This is why you should always plot your data before calculating any statistic. As you will learn in Module 35, some correlation statistics are linear, and some are curvilinear. Using a linear statistic to describe curvilinear data will seriously underestimate the amount of correlation.

CHECK YOURSELF!

How can you tell if a relationship is linear or curvilinear? Why does the shape of the data matter when calculating a correlation?

Outliers and Their Effects

In any relationship, there may be outliers. **Outliers** are scores that fall far outside the trend of the rest of the data—typically 2 or more standard deviations beyond the next closest score. Consider the two scatterplots shown in Figure 34.10. They are identical except for a single score.

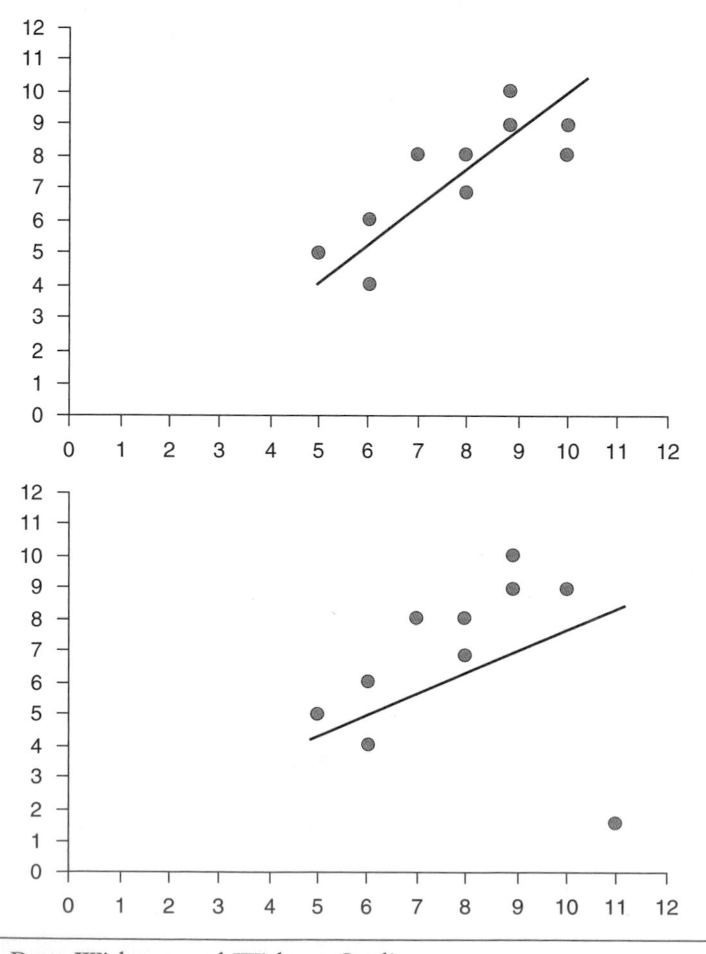

Figure 34.10 Data Without and With an Outlier

For the most part, a straight line fits the data well. However, the outlier "pulls" the line in the direction of the outlier, as shown in the lower example in Figure 34.10. When the line is pulled toward the outlier, the remaining points then fall farther from the line than they otherwise would have. The "fit" is worse; hence, the correlation is lower. Outliers *lower* the amount of correlation from what it would be without the outliers.

Often an outlier is due to error: a mismarked answer paper, a mistake in entering a score in a database, a subject who misunderstood the directions, and so on. You should always seek to understand the cause of an outlying score. If the cause is not legitimate, you should eliminate the outlying score from the analysis. Leaving an incorrect or errant score in the analysis distorts the true correlation.

At other times, outliers are legitimate. One person may legitimately score far beyond everyone else on the variable being measured. Nevertheless, if the outlier seriously distorts interpretation of the remaining data, it is customary to present the data with and without the outlier, state that you are eliminating it, and then conduct further analyses with the outlier removed.

CHECK YOURSELF!

How can you tell if a set of data includes outliers? Why does the presence of outliers matter when calculating a correlation?

Looking Ahead

So far, we have merely estimated the strength and direction of relationships through visual inspection of the data. In the next module, we will calculate a correlation coefficient. Then, we will be able to state the exact strength and direction of a relationship.

PRACTICE

13. Here are the quiz scores of six students on each of three pop quizzes:

	Quiz 1	*Quiz 2*	*Quiz 3*
Ann	11	1	5
Brianna	9	7	9
Chara	8	9	12
Diane	6	10	10
Eliza	4	11	6
Felice	3	12	6

a. Create three separate scatterplots: Quiz 1 and Quiz 2, Quiz 1 and Quiz 3, and Quiz 2 and Quiz 3.

b. Describe the direction and apparent strength of the relationship in each scatterplot. Comment on the effect of outliers, if any.

14. Here is the stress level (scored 1–10, with higher scores = higher stress) for college freshmen during the first week of classes, middle week of classes, and last week of classes:

	First Week of Classes	*Middle Week of Classes*	*Last Week of Classes*
1	5	2	6
2	4	2	5

(Continued)

(Continued)

	First Week of Classes	*Middle Week of Classes*	*Last Week of Classes*
3	5	4	6
4	7	5	7
5	3	3	7
6	6	3	8
7	4	2	5
8	5	4	7
9	7	3	9
10	3	2	5

a. Create three scatterplots: first week with middle week, first week with last week, and middle week with last week.

b. The actual scores show that stress is moderate in the first week, low in the middle week, and high in the last week, yet all plots were positive and strong. Why?

15. Gather data from 10 adults (perhaps from students in your statistics class) for height and shoe size.

   a. Create a scatterplot for the data.

   b. Describe the direction and apparent strength of the data. Comment on the effect of outliers, if any.

16. Gather data from 10 adults (perhaps from students in your statistics class) for the number of hours of television watched on an average day and the typical number of hours of sleep obtained in an average night.

   a. Create a scatterplot for the data.

   b. Describe the direction and apparent strength of the data. Comment on the effect of outliers, if any.

SPSS Connection

Download the file **data_quiz 1 quiz 2.sav** from www.sagepub.com/steinberg2e. These data are used in the textbook example.

Alternatively, manually enter the scores on Quiz 1 and Quiz 2 into the SPSS **Data View** spreadsheet: Set it up like this.

Abby	9	10
Babs	7	8
Clyde	10	7
DeShawn	9	8
Emily	5	5
Fred	6	4
Gino	10	8
Horense	9	6
Ingrid	6	6
Jorge	10	9

Click on the **Variable View** tab to define the variable. Name the first variable **name**, set the Type as **String**, and label the variable as **Name.** Name the second variable **quiz1**, set the decimal at 0, and label the variable as **Quiz 1 (X)**. Name the third variable **quiz2**, set the decimal at 0, and label the variable as **Quiz 2(Y)**.

If the file is not already in **Data View**, click that tab in the lower left of the screen.

In the toolbar at the top of the screen, click on **Graphs**, then **Legacy Dialogs**, then **Scatter/Dot.** Select the **Simple Scatter** example, and click **Define.** Highlight the variable **Quiz 1(X)** in the left window and click on the **arrow** before the **X axis** window (that's the second window, not the top window) on the right, to send the variable into that window. Highlight the variable **Quiz 2(Y)** in the left window and click on the **arrow** before the **Y axis** window (that's the top window, not the second window) on the right, to send the variable into the that window. Click **OK**. This is what you will see.

Graph

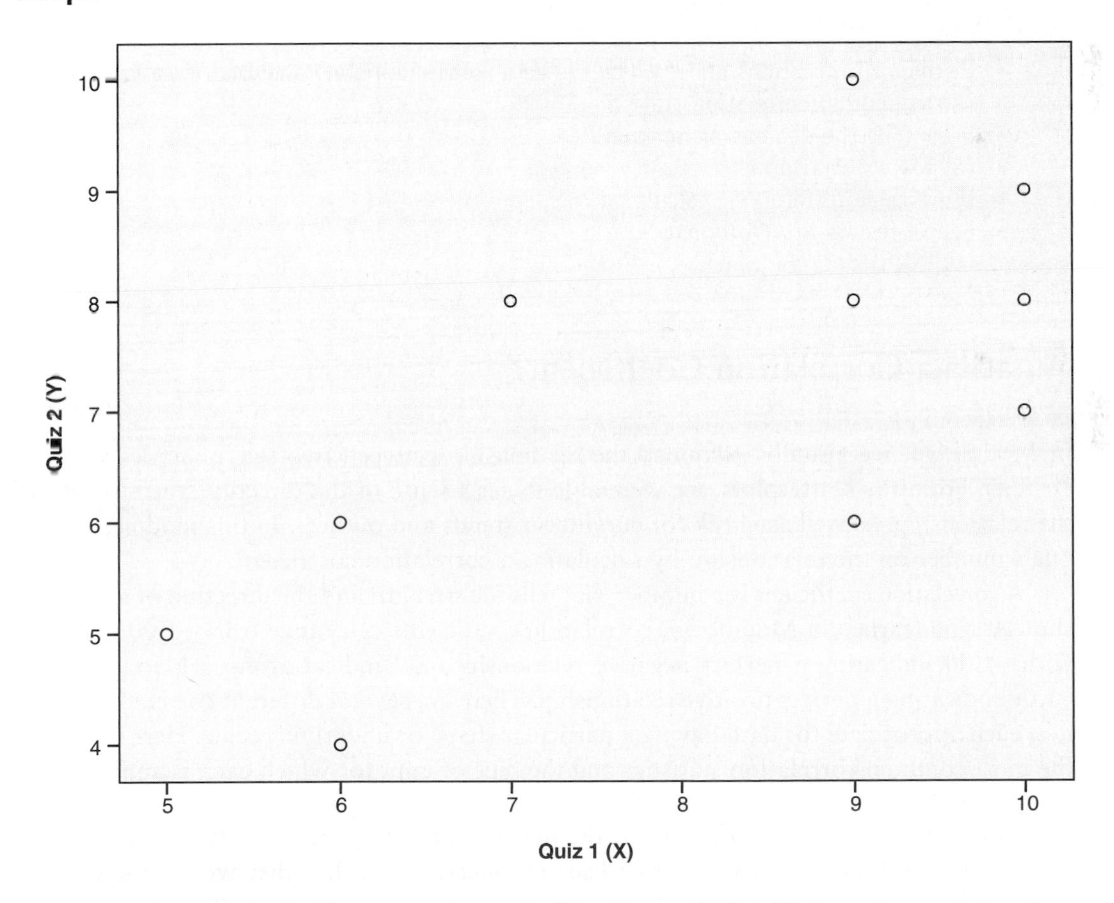

This is the scatterplot shown in the textbook. The strength and direction is not as clear as the one in the textbook because the graph violates the guidelines for graph construction that you learned in Module 4 regarding appropriate axis scaling for best graph interpretation. The complete version of SPSS allows for axis demarcation decisions, but the Student version of SPSS does not.

Visit the study site at www.sagepub.com/steinberg2e for practice quizzes and other study resources.

35

Pearson r

Terms: correlation coefficient, Pearson r

Symbol: r

Learning Objectives:

- Distinguish conditions under which the data are appropriately analyzed via a Pearson r versus another correlation statistic
- Determine the degrees of freedom
- Calculate a Pearson r
- Use a table to interpret calculated r
- Report results in APA format

What Is a Correlation Coefficient?

In Module 34, we visually examined the relationship between two sets of scores via scatterplots. From the scatterplots, we were able to get a sense of the direction and strength of the relationship, as well as check for curvilinear trends and outliers. In this module, we will put a number on the relationship by calculating a correlation coefficient.

A **correlation coefficient** is a number that tells the strength and the direction of a relationship. As you learned in Module 34, correlation coefficients can range from –1.00 to +1.00, with –1.00 indicating a perfect negative relationship, .00 indicating no relationship, and +1.00 indicating a perfect positive relationship. There are several different correlation statistics, each appropriate for data having a particular shape or underlying scale. Here is a list of the most common correlation statistics and the type of data for which each is appropriate:

1. *Pearson r*: This is a measure of the linear relationship between two variables that have both been measured on at least an interval level. In other words, the data on both variables are scores. For example, we could compute a Pearson r statistic between students' SAT score and their freshman GPA, as discussed in Module 34.

2. *Spearman rho*: This is a measure of the linear relationship between two variables that have both been measured on an ordinal level. In other words, the data on both variables are ranks. For example, we could compute a Spearman rho statistic between students' class rank at high school graduation and their class rank at college graduation.

3. *Phi*: This is a measure of the linear relationship between two variables that have both been measured dichotomously. Examples of dichotomous data are pass/fail, yes/no,

high/low, or male/female. For example, we could compute a phi statistic between students' gender and whether or not they pass or fail a particular mathematics course.

4. *Point biserial*: This is a measure of the linear relationship between two variables when one has been measured on at least an interval level and the other has been measured on a dichotomous level. For example, we could compute a point biserial statistic between SAT score and completing college versus dropping out of college. The SAT score would be the interval variable, and dropout status would be the dichotomous variable.

5. *Eta*: This is a measure of the curvilinear relationship between two variables that have both been measured on at least an interval level, but the relationship is known to be nonlinear. For example, eta would be the appropriate statistic for the relationship between test anxiety and test score, as discussed in Module 34.

CHECK YOURSELF!

What two characteristics differentiate the various types of correlation coefficients?

We will not calculate each of these correlation statistics. In this module, we will calculate only a Pearson r. **Pearson r**, symbolized r_{XY} or just r, is appropriate for use with scores whose underlying relationship is linear and that are measured on at least an interval scale. Thus, a Pearson r is not appropriate if the relationship is curvilinear or if the data are ranks or dichotomies.

PRACTICE

1. State whether or not a Pearson r is the appropriate analytic technique for each of the following studies. If it is not, explain why not.

   a. The data take the following form:

	Cola	Orange	Root Beer	Ginger Ale
Boys	18	12	26	9
Girls	14	23	20	6

   b. The data take the following form:

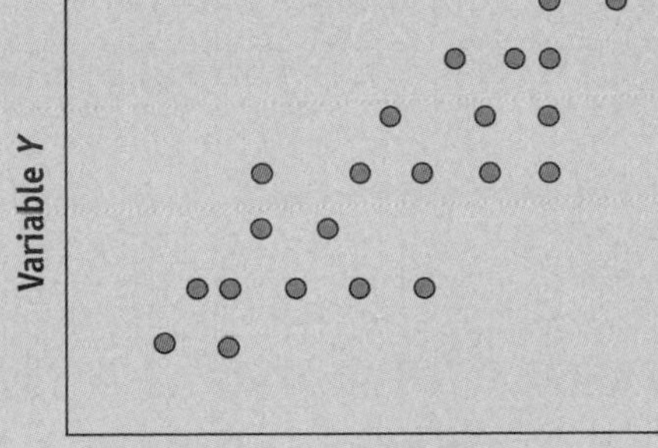

c. The data take the following form:

Subject	*Height*	*Weight*
1	66	124
2	71	165
3	69	160
4	68	182
5	62	133
6	70	149
7	64	144
8	72	187
9	73	161
10	68	140

d. The data take the following form:

Group 1: $N = 30$, $M = 45$

Group 2: $N = 30$, $M = 42$

Group 3: $N = 30$, $M = 46$

Group 4: $N = 30$, $M = 43$

e. The data take the following form:

Subject	*Rank on Quiz 1*	*Rank on Quiz 2*
1	10	10
2	9	8
3	8	6
4	9	9
5	6	7
6	5	4
7	4	5
8	3	2
9	2	3
10	1	1

2. Why is a Pearson r *not* the appropriate statistic for the following studies?
   a. To determine the relationship between incoming college students' class rank in high school and their class rank at the end of their first year of college
   b. To determine if the toy preferences of preschoolers are the same as the toy preferences of elementary school children
   c. To determine if Drug A or Drug B is more effective in relieving disease symptoms

Formulas for Pearson r

There are three different formulas for Pearson r: a z-score formula, a deviation score formula, and a raw score formula. Each of the three formulas is an algebraic derivation from the others, so each leads to the same answer.

z Score formula:

$$r_{XY} = \frac{\sum z_X z_Y}{N}$$

Deviation score formula:

$$r_{XY} = \frac{\sum (X - M_X)(Y - M_Y)}{N s_X s_Y}$$

Raw score formula:

$$r_{XY} = \frac{N\sum XY - \left(\sum X\right)\left(\sum Y\right)}{\sqrt{\left[N\left(\sum X^2\right) - \left(\sum X\right)^2\right]\left[N\left(\sum Y^2\right) - \left(\sum Y\right)^2\right]}}$$

If it were up to you, which of the three formulas would you like to use? You'd like to use the z-score formula, right? That's understandable, given its brevity. However, the brevity of the z-score formula for Pearson r is deceptive. Recall the formula for a z score:

$$z_X = \frac{X - M_X}{s_X} \quad \text{or} \quad z_Y = \frac{Y - M_Y}{s_Y}$$

Thus, before we can plug the z scores for X and Y into the Pearson r formula, we must first calculate the z scores for each raw score. And before we can calculate the z scores, we must first, for both X and Y, calculate the mean and standard deviation, then find the distance of each person's raw scores from those means, and then divide those deviation scores by the standard deviations. That's a lot of work.

The deviation score formula might prove easier to use, so now let's compare the z-score and deviation score formulas. With the deviation score formula, there is no need to find z scores on variables X and Y by dividing the deviation scores by that variable's standard deviation. Thus, the deviation score formula is simpler to use than the z-score version. However, with the deviation score formula, we still must find both the standard deviation and the mean of variables X and Y, as well as the deviation scores. That's still a lot of work.

Well, then, we could use the raw score formula. The raw score formula looks intimidating, doesn't it? However, the formula works directly from the raw scores without first requiring that we calculate means, deviations from means, standard deviations, or z scores. Thus, although the raw score formula looks complicated, it is quite simple to use. It requires nothing more than addition, subtraction, multiplication, division, squares, and square roots.

> The four rules of arithmetic may be regarded as the complete equipment of the mathematician.
>
> —James Clerk Maxwell

Although the z-score formula for Pearson r is the most complicated of the three formulas to use, a scatterplot of z scores conveys the most information about correlational relationships. Therefore, before we calculate a Pearson r correlation coefficient using the raw score formula, let's see what a scatterplot would look like when the data are converted into z scores.

z-Score Scatterplots and r

Let's use a Cartesian coordinate system to create a scatterplot of z scores. A Cartesian coordinate system consists of two perpendicular axes, X and Y, that intersect to form four quadrants.

When graphing z scores, the two axes are the two variables. The intersection point is zero on each axis. Therefore, the upper-right quadrant contains positive values on both variables, the lower-right quadrant contains positive values on variable X and negative values on variable Y, the lower-left quadrant contains negative values on both variables, and the upper-left quadrant contains negative values on variable X and positive values on variable Y. Figure 35.1 shows a typical Cartesian coordinate system.

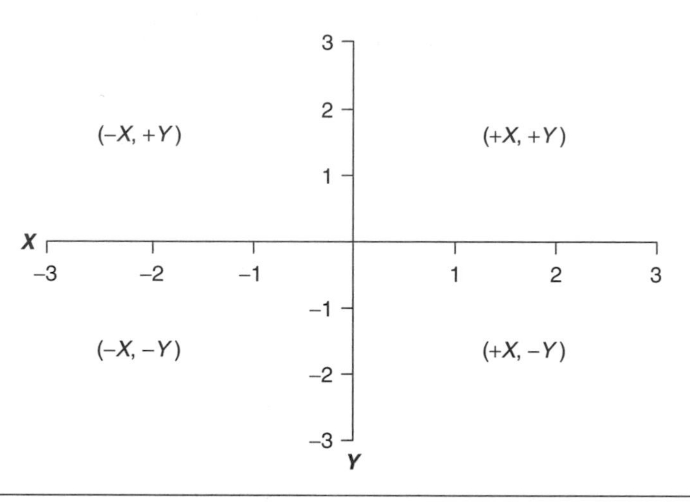

Figure 35.1 Four Quadrants in a Cartesian Coordinate System

Because z scores are either positive or negative and they have a mean of 0, they lie within the coordinate system. Let's look at an example. Continuing with the quiz data from Module 34, here are the z scores for each student's scores on Quiz 1 (X) and Quiz 2 (Y). [*Note:* These z scores are computed using $/N$, not $/n - 1$, in the denominator standard deviation; thus, they are different than the z scores produced by SPSS, which uses $/n - 1$. Only z scores computed with the $/N$ algorithm will produce the correct Pearson r correlation via its definitional z-score formula. See the "Controversy" section at the end of Module 6.]

Student	*X (Quiz 1)*	*Y (Quiz 2)*	z_X	z_Y
Abby	9	10	.496	1.650
Babs	7	8	−.606	.512
Clyde	10	7	1.047	−.057
DeShawn	9	8	.496	.512
Emily	5	5	−1.709	−1.195
Fred	6	4	−1.158	−1.763
Gino	10	8	1.047	.512
Hortense	9	6	.496	−.626
Ingrid	6	6	−1.158	−.626
Jorge	10	9	1.047	1.081
	$M_X = 8.1$	$M_Y = 7.1$		
	$s_X = 1.814$	$s_Y = 1.758$		

Figure 35.2 shows a scatterplot of the raw scores as well as the location of the z scores within a coordinate system. From the pattern and direction in the raw score scatterplot on the top, we can tell that the correlation is positive and moderate: The scores go from the lower left to the upper right, and they fall in an ellipse. However, the z-score scatterplot on the bottom clarifies *why* the relationship is positive: Nearly all the data points fall within the

upper-right and lower-left quadrants. Now, what are those quadrants? In Figure 35.1, we saw that the top-right quadrant contains only cases in which positive (high) scores on X are also positive (high) scores on Y. The bottom-left quadrant contains only cases in which negative (low) scores on X are also negative (low) scores on Y. Recall that a positive relationship is one in which as the values on one variable go up, so do the values on the other variable; or conversely, as the values on one variable go down, so do the values on the other variable. Because the z scores fall primarily in the quadrants in which the signs on X and Y are similar (+, + and –, –), it is clear that the scores on X and Y go up and down together. The z-score formula for Pearson r captures this relationship in its computation.

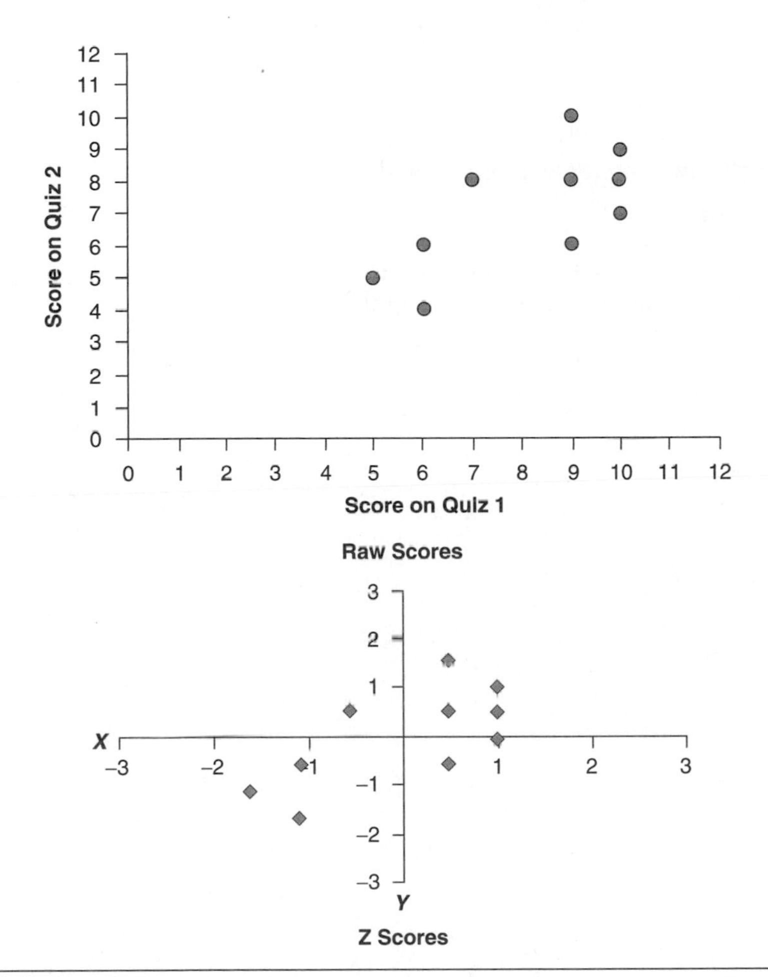

Figure 35.2 Raw Score Scatterplot Versus z Scores in a Cartesian Coordinate System for Quiz 1 and Quiz 2

CHECK YOURSELF!

Now that you know the quadrants in which z scores for a positive relationship fall, in what quadrants would z scores fall if the variables were negatively related? Draw and label a sample z-score scatterplot. In what quadrant(s) would z scores fall if the variables were unrelated? Draw and label a sample z-score scatterplot.

CHECK YOURSELF!

What are the advantages and disadvantages of each of the different formulas for Pearson r? Which one does your own instructor prefer that you use?

Calculating Pearson *r*: Raw Score Method

Now, let's calculate a Pearson *r* for the quiz scores. Looking at the scatterplot for the data in Figure 35.2, we should expect the correlation coefficient to be positive and moderate—say, somewhere between +.50 and +.80. We will use the raw score formula to calculate the actual direction and degree of relationship.

$$r_{XY} = \frac{N\sum XY - \left(\sum X\right)\left(\sum Y\right)}{\sqrt{\left[N\left(\sum X^2\right) - \left(\sum X\right)^2\right]\left[N\left(\sum Y^2\right) - \left(\sum Y\right)^2\right]}}$$

where

X = subject's raw score on variable X,

Y = subject's raw score on variable Y, and

N = sample size.

According to the formula, the values we need are X, Y, XY, X^2, and Y^2, as well as the sums of each of those values. Thus, we set up the data as follows:

Student	*X (Quiz 1)*	*Y (Quiz 2)*	*XY*	X^2	Y^2
Abby	9	10	90	81	100
Babs	7	8	56	49	64
Clyde	10	7	70	100	49
DeShawn	9	8	72	81	64
Emily	5	5	25	25	25
Fred	6	4	24	36	16
Gino	10	8	80	100	64
Hortense	9	6	54	81	36
Ingrid	6	6	36	36	36
Jorge	10	9	90	100	81
$N = 10$	$\Sigma X = 81$	$\Sigma Y = 71$	$\Sigma XY = 597$	$\Sigma X^2 = 689$	$\Sigma Y^2 = 535$

Then, it is just a matter of plugging the above totals into the formula.

$$r_{XY} = \frac{N\sum XY - \left(\sum X\right)\left(\sum Y\right)}{\sqrt{\left[N\left(\sum X^2\right) - \left(\sum X\right)^2\right]\left[N\left(\sum Y^2\right) - \left(\sum Y\right)^2\right.}}$$

$$= \frac{10(597) - (81)(71)}{\sqrt{[(10)(689) - (81)^2][(10)(535) - (71)^2]}}$$

$$= \frac{5970 - 5751}{\sqrt{[(6890) - (6561)][(5350) - (5041)]}}$$

$$= \frac{219}{\sqrt{(329)(309)}}$$

$$= \frac{219}{\sqrt{101661}}$$

$$= \frac{219}{318.843}$$

$$= +.69$$

So our Pearson *r* correlation coefficient is +.69. But what does that mean?

Interpreting a Pearson *r* Coefficient

The correlation coefficient is positive and moderate. In other words, students who score well on Quiz 1 also tend to score well on Quiz 2, with some exceptions. However, if our purpose were to determine whether or not there is, in fact, a correlation between these two quizzes, what would be the null hypothesis? What would be the research hypothesis?

The null hypothesis says that there is no correlation between the two quizzes. The research hypothesis says that there is a correlation between the two quizzes. Note that the research hypothesis does not say how large the correlation is. It says only that there is a correlation—that is, some nonzero number.

Under the null hypothesis, the expected correlation is 0. Our observed correlation is +.69. So are these two quizzes in fact not related but we just happened to get a correlation of +.69? Or is a correlation of +.69 so high that it would be very unlikely to occur if the tests were in fact not related?

Naturally, even if the two quizzes are in fact not related, we do not expect to get a correlation of exactly 0 each time we administer them. Thus, once again, the key question is whether the observed difference from expectation is *a lot*, and therefore probably due to a real underlying correlation between the two quizzes, or whether it is *only a little*, and therefore probably due to mere sampling error among the students in this particular sample.

By now you are so proficient at this that you are probably already searching for a table in which to look it up. Yes, to determine whether or not our calculated correlation is a lot or a little, we must look it up in a table of Pearson *r*. If our calculated correlation meets the tabled value, it is considered to be *a lot*, and we will reject the null hypothesis. If our calculated correlation does not meet the tabled value, it is considered to be due to mere sampling error, and we will not reject the null hypothesis.

CHECK YOURSELF!

What do the entries in a Pearson *r* table tell you?

The Pearson *r* correlation table is found in Appendix G. A portion of that table is reproduced as Table 35.1.

Table 35.1 A Portion of the Pearson *r* Correlation Table

	Levels of Significance for a One-Tailed Test			
	.05	*.025*	*.01*	*.005*
	Levels of Significance for a Two-Tailed Test			
df	*.10*	*.05*	*.02*	*.01*
7	.582	.666	.750	.798
8	.549	.632	.716	.765
9	.521	.602	.685	.735

Notice the one-tailed and two-tailed columns. For a Pearson *r*, a two-tailed (nondirectional) hypothesis indicates that the correlation may be either positive or negative. The one-tailed (directional) hypothesis indicates that the correlation must be positive *or* that it must be negative (one or the other, but not either).

As with previous tables, we enter the table at the correct degrees of freedom (*df*). For a correlational study, $df = N - 2$ (1 *df* is lost for each of the two variables in the study). What is the *df* for this study?

Yes, the *df* for this study is $10 - 2 = 8$. Thus, we enter the table at Row 8.

Type 1 error (α) is listed above each column. Assume that we are willing to make a Type 1 error 5% of the time. For a two-tailed test at $\alpha = .05$, what is the tabled critical *r*? Can we reject the null hypothesis?

The critical *r* is .632; so, yes, we can reject the null hypothesis because the calculated *r* of .69 meets or exceeds the critical *r* of .632. Thus, we can be at least 95% confident that these two quizzes really are related.

Assume that we are willing to make a Type 1 error only 2% of the time. For a two-tailed test at $\alpha = .02$, what is the critical *r*? Can we reject the null hypothesis?

The critical *r* is .716; so, no, we cannot reject the null hypothesis because the calculated *r* of +.69 does not meet or exceed the critical *r* of .716. Thus, we cannot be 98% certain that these two quizzes are really correlated.

In an APA journal, this result would be written as

$$r(8) = +.69,\ p < .05.$$

This is read as "Pearson *r* at 8 degrees of freedom is +.69. There is less than a 5% chance that this correlation is due to mere chance."

Looking Ahead

In this module, you have learned how to calculate and interpret a Pearson *r* correlation coefficient. Unfortunately, correlation coefficients are particularly prone to misinterpretation. In the next module, we will look at the many ways that correlation coefficients can be misinterpreted.

PRACTICE

3. A popular theory says that the more sugar children eat, the more active they become. To test the theory, a researcher receives permission to adjust the diets of 10 preschool children and monitor their activity. Each child is given a snack that looks identical to those of the other children, but the snacks have varying amounts of sugar. The play behavior of the children is then measured at the recess period following the snack. The amount of sugar is coded on a scale of 1 to 20, with 1 indicating *no sugar* and 20 indicating *very high sugar*. The activity level is measured on a scale of 1 to 10, with 1 indicating *inactive* and 10 indicating *extremely active*. The sample consists of five boys and five girls. Two raters agree on the activity ratings. The raters know nothing about the actual amount of sugar the children received until after they have completed the ratings. Here are the results:

Child	*(X) Sugar*	*(Y) Activity*
1	2	5
2	8	4

Child	*(X) Sugar*	*(Y) Activity*
3	4	1
4	10	7
5	16	5
6	6	10
7	12	3
8	20	6
9	14	9
10	18	2

a. Create a scatterplot of the data.
b. From the scatterplot, what direction and approximate strength do you expect the correlation coefficient to take?
c. State the null hypothesis.
d. Calculate Pearson *r*.
e. Interpret the correlation coefficient for a one-tailed test: Can you reject the null hypothesis? If so, with what confidence?
f. Write the result in APA journal format.

4. Here are hypothetical data for a study of the relationship between intelligence and prejudice in adults. For both variables, higher scores indicate more of the measured trait:

Participant	*Intelligence*	*Prejudice*
A	102	22
B	128	18
C	117	26
D	126	20
E	92	30
F	121	19
G	96	26
H	104	23
I	112	22
J	92	28
K	101	25
L	108	25

a. Create a scatterplot of the data.
b. From the scatterplot, what direction and approximate strength do you expect the correlation coefficient to take?
c. State the null hypothesis.
d. Calculate Pearson *r*.
e. Interpret the correlation coefficient for a two-tailed test: Can you reject the null hypothesis? If so, with what confidence?
f. Write the result in APA journal format.

5. The admissions officer of TopNotch University wants to know the relationship between the high school GPAs and college GPAs for the 15 merit scholars in this year's freshman class. Here are the data:

Student	*High School GPA*	*College GPA*
1	3.6	3.7
2	3.9	3.9

(Continued)

(Continued)

Student	*High School GPA*	*College GPA*
3	4.0	3.7
4	3.4	3.5
5	3.8	3.4
6	3.7	3.3
7	3.7	3.3
8	4.0	3.6
9	3.9	3.5
10	3.5	3.2
11	3.3	3.0
12	3.7	3.7
13	3.4	3.2
14	3.6	3.5
15	3.6	3.4

a. Calculate Pearson r.

b. Interpret the correlation coefficient for a one-tailed test: Can you reject the null hypothesis? If so, with what confidence?

6. A physiological psychologist measures the relationship between a rat's weight (in grams) and the time it takes the rat to run a maze (in seconds). Here are the data:

Rat	*Weight*	*Time*
1	302	25
2	290	23
3	281	23
4	299	24
5	307	27
6	286	26
7	293	24
8	275	25
9	301	28
10	295	27

a. Calculate Pearson r.

b. Interpret the correlation coefficient for a one-tailed test: Can you reject the null hypothesis? If so, with what confidence?

7. The data below are for days of rainfall per week over a 12-week period in a resort area and occupancy rate (percentage of beds filled) for a hotel in that same resort area. Compute the Pearson r correlation coefficient.

Week #	*Days Rained*	*Occupancy Rate*
1	3	83
2	0	92
3	5	72
4	1	90
5	3	84

Week #	*Days Rained*	*Occupancy Rate*
6	2	82
7	6	70
8	5	72
9	5	65
10	3	74
11	2	88
12	4	78

8. The following data are for the number of miles per gallon 12 different cars get and the number of miles from home each car owner plans to drive for his or her vacation destination this year. Compute the Pearson *r* correlation coefficient.

Car #	*Miles per Gallon*	*Vacation Distance*
1	30	210
2	16	178
3	23	482
4	27	312
5	48	258
6	31	754
7	21	110
8	17	240
9	23	567
10	25	320
11	46	456
12	24	98

9. Exercise 7 in Module 34 asked you to create a scatterplot for number of mirror looks and level of self-esteem. Now compute the Pearson *r* correlation coefficient. Here, again, are the data.

Girl #	*Level of Self-Esteem*	*Number of Mirror Looks*
1	5	3
2	7	4
3	3	1
4	6	1
5	7	4
6	9	6
7	5	4
8	6	8
9	3	2
10	5	3
11	6	2
12	5	1
13	7	4
14	4	2
15	9	5

(Continued)

(Continued)

10. Exercise 8 in Module 34 asked you to create a scatterplot for number of hours spent playing computer video games and academic average. Now compute the Pearson r correlation coefficient. Here, again, are the data

Student #	*Hours Playing Computer Games*	*Academic Average*
1	1.5	84
2	3.8	76
3	0.5	85
4	2.3	79
5	1.2	88
6	2.9	82
7	1.0	95
8	0.4	93
9	3.0	78
10	4.2	80
11	1.1	90
12	0.3	91

11. Exercise 15 in Module 34 required that you gather height and shoe-size data from 10 adults. Now, calculate and interpret Pearson r on that data.

12. Exercise 16 in Module 34 required that you gather data from 10 adults regarding the number of hours of television watched on an average day and the typical number of hours of sleep obtained in an average night. Now, calculate and interpret Pearson r on that data.

SPSS Connection

Download the file **data_quiz 1 quiz 2.sav** from www.sagepub.com/steinberg2e. These data are used in the textbook example.

Alternatively, manually enter the scores and name the variables as described in Module 34.

If the file is not already in **Data View**, click that tab in the lower left of the screen.

In the toolbar at the top of the screen, click on **Analyze**, then **Correlation**, then **Bivariate.** Highlight the variable **Quiz 1(X)** in the left window and click on the **arrow** before the **Variables** window on the right, to send the variable into that window. Do the same for the **Quiz 2(Y)** variable. Click **OK**. This is what you will see.

Correlations

Correlations

		Quiz 1 (X)	Quiz 2 (Y)
Quiz 1 (X)	Pearson Correlation	1	.687*
	Sig. (2-tailed)		.028
	N	10	10
Quiz 2 (Y)	Pearson Correlation	.687*	1
	Sig. (2-tailed)	.028	
	N	10	10

* Correlation is significant at the 0.05 level (2-tailed).

SPSS correlates each selected variable with each other selected variable twice–once in AB order, and again in BA order. Results are identical and therefore redundant. If you draw a diagonal line through the 1.000 correlations (each variable with itself), you need pay attention only to the coefficients triangle either above or below the diagonal line.

Visit the study site at www.sagepub.com/steinberg2e for practice quizzes and other study resources.

36 Correlation Pitfalls

Terms: heterogeneity, homogeneity, common variance

Symbol: r^2

Learning Objectives:

- Understand the effect that sample size has on the strength of a correlation coefficient
- Distinguish between statistical significance and practical importance
- Understand that guidelines for practical importance differ depending on the use to which the correlation will be put
- Understand the effect that restriction in range has on the strength of a correlation coefficient
- Understand the effect that heterogeneity and homogeneity have on the strength of a correlation coefficient
- Understand the effect that instrument reliability has on the utility of a correlation coefficient
- Distinguish between correlation and common variance
- Distinguish between correlation and causation

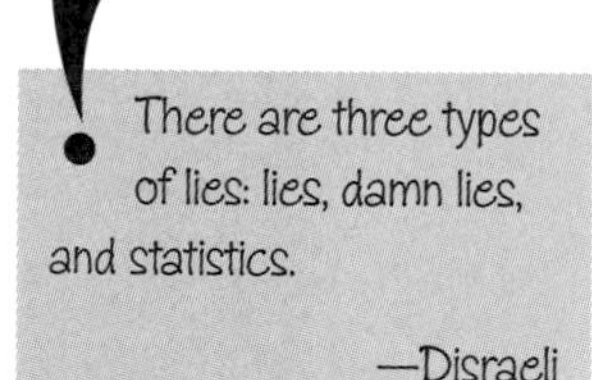

You know how to calculate a correlation coefficient. You also know how to evaluate the coefficient for statistical significance. So is that all? No, that's not all. Although any statistic can be misused or misinterpreted, correlations are particularly prone to these two problems. In this module, we will look at ways in which correlations are often misused or misinterpreted.

Effect of Sample Size on Statistical Significance

Let's revisit our conclusion from Module 35. We were more than 95%, but less than 98%, confident that our correlation of +.69 was significantly different from 0. In other words, between 2 and 5 samples out of every 100 samples of 10 subjects would show a correlation as large as .69 by chance alone, even when the two student quizzes are in fact unrelated.

Are you surprised that a correlation so large would occur by chance so frequently? On the face of it, +.69 seems like a very large correlation to occur by mere chance, doesn't it? The reason such a large correlation does not reach the tabled critical value for rejecting the null hypothesis with more than 98% confidence lies in the sample size.

Consider for a moment a familiar example. You would probably agree that a coin that produces 80 heads and 20 tails on 100 tosses is probably biased (weighted). A 4:1 ratio of heads to tails seems pretty unbalanced. But what if it had not been 100 tosses but only 5 tosses. Suppose that we tossed the coin five times and got 4 heads and 1 tail. That's still a 4:1 ratio of heads to tails. Would you still be willing to conclude that the coin is biased? Or is that result within the realm of reasonable outcomes for a fair coin? Somehow, 4 heads and 1 tail is not as convincing that the coin is biased as 80 heads and 20 tails would be.

It turns out that the number of trials—or in research language, the sample size—also enters into the decision about whether or not the coin is biased. Returning to our correlation, +.69 is not very convincing because the sample consisted of only 10 students.

Now let's see what happens when we increase the sample size. Pretend that we obtained our +.69 correlation coefficient for a sample of 102 students instead of 10 students. The correlation is still +.69, but now the sample size is much larger.

What is the *df* for this study? The $df = 102 - 2 = 100$. Now can we reject the null hypothesis with 98% confidence? How about at 99% confidence?

At 100 *df*, the critical *r* at .01 α is only .254. Our calculated *r* is so much larger than the tabled value that we can reject the null hypothesis not only with 99% confidence but probably also with 99.9% or even 99.99% confidence.

In any study, a larger sample will be more representative of the characteristics of the overall population than a smaller sample will be. Thus, the tabled value for reaching statistical significance for a larger sample is lower than the tabled value for a smaller sample. Conversely, *a smaller sample requires a higher correlation before we can conclude that the correlation really exists* in the overall population from which the sample was drawn.

Statistical Significance Versus Practical Importance

Look at the values along the bottom row of the Pearson *r* table in Appendix G. They are not very large, are they? Yet they are statistically significant. The reality is that if sample size is large enough, practically any nonzero correlation will be statistically significant. However, *statistical significance does not imply practical importance.*

Think about the SAT, for example. The SAT is given to tens of thousands of high school seniors nationwide each year. Suppose we correlate SAT scores with freshman GPA for such a large sample. The table indicates that at the .05 α level and a sample size of 1,000 students, even a .062 correlation between SAT score and freshman GPA would be sufficient to claim that SAT score and freshman GPA are related. For a sample size in the tens of thousands (which the table doesn't show), the value for statistical significance would be even smaller than .062. However, if the SAT is to be used to select students for college admission, we would certainly want the correlation to be greater than that, wouldn't we?

> Attaching significance to invariants is an effort to recognize what is important or significant in what is only trivial or ephemeral.
>
> —H. W. Turnbull

As with other statistics we have studied, a result can be statistically significant and yet be very low in practical value. The question then becomes, is the correlation large enough to be practically useful? Recall that, after you learned about hypothesis testing (t, F, and χ^2), you learned effect size statistics to determine the practical importance of the results. Also recall that with the exception of Cohen's *d*, each of the effect size statistics we used—r, η, ϕ, and V—were correlation coefficients that could vary only between 0.00 and 1.00. But in this module and the previous two modules, our initial test statistic is already a correlation coefficient. Therefore, the correlation coefficient *is* the effect size! No further effect size statistic is needed. We need only to apply the effect size guidelines. For parametric tests on interval score data, those guidelines were as follows:

Small = .25 or less

Medium = .25 to .40

Large = .40 or more

Using those guidelines, the effect size for our +.69 correlation is very large.

A caveat is in order, however. For correlational studies, meaningful correlation size differs depending on the use to which the correlation will be put. Suppose, for example, that a school psychologist is determining which students are too mentally retarded to benefit from classroom instruction. He or she would want to use a screening test whose scores correlate very highly with academic success. That's because the consequence of making a mistake is very great: students might be mistakenly denied appropriate education.

> To isolate mathematics from the practical demands of the sciences is to invite the sterility of a cow shut away from the bulls.
>
> — Pafnuty Chebyshev

On the other hand, suppose an employer wants to use a test to select job applicants. Although the employer might desire a high correlation between test score and subsequent job performance, the truth is that any correlation is better than simply hiring people at random. Thus, the employer can settle for a much lower correlation.

As mentioned in Module 34, correlations are sometimes used for purposes other than prediction. One such use is for establishing the reliability of tests. Correlation values must be much higher when they are used for test reliability rather than for prediction. Let's return to the SAT example. In a study of the predictive relationship between SAT and freshman GPA, a correlation of .50 would be considered large. However, in a study of the reliability of two parallel forms of the SAT, a correlation of .50 would be unacceptably low. Correlations for reliability studies need to be at least in the .80 to .90 range if we are to use the two test forms interchangeably. You will meet this argument again later in this module.

In sum, we must know the use to which a correlation coefficient will be put before deciding what a meaningful value is. Without that information, we cannot effectively interpret a correlation coefficient.

Effect of Restriction in Range

When calculating a correlation coefficient, we include the full range of scores on both variables. Sometimes, however, only part of the range is available for one or both variables. Either the high scores or the low scores are missing from the data set. We call this situation restriction in range. *Restriction in score range decreases the correlation.*

Examine the scatterplot in Figure 36.1 for verbal SAT score and freshman GPA. There appears to be a moderate positive correlation between the two variables. These are the typical data for a nationwide sample.

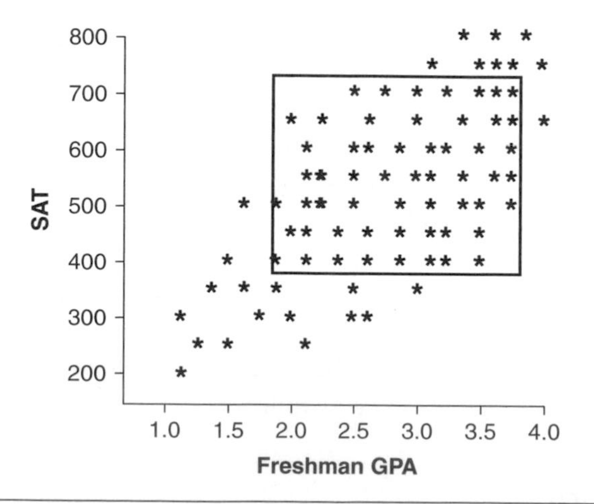

Figure 36.1 Effect of Restricted Range on Correlation

However, suppose that the data are limited to students at a particular college and the verbal SAT scores of students at this college range from 400 to 720. And suppose those same students' GPAs range from 1.8 to 3.8. These ranges on both variables are indicated by the blocked area in Figure 36.1. What does the correlation look like for only the blocked area? It is much lower than for the entire range of scores, isn't it?

Restriction in range is a common situation in real life. It is difficult to justify use of a test that, on the surface, appears to be only marginally related to the other variable. Consider, for example, a lawsuit in which the defendant, a civil service agency, must prove an employment exam's validity. In other words, the agency must prove that scores on the exam are related to job performance. Now, perhaps the exam scores really do correlate highly with success on the job. However, in a civil service system only the top-scoring candidates are hired. The range of test scores is thereby restricted. The agency will never know how the low-scoring candidates would have performed on the job had they been hired. Similarly, because performance evaluations in most government agencies are tied to bonus money, the number of "superior" evaluations that can be given is limited and predetermined. Thus, most agencies give either "satisfactory" or "unsatisfactory" performance ratings. As a result, nearly everyone is rated "satisfactory." With the range of scores for both the civil service exam and the job performance ratings so seriously restricted, the correlation between the two variables cannot be very high. Thus, the agency will have to prove the test's validity some other way.

Effect of Sample Heterogeneity or Homogeneity

Heterogeneity refers to a sample's diversity. Sometimes a sample is inappropriately diverse. For example, assume that the same teacher-made vocabulary and arithmetic tests are given to first, third, and fifth graders. This would be a silly thing to do, of course, because the students should be given grade-appropriate tests for both subjects. But for the sake of argument, suppose the same two tests are given to children in all three grades. Under those conditions, the relationship between scores on the two tests might look like the scatterplot shown in Figure 36.2. The first graders might be expected to score low on both tests because they do not yet know much about either vocabulary or arithmetic; the third graders might be expected to score in the moderate range on both tests because they know somewhat more about vocabulary and arithmetic, and the fifth graders might be expected to score fairly high on both tests because they have learned quite a bit of vocabulary and arithmetic during their 5 years of school. Note that the scatterplots of scores *within* each grade show that the relationship between the two tests is quite low.

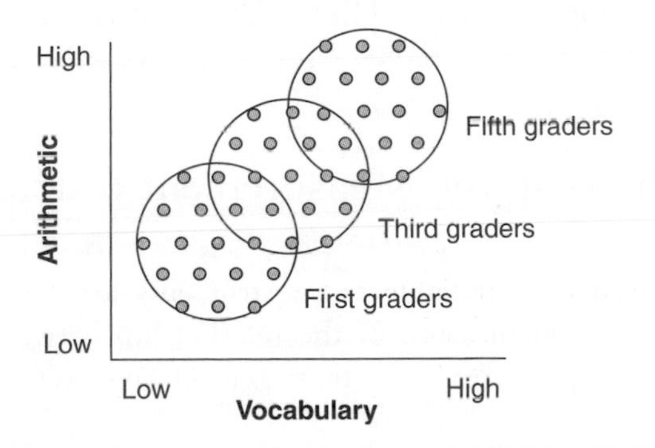

Figure 36.2 Effect of Homogeneity on Correlation

But what does the relationship between the two tests look like across all three grades *combined*? It will look like the scatterplot shown in Figure 36.3.

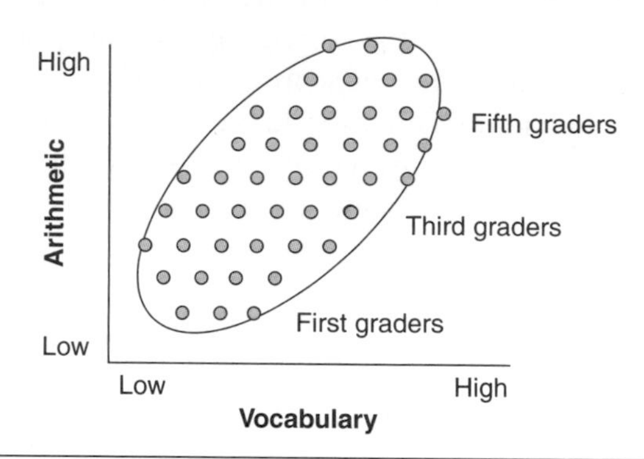

Figure 36.3 Effect of Inappropriate Heterogeneity on Correlation

Combining several distinct subgroups with very different levels of performance gives the impression of a moderate relationship, even when there is little or no relationship within each subgroup. This is because *sample heterogeneity increases the correlation.*

Although giving the same tests to classes at three different grade levels may seem like an unlikely event, issues of sample heterogeneity are common in real life. For example, let's assume that a researcher wants to study the relationship between depression level and activity level. Typically, researchers don't devise their own measuring instruments for their studies; rather, they select preexisting instruments that have already been shown to be good measures of the variables they wish to measure. Now suppose the researcher selects a depression inventory or an activity rating that was designed for use with a more restricted group than the subjects in the researcher's study. For example, the depression inventory and activity ratings may have been designed for use on children, but the people in the researcher's study include both children and adults. This makes the sample in the study inappropriately diverse. The sample's heterogeneity will increase the correlation between the two measured variables—in this case, between depression level and activity level.

This principle works in reverse as well: *Sample homogeneity decreases the correlation.* **Homogeneity** refers to similarity among the subjects. Returning to the SAT and GPA example, many colleges do not attract a diverse nationwide applicant pool. In the boxed area in Figure 36.1, the correlation for the college's own students is quite low because of the relative homogeneity of its students. The College Board, which creates the SAT, claims a much higher correlation than any one college's data will show. This is because the College Board's statistic is based on a diverse nationwide sample that includes the full range of scores—a diversity, it turns out, that does not apply to many of the colleges that actually use the test for admission decisions.

Effect of Unreliability in the Measurement Instrument

Correlations that are based on unreliable test instruments are themselves unreliable. For example, we cannot get a valid measure of the relationship between air temperature and amount of illness in a geographic area if the thermometer gives very different readings from minute to minute.

Again, this may seem like an unlikely scenario. However, such things do happen in real life. For example, when our family moved to a new school district, the district into which my daughter transferred had many programs not available in the small school from which she transferred. One of those features was classes at four levels—below level, on level, above level, and 2 years above level. Not knowing where to place her, the guidance staff gave her a staff-developed 15-min quiz. Obviously, they trusted the correlation between scores on that quiz and future performance in coursework in their school. I, on the other hand, didn't share their confidence. My hunch was that scores on their quiz were prone to error and, hence, not good indicators of subsequent school performance. It was not professionally developed, and there were not enough questions on it. To make a more valid decision, I produced my daughter's scores on a nationally standardized achievement test. As expected, her recommended school placement was different based on her performance on the nationally standardized test than it was based on her performance on the locally developed quiz. The locally developed quiz was not reliable.

Women's magazines frequently include tests of questionable reliability and, hence, questionable validity. I'm sure you've seen them: "Are you honest?" "How socially conscious are you?" "Are you a good lover?" Typically, the scores are intended to indicate the test taker's probable behavior or performance in a related activity. That is, after you take the quiz, the article tells you how your "low" honesty will be interpreted by your boss, how your "high" social consciousness should guide your selection of friends, or how your "characteristics as a lover" match those of your partner. However, *unreliable scores cannot be used to predict other behaviors or outcomes.* As the saying goes, "Garbage in, garbage out!" This principle is known by the acronym GIGO.

Correlation Versus Common Variance

It is tempting to think that if the correlation between two variables is .80, then 80% of the variance in the scores on one variable must be due to variance in the scores on the other variable. It would be tempting to think that, but it would not be correct. This is because correlation is a linear measure, whereas variance is an area measure.

Recall from the discussion of dispersion in Module 6 that variance is a squared, or area, measure. Correlation, on the other hand, is an unsquared, or linear, measure. Just as we had to square the standard deviation (a linear measure) to obtain the variance (an area measure), so we have to square the correlation (a linear measure) to obtain the **common variance** (an area measure) between two sets of scores. Thus, common variance, symbolized r^2, is the square of the correlation.

In the example cited above, the correlation between two tests was .80. Thus, they have $(.80)^2 = .64$, or 64%, of their variance in common. The other 36% of the variance is variance that is unique to each test; it is not shared between the two tests. This situation is depicted in the Venn diagram in Figure 36.4. In the diagram, common variance is indicated by the shaded area, and unique variance is indicated by the unshaded areas.

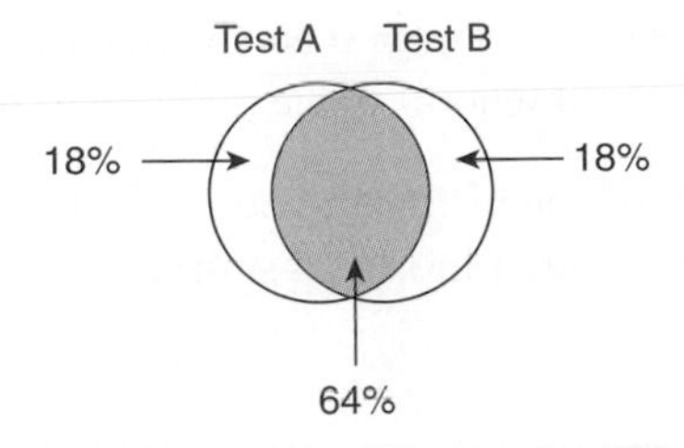

Figure 36.4 Venn Diagram of Common Variance and Unique Variance

Earlier in this module, I mentioned that test reliabilities, which are merely correlations between two different forms or administrations of a test, are required to be at least in the .80 to .90 range. This is because, as the correlation decreases linearly, the common variance decreases exponentially. Thus, two tests that correlate .90 have 81% of their variance in common. But two tests that correlate .70 have only 49% of their variance in common. And tests that correlate .50 have only 25% of their variance in common. Obviously, tests with such low common variance are not testing the same trait or content in the same way. Thus, the tests are not interchangeable.

Correlation Versus Causation

A correlation coefficient tells us about the relationship between variables, but it does not tell us which variable, if either, caused the relationship. With a correlation, *A* can cause *B*; *B* can cause *A*; or a third variable, *C*, can cause both *A* and *B*. In any of these cases, the correlation will be the same.

Suppose a researcher finds a strong negative relationship between the frequency of changes in administrative staff and the level of morale in a school system. That's certainly a believable finding. However, does that mean that frequent changes in administrative staff cause poor morale? Or does it mean that poor morale causes frequent changes in administrative staff? Or does some third variable—such as violent students or budget restrictions—cause both the low morale and the frequent changes in administrative staff? A correlation coefficient, no matter how high, does not indicate where the cause lies.

Similarly, there is a moderate positive correlation between the amount of ice cream eaten in a given geographic area and the number of drownings in that same area. However, certainly no one believes that eating ice cream causes people to drown. Nor does anyone think that preventing people from eating ice cream will result in fewer drownings. Why? It turns out that the relationship between eating ice cream and drowning is an artifact of a third variable: outside air temperature. In warm weather, more people eat ice cream, and in warm weather more people go swimming and, hence, more people drown.

There is probably no more common error in research interpretation than this: confusing correlation with causation. However, *correlation is not causation.* This error in causal reasoning shows up regularly in newspapers, on the nightly news, and in other lay sources of information. Consider the following:

- "Smokers' children have lower IQs," the headline said. Maybe so. But does that mean that parents can increase their babies' IQs simply by decreasing the amount they smoke? Probably not. Maybe people who smoke have lower IQs themselves. Or maybe smokers provide a less stimulating home environment. Note that smoking is irrelevant in either explanation.
- "Standardized test scores are higher in schools that allow prayer," claimed a recent magazine article. Personally, I would like to believe that there is a causal connection between prayer and academic achievement. However, this study did not prove such a connection. Perhaps parents who strongly encourage their children to achieve also choose to send their children to religiously affiliated schools. Note that prayer is irrelevant in this explanation.

Be on the lookout for each of the pitfalls mentioned in this module. Survey research, in particular, is frequently misrepresented in media sources.

Repeat the eight pitfalls learned in this module.

PRACTICE

1. You conduct a correlational study and obtain a low correlation coefficient. Assume that you did not make a calculation error. Also assume that there really is a high correlation between the two variables. List and explain two reasons why the correlation might be low for your study.

2. Will the following study flaws lead to a Type 1 error or to a Type 2 error? As a reminder, here are the definitions from Module 13 (modified to apply to correlations):

   Type 1: There really is not a significant correlation, but you find one.

   Type 2: There really is a significant correlation, but you do not find one.

   _____ The sample size is too small.

   _____ The study uses only a restricted portion of the real score range.

   _____ The sample is too heterogeneous.

3. Will the following study flaws lead to a Type 1 error or to a Type 2 error? As a reminder, here are those definitions from Module 13 (modified to apply to correlations):

   Type 1: There really is not a significant correlation, but you find one.

   Type 2: There really is a significant correlation, but you do not find one.

   _____ The sample size is too large.

   _____ The sample is too homogeneous.

   _____ Even a very small effect size is acceptable.

4. What is the common variance for the following correlations? Sketch the Venn diagram for each common variance:

   a. +1.00

   b. −.60

   c. +.40

5. What is the common variance for the following correlations? Sketch the Venn diagram for the common variance"

   a. −0.90

   b. +0.25

   c. −1.00

6. In Modules 34 and 35, several exercises required that you gather data from 10 people (perhaps students in your statistics class) for two variables and compute the correlation relationship between the variables. For example, one exercise required that you calculate the relationship between adults' height and shoe size. What effect did the small sample size ($N = 10$) have on the tabled value needed for statistical significance?

7. Assume that you repeat the study of the relationship between height and shoe size, but this time you gather data from children as well as adults. What effect will the increased diversity of the sample have on the calculated value of the correlation coefficient?

8. Assume that you repeat the study of the relationship between height and shoe size, but this time you estimate people's heights and shoe sizes visually rather than use accurate measuring devices. What effect, if any, will the sloppy data collection have on the calculated correlation coefficient?

9. Read the Dear Abby column shown below. A reader reports several studies showing that the more people volunteer, the longer they live. The reader then implies that people

should volunteer more because it will make them live longer. Abby agrees and urges readers to get out there and volunteer because, in Abby's words, "Those who give, *get!*"

> **Dear Abby:** In a recent column, a writer stated that it was foolish to work for nothing as a hospital volunteer. Thanks for saying, "The rewards are far more valuable than money." You are so right, Abby.
>
> In 1988, a study was done by the University of Michigan Survey Research Center. They followed 2,700 people in Tecumseh, Michigan, over a 10-year period to determine the impact of social relationships on health. They found that regular volunteer work, more than any other activity, dramatically increased life expectancy.
>
> This was especially significant for men. Men who did no volunteer work were 2½ times more likely to die during the course of the 10-year study than those who volunteered at least once a week.
>
> Research at Yale, the University of California, Johns Hopkins, the National Institute of Mental Health, and Ohio State support these findings.
>
> Longtime Volunteer Leader
> Mechanicsville, VA
>
> **Dear Leader:** So what else is new? People who spend their time doing for others feel useful, productive, and good about themselves. Volunteers, particularly those who work in hospitals, hospices, and nursing homes, are too busy to dwell on their own troubles or feel depressed. Those who give—get!
>
>

a. Suggest a plausible explanation why volunteering may cause people to live longer.

b. Suggest a plausible explanation why living longer may cause people to volunteer.

c. Suggest a plausible third variable that may account for people's tendency to volunteer as well as their tendency to live longer.

10. A study was done of diet and activity level. The study found that, as the amount of meat in the diet decreased, the amount of physical activity increased. Vegetarians were the most active of all. The nightly news reported on the study: "Eating a vegetarian diet leads to a more active lifestyle." Is that a valid conclusion? Why or why not?

11. A study was done of diet and activity level. The study found that, as the amount of meat in the diet decreased, the amount of physical activity increased. The correlation was +.20. The nightly news reported on the study: "Twenty percent of the variation in activity level is related to meat in the diet." Is that a valid conclusion? Why or why not?

12. A study was done of diet and activity level. The study found that, as the amount of meat in the diet decreased, the amount of physical activity increased. The correlation was +.20, which was statistically significant at the .0001 level (highly significant). The nightly news reported on the study: "Study finds that meat is an important dietary contributor to physical activity." Is that a valid conclusion? Why or why not?

Visit the study site at www.sagepub.com/steinberg2e for practice quizzes and other study resources.

PART XVI

Linear Prediction

"You're So Predictable"

37

Linear Prediction

Terms: predictor, criterion, prediction line, regression line, line of best fit, prediction error, least squares, slope, Y intercept

Symbols: X, Y, Y', M_X, M_Y

Learning Objectives:

- Understand why a line of best fit minimizes errors of prediction
- Find the equation for the prediction line in a set of bivariate data
- Predict scores on variable Y from known scores on variable X

Correlation Permits Prediction

> Prediction is difficult, especially of the future.
>
> —Niels Bohr, physicist

In Module 36, you learned that correlation, even a high one, does not indicate causation. Nevertheless, whenever there is a correlation between two variables, we can make predictions based on the correlation. This is because, when there is a correlation between two sets of scores, a person's score on the first variable gives some indication of that person's probable score on the second variable. A few examples are given below:

- There is a positive correlation between SAT score and subsequent freshman GPA in college. Therefore, because the scores predict academic success, college admissions officers use SAT scores in their selection process.
- There is a negative correlation between outside air temperature and illness. Therefore, because cooler air temperatures predict more illness, employers expect to experience greater illness-related employee absences in colder months than in warmer months.
- There is a positive correlation between hours spent practicing a sport and subsequent skill level in that sport. Therefore, because time spent practicing predicts athletic performance; professional athletes spend a substantial amount of time practicing in preparation for competitive games.

The variable from which a prediction is made is called the **predictor**, and the variable being predicted is called the **criterion.**

CHECK YOURSELF!

What is the predictor and what is the criterion in the sports example above?

When the scores on any two linearly related variables are perfectly correlated ($r = 1.00$), the scores will fall along a single straight line. Let's return to the Celsius-Fahrenheit example depicted in Figure 37.1.

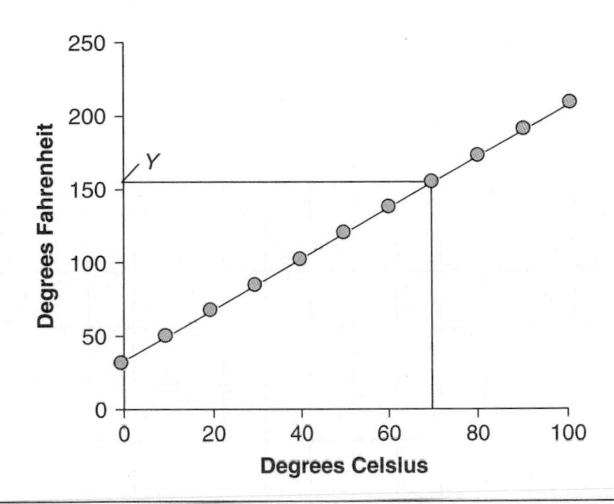

Figure 37.1 Linear Relationship Between Degrees Celsius and Degrees Fahrenheit

The predictor variable is designated as *X*, and it is found along the horizontal *X*-axis. The criterion variable is designated as *Y*, and it is found along the vertical *Y*-axis. The predicted score on *Y* is designated as *Y′*, pronounced "*Y* prime." When two variables correlate perfectly, as these two do, we can predict exactly the score on the criterion variable simply by knowing the score on the predictor variable. To find *Y′*, locate *X*, proceed vertically to the line that describes the relationship between the two variables, and then read horizontally to the value on the *Y*-axis. In the graph above, a score of 70°C predicts a score of 158°F, indicated by *Y′* at the arrow.

CHECK YOURSELF!

What is the predicted temperature in degrees Fahrenheit for a temperature of 40°C? What is it for 0°C? What is it for 100°C?

Logic of a Prediction Line

But how did we get that line? In Module 34, you learned to graph a line by plotting paired *X*, *Y* values. Recall that to place a dot, you (1) find the *X* score on the *X*-axis and find the *Y* score on the *Y*-axis, (2) run imaginary perpendicular lines to find where these two scores meet, and (3) place a dot at that spot.

Plot the *X*, *Y* values on the graph in Figure 37.2. Once you have plotted the values, connect the dots to form a straight line.

°C	°F
0	32
10	50
20	68
30	86
40	104
50	122
60	140
70	158
80	176
90	194
100	212

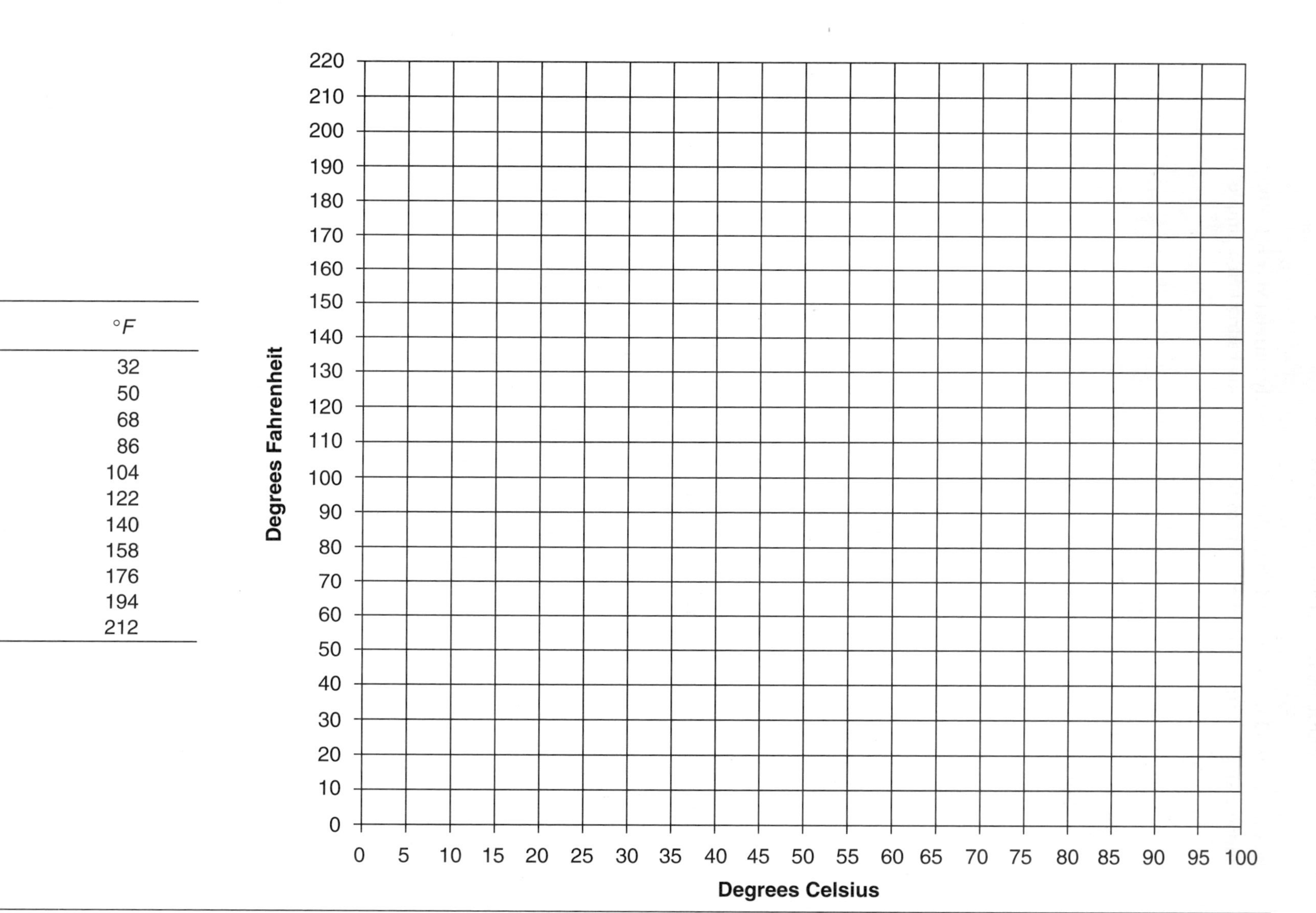

Figure 37.2 Grid for Temperature Plot

Unlike Celsius and Fahrenheit, social science variables are rarely perfectly related. For example, the correlation between SAT and freshman GPA is typically only moderate, as shown in Figure 37.3.

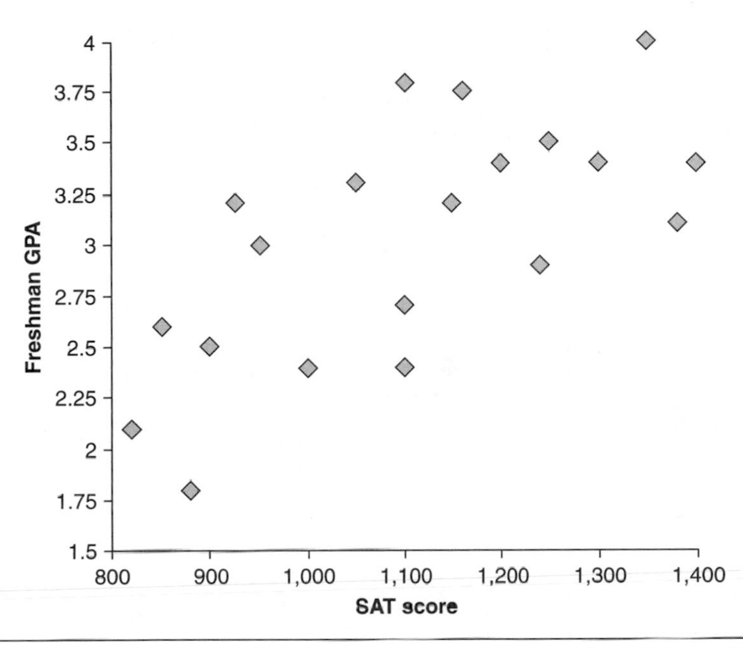

Figure 37.3 Moderate Relationship Between SAT Score and Freshman GPA

When a correlation is less than perfect, the data do not all fall along a straight line. Thus, for any given score on X, there are many scores on Y. In Figure 37.4, you can see that when SAT score = 1,100, freshman GPA can be anything between 2.30 and 3.80.

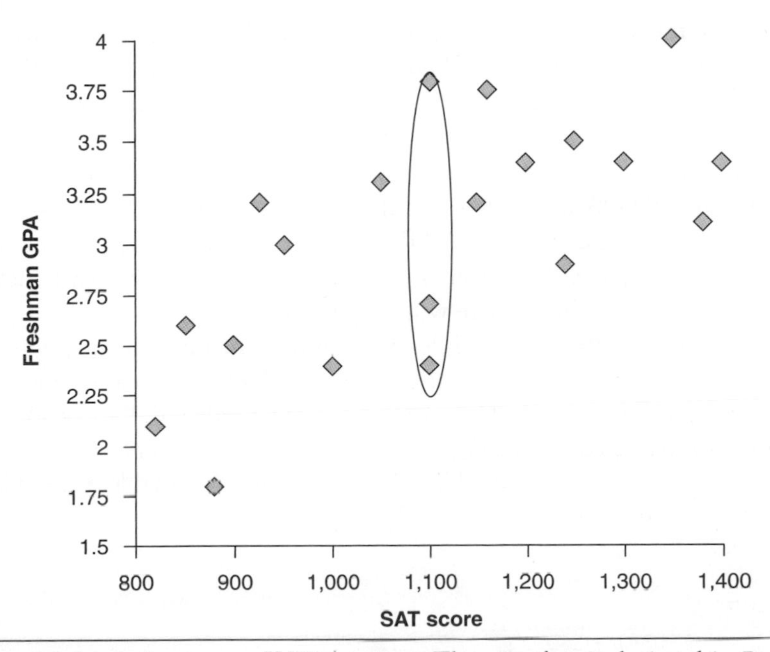

Figure 37.4 Multiple Scores on *Y* With a Less Than Perfect Relationship Between *X* and *Y*

Q: What was Sigmund Freud's favorite statistical procedure?

A: Regression.

The question then becomes, "Which of the many Y scores should we predict for a given X score?" We must place a line through the data that will predict the most probable single score, rather than a range of scores, for Y. The line through the data that we use for predicting scores on Y from scores on X is called a **prediction line.** It is also called a **regression line** or a **line of best fit.** Figure 37.5 shows one such line.

In this example, the predicted GPA for all freshmen having a SAT score of 1,100 will fall at the point indicated, Y'—about 3.05. However, the actual

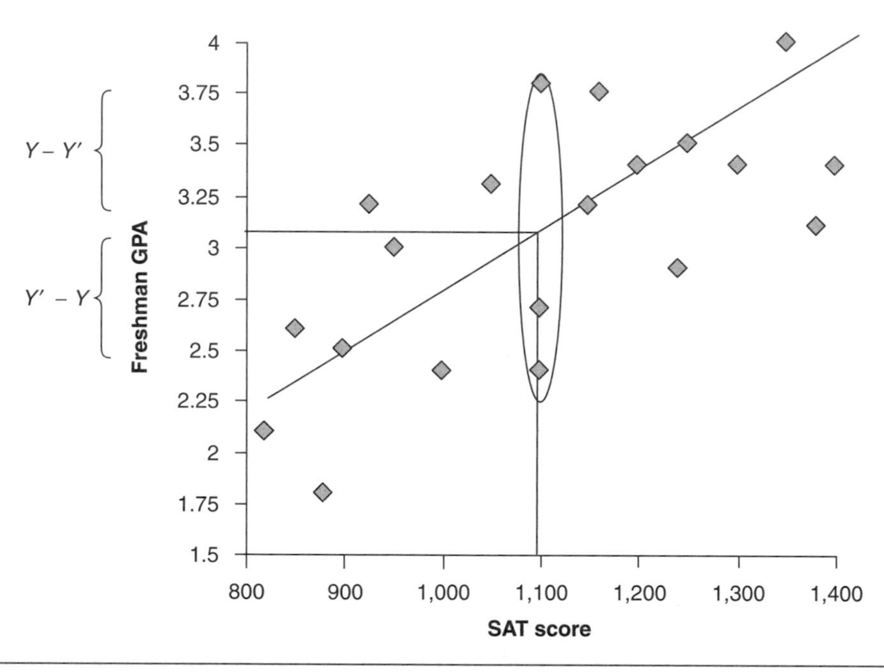

Figure 37.5 Prediction Line for SAT-GPA Data

Honorary degrees are given generally to people whose SAT scores were too low to get them into schools the regular way.

—playwright Neil Simon, on receiving an honorary degree, 1984

GPA for students having a SAT score of 1,100 can be anywhere above or below the prediction line directly over the X score of 1,100—that is, between 2.30 and 3.80 on the Y-axis. Thus, sometimes the student's actual GPA will be lower than the predicted 3.05 GPA, and at other times the student's actual GPA will be higher than the predicted 3.05 GPA. The amount by which the predicted GPA misses the actual GPA is indicated as $Y - Y'$ and by $Y' - Y$ in Figure 37.5. The deviation of the actual score from the predicted score is called **prediction error.** Whenever the correlation between two variables is less than perfect, as it is here, there will be prediction error.

Now recall from Modules 5 and 6 that the mean is the point where deviations below it balance deviations above it. This is the case for a prediction line as well. For any given score on X, deviations above the predicted Y score $(Y - Y')$ balance deviations below the predicted Y score $(Y' - Y)$. Therefore, a prediction line is nothing more than a series of means—a mean score on variable Y for every score on variable X. See Figure 37.6 for a depiction of this point.

CHECK YOURSELF!

Explain how a prediction line is like a mean.

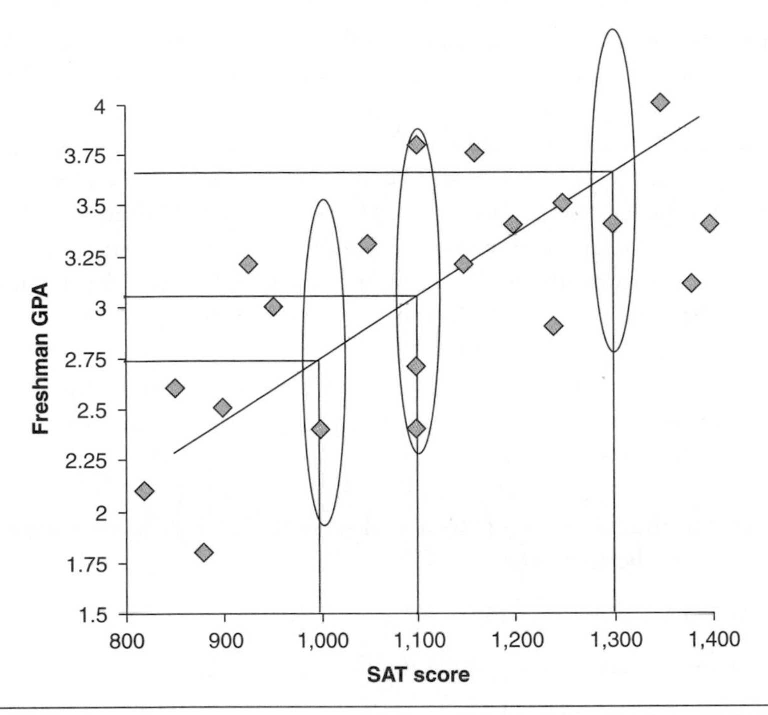

Figure 37.6 Prediction Line as a Series of Mean *Y* Scores for Various *X* Scores

Concept of Best-Fitting Line

Although I placed a line through the data for demonstration purposes, there are many possible lines through the data. Look at the graph in Figure 37.7. Each of these lines gives different predicted *Y* scores for the same *X* score. Obviously, only one of these lines is the correct line for the data. Which line is the correct line for predicting *Y* from *X*?

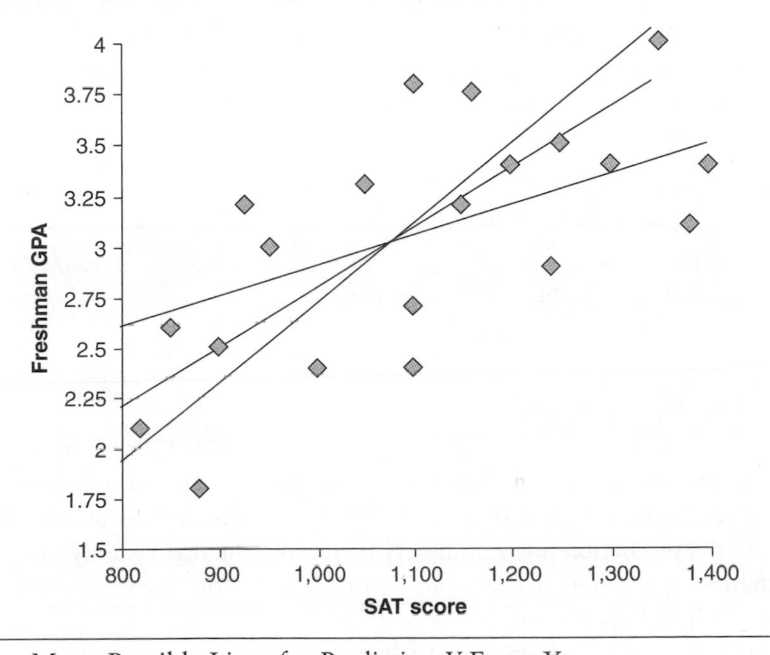

Figure 37.7 Many Possible Lines for Predicting *Y* From *X*

We want whichever line minimizes the $Y - Y'$ prediction error. Which line is that? To find that line, we could average all the actual Y scores for every X score. In Figure 37.6, all Y scores are encircled for a single X score of 1,100. We could find the total amount of error (the $Y - Y'$) for that X score. We could then do that for every possible X score. We could sum all the $Y - Y'$ errors for every X score. Then, we could do that for every possible line to see which line gives the least prediction error. However, if we did that, what would we get? We would get 0! As you learned when you studied central tendency and dispersion in Modules 5 and 6, deviations above the mean and deviations below the mean balance, and so their sum is 0. Because the sums of deviations for every line would be 0, this would not be very helpful in selecting the best-fitting line.

Recall from Module 6 that we solved this problem by squaring the deviation scores before summing them. Now we follow that same process for finding the best-fitting prediction line. Here are the steps:

1. First, we find the deviation of the actual score from the predicted score. That is, we find $Y - Y'$ (prediction error).
2. Then, we square the prediction error.
3. We do this for every X score along the line.
4. Then, we sum those squared prediction errors.
5. Finally, we repeat this process for every possible line.
6. We then choose the line for which the sum of the squared prediction errors (Step 4) is the smallest.

Because this method for selecting the best-fitting line minimizes *squared* prediction error, it is called the **least squares** method. With this best-fitting line, we can predict Y' for any given X with the least amount of error.

The process I have described above for finding the best-fitting line is conceptual only. Fortunately, we do not actually follow those steps for every possible line to find the best-fitting line. Instead, we use a mathematical equation. However, the process I described underlies the mathematical equation for any prediction line. Because the equation incorporates the least squares approach, prediction error is minimized when we find the prediction line mathematically rather than merely by placing a line visually.

CHECK YOURSELF!

Why do we square the $Y - Y'$ prediction errors before summing and averaging them? Why does this process give us the "best-fitting" line?

Equation for Best-Fitting Line

In Modules 34 and 35, we first depicted a relationship visually in a scatterplot and then calculated a correlation coefficient to place a numerical value on the relationship. Similarly, we now must find a prediction line not merely by placing the line visually but by calculating it from an equation. We calculate the equation for a prediction line to find the exact numerical values for predicted scores.

You may recall from high school algebra that the equation for a straight line is $Y = mX + b$. In words, to get the score on variable Y, multiply the score on variable X by m and then add b. In the equation, m (the variable modifying X) is the **slope.** You may have learned slope as "rise over run." Slope is how many units the line goes up on the Y-axis for every unit it goes over on the X-axis. The constant at the end of the equation, b, is the **y intercept.** It is the point at which the line crosses the Y-axis.

Unfortunately, statisticians use symbols for a straight line different from those used in algebra. For statisticians, the statistical equation for a straight line is $Y = bX + a$. As you can see, the letters used for the slope and the y intercept differ in algebra and statistics. Of course, it doesn't matter what letters we use to indicate the terms in the equation. After all, a line by any other symbols is still a line. (Did William Shakespeare say that? No, I guess not.) The lack of consistent symbols between the mathematical fields of algebra and statistics probably falls in the same category as the change from "variance" to "mean square" terminology that we encountered in Module 24: different mathematicians working separately in different areas and at different times. However, for the sake of consistency within the statistical field, we will use the statisticians' symbols rather than the algebra symbols to define a straight line. Therefore, our equation for a straight line will be $Y = bX + a$. In this equation, b is the *slope* and a is the *y intercept.*

> That which we call a rose, by any other name would smell as sweet.
>
> —William Shakespeare

Returning to the Fahrenheit-Celsius example, the equation defining the prediction line for conversion of Fahrenheit to Celsius is °F = 9/5°C + 32. In this equation, the slope is 9/5 and the y intercept is +32. To convert from Celsius to Fahrenheit, we multiply degrees Celsius by 9/5 and then add 32. As depicted in Figure 37.8, if the air temperature is 60°C, it is 140°F.

$$
\begin{aligned}
°F &= \frac{9}{5}°C + 32 \\
&= 1.8(60) + 32 \\
&= 108 + 32 \\
&= 140
\end{aligned}
$$

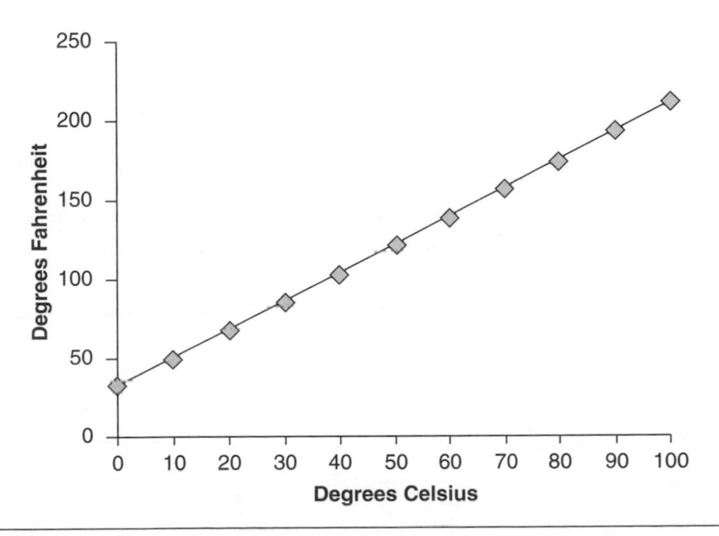

Figure 37.8 Linear Conversion of Degrees Celsius to Degrees Fahrenheit

"THERE'S A 70% CHANCE OF RAIN TOMORROW.
... THAT'S 21% CELSIUS."

SOURCE: Reprinted with permission of the Carolina Biological Supply Co.

Because any straight line is defined by $Y = bX + a$, and because a prediction line is a straight line, a prediction line is also defined by the equation $Y = bX + a$. The formula for a prediction line is

$$Y' = r_{XY}\left(\frac{s_Y}{s_X}\right)X - r_{XY}\left(\frac{s_Y}{s_X}\right)M_X + M_Y$$

where

r_{XY} = correlation between X and Y,

s_Y = standard deviation of Y,

s_X = standard deviation of X,

M_X = mean of X, and

M_Y = mean of Y.

Wait! Didn't I just say that the equation for a prediction line is $Y = bX + a$? So what is this other formula? This equation, while it looks formidable, is merely an expanded version of the $Y = bX + a$ equation. Compare the two equations. Arrows indicate comparable parts of the two equations (Figure 37.9).

$$Y \quad = \quad b \quad X \quad + \quad a$$

$$\downarrow \quad \downarrow \quad \swarrow\searrow \quad \downarrow \quad \downarrow \quad \swarrow\searrow$$

$$Y' \quad = \quad r_{xy}\left(\frac{S_y}{S_x}\right) \quad X \quad - \quad r_{xy}\left(\frac{S_y}{S_x}\right) M_x + M_y$$

Figure 37.9 The Linear Prediction Equation

The longer equation applies to any prediction line. It takes into account the two variables' central tendencies (M_X and M_Y), dispersions (s_X and s_Y), and degree of relationship (r). However, once we plug in these values for any given set of data, the equation reduces to the shorter version.

Let's look at an example. Assume the following values for SAT and GPA for a particular college:

$r_{XY} = .43$

$M_X = 1{,}050$ (mean SAT, quantitative and verbal combined)

$s_X = 180$ (SAT standard deviation)

X = SAT score for an applicant

$M_Y = 2.52$ (mean GPA)

$s_Y = 0.57$ (GPA standard deviation)

Y' = predicted GPA for the applicant

Plugging these values into the prediction equation,

$$\begin{aligned}
Y' &= r_{XY}\left(\frac{s_Y}{s_X}\right)X - r_{XY}\left(\frac{s_Y}{s_X}\right)M_X + M_Y \\
&= (.43)(0.57/180)X - (.43)(0.57/180)(1050) + 2.52 \\
&= (.43)(0.00317)X - (.43)(0.00317)(1050) + 2.52 \\
&= 0.00136X - (.43)(0.00317)(1050) + 2.52 \\
&= 0.00136X - (0.00136)(1050) + 2.52 \\
&= 0.00136X - 1.428 + 2.52 \\
&= 0.00136X + 1.09
\end{aligned}$$

Notice that, as promised, the equation is now in the short form of $Y = bX + a$. Specifically, the prediction equation for this college is $Y = 0.00136X + 1.09$.

SOURCE: Peanuts: © United Feature Syndicate, Inc.

Using a Prediction Equation to Predict Scores on *Y*

The purpose of a prediction equation is, of course, to predict. Using the prediction equation for the relationship between SAT and GPA, a college admissions officer can predict freshman GPA for prospective students from their SAT scores. For any given applicant, the admissions officer multiplies the SAT score, designated X, by 0.00136 and then adds 1.09. This gives that applicant's predicted freshman GPA. If the predicted GPA is high, the admissions officer will probably invite the student to become a member of the incoming class. The admissions officer also might deny admission to an applicant whose predicted GPA falls short of the college's standards.

> It was my SAT scores that led me into my present vocation in life, comedy.
>
> —playwright Neil Simon, on receiving an honorary degree, 1984

Let's see how this works for two applicants. Recall from Module 9 that scores on any single part of the SAT range from 200 to 800, with a mean of 500 and a standard deviation of 100. Taking quantitative and verbal scores together, the range is 400 to 1,600, with a mean of 1,000 and a standard deviation of 200. Let's take two applicants, one with a combined SAT score of 700 and the other with a combined SAT score of 1,300.

For a combined SAT score of 700,

$$\begin{aligned} Y &= (0.00136)\,(700) + 1.09 \\ &= 0.952 + 1.09 \\ &= 2.042 \end{aligned}$$

For a combined SAT score of 1,300,

$$\begin{aligned} Y &= (0.00136)\,(1300) + 1.09 \\ &= 1.768 + 1.09 \\ &= 2.858 \end{aligned}$$

Are you surprised at the predicted GPA for each applicant? On the surface, it seems as if a combined SAT score of 700 should predict a GPA lower than 2.042 and that a combined SAT score of 1,300 should predict a GPA higher than 2.858. However, several factors are operating to counteract these expectations.

- The correlation between SAT and GPA at this college is only modest (.43). Recall from Module 36 that the amount of variance that can be predicted in one variable from variance in the other variable is r^2. Thus, the common variance between SAT and GPA at this college is $(.43)^2$—or only .18. This is why SAT scores are typically only one part of the basis for a college admissions officer's decision. Other variables such as high school grades, amount of extracurricular leadership, parental level of education, and quality of letters of recommendation are also predictive of academic success and often contribute to the admissions decision.

> Our federal income tax law defines the tax y to be paid in terms of the income x; it does so in a clumsy enough way by pasting several linear functions together, each valid in another interval or bracket of income.
>
> —Hermann Wehl, 1940

- The average GPA at this college is 2.52, not the standard 2.00 ("C"). Thus, predicted GPAs of 2.042 and 2.858 are, indeed, below average and above average, respectively, for this college.
- The 2.52 average GPA at this college indicates that the grades are rather high. Using terminology from Modules 4 and 5, we can say that GPAs at this college are negatively skewed. When negatively skewed scores are correlated with normally distributed scores, the prediction line is not described by a single line. Rather, one line predicts for the lower scores and a second, more sloped line predicts for the higher scores. This situation is depicted in Figure 37.10. Note the underprediction of the higher GPAs when a single prediction line is imposed.

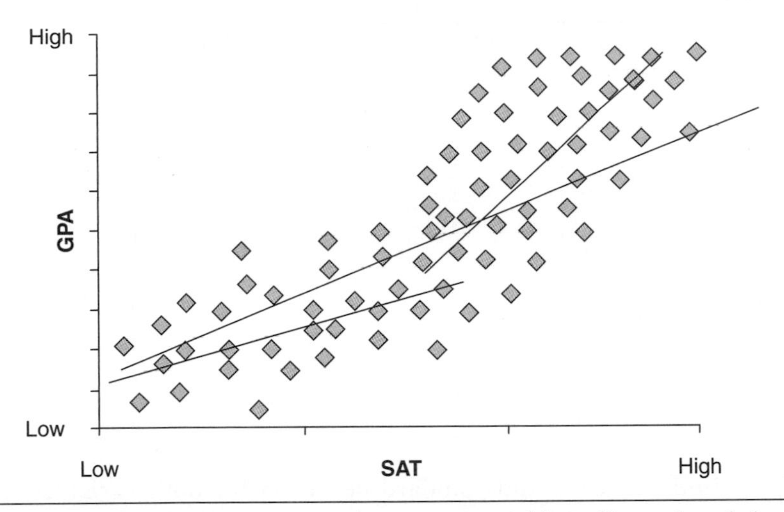

Figure 37.10 Double Prediction Lines When One Variable Is Skewed and the Other Variables Are Normally Distributed

CHECK YOURSELF!

What are some of the factors that influence how accurately a predictor will be able to predict a criterion?

Another Calculation Example

Now that I have completed the instruction for calculating and using a prediction equation, it is appropriate to remark further on the process. Fitting a prediction line is a very different process from the hypothesis-testing procedures discussed throughout much of this textbook. With prediction, there are no independent or dependent variables, only a predictor and criterion. Causation is not implied, only relationship. We do still calculate means and standard deviations, but they are not used for scaling of differences between groups, because there is only one group of subjects. Rather, they are used to create the equation that predicts one variable from another. Prediction also requires an additional statistic, the correlation coefficient, which we calculated in Module 35.

Because the entire process has taken several modules to explain, it would be helpful to calculate another example, this time starting with the raw scores, progressing through correlation, and continuing on to the prediction equation. This will better fix the steps in your mind. For students using SPSS in the course, it also will provide a fully worked out example to compare with the SPSS output.

Here is the new example. A criminologist is investigating the relationship between the number of convicted felons' prior arrests and the number of years of their prison sentences. He hypothesizes that prior arrest record is a good predictor of time to be served. Here are the data.

Number of Years in Sentence (Y)	*Number of Prior Arrests (X)*
12.2	3
8.4	2
8.0	1
14.2	5
6.0	1
4.4	0
6.2	4
16.1	7
9.3	4
7.7	3

1. First, we find the mean and standard deviation for both variables. *Note:* When computing the standard deviation, we will need both the descriptive (N divisor) and inferential ($n - 1$ divisor) values. This is because the correlation coefficient uses the descriptive standard deviation, but the prediction equation uses the inferential standard deviation.

Number of Prior Arrests (X)	$X - M$	$(X - M)^2$	*Number of Years in Sentence (Y)*	$Y - M$	$(Y - M)^2$
3	0.00	0.00	12.2	2.95	8.702
2	–1.00	1.00	8.4	–0.85	0.722
1	–2.00	4.00	8.0	–1.25	1.562
5	2.00	4.00	14.2	4.95	24.502
1	–2.00	4.00	6.0	–3.25	10.562
0	–3.00	9.00	4.4	–4.85	23.522
4	1.00	1.00	6.2	–3.05	9.302
7	4.00	16.00	16.1	6.85	46.922
4	1.00	1.00	9.3	0.05	0.002
3	0.00	0.00	7.7	–1.55	2.402
$\Sigma X = 30$		$\Sigma(X - M)^2 = 40.00$	$\Sigma Y = 92.5$		$\Sigma(Y - M)^2 = 128.20$
$M_X = \Sigma X/N$ $= 30/10$ $= 3.00$		$s^2 = \Sigma(X - M)^2/N$ $= 40.00/10$ $= 4.00$ (or 4.44, using $n - 1$)	$M_Y = \Sigma Y/N$ $= 92.5/10$ $= 9.25$		$s^2 = \Sigma(Y - M)^2/N$ $= 128.20/10$ $= 12.82$ (or 14.24, using $n - 1$)
		$s = \sqrt{s^2}$ $= \sqrt{4.00}$ $= 2.00$ (or 2.11, using $n - 1$)			$s = \sqrt{s^2}$ $= \sqrt{12.82}$ $= 3.58$ (or 3.77, using $n - 1$)

2. Next, we calculate the correlation coefficient. The raw score method is shown here.

Number of Prior Arrests (X)	*Number of Years in Sentence (Y)*	XY	X^2	Y^2
3	12.2	36.6	9	148.84
2	8.4	16.8	4	70.56
1	8.0	8.0	1	64.00
5	14.2	71.0	25	201.64
1	6.0	6.0	1	36.00
0	4.4	0.0	0	19.36
4	6.2	24.8	16	38.44
7	16.1	112.7	49	259.21
4	9.3	37.2	16	86.49
3	7.7	23.1	9	59.29
$\Sigma X = 30$	$\Sigma Y = 92.5$	$\Sigma XY = 336.2$	$\Sigma X^2 = 130$	$\Sigma Y^2 = 983.83$

$$
\begin{aligned}
r_{XY} &= \frac{N\sum XY - (\sum X)(\sum Y)}{\sqrt{[N(\sum X^2) - (\sum X)^2][N(\sum Y^2) - (\sum Y)^2]}} \\
&= \frac{10(336.2) - (30)(92.5)}{\sqrt{[10(130) - (30)^2][10(983.83) - (92.5)^2]}} \\
&= \frac{3362 - 2775}{\sqrt{(1300 - 900)(9838.3 - 8556.25)}} \\
&= \frac{587}{\sqrt{(400)(1282.05)}} \\
&= \frac{587}{\sqrt{512820}} \\
&= \frac{587}{716.1145} \\
&= +.819 \text{ or } .82
\end{aligned}
$$

3. Next we create the prediction equation. *Note:* Prediction is an inferential process. That is, the equation will be applied to subjects other than those in this data set. Therefore, we use the inferential standard deviations ($n - 1$ divisor).

$$
\begin{aligned}
Y' &= r_{XY}\frac{(s_Y)}{(s_X)}X - r_{XY}\frac{(s_Y)}{(s_X)}M_X + M_Y \\
&= (.82)\frac{(3.77)}{(2.11)}X - (.82)\frac{(3.77)}{(2.11)}(3.00) + 9.25 \\
&= (.82)(1.79)X - (.82)(1.79)(3.00) + 9.25 \\
&= 1.4678X - (1.4678)(3.00) + 9.25 \\
&= 1.4678X - 4.4034 + 9.25 \\
&= 1.4678X + 4.8466
\end{aligned}
$$

4. Finally, we use the prediction equation to predict number of years in prison sentence from number of prior arrests. For our example, let's assume that a convicted felon has six arrests at the time of sentencing.

$$Y' = 1.4678\ (X) + 4.8466$$

$$= 1.4678\ (6) + 4.8466$$

$$= 8.8068 + 4.8466$$

$$= 13.65 \text{ years}$$

Now you have a complete example to follow for similar problems. The exercises in this module and in the next each start with raw data from Module 35, for which you already calculated correlation coefficients. Each exercise below is the "next step" in those exercises. Therefore, for the exercises that follow, I remind you of the correlation coefficient value, and I provide you with the means and standard deviations. With those preliminary statistics, you can calculate the prediction equations.

PRACTICE

1. Change the SAT-GPA example in this module to a correlation of +.72 and an average GPA of 2.14.
   a. What is the new prediction equation?
   b. What is the new expected GPA for the applicant with a combined SAT of 700?
   c. What is the new expected GPA for the applicant with a combined SAT of 1,300?

2. Change the SAT-GPA example in this module to a correlation of +.45 and an average GPA of 2.40.
   a. What is the new prediction equation?
   b. What is the new predicted GPA for the applicant with a combined SAT score of 700?
   c. What is the new predicted GPA for the applicant with a combined SAT score of 1,300?

3. In Module 35, we calculated the correlation between two quizzes for 10 students:

Student	*(X) Quiz 1*	*(Y) Quiz 2*
Abby	9	10
Babs	7	8
Clyde	10	7
DeShawn	9	8
Emily	5	5
Fred	6	4
Gino	10	8
Hortense	9	6
Ingrid	6	6
Jorge	10	9

The correlation was +.69. The means and standard deviations necessary for prediction are as follows:

$M_X = 8.1$ $M_Y = 7.1$

$s_X = 1.912$ (using/$n - 1$) $s_Y = 1.853$ (using/$n - 1$)

a. What is the equation for predicting Quiz 2 (*Y*) from Quiz 1 (*X*)?
b. Assume that a new student, Kate, takes Quiz 1 and scores 7. What is Kate's predicted score on Quiz 2?

4. Exercise 4 in Module 35 required that you calculate the correlation between intelligence and prejudice in adults. The correlation was −.83. The means and standard deviations necessary for prediction are as follows:

IQ (X)	*Prejudice (Y)*
$M_X = 108.25$	$M_y = 23.667$
$s_X = 12.600$ (using/$n - 1$)	$s_Y = 3.651$ (using/$n - 1$)

a. What is the equation for predicting prejudice (*Y*) from intelligence (*X*)?
b. What is the expected prejudice score for someone with an IQ of 115?

5. Exercise 5 in Module 35 required that you calculate the correlation between high school GPA and college GPA for a group of National Merit Scholars. The correlation was +.66. The means and standard deviations necessary for prediction are as follows:

High School GPA (X)	*College GPA (Y)*
$M_X = 3.673$	$M_Y = 3.460$
$s_X = 0.219$ (using/$n - 1$)	$s_Y = 0.238$ (using/$n - 1$)

a. What is the equation for predicting college GPA (*Y*) from high school GPA (*X*) for this group of National Merit Scholars?
b. What is the expected college GPA for a National Merit Scholar whose high school GPA is 3.8?

6. Exercise 6 in Module 35 required that you calculate the correlation between rats' weight (in grams) and their running speed (in seconds). The correlation was +.46. The means and standard deviations necessary for prediction are as follows:

Weight (X)	*Running Speed (Y)*
$M_X = 292.90$	$M_Y = 25.20$
$s_X = 10.04$ (using/$n - 1$)	$s_Y = 1.75$ (using/$n - 1$)

a. What is the equation for predicting running speed in seconds (*Y*) from weight (*X*)?
b. What is the expected running speed in seconds for a rat that weighs 290 g?

7. Exercise 7 in Module 35 required you to calculate the correlation between days of weekly rainfall and hotel occupancy rate in a resort area. The correlation was +0.913. The means and standard deviations necessary for prediction are as follows:

Days of Weekly Rainfall (X)	*Occupancy Rate (Y)*
$M_X = 3.250$	$M_Y = 79.167$
$s_X = 1.815$ (using/$n - 1$)	$s_Y = 8.643$ (using/$n - 1$)

a. What is the equation for predicting hotel occupancy rate (*Y*) from days of weekly rainfall (*X*) for the group of vacationers?

b. What is the expected weekly occupancy rate for a week in which it rains on 3 days?

8. Exercise 8 in Module 35 required you to calculate the correlation between the number of miles per gallon cars get and the distance that the car owners planned to drive on the current year's vacation. The correlation was +0.239. The means and standard deviations necessary for prediction are as follows:

Miles Per Gallon (X)	*Planned Vacation Driving Distance (Y)*
$M_X = 27.583$	$M_y = 332.083$
$s_X = 10.113$ (using/$n - 1$)	$s_Y = 197.325$ (using/$n - 1$)

a. What is the equation for predicting planned driving distance (*Y*) from miles per gallon (*X*) for this group of car owners?

b. What is the expected vacation driving distance for someone whose car gets 26 miles per gallon?

9. Exercise 9 in Module 35 required you to calculate the correlation between the number of times per day that teenage girls look at themselves in a mirror and those girls' level of self-esteem. The correlation was +0.612. The means and standard deviations necessary for prediction are as follows:

Self-Esteem (X)	*Mirror Looks (Y)*
$M_X = 5.80$	$M_Y = 3.333$
$s_X = 1.821$ (using/$n - 1$)	$s_Y = 1.988$ (using/$n - 1$)

a. What is the equation for predicting mirror looks (*Y*) from self-esteem (*X*) for this group of teenage girls?

b. What is the expected number of mirror looks for a teenage girl whose level of self-esteem is 4?

10. Exercise 10 in Module 35 required you to calculate the correlation between the daily hours spent playing video computer games (*X*) and academic average (*Y*). The correlation was −0.834. The means and standard deviations necessary for prediction are as follows:

Video Game Hours (X)	*Academic Average (Y)*
$M_X = 1.850$	$M_Y = 85.083$
$s_X = 1.352$ (using/$n - 1$)	$s_Y = 6.288$ (using/$n - 1$)

a. What is the equation for predicting academic average (*Y*) from video game hours (*X*) for this group of students?

b. What is the expected academic average for a student who plays video games for 2.3 hr daily?

SPSS Connection

Download the file **data_years in sentence and prior arrests.sav** from www.sagepub.com/steinberg2e. These data are used in the textbook example.

Alternatively, click on the **Data View** tab and manually enter the scores into the spreadsheet. Set it up like this.

12.2	3
8.4	2
8.0	1
14.2	5
6.0	1
4.4	0
6.2	4
16.1	7
9.3	4
7.7	3

Click on the **Variable View** tab to define the variable. Name the first variable **yearssen,** set the decimal at **1** (not at 0), and label the variable as **Years in Sentence.** Name the second variable **prioarr,** set the decimal at **0,** and label the variable as **Prior Arrests.**

If the file is not already in **Data View,** click that tab in the lower left of the screen.

In the toolbar at the top of the screen, click on **Analyze,** then **Regression,** then **Linear.** Highlight the variable **Years in Sentence** in the left window and click on the **arrow** before the **Dependent** window on the right to send the variable into that window. In a simple linear prediction data set without a causal variable, this is the criterion variable, although SPSS refers to it as dependent variable. Highlight the **Prior Arrests** variable, click on the **arrow** before the **Independent** window on the right to send the variable into that window. In a simple linear prediction data set without a causal variable, this is the predictor variable, although SPSS refers to it as independent variable. Click **OK.** This is what you will see.

Regression

Variables Entered/Removed[b]

Model	Variables Entered	Variables Removed	Method
1	Prior Arrests[a]		Enter

a. All requested variables entered.

b. Dependent Variable: Years in Sentence

Model Summary

Model	R	R Square	Adjusted R Square	Std. Error of the Estimate
1	.820[a]	.672	.631	2.2930

a. Predictors: (Constant), Prior Arrests

ANOVA[b]

Model		Sum of Squares	df	Mean Square	F	Sig.
1	Regression	86.142	1	86.142	16.384	.004[a]
	Residual	42.063	8	5.258		
	Total	128.205	9			

a. Predictors: (Constant), Prior Arrests

b. Dependent Variable: Years in Sentence

(Continued)

(Continued)

Coefficients[a]

Model		Unstandardized Coefficients		Standardized Coefficients	t	Sig.
		B	Std. Error	Beta		
1	(Constant)	4.848	1.307		3.708	.006
	Prior Arrests	1.468	.363	.820	4.048	.004

a. Dependent Variable: Years in Sentence

The statistics needed to create the prediction equation are found in the final table labeled **Coefficients.** The column labeled **B** gives the slope, now referred to as the multiplier or weight, for each equation entry. The first equation entry is our predictor variable, which was **Prior Arrests.** This value, 1.468, is our prediction equation's "b." The second equation entry is our *Y* intercept, which is a **Constant.** This value, 4.848, is our prediction equation's "a." Thus, the prediction equation is $Y' = 1.468(X) + 4.848$. This equation agrees with the textbook calculations, within rounding error.

Other tables on the output reflect SPSS's regression algorithm. I will defer discussion of those tables until after I have discussed multiple regression in Module 39.

Note: As explained in Modules 6 and 8, SPSS uses the $n - 1$ divisor for a descriptive standard deviation, which it then uses in calculating *z* scores. Thus, the standard deviations and *z* scores produced through SPSS differ from those in the textbook, which maintains *N* as the divisor for descriptive standard deviations.

As explained in Module 35, the definitional formula for a correlation coefficient is the average of the cross products of the *z* scores, and those *z* scores must be computed from a standard deviation with *N* in the divisor. Thus, if we use the *z* scores computed by SPSS to compute the correlation coefficient, we will get an incorrect correlation coefficient. Despite this, SPSS produces the correct correlation coefficient, not an incorrect one. Clearly, it does not do so by using its own incorrect *z* scores. Apparently, SPSS either (1) recomputes the standard deviations using *N* in the divisor before computing the correlation coefficient via the cross-products method or (2) uses the raw score formula for the correlation coefficient, thereby avoiding its incorrect *z* scores altogether.

The current linear prediction module, as well as the standard error of prediction module to follow, requires standard deviations to calculate the necessary equations. SPSS uses the $n - 1$ divisor for standard deviations when calculating both the prediction equation and the standard error of prediction. This is appropriate, in that sample data are being used to inferentially predict the outcome for subjects who were not part of the original data set from which the prediction equation was derived.

This is why both standard deviation values are given in these textbook modules. You need the *N* divisor to compute the descriptive standard deviation, the *z* score, and the Pearson correlation coefficient via its definitional formula. But you need the $n - 1$ divisor to compute inferential statistics, including *t*, *F*, and the prediction equation and standard error of prediction found in this module and the next.

Visit the study site at www.sagepub.com/steinberg2e for practice quizzes and other study resources.

38 Standard Error of Prediction

Terms: point estimate, confidence interval, prediction error, standard error of prediction, standard error of estimate

Symbols: s_{YX}, CI

Learning Objectives:

- Know the shape of a distribution of prediction errors
- Calculate a standard error of prediction
- Establish intervals having various probabilities of containing the predicted score
- Understand the effect of sample size on the size of the standard error of prediction
- Understand the effect of variability among predicted scores on the size of the standard error of prediction

What Is a Confidence Interval?

As you learned in Module 37, when correlation is perfect—that is, when all data fall on the prediction line—the actual score will precisely equal the predicted score. To the extent that a correlation is not perfect, the actual score may be either higher or lower than the predicted score. Thus, when predicting scores on one variable from scores on another variable, statisticians typically state not only the predicted score but also the range of scores within which the actual score probably will fall. The predicted score, which you learned how to calculate in Module 37, is called a **point estimate**. The range within which the actual score will probably fall is called a **confidence interval** (CI). In this module, we will calculate the confidence interval within which we expect the actual score to lie.

Correlation and Prediction Error

The stronger the correlation, the less the data deviate from the prediction line and, hence, the closer the actual scores will be to the predicted scores. In contrast, the weaker the correlation, the more the data deviate from the prediction line and, hence, the farther the actual scores will be from the predicted scores (see Figure 38.1).

> Errors using inadequate data are much less than those using no data at all.
>
> —Charles Babbage

As shown in the two graphs in Figure 38.1, the correlation between SAT score and freshman GPA is moderate, but the correlation between IQ and shoe size is zero. As depicted, someone with a SAT score of 1,100 might earn a GPA anywhere between 2.3 and 3.8 but would not likely earn a GPA below 2.3 or above 3.8. Because of the correlation between SAT score and freshman GPA, the range of GPA within which someone with a given SAT score will fall is limited. Thus, if we predict GPA based on the prediction line, our possible error—the range of actual GPA scores for that SAT—will be restricted.

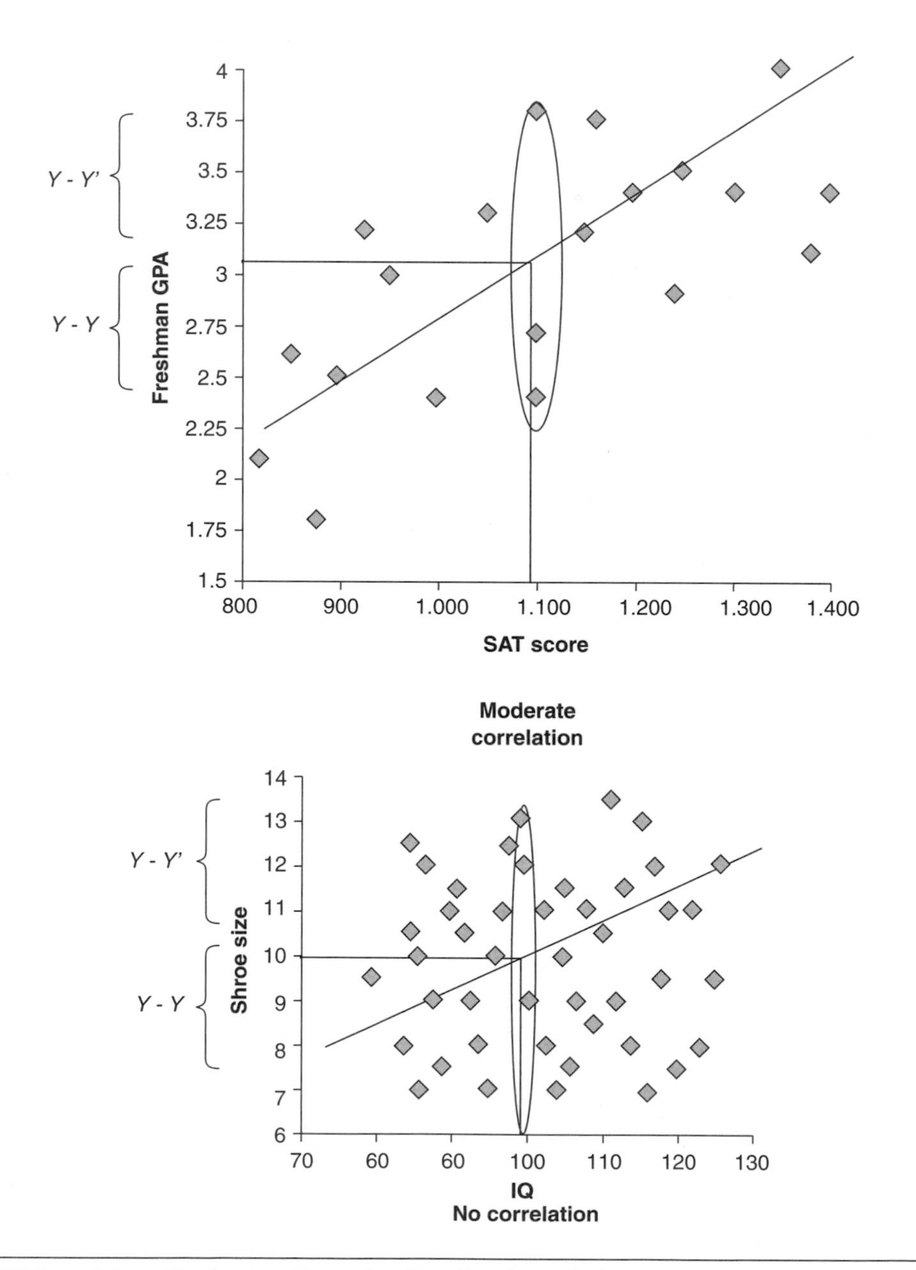

Figure 38.1 More Prediction Error for Smaller Correlations

In contrast, knowing someone's IQ gives us no information about what that person's shoe size might be. Because there is no correlation between IQ and shoe size, someone with an IQ of 100 could have any of the shoe sizes shown in the graph. Thus, we expect much more error in predicting shoe size from IQ score than in predicting freshman GPA from SAT score.

CHECK YOURSELF!

Which would you expect to have the larger prediction error: predicting weight from height or predicting weight from intelligence? Why?

Distribution of Prediction Error

An actual score's deviation from the predicted score is called **prediction error.** How are these deviations distributed? As you learned when you studied probability in Module 11, and again when you studied the standard error of the mean in Module 15, smaller deviations occur more often than larger ones. Also, deviations above the predicted score are neither more likely nor less likely to occur than deviations below the predicted score. Thus, deviations of actual *Y* scores around the predicted *Y′* score will be normally distributed.

Prediction errors are deviations from expectation. Because they are normally distributed, their distribution, like that of any normal curve, can be divided into standard deviation units. The standard deviation of prediction errors is called a **standard error of prediction,** symbolized s_{YX}. It is also called a **standard error of estimate.**

Like any standard deviation, the standard error of prediction is the average amount of linear dispersion among the units being plotted—in this case, among the errors of prediction. From your study of the normal curve in Module 7 and again in various modules covering hypothesis testing, you know that nearly all of a normal distribution is accounted for by ±3 standard deviation units. Thus, the distribution of errors around the prediction line—that is, around *Y′*—looks like Figure 38.2. Most scores fall close to the predicted score. The farther we go from the predicted score (either higher or lower), the fewer the number of scores.

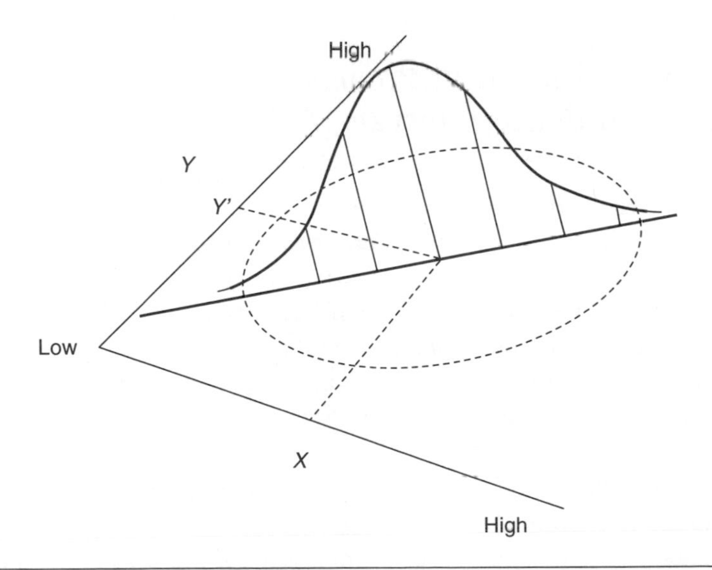

Figure 38.2 Normal Distribution of Prediction Errors Around the Prediction Line

CHECK YOURSELF!

Explain why a distribution of prediction errors will be normally distributed.

Calculating the Standard Error of Prediction

How big is the standard error of prediction? From the discussion above, the value of the standard error of prediction will be larger for lower correlations than for higher correlations. We find the size of the standard error of prediction via the following formula:

$$s_{YX} = s_Y\sqrt{1 - r_{XY}^2}$$

where

s_Y = standard deviation of the variable Y and

r_{XY} = correlation between the variables X and Y.

For the SAT-GPA example with which we worked in Module 37, the correlation was +.43. The standard deviation for the variable Y (GPA) was 0.57. Now let's compute the standard error of prediction for those data.

$$\begin{aligned} s_{YX} &= 0.57\sqrt{1 - (.43)^2} \\ &= 0.57\sqrt{1 - .1849} \\ &= 0.57\sqrt{.8151} \\ &= 0.57(.9028) \\ &= 0.515 \end{aligned}$$

So s_{YX} is 0.515. But what does that mean?

Using the Standard Error of Prediction to Calculate Confidence Intervals

Because the standard error of prediction is simply a standard deviation, it can be interpreted just like any standard deviation: It is the average amount of linear dispersion in the errors of prediction. When you studied the normal curve in Module 7, you learned that approximately 68% of all scores fall within ±1 standard deviation from the mean, approximately 95% within ±2 standard deviations from the mean, and approximately 99% within ±3 standard deviations from the mean. Because actual Y scores are normally distributed around Y', the same percentages apply. Thus, we can say that

- approximately 68% of actual Y scores for a given X score will fall within ±1 standard error of prediction around Y',
- approximately 95% of actual Y scores for a given X score will fall within ±2 standard errors of prediction around Y', and
- approximately 99% of actual Y scores for a given X score will fall within ±3 standard errors of prediction around Y'.

This is shown in Figure 38.3, which is the same as Figure 38.2 but with percentages added.

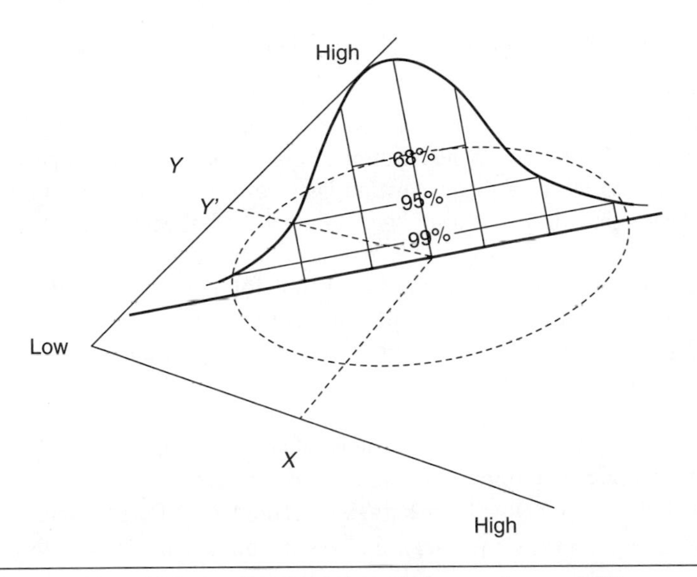

Figure 38.3 Percentage of Normally Distributed Scores Around Y'

Recall from Module 37 that the predicted freshman GPA for a student with a combined SAT score of 700 was 2.042. Therefore, with a standard error of prediction of 0.515, the admissions officer can be

- approximately 68% confident that this student's GPA will fall at 2.042 ± *one* unit of 0.515, which is a GPA between 1.53 and 2.56;
- approximately 95% confident that this student's GPA will fall at 2.042 ± *two* units of 0.515, which is 2.042 ± 1.030, which is a GPA between 1.012 and 3.072; and
- approximately 99% confident that this student's GPA will fall at 2.042 ± *three* units of 0.515, which is 2.042 ± 1.545, which is a GPA between 0.497 and 3.587.

SOURCE: By permission of John L. Hart FLP and Creators Syndicate, Inc. (B.C.)

CHECK YOURSELF!

What value will always fall in the middle of the confidence interval for predicted scores?

PRACTICE

1. The predicted resting heart rate for Diego based on his score on a physical fitness test is 78.5 beats per minute. The standard error of prediction is 2.4 beats per minute. His physician is about to measure Diego's resting heart rate. The physician can be approximately 95% confident that Diego's resting heart rate will be between which two values?

2. Using a prediction equation, the predicted time for John, a member of the college soccer team, to run a mile is 355 s. The standard error of prediction for members of the team is 3.4 s. The coach can be about 68% confident that John's mile running time will be between which two times?

3. Exercise 3 in Module 37 required that you calculate the prediction equation for 10 students who took two quizzes. The same exercise also required that you determine the predicted score on Quiz 2 for Kate, who scored 7 on Quiz 1. Relevant statistics are $r_{XY} = +.69$, $s_Y = 1.853$ (using $n - 1$), and Y' for Quiz 2 = 6.364.
   a. Calculate the standard error of prediction (s_{YX}) for the two quizzes.
   b. You can be approximately 68% confident that Kate's actual score on Quiz 2 (Y) will fall between which two values?

4. Exercise 4 in Module 37 required that you calculate the equation for predicting level of prejudice from intelligence level. The same exercise also required that you determine the level of prejudice for someone with an IQ of 115. Relevant statistics are $r_{XY} = -.83$, $s_Y = 3.651$ (using $n - 1$), and Y' for level of prejudice = 22.043.
   a. Calculate the standard error of prediction (s_{YX}) for the same data.
   b. You can be approximately 95% confident that this person's level of prejudice (Y) will fall between which two values?

5. Exercise 5 in Module 37 required that you calculate the equation for predicting college freshman GPA from high school GPA for a group of National Merit Scholars. The same exercise also required that you determine the predicted college GPA of a National Merit Scholar whose high school GPA is 3.8. Relevant statistics are $r_{XY} = +.66$, $s_Y = 0.238$ (using $n - 1$), and Y' for college freshman GPA = 3.539.
   a. Calculate the standard error of prediction (s_{YX}) for the same data.
   b. You can be approximately 99% confident that this student's college GPA (Y) will fall between which two values?

6. Exercise 6 in Module 37 required that you calculate the equation for predicting rats' running speed from their weights. The same exercise also required that you determine the running speed for a rat that weighs 290 g. Relevant statistics are $r_{XY} = +.46$, $s_Y = 1.75$ (using $n - 1$), and Y' for running speed = 24.967.
   a. Calculate the standard error of prediction (s_{YX}) for the same data.
   b. You can be approximately 68% confident that this rat's running speed (Y) will fall between which two values?

7. Exercise 7 in Module 37 required that you calculate the equation to predict hotel occupancy rate from days of rainfall. That same exercise required that you predict hotel living occupancy rate for a week having 3 days of rainfall. Relevant statistics are $r_{XY} = -0.913$, $s_Y = 8.643$, and Y' for hotel occupancy = 80.25% occupancy.
   a. Calculate the standard error of prediction (s_{YX}) for the same data.
   b. You can be approximately 95% confident that the predicted week's occupancy rate will be between which two values?

8. Exercise 8 in Module 37 required that you calculate the equation to predict planned vacation driving distance from cars' miles/gallon. That same exercise required that you predict planned vacation driving distance for a person whose car gets 26 miles/gallon. Relevant statistics are $r_{XY} = +0.239$, $s_Y = 197.325$, and Y' for this vacationer = 324.70 miles.
   a. Calculate the standard error of prediction (s_{YX}) for the same data.
   b. You can be approximately 68% confident that this vacationer's driving distance will be between which two values?

9. Exercise 9 in Module 37 required that you calculate the equation to predict number of mirror looks from level of self-esteem. That same exercise required that you predict number of mirror looks for a girl with a self-esteem level of 4. Relevant statistics are $r_{XY} = +0.612$, $s_Y = 1.988$, and Y' for this girl = 2.21 mirror looks.
   a. Calculate the standard error of prediction (s_{YX}) for the same data.
   b. You can be approximately 68% confident that this girl will look in the mirror today between which two number of times?

10. Exercise 10 in Module 37 required that you calculate the equation to predict academic average from daily hours spent playing computer video games. That same exercise required that you predict academic average for a student who averages 2.3 hr of daily video game playing. Relevant statistics are $r_{XY} = -0.834$, $s_Y = 6.288$, and Y' for this student = 83.34% academic average.
    a. Calculate the standard error of prediction (s_{YX}) for the same data.
    b. You can be approximately 95% confident that this student's academic average will be between which two values?

Factors Influencing the Standard Error of Prediction

Let's look again at the formula for s_{YX}.

$$s_{YX} = s_Y\sqrt{1 - r_{XY}^2}$$

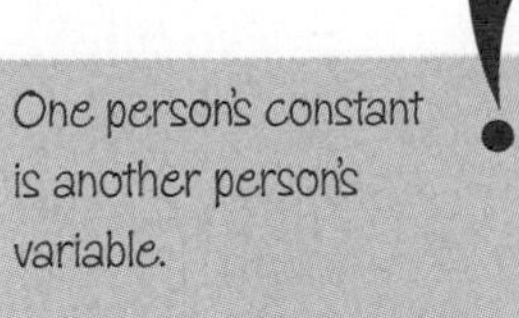

According to the formula, only two variables can influence the size of the standard error of prediction (s_{YX}). What are those two variables?

In the formula, the numeral 1 is a constant. Because it does not vary, it cannot affect the size of the standard error of prediction. The size of the standard error of prediction is determined only by the two variables—the size of the standard deviation within the set of Y scores (s_Y) and the size of the correlation coefficient between the X and Y scores (r_{XY}). Let's look more closely at the effect of those two variables on the standard error of prediction.

Returning to out SAT and GPA example, the standard deviation of GPA scores (Y) is 0.57. Here's what happens to the standard error of prediction when the correlation between SAT and GPA (X and Y) is perfect (i.e., when $r_{XY} = 1.00$).

$$\begin{aligned} s_{YX} &= 0.57\sqrt{1 - (1.00)^2} \\ &= 0.57\sqrt{0} \\ &= 0.57(0) \\ &= 0 \end{aligned}$$

Thus, when the correlation is perfect, there is no variation in predicted Y' scores. Recall that this was the case with Celsius to Fahrenheit conversions, in which the correlation was perfect. For a given Celsius temperature, we did not have to estimate what the Fahrenheit temperature would be. We knew exactly what it would be.

"LET'S GO OVER TO CELSIUS'S PLACE. I HEAR IT'S ONLY 36° OVER THERE."

SOURCE: Reprinted with permission from Sydney Harris, ScienceCartoonsPlus.com.

Now let's see what happens when there is no correlation between SAT and GPA (i.e., when $r_{XY} = .00$).

$$\begin{aligned} s_{YX} &= 0.57\sqrt{1 - (.00)^2} \\ &= 0.57\sqrt{1} \\ &= 0.57(1) \\ &= 0.57 \end{aligned}$$

Thus, when the correlation is 0, there is a great deal of variation in predicted Y scores. More precisely, when the correlation is 0, the size of the standard error of prediction is equal to the standard deviation of the Y scores (which, in this case, was 0.57). To put it another way, the range of predicted Y scores for any given score on X is as large as the range of all Y scores.

Figure 38.4 shows the diagrams of this principle. Note that, for any given score on X, the range of the predicted scores on Y increases as the size of the correlation decreases.

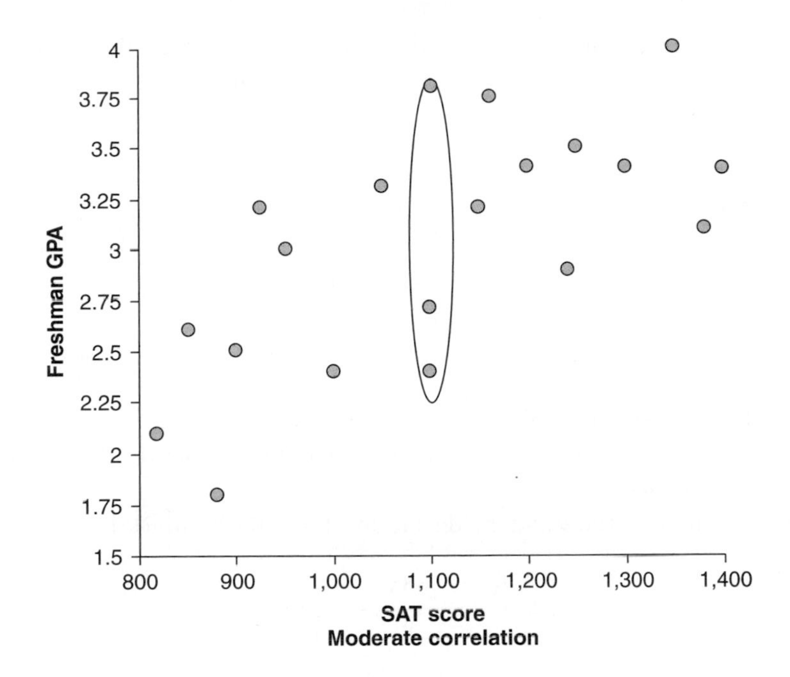

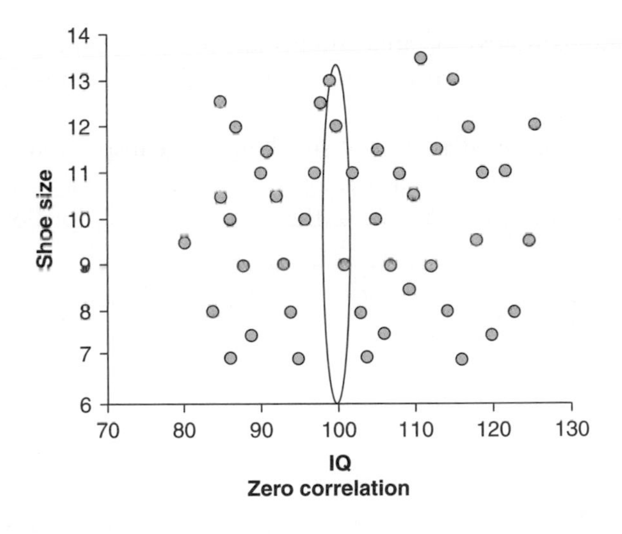

Figure 38.4 Range in Predicted Scores on *Y* Under Different Correlation Values Between *X* and *Y*

CHECK YOURSELF!

As the value of the correlation coefficient changes, the size of the standard error of prediction ranges between which two values?

In like manner, the size of the standard deviation of Y (s_Y) affects the size of the standard error of prediction (s_{YX}). First, let's assume that the standard deviation of Y is 0.46 rather than 0.57.

$$\begin{aligned} s_{YX} &= 0.46\sqrt{1-(.43)^2} \\ &= 0.46\sqrt{1-.1849} \\ &= 0.46\sqrt{.8151} \\ &= 0.46(.9028) \\ &= 0.415 \end{aligned}$$

Note that the standard error of prediction is smaller than it previously was. The smaller the dispersion in the entire set of scores being predicted, the smaller the dispersion in predicting any single person's score.

Now let's assume that the standard deviation of Y is 0.68 rather than 0.57:

$$\begin{aligned} s_{YX} &= 0.68\sqrt{1-(.43)^2} \\ &= 0.68\sqrt{1-.1849} \\ &= 0.68\sqrt{.8151} \\ &= 0.68(.9028) \\ &= 0.614 \end{aligned}$$

Note that the standard error of prediction is larger than it previously was. The greater the dispersion in the entire set of scores being predicted, the greater the dispersion in predicting any single person's score.

To illustrate this principle, diagrams of small and large dispersion in a set of scores are given in Figure 38.5. Bob's score is indicated by an X. Note that, as the size of the dispersion within the set of scores increases, the size of the score dispersion for Bob also increases.

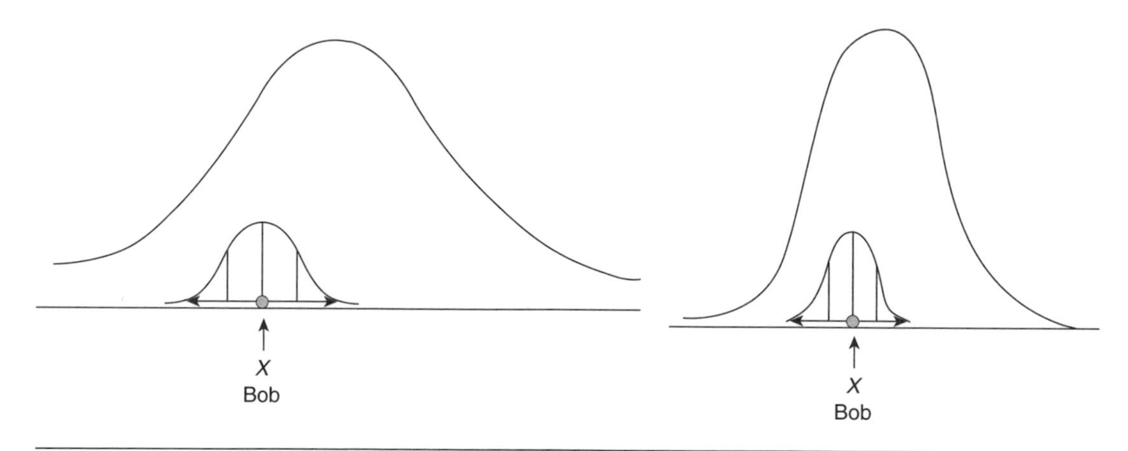

Figure 38.5 Size of the Standard Error of Prediction When the Standard Deviation of Y Is Large and When the Standard Deviation of Y Is Small

PRACTICE

11. Assume that the correlation between toddlers' ages and their play distances from their mothers is +.90. This correlation is very high. What effect will such a high correlation have on the standard error of prediction for this study?

12. Assume that the psychologist who studied the relationship of toddlers' age to their play distance now includes additional toddlers who are older than those listed. And assume that these additional toddlers, being older, play farther from their mothers. What effect will this additional spread of *Y* scores have on the standard error of prediction for this study?

13. In Module 37, we predicted sentence length from number of prior arrests. Now let's assume that we found a correlation of +0.54 instead of the +0.82 correlation that we found. What effect will this decreased correlation have on the standard error of prediction for this study?

14. In Module 37, we predicted sentence length from number of prior arrests. Now let's assume that we had a much smaller range of sentences. What effect will this smaller range of sentences have on the standard error of prediction for this study?

Another Calculation Example

I conclude this module with another example. At the end of Module 37, we predicted years of prison sentence from number of prior arrests. After calculating the prediction equation and applying it to a convicted felon with six prior arrests, we found the predicted prison term to be 13.65 years. Other relevant statistics from that example were $r_{XY} = +.82$ and $s_Y = 3.77$ (using $n - 1$). Module 37 included calculation steps 1 through 4. Let's complete that example by finding the standard error of prediction and confidence interval. We pick up at Step 5.

5. Find the standard error of prediction.

$$
\begin{aligned}
s_{YX} &= s_Y\sqrt{1 - r_{XY}^2} \\
&= 3.77\sqrt{1 - (.82)^2} \\
&= 3.77\sqrt{1 - .672} \\
&= 3.77\sqrt{0.328} \\
&= 3.77(.573) \\
&= 2.16
\end{aligned}
$$

6. Find the confidence interval for the predicted prison term. First, we must decide what level of confidence we would like. I will show both 95% and 68% confidence.

$$
\begin{aligned}
95\%\text{CI} &= 13.65 \pm 2(2.16) \\
&= 13.65 \pm 4.32 \\
&= 13.65 + 4.32 \text{ and } 13.65 - 4.32 \\
&= 17.97 \text{ and } 9.33
\end{aligned}
$$

We can be approximately 95% certain that this felon's sentence will be between 9.33 years and 17.97 years.

$$
\begin{aligned}
68\%\text{CI} &= 13.65 \pm 1(2.16) \\
&= 13.65 \pm 2.16 \\
&= 13.65 + 2.16 \text{ and } 13.65 - 2.16 \\
&= 15.81 \text{ and } 11.49
\end{aligned}
$$

We can be approximately 68% certain that this felon's sentence will be between 11.49 years and 15.81 years.

Visit the study site at www.sagepub.com/steinberg2e for practice quizzes and other study resources.

39 Introduction to Multiple Regression

Terms: regression, weight, beta, common variance, residual, model (full, reduced, and testing), General Linear Model (GLM)

Symbols: b, B, R, R^2

Learning Objectives:

- Break total variance into its components
- State the advantage of multiple predictors over a single predictor
- Use predictor scores and their coefficients to predict outcomes
- Explain the logical difference between experimental and regression analytic methods
- Locate and interpret key statistics found on multiple regression software output

What Is Regression?

In Module 37, we predicted GPA from SAT score, and we predicted length of prison sentence from number of prior arrests. When we use one variable to predict another, we are said to regress the criterion or dependent variable onto the independent variable. In our examples, we regressed GPA onto SAT score, and we regressed length of prison sentence onto number of prior arrests. The process of predicting the dependent variable from the independent variable is called **regression.** Thus, we could as well have referred to the process in Module 37 as linear regression rather than linear prediction. The two terms are synonymous.

Prediction Error, Revisited

Unless correlation is perfect ($r = \pm 1.00$), prediction is not perfect, and hence the need in Module 38 to calculate the standard error of prediction, also known as the standard error of estimate. We used the standard error of prediction to create a confidence interval within which the predicted score probably fell.

However, not only is prediction not perfect when correlation is not perfect, but a decrease in correlation *exponentially* lowers the prediction accuracy. This is because, as you learned in Module 36, the common variance between two variables is not the correlation coefficient (r) itself, but rather the *square* of the correlation coefficient (r^2). For example, when the correlation is .90, the common variance is $(.90)^2$, which is .81; when the correlation is .80, the common variance is $(.80)^2$, which is .64; when the correlation is .70, the common variance is $(.70)^2$, which is .49; and so on. In Module 36, we defined this common variance as the amount of variance in one variable that can be accounted for by the other variable. Now we see that common variance is the proportion of the total variance that is predictable. Consequently, prediction accuracy decreases dramatically with even a modest decrease in the correlation between the two variables.

Total variance consists of predictable variance plus unpredictable variance or error variance. In mathematical terms, $T = P + E$. In mathematical terms, $1.00 = r^2 + \text{error}$. In regression, the error variance is called **residual**, because it is what is "left over" after all predictable variance has been accounted for. Thus, in simple linear regression we see results presented as r^2 and residual. By manipulating the variables algebraically, $E = T - P$, or residual = $1.00 - r^2$. Plugging in numbers from the above examples,

when r is .90, r^2 = .81 and residual = .19

when r is .80, r^2 = .64 and residual = .36

when r is .70, r^2 = 49 and residual = .51.

CHECK YOURSELF!

What three types of variance make up a complete variance equation? Which portion is predictable and which is not? What happens to the relative values of the portions as the correlation between the two variables decreases?

Why Multiple Regression?

As we have just seen, even moderate correlations leave quite a bit of variance in the criterion (or dependent variable) variable unaccounted for. From Module 37, the correlation between SAT score and GPA was +.43, and the correlation between number of prior arrests and prison sentence length was +.82. This means that the residual for GPA prediction is $1.00 - (+.43)^2 = 1.00 - .18 = .82$. And this means that the residual for prison sentence prediction is $1.00 - (.82)^2 = 1.00 - .67 = .33$. Clearly, quite a bit of the criterion variance is *not* accounted for in either example.

What can we do about that? One thing we can do is include additional predictor variables. For example, while SAT score provides some predictive information about incoming students' probable GPA, many other variables also provide predictive information. These variables include time spent studying, motivation to achieve, and parental level of education, among others. Including these variables in the prediction equation along with SAT score will greatly increase the accuracy of the predicted GPA. Similarly, while number of prior arrests provides some predictive information about prison sentence length, many other variables also provide predictive information. These variables include age at first offense and family income level, among others. Including these variables in the prediction equation along with number of prior arrests will greatly increase the accuracy of the predicted prison sentence length.

Predicting a criterion outcome based on more than one predictor variable simultaneously is called **multiple regression**. This type of research is very common. After you learned t tests and one-way ANOVA, each of which included a single independent variable, you learned factorial ANOVA, which included more than one independent variable. At the time, I told you that such designs were more realistic than those with single independent variables because real-life relationships are seldom univariate. Rather, it is usually the case that multiple independent variables influence a dependent variable, and so it is best to include those additional variables in the study from the outset. The same is true of prediction studies. Rarely is an outcome adequately predicted by a single predictor. Therefore, prediction accuracy is increased by including other relevant predictor variables in the design from the outset.

CHECK YOURSELF!

What is the major advantage of multiple regression over simple regression?

The Multiple Regression Equation

Recall the simple prediction equation: $Y' = bX + a$. In this equation, Y' was the predicted criterion score, X was the score on the predictor variable, "b" (the slope) was the multiplication factor for X, and "a" was the constant.

The equation for multiple regression follows the format for simple linear prediction except that it includes additional predictor variables. Here is the equation, where n = number of predictor variables. Each of the X's is a predictor variable, and each of the b's is the multiplier given to that predictor variable.

$$Y' = b_1X_1 + b_2X_2 + b_3X_3 + \ldots + b_nX_n + a$$

By convention, the b's are now called **weights**. Thus, we weight (or multiply) the first predictor variable's score by b_1, the second predictor variable's score by b_2, the third predictor variable's score by b_3, and so on. The weight tells the relative importance of that variable in predicting the criterion's variance. Here is an example:

Predictor Variable	*Score*	*Weight*
X_1	10	.30
X_2	20	.20
X_3	50	.10
X_4	50	.05
Constant = 5.5		

To obtain the predicted criterion score, we simply sum the weighted predictor scores, along with the constant, as indicated in the equation.

$$\begin{aligned} Y' &= b_1X_1 + b_2X_2 + b_3X_3 + b_4X_4 + a \\ &= .30(10) + .20(20) + .10(50) + .05(20) + 5.5 \\ &= 3 + 4 + 5 + 1 + 5.5 \\ &= 18.5 \end{aligned}$$

Coefficient weights can be either unstandardized or standardized (via z scores). It is the *un*standardized weights that are used in the equation to predict individual outcome scores, as above. The unstandardized weights have the same meaning as those that we derived in Module 37 for our single predictor.

On the other hand, it is the *standardized* coefficients that are used to compare the relative contribution of individual predictors. Standardized weights are called **beta** weights and are symbolized by a capital **B**. Standardized beta weights have many advantages. Without standardization, the weights are highly dependent on the underlying scales of the predictors. For example, when predicting the number of resting heart beats from time spent exercising and from calories consumed, we would expect very different weights per resting heart beat when the time spent exercising is in hours than when it is in minutes, and when the calories consumed are in thousand-calorie units than in single-calorie units. Hence, comparisons between the weights assigned to exercise and calories when predicting resting heart beat are meaningless unless the scales of the two variables are equated. Similarly, when predicting college GPA from high school GPA and from time spent studying, we would expect very different weights for predicting college GPA when high school GPAs are reported on a 0% to 100% scale than when they are reported on a 0.00 to 4.00 scale, and for time spent studying when it is reported in hours than when it is reported in minutes. Hence, comparisons between the weights assigned to high school average and time spent studying are meaningless unless the scales of the two variables are equated. For this reason, neither can we directly compare the relative contributions of the variables to the equation unless their weights have been standardized. Therefore, we will look briefly at the standardized beta weights when I discuss the relative importance of individual predictors.

CHECK YOURSELF!

A prediction study produces the following equation: Outcome = .54(VarA) + .18(VarB) + .06(VarC) + *a*. How many predictor variables are there in this study?

Multiple Regression and Predicted Variance

In simple linear regression, there is only one predictor variable (X) correlated with the criterion variable (Y). There is no need to weight the predictor's contribution to the criterion because the single correlation is the whole of the relationship between X and Y. Thus, the weight and the correlation are the same.

Also in simple linear regression, the square of the correlation coefficient (r^2) is the **common variance**. That is, if we square the correlation coefficient between the predictor variable X and the criterion variable Y, we have all of the variance that is in common between X and Y. Whatever is left is error variance or residual.

It is tempting to think that we could follow this same process with multiple regression—merely adding together the correlations between each of the predictor variables and the criterion to obtain the overall correlation between the set of predictors and the criterion; and merely adding together the squared correlations between each of the predictor variables and the criterion to obtain the overall common variance for the set of predictors with the criterion. However, we cannot do that. Let's see what happens if we simply follow our intuition. Assume the following correlations between the predictors and the criterion for the prison sentence example.

r between prior arrests and sentence length	= +.82	$r^2 = (+.82)^2 = .67$
r between age at first arrest and sentence length	= −.71	$r^2 = (-.71)^2 = .50$
r between family income and sentence length	= −.52	$r^2 = (-.52)^2 = .27$
Total r (absolute value)	= 2.05	Total r square = 1.44

Of course, the highest possible correlation is 1.00, so how can we get a total correlation of 2.05? We cannot. Similarly, total variance (or total *anything*) is 1 unit, so how can we get a total value of 1.44 unit? We cannot. Obviously, something is wrong. What is wrong?

Just as the independent variables in factorial ANOVA often *interact* to produce an effect in addition to their independent contributions, so the predictors in multiple regression often are *intercorrelated* to produce effects in addition to their independent contributions. That is, the predictors not only are correlated with the criterion, but they also are correlated with the other predictors. In our example, the number of prior arrests, age at first arrest, and family income each are intercorrelated, in addition to being individually correlated with prison sentence length. Here is the situation shown visually via a Venn diagram:

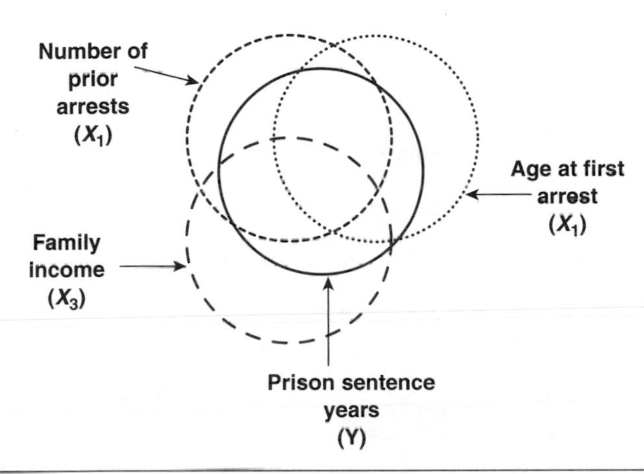

Figure 39.1 Venn Diagram Demonstrating Predictor Intercorrelation in Multiple Regression

Unlike ANOVA, in which the effects of the independent variables and their interactions are already additively partitioned into the total variance, this intercorrelation is not accounted for in a set of simple correlations between each predictor with the criterion. Thus, if we simply add the correlations between each predictor with the criterion (or even the squares of each predictor variable with the criterion), the amount by which the variables are intercorrelated is counted multiple times. We get an inflated total value.

Multiple regression removes the effect of intercorrelation. It does this during the process of deriving the *b* weights. One method for removing the effect of predictor correlation is to find the common variance between the first predictor and the criterion; then find only that portion of the common variance between the second predictor and the criterion that is not already taken by the first predictor; then find only that portion of the common variance between the third predictor and the criterion that is not already taken by the first or second predictors; and so on. This process is illustrated in Figure 39.2.

There are other ways of partitioning the common variance between the predictors and the criterion (e.g., dividing the shared variance between predictors). However, all these methods remove correlation between predictors when determining total predicted variance, and that was our major concern.

While multiple regression is conceptually simple, it is mathematically tedious, partly because of the number of variables and partly because of the intercorrelation among the predictors. To find the correct predicted variance by hand calculation for a set of predictors, we would have to compute the correlation between each predictor and the criterion, and also between each predictor and every other predictor. Then we would have to subtract the intercorrelations through a

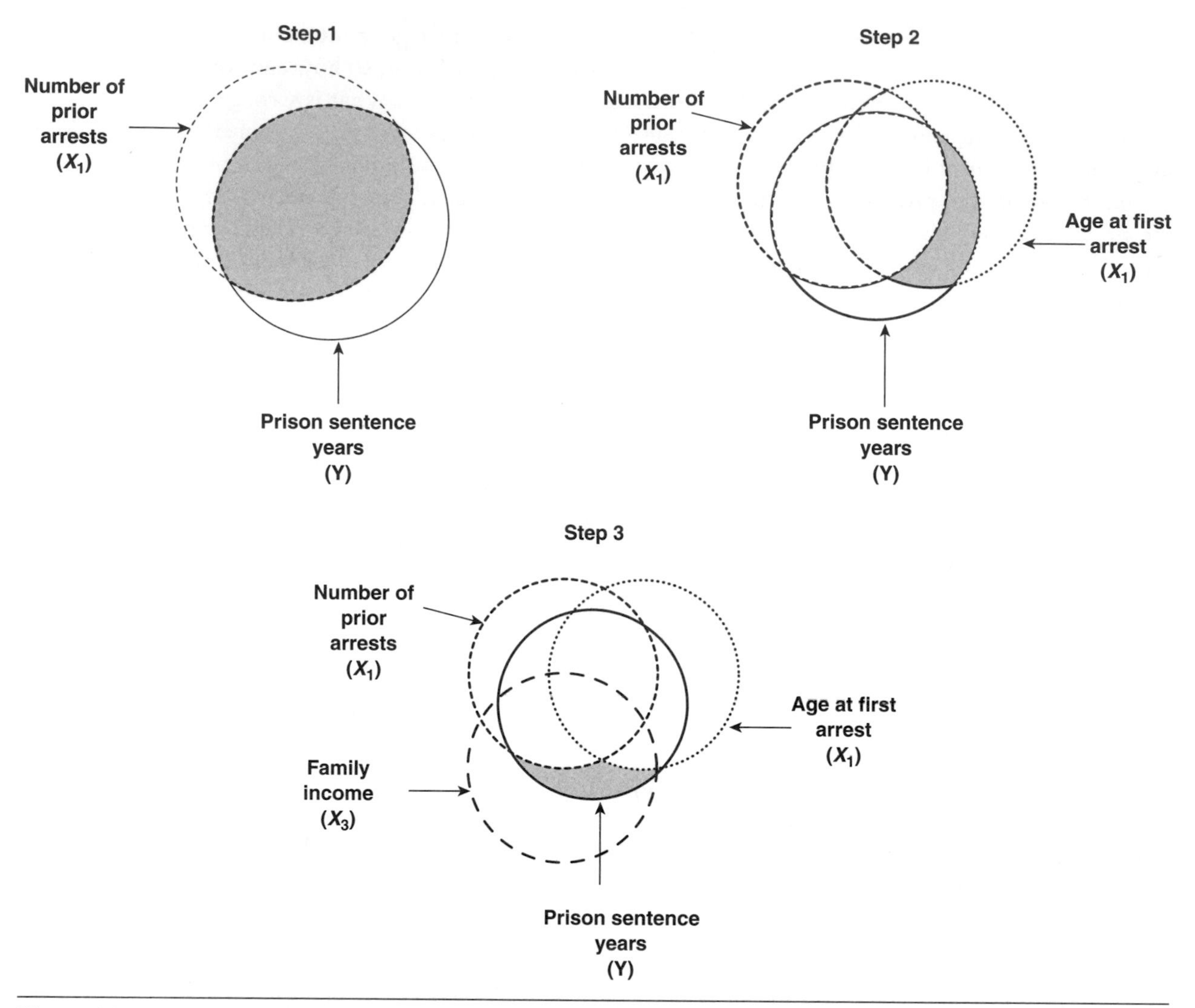

Figure 39.2 Venn Diagram Demonstrating a Method for Partitioning Common Variance Among Predictors in Multiple Regression

complicated formula. We would have to square values, adjust for degrees of freedom, and many other steps. Because of the time required and the potential for error in such lengthy calculations, few people hand-calculate multiple regression equations. Instead, they use statistical software. Thus, we will use the same. My discussion from this point forward will be limited to interpreting output obtained via such software.

There are several ways of entering variables into a multiple regression equation. For example, we can pre-decide the variables' entry order. However, the predictors' weights will differ with each reordering of the variables. This is because subsequent variables are allotted only that portion of the criterion variance that is not already accounted for by the predictors already entered into the equation. Because of the changing weights, we should not give much credence to their size when we force the variables' order. Another entry option, called "stepwise," sets statistical criteria for the variables' inclusion and then drops variables that don't meet those criteria. This method is controversial because it capitalizes on the random error inherent in any study and thus drops variables and sets weights that are not likely to reproduce in subsequent studies.

A full discussion of the options for variables' entry order is beyond the scope of this introductory textbook. The default option in most software programs is to include all variables in the model and let the computer enter them in the order of their shared criterion variance—selecting first whichever variable accounts for most criterion variance, then whichever variable accounts for most of the remaining criterion variance, and so forth. In that way, all variables that we intend to be part of the study are included and we are not giving undue importance to random error in determining the weights. This is the process we will assume here.

Regardless of the order in which the predictor variables are entered, however, the coefficients *together* will predict the same amount of criterion variance and, hence, predict the same criterion score for individual cases. Thus, the important statistic is the R^2, which is the total criterion variance accounted for by a set of predictors (more about that in a bit). Think of it this way. If three people together eat 9 of the 12 slices in a pizza, then three quarters of the pizza has been eaten (our predicted variance) and one quarter of the pizza remains (our residual variance). It does not matter if the three people each ate 3 slices, or if two of the people ate 4 slices and one person ate only 1 slice, or if one of the people ate 5 slices and the other two people ate 2 slices apiece. The same amount of the pizza, 9 slices, was eaten, and the same amount of pizza, 3 slices, remains uneaten. It is merely apportioned differently among the eaters. The total amount eaten will change only if we change the number of people or the identity of the people eating the pizza. So it is with regression equation coefficients. Changing the order of the variables will change their weights, but the total criterion variance accounted for by the reallocated weights will remain the same. Only if we change the number or nature of the predictors will their total predicted variance change.

CHECK YOURSELF!

R^2 for a study with two predictors is .68. If you switch the order of the two predictors, what will be the new R^2?

Hypothesis Testing in Multiple Regression

In addition to using the prediction equation to predict scores for particular subjects, we can answer two types of research questions through multiple regression: Does the set of predictors taken together predict a significant amount of criterion variance? And do the individual predictors predict a significant amount of criterion variance?

As you probably realize by now, we test the overall set of predictors via R^2. If the set of predictors is doing a good job at collectively predicting criterion outcome, the R^2 will be high. But how high should it be? A general rule of thumb, based on Cohen, is that a correlation greater than .30 is medium and one of .40 is large. This translates into a medium R^2 of about .10, and a large R^2 of about .15. Many sets of predictors do much better than this, however.

The software also tests the statistical significance of the R^2. It does this by reverting to an ANOVA approach. When we calculated ANOVAs we formed an F ratio of between groups variance to within groups variance. What is the equivalent in multiple regression? Recall from the discussion early in this module that R^2 and the residual together make up all of the variance. Therefore, the residual, or error, is $1 - R^2$. Knowing the R^2 and the degrees of freedom by which to divide (based on number of predictors and sample size), we have enough information to form an F ratio of R^2 to error. We then test this F ratio by the appropriate degrees of freedom. As I said earlier, I will not calculate this statistic here. See the "Example" section to follow, for guidance in interpreting that portion of statistical software output.

To test individual predictors, we examine the standardized beta weights. Once standardized, they can be interpreted as simple correlations between the predictor and the criterion.

Thus, the same criterion applies as above: .30 or higher is medium, and .40 or higher is large. Because they are standardized on a common scale (the *z*-score scale), we also can compare the relative strength of the predictors. The largest beta weight is the strongest predictor, the second highest beta weight is the second strongest predictor, and so on.

We also test the individual predictors via *t* tests. Degrees of freedom are complicated for the predictors, and I again will not explain them here. See the "Example" section to follow for guidance in interpreting that portion of typical software output.

A final way of testing the contribution of individual predictors is to compare the R^2 with and without a specific predictor. The software allows us to determine predicted variance for subsets of predictors—say, for number of prior arrests and age at first arrest, but not family income; or, for family income and age at first arrest, but not number of prior arrests; and so on. Each set of predictors is called a **model**. The software then compares the **full model** (all predictors) to the **reduced model** (fewer than all predictors) to see if the eliminated predictor added any significant variance. This is called **model testing**. Software programs delight in calculating and comparing all possible models. People like you and me, in contrast, do not. This is another reason for relying on the software. Model testing would require more instruction in both multiple regression theory and software use than is appropriate for this introductory course. Therefore, I will not discuss it even via the software output. Model testing is typically an important part of a next course in statistics, however, so this brief introduction should help you begin thinking of hypothesis testing in multiple regression terms.

CHECK YOURSELF!

Which model–full or reduced–should predict the greater amount of variance? Why?

PRACTICE

1. R^2 for a study is .74. What is the residual?
2. The residual for a study is .19. What is R^2?
3. A researcher reports an R^2 of .88 and a residual of 14. Comment on this.
4. Which of these is preferred—a study with an R^2 of .73, or a study with a residual of .27?
5. The R^2 for a study with three predictors is .64. If the researcher drops one of the predictors and re-runs the study, what should happen to the R^2?
6. The R^2 for a study with three predictors is .64. If the researcher adds a fourth predictor, what should happen to the R^2?
7. The R^2 for a study with three predictors is .64. If the researcher changes the order of the three predictors, what should happen to the R^2?

An Example

Here is the example in Module 37 predicting prison sentence length from number of prior arrests. I have added data for age at first arrest and family income, as discussed above. Thus, we now have three predictors and one criterion.

First, a word about sample size. A rule of thumb for multiple regression studies is at least 10 subjects per predictor variable. Fewer subjects are needed when the predictors are highly correlated with the criterion and when the predictors are uncorrelated with each other. As these assumptions are violated, more subjects are needed. When sample size is insufficient, the weights are very unstable and are not likely to reappear on a different sample of subjects. Obviously, our textbook example is *very* deficient in sample size. The sample size was kept small for calculation purposes (recall that we hand-calculated the simple linear portion in Module 37). Real research would include many more subjects than this textbook example.

Number of Years in Sentence (Y)	*Number of Prior Arrests* (X_1)	*Age at First Arrest* (X_2)	*Family Income (in thousand $)*
12.2	3	18	40
8.4	2	23	42
8.0	1	20	58
14.2	5	15	18
6.0	1	33	44
4.4	0	21	76
6.2	4	25	21
16.1	7	17	27
9.3	4	24	72
7.7	3	25	39

Most software programs produce three sets of statistics: predicted variance, equation coefficients, and ANOVA table. We will examine these outputs one at a time.

Table 39.1 shows the correlation and the predicted variance between the set of predictors and the criterion. **R** is the correlation. The R is capitalized to signify that the correlation is based on multiple predictors taken simultaneously. This is in contrast to the small r of simple linear regression. $\mathbf{R^2}$ is the predictable variance for the set of predictors with the criterion. Again, this is in contrast to the small r^2 of simple linear regression. The higher the R^2, the more of the criterion variance accounted for by the tested set of predictors. Here is the pertinent SPSS output for our data. On this output, R^2 is written as R Square.

Table 39.1 SPSS Multiple Regression Output Showing Predicted Variance

Model Summary

Model	R	R Square	Adjusted R Square	Std. Error of the Estimate
1	.901[a]	.811	.717	2.0073

a. Predictors: (Constant), Family Income (thousands), Age at First Arrest, Prior Arrests

From this output we can see that the correlation between the set of predictors and the criterion is .901. Recall from Module 37 that the correlation between prior arrests alone and sentence length was .82. Although this correlation already was quiet high, by adding the two additional predictors—age at first arrest and family income—we were able to increase the correlation by about 8 points.

From this output, we also can see that the predicted variance (R Square) is .811. Recall from Module 37 that the predictable variance using only the number of prior arrests as a single predictor was .67. Thus, by adding the two additional predictors we were able to increase the predicted variance by about 14 percentage points.

Table 39.2 shows the coefficients for the regression equation. Coefficients are weights, as previously discussed. They are shown both as unstandardized values and as standardized (*z* score) values.

Table 39.2 SPSS Multiple Regression Output Showing Coefficients for the Prediction Equation

Coefficients[a]

Model		Unstandardized Coefficients		Standardized Coefficients		
		B	Std. Error	Beta	t	Sig.
1	(Constant)	13.431	4.994		2.690	.036
	Number of Prior Arrests	1.053	.453	.588	2.325	.059
	Age at First Arrest	-.310	.147	-.424	-2.107	.080
	Family Income (thousands)	-.011	.044	-.058	-.253	.809

a. Dependent Variable: Years in Sentence

The weight for our constant and for our three predictor variables are listed in the left-most column. SPSS unfortunately uses the capital B designation, but these are the unstandardized weights, not the standardized weights. Notice how the coefficient for Prior Arrests has changed since Module 37. When Number of Prior Arrests was the only predictor, its coefficient was 1.468. Now its weight is less, only 1.053, because the other two predictors have acquired some of its predictive function, along with some of their own unique predictive functions.

Let's apply the tabled unstandardized coefficients to an example. Assume that a convict has three prior arrests, was first arrested at age 21, and has a family income of $42,000. Plugging the coefficients and this subject's data into the prediction equation, this convict's predicted prison sentence is 9.618 years, as follows.

$$\begin{aligned}\text{Predicted sentence} &= b_1(\text{prior arrests}) + b_2(\text{age at first arrest}) + b_3(\text{family income in thousand dollars}) + a \\ &= 1.053\ (3) + -.310\ (21) + -.011(42) + 13.431 \\ &= 3.159 - 6.51 - 0.462 + 13.431 \\ &= 9.618\end{aligned}$$

The unstandardized coefficients cannot tell us anything about the variables' relative importance in predicting the criterion because they are measured on different scales. To judge the predictors' relative importance, we instead must look at the values in the standardized coefficients, also called Beta, column. Here we can see that both Number of Prior Arrests and Age at First Arrest contribute heavily toward predicting prison sentence length (.588 and –.424, respectively), while Family Income contributes very little (–.058) to the prediction.

This interpretation is confirmed in the last two columns, where the individual predictors are subjected to *t* tests. Number of Prior Arrests and Age at First Arrest do predict prison sentence length, but only if we drop our confidence level to 94% and 92%, respectively ($\alpha = .059$ and .080). Clearly, the small sample size is hurting our ability to find statistical significance.

Family Income is not contributing to the prediction at all (α = .809). We might even consider dropping this variable in future studies, since it predicts so poorly. A word about dropping variables is appropriate. Recall that the criterion variance available to the third predictor is only that variance that was not already accounted for by the first two variables. It is the beauty of multiple regression that it removes intercorrelation for us. However, it is easy to think that Family Income, therefore, is not highly correlated with length of prison sentence and so makes a poor predictor of prison sentence. That would be a false conclusion from these data. It makes a poor predictor with these other two predictors present. But that might be because it highly correlates with one or both of those predictors and so its contribution to prison sentence length has already been accounted for by the time it is entered into the equation. The moral of this story is that predictors chosen for equation entry ought to be theory driven and, therefore, ought not to be dropped without further investigation. A simple Pearson r correlation between Family Income and prison sentence length would tell if Family Income has the potential to predict, as would including Family Income with a different set of predictors with which it might be less correlated.

Table 39.3 gives yet another way of looking at the data—an ANOVA table. The table format is the same as the one we worked with in Module 25. Here we see the familiar ANOVA variance allocations, except that the terms have been changed to reflect a regression approach. The "Between" sum of squares is now called "Regression" sum of squares, and the "Within" sum of squares is now called "Residual" sum of squares. Also, *df* refers to between the estimators, rather than between groups. Estimators include the predictors and the constant. The *df* is the usual $n - 1$. With three predictors and one constant, we have four estimators and thus three *df*. Because there is always one constant in multiple regression, the regression *df* ends up being the same as the number of predictor variables.

Table 39.3 SPSS Multiple Regression Output Showing ANOVA Summary Table

ANOVA

Model		Sum of Squares	Df	Mean Square	F	Sig.
1	Regression	104.029	3	34.676	8.606	.014[a]
	Residual	24.176	6	4.029		
	Total	128.205	9			

a. Predictors: (Constant), Family Income (thousands), Age at First Arrest, Prior Arrests

The F value for this set of predictors is 8.606. According to the final column, there is only a 1.4% chance that these predictors account for the large amount of criterion variance that they do by chance alone. Said another way, we can be 98.6% confident that the criterion variance predicted by this set of predictors is valid and reproducible.

Here's a final check on the ANOVA presentation. Recall that common variance is the proportion of criterion variance accounted for by the predictors. That is the same definition we gave for R^2. Thus, the two values ought to agree. Let's see if they do.

$$\begin{aligned}\text{Common Variance} &= \frac{\text{Regression } SS}{\text{Total } SS} \\ &= \frac{104.029}{128.205} \\ &= .811\end{aligned}$$

Now refer back to Table 39.1. We see that R squared is .811. Thus, ANOVA and regression give the same result.

CHECK YOURSELF!

Which gives the more accurate results–ANOVA or multiple regression?

PRACTICE

8. Examine the following output.

Model Summary

Model	R	R Square	Adjusted R Square	Std. Error of the Estimate
1	.842	.709	.688	1.7403

a. What proportion of the study's total variance do the predictors account for?
b. Which cells in an ANOVA summary table would give us the same information as we find in the R Square cell of this table?
c. What is the residual for this study?
d. Assume that we add an additional predictor. What might be a reasonable R Square to expect with the additional predictor in the full model? Suggest a value and defend your choice.

9. Examine the following output.

Model Summary

Model	R	R Square	Adjusted R Square	Std. Error of the Estimate
Full	.623	.388	.364	1.8061

a. What proportion of the study's total variance do the predictors account for?
b. Which cells in an ANOVA summary table would give us the same information as we find in the R Square cell of this table?
c. What is the residual for this study?
d. The R Square value for this study is rather low. Suggest reason(s) why that might be so.

10. Examine the following partial output.

Coefficients

	Unstandardized Coefficients		Standardized Coefficients		
Model	B	Std. Error	Beta	t	Sig.
1 (Constant)	.448				
Same Sex Parent's Longevity	1.018		.523		.013
Cigarettes Smoked Daily	-.269		-.228		.044
Minutes of Exercise Daily	.112		.106		.287

Dependent Variable: Longevityin Years

a. Which predictor variable best predicts longevity?

b. In this study, exercise is not a statistically significant predictor of longevity. Yet most research has shown exercise to be a significant predictor of longevity. Assume that the sample size on this study is adequate, so sample size cannot explain the result. What aspect of multiple regression research might explain this result?

c. Joe's father died at age 77. On an average day, Joe smokes 20 cigarettes and exercises for 30 minutes. What is Joe's predicted longevity in years?

11. Examine the following partial output.

Coefficients

Model	Unstandardized Coefficients		Standardized Coefficients	t	Sig.
	B	Std. Error	Beta		
1 (Constant)	78.241				
Number of Students in Class—beyond class size of 20	-.456		-.115		
Teacher's Years of Teaching Experience	.232		.362		
Budget Expenditure per Student—above $4,000(in hundred dollars)	.129		.062		

Dependent Variable: Class Averageon National Exam

a. Which predictor variable among this set best predicts class score on the National Exam?

b. The beta weight for class size is negative. What does this tell us about class size and student achievement?

c. The unstandardized weight for teaching experience is only about half that (in absolute value) of class size. Does this mean that teaching experience is only half as important as class size in predicting student success on the National Exam? Why or why not? What aspect of multiple regression might explain these relative weights?

d. A class has 27 students (that's 7 more than the 20 listed). The teacher has 12 years of teaching experience. The district expends $6,400 per student (that's $2,400 more than $4,000 listed, which is 24 of the hundred-dollar units more than $4,000 listed). What is this class's expected score on the National Exam?

12. Examine the following output.

ANOVA

Model		Sum of Squares	df	Mean Square	F	Sig.
1	Regression	378.55	3	126.183	15.572	.000
	Residual	210.67	26	8.103		
	Total	589.22	29			

a. What is R^2 for this study?

b. What is R for this study?

c. How many predictors were in this study?

d. How many subjects were in this study? Comment on this sample size.

e. Does this set of predictors predict a significant amount of criterion variance?

13. Examine the following output.

ANOVA

Model		Sum of Squares	df	Mean Square	F	Sig.
1	Regression	125.74	2	62.87	3.949	.055
	Residual	143.27	9	15.92		
	Total	269.01	11			

a. What is R^2 for this study?
b. What is R for this study?
c. How many predictors were in this study?
d. How many subjects were in this study? Comment on this sample size.
e. Does this set of predictors predict a significant amount of criterion variance?

The General Linear Model

In the last section we saw that ANOVA and Regression give the same result. This leads us to consider: Can we always use regression in lieu of ANOVA? Should we?

You probably didn't realize it, but we have been doing linear prediction all along. We just didn't call it that, nor did we set up the data to look linear. Nevertheless, ANOVA and *t* tests are each just special cases of linear prediction. Here is the explanation.

Throughout the course, I drilled home the logic of experimental hypothesis testing. Repeatedly, I asked, "What did you get?" versus "What did you expect to get?" The obtained difference from expectation went into the numerator of our test statistic. Because that difference from expectation was supposedly due to the independent variable, it was our Treatment effect. Then I asked, "Is that difference from expectation a lot? Or is it only a little?" To judge the magnitude of our obtained treatment effect, we had to scale it against some standardized unit, so that we could say how many of those standardized units our observed difference comprised. That standardized unit was a random error unit—standard deviation, standard error of the mean, standard error of the difference between the means, or within-group variance. In other words, it was our Error term. Thus, each of our test statistics was set up as a ratio of Variance due to Treatment over Variance due to Error.

$$\frac{\text{Treatment}}{\text{Error}}$$

But surely the Treatment variance is Predicted variance. It is not a random occurrence, but a very predictable occurrence, due to the applied treatment. Thus, we could have as well have written the ratio as

$$\frac{\text{Prediction}}{\text{Error}}$$

In correlation and linear prediction, in contrast, we have asked what proportion of the total score dispersion is accounted for by the predictors. The total is always one unit. The portion predicted by the set of predictors is a percentage of that unit. Thus, this approach scaled Predicted Variance over Total Variance rather than over Error Variance. However, we also learned that Total Variance consists of two portions: Predicted Variance and Error Variance. Thus, our scaling was as follows:

$$\frac{\text{Prediction}}{\text{Total}} = \frac{\text{Prediction}}{\text{Prediction} + \text{Error}}$$

Already, our equations are looking similar. But I can make it even clearer. In the ANOVA approach, the dependent variable is a *difference in dependent variable means between independent variable groups*. Do medicated depressives experience a different relief level than counseled depressives? Do smokers have a different number of health problems than nonsmokers? Our research design looks like this.

Number of Health Problems	
Smoke	*Don't Smoke*
8	3
6	4
5	2
3	3
7	1
$\text{Mean}_{\text{Smoke Group}}$	$\text{Mean}_{\text{Don't Smoke Group}}$

You know how to analyze these data. We test the difference in group means for statistical significance. Here is the result: $t = 3.20$, $\alpha = .013$.

Now let's keep the study the same and simply consider *group membership itself as a predictor*. We will phrase the question differently. Does the treatment group of which one is a member (medication vs. counseling) predict depression relief? Does smoking status (smokes or does not smoke) predict number of health problems? To answer these questions, we would collect data in the same way as we did above, but we would record it differently. Using the same data as above, here is the data setup, where 1 = *Smokes* and 0 = *Does Not Smoke.*

Group Membership	*Number of Health Problems*
1	8
1	6
1	5
1	3
1	7
0	3
0	4
0	2
0	3
0	1

You know how to analyze these data, as well. Here is the regression result: $R = .749$, $R^2 = .561$, t for the one predictor = 3.20, $F = 10.240$, $\alpha = .013$.

Not only do the two types of analyses—*t* test and linear regression—give the same result, but, if you have been using SPSS, you might recognize the data set-up. Recall that, even for *t* tests and ANOVA, SPSS required that we code group membership and set up our data in the column format above. At the time, the data setup was jarringly different from what you expected, given the textbook examples. SPSS required it that way because it analyzed the data through the regression approach regardless of which statistic we asked it for. It altered the output to fit our request, but it analyzed it linearly regardless of what we requested.

This short demonstration and discussion proves that experimental approaches to data analysis—*t* tests, ANOVA, and the like—are merely special cases of a linear regression model. The whole set of linear approaches is called the **General Linear Model**, usually abbreviated as

GLM. All of the GLM statistics scale treatment variance over total variance, rather than treatment variance over error variance. Thus, all of them report outcomes as proportions of unit variance accounted for, rather than as standardized differences from a mean.

The GLM is so flexible that we can use it to analyze data from true experiments, post hoc studies, and prediction studies; we can use it on ratio, interval, ordinal, or nominal-level data; we can use it with continuously scored or dichotomously scored data; and we can use it with single or multiple predictors (IVs) and single or multiple criteria (DVs). In short, we can virtually always use it. It also is methodologically more straightforward once you master it, requiring only variations on the theme for differences between designs, rather than whole different statistics for each different type of design. For this reason, it is the most widely used analytic method today. You can expect your next statistics course, if you take one, to be largely about the General Linear Model, merely adding in computational variations as the underlying research designs become more complicated.

SPSS Connection

Download the file **data_years in sentence and prior arrest age at first arrest family income.sav** from www.sagepub.com/steinberg2e. These data are used in the textbook example.

Alternatively, manually enter the three scores for the 10 subjects in Module 39 into the SPSS **Data View** spreadsheet. Data entry for a multiple regression follows the output explained in this module: one column for scores on the criterion variable and three columns for scores on the three predictor variables. Thus, enter the data as follows:

12.2	3	18	40
8.4	2	23	42
8.0	1	20	58
14.2	5	15	18
6.0	1	33	44
4.4	0	21	76
6.2	4	25	21
16.1	7	17	27
9.3	4	24	72

Click on the **Variable View** tab to define the variable. Name the first variable **yearssen**, set the decimal at **1** (not at 0), and label the variable as **Years in Sentence**. Name the second variable **prioarr**, set the decimal at **0**, and label the variable as **Number of Prior Arrests**. Name the third variable **agearres**, set the decimal at **0**, and label the variable as **Age at First Arrest**. Name the fourth variable **famincom**, set the type as **dollar**, set the width at **4**, and label the variable **Family Income**.

If the file is not already in **Data View**, click that tab in the lower left of the screen.

In the toolbar at the top of the screen, click on **Analyze**, then **Regression**, then **Linear**. Highlight the variable **Years in Sentence** in the left window and click on the **arrow** before the **Dependent** window on the right to send the variable into that window. In a prediction study without a causal variable, this is the criterion variable, although SPSS refers to it as a dependent variable. Multiple regression can be used for experimental studies as well; in that case, the criterion variable would be a dependent variable. Highlight the **Prior Arrests** variable, click on the **arrow** before the **Independent** window on the right to send the variable into that window. Do the same with the **Age at First Arrest** and **Family Income** variables. In a prediction study without a causal variable, these are the predictor variables, although SPSS refers to them as independent variables. Multiple regression can be used for experimental studies as well; in that case, the predictor variables would be independent variables. Keep the method at the default **Enter**. Click **OK**. This is what you will see.

Regression

Variables Entered/Removed[b]

Model	Variables Entered	Variables Removed	Method
1	Family Income (thousands), Age at First Arrest, Number of Prior Arrests[a]		Enter

a. All requested variables entered.

b. Dependent Variable: Years in Sentence

Model Summary

Model	R	R Square	Adjusted R Square	Std. Error of the Estimate
1	.901[a]	.811	.717	2.0073

a. Predictors: (Constant), Family Income (thousands), Age at First Arrest, Number of Prior Arrests

ANOVA[b]

Model		Sum of Squares	df	Mean Square	F	Sig.
1	Regression	104.029	3	34.676	8.606	.014[a]
	Residual	24.176	6	4.029		
	Total	128.205	9			

a. Predictors: (Constant), Family Income (thousands), Age at First Arrest, Number of Prior Arrests

b. Dependent Variable: Years in Sentence

Coefficients[a]

Model		Unstandardized Coefficients		Standardized Coefficients		
		B	Std. Error	Beta	t	Sig.
1	(Constant)	13.431	4.994		2.690	.036
	Number of Prior Arrests	1.053	.453	.588	2.325	.059
	Age at First Arrest	-.310	.147	-.424	-2.107	.080
	Family Income (thousands)	-.011	.044	-.058	-.253	.809

a. Dependent Variable: Years in Sentence

Unlike prior modules, discussion of this SPSS output took place within the module itself. Therefore, no further explanation is needed here.

Visit the study site at www.sagepub.com/steinberg2e for practice quizzes and other study resources.

PART XVII

Review

"Say It Again, Sam"

40

Selecting the Appropriate Analysis

You have learned how to display and interpret data in tables and graphs, calculate and interpret various descriptive statistics, develop and word hypotheses, sample subjects, compute probabilities, calculate and interpret inferential statistics for data of various scales and having differing numbers of variables and treatment groups, determine the degree and direction of the relationship between two variables, and predict scores on one variable from scores on another variable. You have certainly learned a lot! It is, of course, impossible to summarize in a single module all that you have learned throughout this textbook. It is appropriate, however, to review which displays and statistics to use under which conditions. This module provides an organizational structure for making that decision.

The easiest way to arrive at the correct display or statistic for a particular study is through flowcharts. However, to use the flowcharts, you must first answer several questions about your study. These include the following:

- What is my study's purpose? Am I seeking to display or report data, establish causation, or establish a relationship?
- What is my data's measurement scale?
- How many groups of subjects do I have?
- How many variables do I have?

Review of Descriptive Methods

Descriptive methods are used to simplify a set of scores for reporting purposes. This simplification may be accomplished via tables and graphs or via summary statistics. It is appropriate and helpful to present data descriptively even when a study is inferential. Figure 40.1 is a flowchart for selecting the appropriate descriptive display or statistic for a particular study. Narrative guidelines follow the flowchart.

Tables and Graphs

Frequency Tables

Frequency tables show how many cases fall at each score. They provide a view of the entire set of scores while at the same time indicating which scores are the most and the least frequent. Score frequencies may be reported as numbers or as percentages and for single

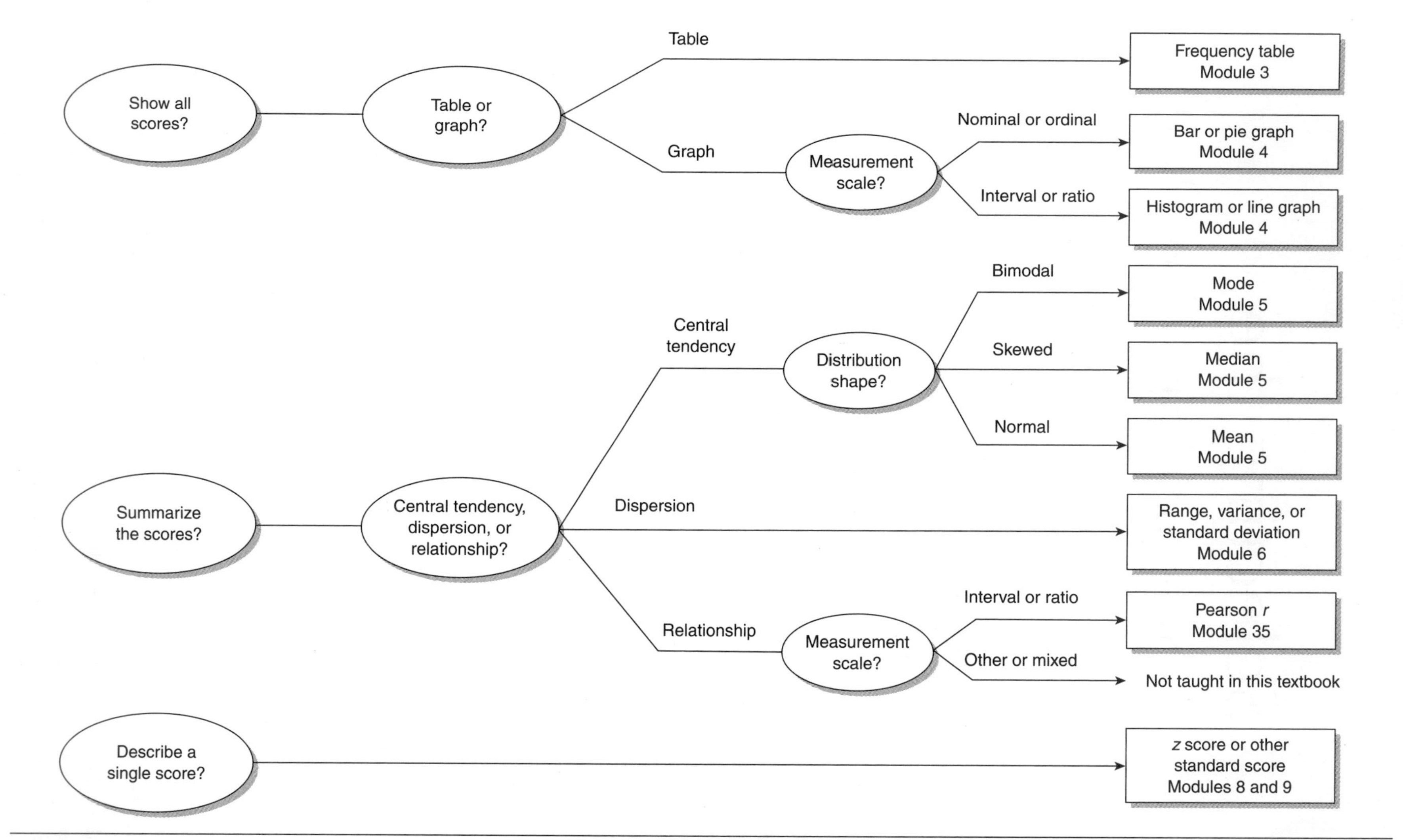

Figure 40.1 Flowchart for Selecting a Descriptive Display or Statistic

scores or for grouped scores. When grouping scores, the deciding principle is to achieve clarity of the data so as not to obscure or exaggerate the data trends.

Graphs

Graphs give a visual picture of the data. This allows us to see not only frequencies but also the general shape of the data. We can also see if there are outliers. Shape is important because some statistics are appropriate only if the data assume a normal curve. For example, skewed or multimodal data distort the mean.

Most univariate graphs plot scores on the *X*-axis and frequencies on the *Y*-axis. Histograms or frequency polygons are appropriate for continuously scored data. Bar graphs or pie graphs are appropriate for discretely scored or nominally scaled data.

Bivariate graphs, also called scatterplots, show scores on Variable 1 on the *X*-axis and scores on Variable 2 on the *Y*-axis. Frequency is not shown. Individual cases are depicted within the interior of the graph as dots or symbols where the scores on *X* and *Y* intersect.

Descriptive Statistics

Central Tendency

Measures of central tendency summarize in a single number the general location of the scores. Measures include the mode, median, and mean. The mode is the most frequent score. It is the easiest to calculate but also the least stable from one sample to the next. Always report the modes if the data are bimodal. The median is the midpoint. It is a more appropriate measure to report than the mean when the data are seriously skewed. The mean, or average, is the balance point of the data. It is appropriate to report the mean as long as the data take on a roughly normal distribution. The mean is a necessary component in many parametric inferential tests.

Dispersion

Measures of dispersion summarize in a single number the amount of spread in the scores. Measures include the range, variance, and standard deviation. The range is the difference between the highest and the lowest scores. It is not only easy to calculate but also the least stable from one sample to the next. The variance is the average area distance from the mean. Because it is not a linear measure, it is not very useful in descriptive statistics. However, it is useful for depicting the common variance (overlap) between two sets of scores. It is necessary to first calculate the variance in order to calculate the standard deviation. The standard deviation is the average linear distance from the mean. It is appropriate to calculate the standard deviation if there are at least 30 subjects and the data are approximately normally distributed. A normal distribution for an infinite sample size consists of roughly 6 standard deviation units. Either the variance or the standard deviation is a necessary component in many parametric inferential tests of differences between groups.

Standard Scores

Standard scores rescale raw scores into standardized units. A z score expresses the score in standard deviation units. Thus, it tells how many standard deviation units a particular score is above or below the mean. Once raw scores are converted to z scores, we can use the normal curve table to find the percentage of scores falling above or below any given score. Many other standard scores—T scores, deviation IQ scores, and CEEB (College Entrance Examination

Board) scores—are linear transformations of z scores. Such scores are relative to a new mean and standard deviation but fall at the same locations within a normal curve as the original z scores fell. We use standard scores to describe the location of a particular score.

Correlation

A correlation coefficient tells the amount and direction of relationship between two sets of scores—that is, between two variables. It is the calculated statistic for a bivariate scatterplot. A correlation coefficient can be interpreted either descriptively (its value and direction for this particular set of subjects) or inferentially (its statistical significance given a hypothesized null relationship within the populations from which these subjects were drawn). A Pearson r is the appropriate coefficient to use for data measured on at least an interval scale. For data measured nominally or ordinally, other correlation coefficients are more appropriate. Once the mean and standard deviation are calculated for each set of scores and the correlation between the two sets of scores is calculated, those values can be used to predict the future score on one variable given a known score on the other variable. However, correlation, even a high one, does not imply causation.

Review of Inferential Methods

Inferential methods are used to draw a probabilistic conclusion about a population or populations based on sample data. That is, we draw a conclusion about a larger group of subjects than those actually in the study. Typically, the population about which we draw a conclusion received a treatment, while another population did not. The conclusion, therefore, is not about the members of the population but about the treatment in that population.

In the social sciences, the treatment of interest cannot always be applied; sometimes it preexists. In those cases, we use the same inferential statistics but control for the influence of extraneous variables by matching subjects or by more sophisticated statistical methods than those covered in this textbook. Figure 40.2 is a flowchart for selecting the appropriate inferential statistic for a particular study. Narrative guidelines are described in the following sections.

Parametric Test Statistics

Parametric tests compare sample statistics with population parameters. The data must be on at least an interval scale.

One-Sample Tests

In a one-sample test, we compare a sample statistic (usually the mean) with the corresponding known or hypothesized population parameter (again, usually the mean). If we know the standard deviation of the population, we calculate a normal deviate Z test. If we don't know the population standard deviation or if the sample size is small (say, less than 30), we calculate a one-sample t test. In either case, if the difference between the sample mean and the population mean is greater than mere random error, we reject the null hypothesis of no difference between their means and conclude instead that the sample is significantly different from the population (and, hence, probably a representative sample from a population other than the one tested).

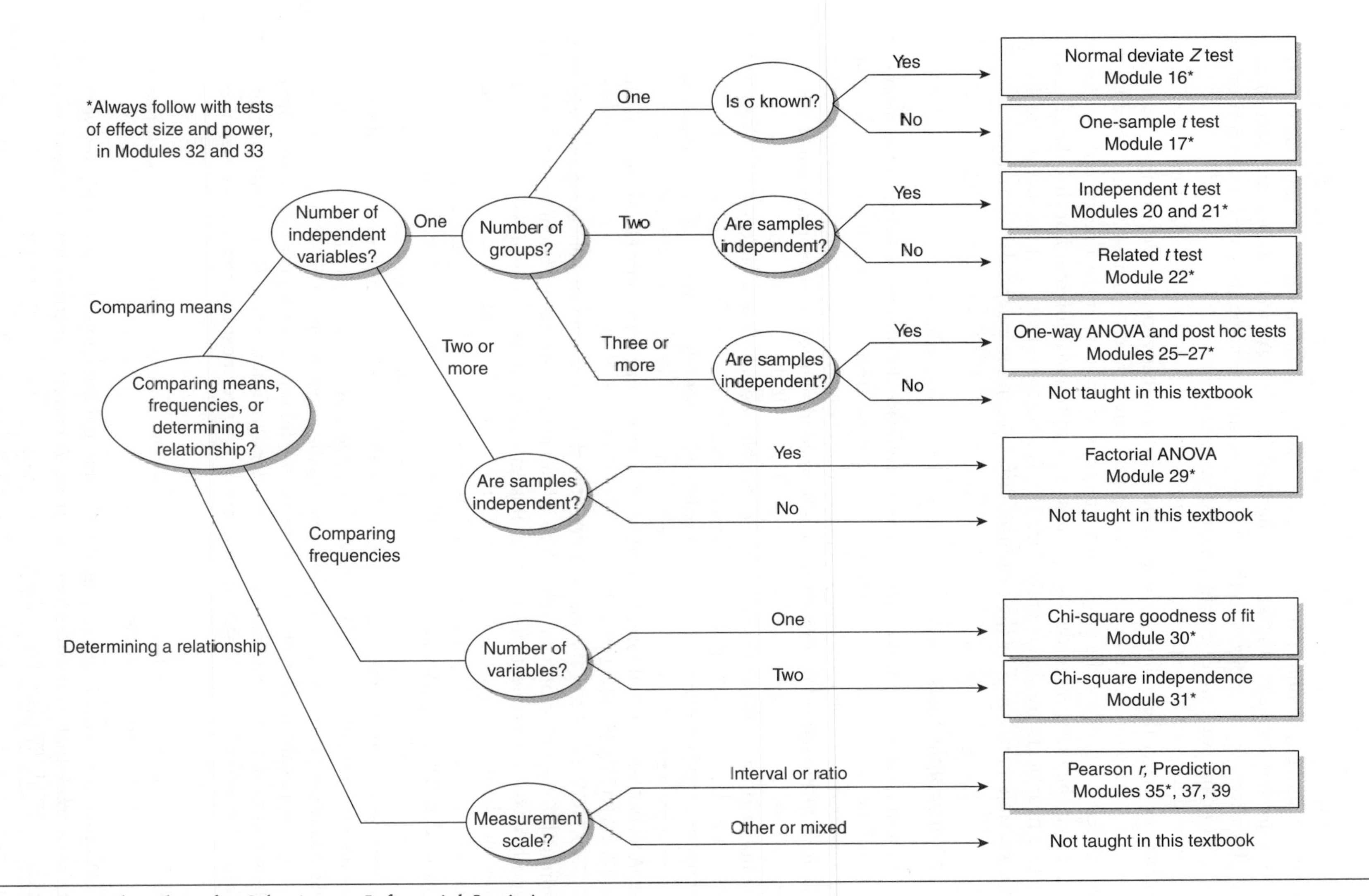

Figure 40.2 Flowchart for Selecting an Inferential Statistic

Two-Sample Tests

In a two-sample test, we compare the difference between the means of two samples (usually a treated group and an untreated group) with the difference expected between their corresponding populations (which, under the null hypothesis, is 0). If the subjects in the two groups are not the same and are not matched in any way, we calculate an independent-samples t test. If the subjects in the two groups are the same (repeated measures) or are matched in any way, we calculate a related-samples t test.

If the difference between the sample means is farther from 0 than random error would predict, we reject the null hypothesis of no difference between the two groups' means and conclude instead that the two groups are significantly different from each other. Because the only difference between the groups in a true experiment is their treatments, our conclusion is not that the members of the two groups are significantly different but that the effects of the two treatments are significantly different. We conclude that the difference in the dependent variable was caused by the treatment difference in the independent variable.

Multisample Tests

In a multisample test, we simultaneously compare the difference between the means of several differently treated samples (i.e., three or more treatment groups) with the difference expected between their corresponding populations (whose difference, under the null hypothesis, is 0). If the several treatments are levels or conditions of a single independent variable, we calculate a one-way ANOVA (one-way F test). If the several treatments are levels or conditions of more than one independent variable, we calculate a factorial ANOVA. If the subjects in the several groups are not the same and are not matched in any way, we calculate the one-way or factorial ANOVA as an independent-samples ANOVA. If the subjects in the two groups are the same or are matched in any way, we calculate the one-way or factorial ANOVA as a related-samples ANOVA. (You did not learn how to calculate a related-samples ANOVA in this textbook.)

In ANOVA, we test the mean differences through a ratio of between-group variance (summed across all treatment groups) to within-group variance (random error). If the null hypothesis of no treatment effect is true, this ratio of treatment to error will be +1.00. If the difference between the sample means is farther from +1.00 than random error would predict, we reject the null hypothesis of no difference between the groups' means and conclude instead that the groups are significantly different from each other. Because the only difference between the groups in a true experiment is their treatments, our conclusion is not that the members of the several groups are significantly different but that the effects of the several treatments are significantly different. We conclude that the difference in the dependent variable was caused by the treatment difference in the independent variable.

If a significant treatment effect is found, we follow up with post hoc tests to find out which pair or pairs of treatments account for the significant overall F. A Scheffé post hoc test is more conservative in its distribution of error. A Tukey post hoc test is less conservative in its distribution of error.

Correlation and Prediction

A correlation coefficient tells us the amount and direction of relationship between two sets of scores—that is, between two variables. A correlation coefficient can be interpreted either descriptively (its value and direction for this particular set of subjects) or inferentially (its statistical significance given a hypothesized null relationship within the populations from which these subjects were drawn). A Pearson r is the appropriate coefficient to use for data measured on at least an interval scale. For data measured nominally or ordinally, other correlation coefficients are more appropriate. You did not learn how to calculate these other correlation coefficients in this textbook.

Correlation coefficients range from –1.00 to +1.00. The null hypothesis states that there is no relationship—that is, the correlation is 0.00. If the correlation coefficient is farther from 0.00 than random error would predict, we reject the null hypothesis of no correlation between the two variables and conclude instead that the two variables are related.

Correlation coefficients are useful for prediction. In simple prediction, also known as simple regression, a single predictor variable is used to predict the criterion. Results are reported as the proportion of the criterion variance accounted for by the predictor variable. The residual is the unpredicted portion.

In multiple regression, more than one predictor variable is used to predict the criterion. Results are reported as the proportion of criterion variance accounted for by the entire set of predictor variables. Model testing is the process of comparing predicted criterion variance for each predictor/independent variable to the predicted criterion variance for the entire set of predictors, to see which variables are making significant contributions.

Multiple regression is part of the General Linear Model (GLM), which can be used to report results for data on any measurement scale and for not only correlational studies but also experimental studies. In experimental studies, results are the proportion of the dependent variable accounted for by the entire set of independent variables. Because of its generality, the GLM is the preferred analytical method for most research and, hence, is the basis for most upper-level statistics courses.

Nonparametric Test Statistics

Nonparametric tests compare the frequencies, ranks, shapes, or patterns of sample data with expected or hypothesized frequencies, ranks, shapes, or patterns. There are no population parameters; hence, we calculate no summary statistics for comparison. There are different nonparametric statistics for data measured on different measurement scales, having different numbers of groups, having different numbers of variables, and asking different research questions. Examples of these questions are as follows: Are the ranks of subjects on two variables similar? Is the shape of the data in two groups similar? Is the pattern of responses in two groups similar? Are the frequencies of responses of two groups similar?

In this textbook, you learned only one nonparametric test statistic, the chi-square, which answers only one of those questions. A chi-square test determines if the observed frequencies in a sample are different from the expected frequencies. Data are categorical—that is, nominal. Frequencies are tallied for each nominal category.

Chi-Square Goodness of Fit

A chi-square (χ^2) goodness-of-fit test determines whether the frequencies for a single variable in a sample are representative of—or a good "fit" with—the frequencies for that same variable in another group. The null hypothesis is that the observed frequencies in the sample are as expected across all categories of the variable. Because total numbers across all categories typically differ for the focal sample and the comparison group, expected frequencies in the focal group are found by theory, which is sometimes by precedent but is typically by the percentage in the comparison group. If the observed frequencies in the focal group are significantly different than expected, we reject the null hypothesis and conclude that the focal sample is not a good fit with—and, hence, probably was not drawn from—the comparison group.

Chi-Square Test of Independence

A chi-square (χ^2) test of independence determines whether two variables operate independently. The null hypothesis is that the two variables operate independently. Expected frequencies for cells within the two-variable matrix are found by multiplying row and column totals and dividing by total *N*. This allots expected frequencies in each variable independent of the other variable. If the observed and expected frequencies are significantly different, we reject the null hypothesis and conclude that the two variables do not operate independently.

Effect Size and Power

Finally, remember that statistical significance is necessary but not sufficient for accepting the utility of a treatment. Always follow a test of a hypothesis with the appropriate measures of the study's effect size and power. Different effect-size statistics are appropriate for each hypothesis-testing statistic. Power can be calculated with a formula, but is typically determined by consulting a table of calculated values.

PRACTICE

1. Assume that you are studying the effect of noise level on memory. Your study involves manipulating the level of noise while subjects try to memorize a list of words and then measuring the number of words the subjects correctly remember. For *each* statistical method listed below: (a) state the necessary or appropriate number of groups in the study, (b) state the necessary or appropriate measurement scale for the data, and (c) draw a diagram of one possible design for the study:

   A Pearson *r*

   A *t* test

   An *F* test

   A χ^2 test

2. Read the description of each of the following research studies and then decide which statistic would be the best one to determine the statistical significance of the results. Select from the following choices: Pearson *r*, *t* test, *F* test, and χ^2.
   a. One group of children classified as having behavioral problems is put into a behavioral change program for a month. A second group of children with similar behavioral problems is not put into any special treatment program. You want to see whether or not the behavioral change program affects the number of children who exhibit socially unacceptable behavior at the end of the study.
   b. While waiting for a friend in a restaurant lobby, a bored statistician recorded the number of male and female customers who helped themselves to a bowl of green mint candies or a bowl of pink mint candies while leaving the restaurant. His goal was to see if men and women differed in their choice of candy color.

c. A researcher wonders if height is related to running speed, so he obtains the running speeds and the heights of all runners in a national marathon race.

d. Two randomly formed groups are asked to read a number of passages. One group is then taught strategies for remembering facts and concepts, while the other group is not. At the end of the study, both groups are tested on their retention of the material in the reading passages, and their retention level scores are compared to determine whether having been taught the memory strategies helped their retention.

e. Children who are small for their age are grouped and treated in different ways: One group receives a growth hormone drug, one group receives nutritional supplements (vitamins, minerals, etc.), one group receives physical therapy and stretching exercises, and the last group receives no special treatment. You want to see whether or not the type of treatment received affects height increase.

Visit the study site at www.sagepub.com/steinberg2e for practice quizzes and other study resources.

Appendix A

Normal Curve Table

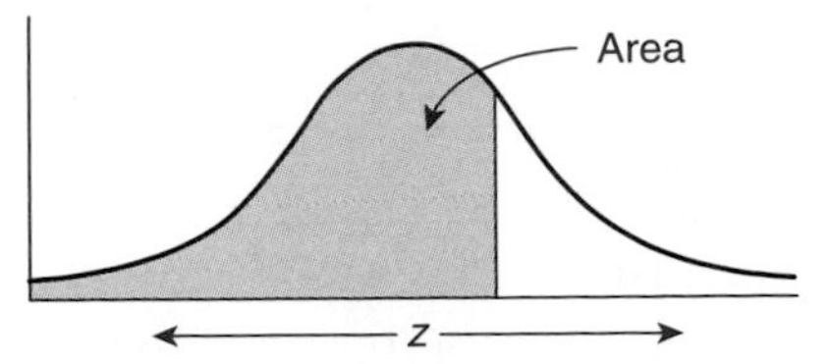

z	*Area*	*z*	*Area*	*z*	*Area*	*z*	*Area*
–3.00	.0013						
–2.99	.0014	–2.64	.0041	–2.29	.0110	–1.94	.0262
–2.98	.0014	–2.63	.0043	–2.28	.0113	–1.93	.0268
–2.97	.0015	–2.62	.0044	–2.27	.0116	–1.92	.0274
–2.96	.0015	–2.61	.0045	–2.26	.0119	–1.91	.0281
–2.95	.0016	–2.60	.0047	–2.25	.0122	–1.90	.0287
–2.94	.0016	–2.59	.0048	–2.24	.0125	–1.89	.0294
–2.93	.0017	–2.58	.0049	–2.23	.0129	–1.88	.0301
–2.92	.0018	–2.57	.0051	–2.22	.0132	–1.87	.0307
–2.91	.0018	–2.56	.0052	–2.21	.0136	–1.86	.0314
–2.90	.0019	–2.55	.0054	–2.20	.0139	–1.85	.0322
–2.89	.0019	–2.54	.0055	–2.19	.0143	–1.84	.0329
–2.88	.0020	–2.53	.0057	–2.18	.0146	–1.83	.0336
–2.87	.0021	–2.52	.0059	–2.17	.0150	–1.82	.0344
–2.86	.0021	–2.51	.0060	–2.16	.0154	–1.81	.0351
–2.85	.0022	–2.50	.0062	–2.15	.0158	–1.80	.0359
–2.84	.0023	–2.49	.0064	–2.14	.0162	–1.79	.0367
–2.83	.0023	–2.48	.0066	–2.13	.0166	–1.78	.0375
–2.82	.0024	–2.47	.0068	–2.12	.0170	–1.77	.0384
–2.81	.0025	–2.46	.0069	–2.11	.0174	–1.76	.0392
–2.80	.0026	–2.45	.0071	–2.10	.0179	–1.75	.0401
–2.79	.0026	–2.44	.0073	–2.09	.0183	–1.74	.0409
–2.78	.0027	–2.43	.0075	–2.08	.0188	–1.73	.0418
–2.77	.0028	–2.42	.0078	–2.07	.0992	–1.72	.0427
–2.76	.0029	–2.41	.0080	–2.06	.0197	–1.71	.0436
–2.75	.0030	–2.40	.0082	–2.05	.0202	–1.70	.0446
–2.74	.0031	–2.39	.0084	–2.04	.0207	–1.69	.0455
–2.73	.0032	–2.38	.0087	–2.03	.0212	–1.68	.0465
–2.72	.0033	–2.37	.0089	–2.02	.0217	–1.67	.0475
–2.71	.0034	–2.36	.0091	–2.01	.0222	–1.66	.0485
–2.70	.0035	–2.35	.0094	–2.00	.0228	–1.65	.0495
–2.69	.0036	–2.34	.0096	–1.99	.0233	–1.64	.0505
–2.68	.0037	–2.33	.0099	–1.98	.0239	–1.63	.0516
–2.67	.0038	–2.32	.0102	–1.97	.0244	–1.62	.0526
–2.66	.0039	–2.31	.0104	–1.96	.0250	–1.61	.0537
–2.65	.0040	–2.30	.0107	–1.95	.0256	–1.60	.0548

(Continued)

(Continued)

z	Area	z	Area	z	Area	z	Area
−1.59	.0559	−1.19	.1170	−0.79	.2148	−0.39	.3483
−1.58	.0571	−1.18	.1190	−0.78	.2177	−0.38	.3520
−1.57	.0582	−1.17	.1210	−0.77	.2206	−0.37	.3557
−1.56	.0594	−1.16	.1230	−0.76	.2236	−0.36	.3594
−1.55	.0606	−1.15	.1251	−0.75	.2266	−0.35	.3632
−1.54	.0618	−1.14	.1271	−0.74	.2296	−0.34	.3669
−1.53	.0630	−1.13	.1292	−0.73	.2327	−0.33	.3707
−1.52	.0643	−1.12	.1314	−0.72	.2358	−0.32	.3745
−1.51	.0655	−1.11	.1335	−0.71	.2389	−0.31	.3783
−1.50	.0668	−1.10	.1357	−0.70	.2420	−0.30	.3821
−1.49	.0681	−1.09	.1379	−0.69	.2451	−0.29	.3859
−1.48	.0694	−1.08	.1401	−0.68	.2483	−0.28	.3897
−1.47	.0708	−1.07	.1423	−0.67	.2514	−0.27	.3936
−1.46	.0721	−1.06	.1446	−0.66	.2546	−0.26	.3974
−1.45	.0735	−1.05	.1469	−0.65	.2578	−0.25	.4013
−1.44	.0749	−1.04	.1492	−0.64	.2611	−0.24	.4052
−1.43	.0764	−1.03	.1515	−0.63	.2643	−0.23	.4090
−1.42	.0778	−1.02	.1539	−0.62	.2676	−0.22	.4129
−1.41	.0793	−1.01	.1562	−0.61	.2709	−0.21	.4168
−1.40	.0808	−1.00	.1587	−0.60	.2743	−0.20	.4207
−1.39	.0823	−0.99	.1611	−0.59	.2776	−0.19	.4247
−1.38	.0838	−0.98	.1635	−0.58	.2810	−0.18	.4286
−1.37	.0853	−0.97	.1660	−0.57	.2843	−0.17	.4325
−1.36	.0869	−0.96	.1685	−0.56	.2877	−0.16	.4364
−1.35	.0885	−0.95	.1711	−0.55	.2912	−0.15	.4404
−1.34	.0901	−0.94	.1736	−0.54	.2946	−0.14	.4443
−1.33	.0918	−0.93	.1762	−0.53	.2981	−0.13	.4483
−1.32	.0934	−0.92	.1788	−0.52	.3015	−0.12	.4522
−1.31	.0951	−0.91	.1814	−0.51	.3050	−0.11	.4562
−1.30	.0968	−0.90	.1841	−0.50	.3085	−0.10	.4602
−1.29	.0985	−0.89	.1867	−0.49	.3121	−0.09	.4641
−1.28	.1003	−0.88	.1894	−0.48	.3156	−0.08	.4681
−1.27	.1020	−0.87	.1922	−0.47	.3192	−0.07	.4721
−1.26	.1038	−0.86	.1949	−0.46	.3228	−0.06	.4761
−1.25	.1056	−0.85	.1977	−0.45	.3264	−0.05	.4801
−1.24	.1075	−0.84	.2005	−0.44	.3300	−0.04	.4840
−1.23	.1093	−0.83	.2033	−0.43	.3336	−0.03	.4880
−1.22	.1112	−0.82	.2061	−0.42	.3372	−0.02	.4920
−1.21	.1131	−0.81	.2090	−0.41	.3409	−0.01	.4960
−1.20	.1151	−0.80	.2119	−0.40	.3446	0.00	.5000

z	*Area*	*z*	*Area*	*z*	*Area*	*z*	*Area*
0.01	.5040	0.41	.6591	0.81	.7910	1.21	.8869
0.02	.5080	0.42	.6628	0.82	.7939	1.22	.8888
0.03	.5120	0.43	.6664	0.83	.7967	1.23	.8907
0.04	.5160	0.44	.6700	0.84	.7995	1.24	.8925
0.05	.5199	0.45	.6736	0.85	.8023	1.25	.8944
0.06	.5239	0.46	.6772	0.86	.8051	1.26	.8962
0.07	.5279	0.47	.6808	0.87	.8078	1.27	.8980
0.08	.5319	0.48	.6844	0.88	.8106	1.28	.8997
0.09	.5359	0.49	.6879	0.89	.8133	1.29	.9015
0.10	.5398	0.50	.6915	0.90	.8159	1.30	.9032
0.11	.5438	0.51	.6950	0.91	.8186	1.31	.9049
0.12	.5478	0.52	.6985	0.92	.8212	1.32	.9066
0.13	.5517	0.53	.7019	0.93	.8238	1.33	.9082
0.14	.5557	0.54	.7054	0.94	.8264	1.34	.9099
0.15	.5596	0.55	.7088	0.95	.8289	1.35	.9115
0.16	.5636	0.56	.7123	0.96	.8315	1.36	.9131
0.17	.5675	0.57	.7157	0.97	.8340	1.37	.9147
0.18	.5714	0.58	.7190	0.98	.8365	1.38	.9162
0.19	.5753	0.59	.7224	0.99	.8389	1.39	.9177
0.20	.5793	0.60	.7257	1.00	.8413	1.40	.9192
0.21	.5832	0.61	.7291	1.01	.8438	1.41	.9207
0.22	.5871	0.62	.7324	1.02	.8461	1.42	.9222
0.23	.5910	0.63	.7357	1.03	.8485	1.43	.9236
0.24	.5948	0.64	.7389	1.04	.8508	1.44	.9251
0.25	.5987	0.65	.7422	1.05	.8531	1.45	.9265
0.26	.6026	0.66	.7454	1.06	.8554	1.46	.9279
0.27	.6064	0.67	.7486	1.07	.8577	1.47	.9292
0.28	.6103	0.68	.7517	1.08	.8599	1.48	.9306
0.29	.6141	0.69	.7549	1.09	.8621	1.49	.9319
0.30	.6179	0.70	.7580	1.10	.8643	1.50	.9332
0.31	.6217	0.71	.7611	1.11	.8665	1.51	.9345
0.32	.6255	0.72	.7642	1.12	.8686	1.52	.9357
0.33	.6293	0.73	.7673	1.13	.8708	1.53	.9370
0.34	.6331	0.74	.7704	1.14	.8729	1.54	.9382
0.35	.6368	0.75	.7734	1.15	.8749	1.55	.9394
0.36	.6406	0.76	.7764	1.16	.8770	1.56	.9406
0.37	.6443	0.77	.7794	1.17	.8790	1.57	.9418
0.38	.6480	0.78	.7823	1.18	.8810	1.58	.9429
0.39	.6517	0.79	.7852	1.19	.8830	1.59	.9441
0.40	.6554	0.80	.7881	1.20	.8849	1.60	.9452

(Continued)

(Continued)

z	*Area*	*z*	*Area*	*z*	*Area*	*z*	*Area*
1.61	.9463	1.96	.9750	2.31	.9896	2.66	.9961
1.62	.9474	1.97	.9756	2.32	.9898	2.67	.9962
1.63	.9484	1.98	.9761	2.33	.9901	2.68	.9963
1.64	.9495	1.99	.9767	2.34	.9904	2.69	.9964
1.65	.9505	2.00	.9772	2.35	.9906	2.70	.9965
1.66	.9515	2.01	.9778	2.36	.9909	2.71	.9966
1.67	.9525	2.02	.9783	2.37	.9911	2.72	.9967
1.68	.9535	2.03	.9788	2.38	.9913	2.73	.9968
1.69	.9545	2.04	.9793	2.39	.9916	2.74	.9969
1.70	.9554	2.05	.9798	2.40	.9918	2.75	.9970
1.71	.9564	2.06	.9803	2.41	.9920	2.76	.9971
1.72	.9573	2.07	.9808	2.42	.9922	2.77	.9972
1.73	.9582	2.08	.9812	2.43	.9925	2.78	.9973
1.74	.9591	2.09	.9817	2.44	.9927	2.79	.9974
1.75	.9599	2.10	.9821	2.45	.9929	2.80	.9974
1.76	.9608	2.11	.9826	2.46	.9931	2.81	.9975
1.77	.9616	2.12	.9830	2.47	.9932	2.82	.9976
1.78	.9625	2.13	.9834	2.48	.9934	2.83	.9977
1.79	.9633	2.14	.9838	2.49	.9936	2.84	.9977
1.80	.9641	2.15	.9842	2.50	.9938	2.85	.9978
1.81	.9649	2.16	.9846	2.51	.9940	2.86	.9979
1.82	.9656	2.17	.9850	2.52	.9941	2.87	.9979
1.83	.9664	2.18	.9854	2.53	.9943	2.88	.9980
1.84	.9671	2.19	.9857	2.54	.9945	2.89	.9981
1.85	.9678	2.20	.9861	2.55	.9946	2.90	9981
1.86	9686	2.21	.9864	2.56	.9948	2.91	.9982
1.87	.9693	2.22	.9868	2.57	.9949	2.92	.9982
1.88	.9699	2.23	.9871	2.58	.9951	2.93	.9983
1.89	.9706	2.24	.9875	2.59	.9952	2.94	.9984
1.90	.9713	2.25	.9878	2.60	.9953	2.95	.9984
1.91	.9719	2.26	.9881	2.61	.9955	2.96	.9985
1.92	.9726	2.27	.9884	2.62	.9956	2.97	.9985
1.93	.9732	2.28	.9887	2.63	.9957	2.98	.9986
1.94	.9738	2.29	.9890	2.64	.9959	2.99	.9986
1.95	.9744	2.30	.9893	2.65	.9960	3.00	.9987

Appendix B

Binomial Table

N	No. of p or q Events	p or q									
		.05	.10	.15	.20	.25	.30	.35	.40	.45	.50
1	0	.9500	.9000	.8500	.8000	.7500	.7000	.6500	.6000	.5500	.5000
	1	.0500	.1000	.1500	.2000	.2500	.3000	.3500	.4000	.4500	.5000
2	0	.9025	.8100	.7225	.6400	.5625	.4900	.4225	.3600	.3025	.2500
	1	.0950	.1800	.2550	.3200	.3750	.4200	.4550	.4800	.4950	.5000
	2	.0025	.0100	.0225	.0400	.0625	.0900	.1225	.1600	.2025	.2500
3	0	.8574	.7290	.6141	.5120	.4219	.3430	.2746	.2160	.1664	.1250
	1	.1354	.2430	.3251	.3840	.4219	.4410	.4436	.4320	.4084	.3750
	2	.0071	.0270	.0574	.0960	.1406	.1890	.2389	.2880	.3341	.3750
	3	.0001	.0010	.0034	.0080	.0156	.0270	.0429	.0640	.0911	.1250
4	0	.8145	.6561	.5220	.4096	.3164	.2401	.1785	.1296	.0915	.0625
	1	.1715	.2916	.3685	.0496	.4219	.4116	.3845	.3456	.2995	.2500
	2	.0135	.0486	.0975	.1536	.2109	.2646	.3105	.3456	.3675	.3750
	3	.0005	.0036	.0115	.0256	.0469	.0756	.1115	.1536	.2005	.2500
	4	.0000	.0001	.0005	.0016	.0039	.0081	.0150	.0256	.0410	.0625
5	0	.7738	.5905	.4437	.3277	.2373	.1681	.1160	.0778	.0503	.0312
	1	.2036	.3280	.3915	.4096	.3955	.3602	.3124	.2592	.2059	.1562
	2	.0214	.0729	.1382	.2048	.2637	.3087	.3364	.3456	.3369	.3125
	3	.0011	.0081	.0244	.0512	.0879	.1323	.1811	.2304	.2757	.3125
	4	.0000	.0004	.0022	.0064	.0146	.0284	.0488	.0768	.1128	.1562
	5	.0000	.0000	.0001	.0003	.0010	.0024	.0053	.0102	.0185	.0312
6	0	.7351	.5314	.3771	.2621	.1780	.1176	.0754	.0467	.0277	.0156
	1	.2321	.3543	.3993	.3932	.3560	.3025	.2437	.1866	.1359	.0938
	2	.0305	.0984	.1762	.2458	.2966	.3241	.3280	.3110	.2780	.2344
	3	.0021	.0146	.0415	.0819	.1318	.1852	.2355	.2765	.3032	.3125
	4	.0001	.0012	.0055	.0154	.0330	.0595	.0951	.1382	.1861	.2344
	5	.0000	.0001	.0004	.0015	.0044	.0102	.0205	.0369	.0609	.0938
	6	.0000	.0000	.0000	.0001	.0002	.0007	.0018	.0041	.0083	.0156
7	0	.6983	.4783	.3206	.2097	.1335	.0824	.0490	.0280	.0152	.0078
	1	.2573	.3720	.3960	.3670	.3115	.2471	.1848	.1306	.0872	.0547
	2	.0406	.1240	.2097	.2753	.3115	.3177	.2985	.2613	.2140	.1641
	3	.0036	.0230	.0617	.1147	.1730	.2269	.2679	.2903	.2918	.2734
	4	.0002	.0026	.0109	.0287	.0577	.0972	.1442	.1935	.2388	.2734
	5	.0000	.0002	.0012	.0043	.0115	.0250	.0466	.0774	.1172	.1641
	6	.0000	.0000	.0001	.0004	.0013	.0036	.0084	.0172	.0320	.0547
	7	.0000	.0000	.0000	.0000	.0001	.0022	.0006	.0016	.0037	.0078

N	No. of *p or q* Events	*p or q*									
		.05	*.10*	*.15*	*.20*	*.25*	*.30*	*.35*	*.40*	*.45*	*.50*
8	0	.6634	.4305	.2725	.1678	.1001	.0576	.0319	.0168	.0084	.0039
	1	.2793	.3826	.3847	.3355	.2607	.1977	.1373	.0896	.0548	.0312
	2	.0515	.1488	.2376	.2936	.3115	.2965	.2587	.2090	.1569	.1094
	3	.0054	.0331	.0839	.1468	.2076	.2541	.2786	.2787	.2568	.2188
	4	.0004	.0046	.0185	.0459	.0865	.1361	.1875	.2322	.2627	.2734
	5	.0000	.0004	.0026	.0092	.0231	.0467	.0808	.1239	.1719	.2188
	6	.0000	.0000	.0002	.0011	.0038	.0100	.0217	.0413	.0703	.1094
	7	.0000	.0000	.0000	.0001	.0004	.0012	.0033	.0079	.0164	.0312
	8	.0000	.0000	.0000	.0000	.0000	.0001	.0002	.0007	.0017	.0039
9	0	.6302	.3874	.2316	.1342	.0751	.0404	.0207	.0101	.0046	.0020
	1	.2985	.3874	.3679	.3020	.2253	.1556	.1004	.0605	.0339	.0176
	2	.0629	.1722	.2597	.3020	.3003	.2668	.2162	.1612	.1110	.0703
	3	.0077	.0446	.1069	.1762	.2336	.2668	.2716	.2508	.2119	.1641
	4	.0006	.0074	.0283	.0661	.1168	.1715	.2194	.2508	.2600	.2461
	5	.0000	.0008	.0050	.0165	.0389	.0735	.1181	.1672	.2128	.2461
	6	.0000	.0001	.0006	.0028	.0087	.0210	.0424	.0743	.1160	.1641
	7	.0000	.0000	.0000	.0003	.0012	.0039	.0098	.0212	.0407	.0703
	8	.0000	.0000	.0000	.0000	.0001	.0004	.0013	.0035	.0083	.0176
	9	.0000	.0000	.0000	.0000	.0000	.0000	.0001	.0003	.0008	.0020
10	0	.5987	.3487	.1969	.1074	.0563	0282	.0135	.0060	.0025	0010
	1	.3151	.3874	.3474	.2684	.1877	.1211	.0725	.0403	.0207	.0098
	2	.0746	.1937	.2759	.3020	.2816	.2335	.1757	.1209	.0763	.0439
	3	.0105	.0574	.1298	.2013	.2503	.2668	.2522	.2150	.1665	.1172
	4	.0010	.0112	.0401	.0881	.1460	.2001	.2377	.2508	.2384	.2051
	5	.0001	.0015	.0085	.0264	.0584	.1029	.1536	.2007	.2340	.2461
	6	.0000	.0001	.0012	.0055	.0162	.0368	.0689	.1115	.1596	.2051
	7	.0000	.0000	.0001	.0008	.0031	.0090	.0212	.0425	.0746	.1172
	8	.0000	.0000	.0000	.0001	.0004	.0014	.0043	.0106	.0229	.0439
	9	.0000	.0000	.0000	.0000	.0000	.0001	.0005	.0016	.0042	.0098
	10	.0000	.0000	.0000	.0000	.0000	.0000	.0000	.0001	.0003	.0010

Appendix C

t Table

	Level of Significance for One-Tailed Test (%)			
	5	*2.5*	*1*	*.5*
	Level of Significance for Two-Tailed Test (%)			
df	*10*	*5*	*2*	*1*
1	6.3138	12.7062	31.8207	63.6574
2	2.9200	4.3027	6.9646	9.9248
3	2.3534	3.1824	4.5407	5.8409
4	2.1318	2.7764	3.7469	4.6041
5	2.0150	2.5706	3.3649	4.0322
6	1.9432	2.4469	3.1427	3.7074
7	1.8946	2.3646	2.9980	3.4995
8	1.8595	2.3060	2.8965	3.3554
9	1.8331	2.2622	2.8214	3.2498
10	1.8125	2.2281	2.7638	3.1693
11	1.7959	2.2010	2.7181	3.1058
12	1.7823	2.1788	2.6810	3.0545
13	1.7709	2.1604	2.6503	3.0123
14	1.7613	2.1448	2.6245	2.9768
15	1.7531	2.1315	2.6025	2.9467
16	1.7459	2.1199	2.5835	2.9208
17	1.7396	2.1098	2.5669	2.8982
18	1.7341	2.1009	2.5524	2.8784
19	1.7291	2.0930	2.5395	2.8609
20	1.7247	2.0860	2.5280	2.8453
21	1.7207	2.0796	2.5177	2.8314
22	1.7171	2.0739	2.5083	2.8188
23	1.7139	2.0687	2.4999	2.8073
24	1.7109	2.0639	2.4922	2.7969
25	1.7081	2.0595	2.4851	2.7874
26	1.7056	2.0555	2.4786	2.7787
27	1.7033	2.0518	2.4727	2.7707
28	1.7011	2.0484	2.4671	2.7633
29	1.6991	2.0452	2.4620	2.7564
30	1.6973	2.0423	2.4573	2.7500
35	1.6869	2.0301	2.4377	2.7238
40	1.6839	2.0211	2.4233	2.7045
45	1.6794	2.0141	2.4121	2.6896
50	1.6759	2.0086	2.4033	2.6778

	Level of Significance for One-Tailed Test (%)			
	5	*2.5*	*1*	*.5*
	Level of Significance for Two-Tailed Test (%)			
df	*10*	*5*	*2*	*1*
60	1.6706	2.0003	2.3901	2.6603
70	1.6669	1.9944	2.3808	2.6479
80	1.6641	1.9901	2.3739	2.6387
90	1.6620	1.9867	2.3685	2.6316
100	1.6602	1.9840	2.3642	2.6259
110	1.6588	1.9818	2.3607	2.6213
120	1.6577	1.9799	2.3598	2.6174
∞	1.6449	1.9600	2.3263	2.5758

Appendix D

F Table (ANOVA)

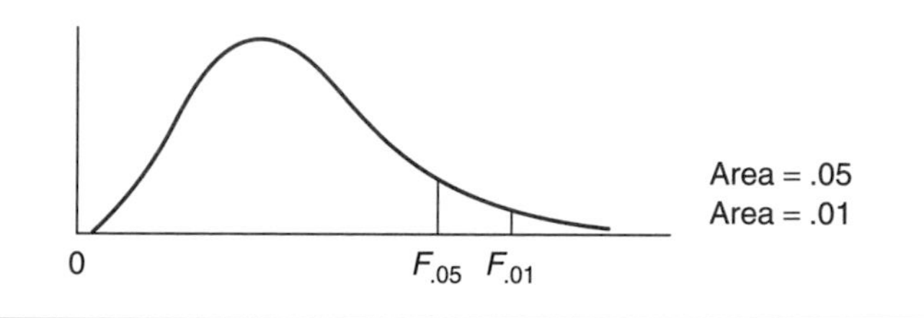

Degrees of Freedom: Denominator	*Degrees of Freedom: Numerator* 1	2	3	4	5	6	7	8	9	10	11	12	14	16	20	24	30	40	50	75	100	200	500	∞
1	161	200	216	225	230	234	237	239	241	242	243	244	245	246	248	249	250	241	252	255	253	254	254	254
	4052	4999	5403	5625	5764	5859	5928	5981	6022	6056	6082	6106	6142	6169	6208	6234	6258	6286	6302	6323	6334	6352	6361	6366
2	18.51	19.00	19.16	19.25	19.30	19.33	19.36	19.37	19.38	19.39	19.40	19.41	19.42	19.43	19.44	19.45	19.46	19.47	19.47	19.48	19.49	19.49	19.50	19.50
	98.49	99.00	99.17	99.25	99.30	99.33	99.34	99.36	99.38	99.40	99.41	99.42	99.43	99.44	99.45	99.46	99.47	99.48	99.48	99.49	99.49	99.49	99.50	99.50
3	10.13	9.55	9.28	9.12	9.01	8.94	8.88	8.84	8.81	8.78	8.76	8.74	8.71	8.69	8.66	8.64	8.62	8.60	8.58	8.57	8.56	8.54	8.54	8.53
	34.12	30.82	29.46	28.71	28.24	27.91	27.67	27.49	27.34	27.23	27.13	27.05	26.92	26.83	26.69	26.60	26.50	26.41	26.35	26.27	26.23	26.18	26.14	26.12
4	7.71	6.94	6.59	6.39	6.26	6.16	6.09	6.04	6.00	5.96	5.93	5.91	5.87	5.84	5.80	5.77	5.74	5.71	5.70	5.68	5.66	5.65	5.64	5.63
	21.20	18.00	16.69	15.98	15.52	15.21	14.98	14.80	14.66	14.54	14.45	14.37	14.24	14.15	14.02	13.93	13.83	13.74	13.69	13.61	13.57	13.52	13.48	13.46
5	6.61	5.79	5.41	5.19	5.05	4.95	4.88	4.82	4.78	4.74	4.70	4.68	4.64	4.60	4.56	4.53	4.50	4.46	4.44	4.42	4.40	4.38	4.37	4.36
	16.26	13.27	12.06	11.39	10.97	10.67	10.45	10.27	10.15	10.05	9.96	9.89	9.77	9.68	9.55	9.47	9.38	9.29	9.24	9.17	9.13	9.07	9.04	9.02
6	5.99	5.14	4.76	4.53	4.39	4.28	4.21	4.15	4.10	4.06	4.03	4.00	3.96	3.92	3.87	3.84	3.81	3.77	3.75	3.72	3.71	3.69	3.68	3.67
	13.74	10.92	9.78	9.15	8.75	8.47	8.26	8.10	7.98	7.87	7.79	7.72	7.60	7.52	7.39	7.31	7.23	7.14	7.09	7.02	6.99	6.94	6.90	6.88
7	5.59	4.47	4.35	4.12	3.97	3.87	3.79	3.73	3.68	3.63	3.60	3.57	3.52	3.49	3.44	3.41	3.38	3.34	3.32	3.29	3.28	3.25	3.24	3.23
	12.25	9.55	8.45	7.85	7.46	7.19	7.00	6.84	6.71	6.62	6.54	6.47	6.35	6.27	6.15	6.07	5.98	5.90	5.85	5.78	5.75	5.70	5.67	5.65
8	5.32	4.46	4.07	3.84	3.69	3.58	3.50	3.44	3.39	3.34	3.31	3.28	3.23	3.20	3.15	3.12	3.08	3.05	3.03	3.00	2.98	2.96	2.94	2.93
	11.26	8.65	7.59	7.01	6.63	6.37	6.19	6.03	5.91	5.82	5.74	5.67	5.56	5.48	5.36	5.28	5.20	5.11	5.06	5.00	4.96	4.91	4.88	4.86
9	5.12	4.26	3.86	3.63	3.48	3.37	3.29	3.23	3.18	3.13	3.10	3.07	3.02	2.98	2.93	2.90	2.86	2.82	2.80	2.77	2.76	2.73	2.72	2.71
	10.56	8.02	6.99	6.42	6.06	5.80	5.62	5.47	5.35	5.26	5.18	5.11	5.00	4.92	4.80	4.73	4.64	4.56	4.51	4.45	4.41	4.36	4.33	4.31
10	4.96	4.10	3.71	3.48	3.33	3.22	3.14	3.07	3.02	2.97	2.94	2.91	2.86	2.82	2.77	2.74	2.70	2.67	2.64	2.61	2.59	2.56	2.55	2.54
	10.04	7.56	6.55	5.99	5.64	5.39	5.21	5.06	4.95	4.85	4.78	4.71	4.60	4.52	4.41	4.33	4.25	4.17	4.12	4.05	4.01	3.96	3.93	3.91
11	4.84	3.98	3.59	3.36	3.20	3.09	3.01	2.95	2.90	2.86	2.82	2.79	2.74	2.70	2.65	2.61	2.57	2.53	2.50	2.47	2.45	2.42	2.41	2.40
	9.65	7.20	6.22	5.67	5.32	5.07	4.88	4.74	4.63	4.54	4.46	4.40	4.29	4.21	4.10	4.02	3.94	3.86	3.80	3.74	3.70	3.66	3.62	3.60
12	4.75	3.88	3.49	3.26	3.11	3.00	2.92	2.85	2.80	2.76	2.72	2.69	2.64	2.60	2.54	2.50	2.46	2.42	2.40	2.36	2.35	2.32	2.31	2.30
	9.33	6.93	5.95	5.41	5.06	4.82	4.65	4.50	4.39	4.30	4.22	4.16	4.05	3.98	3.86	3.78	3.70	3.61	3.56	3.49	3.46	3.41	3.38	3.36
13	4.67	3.80	3.41	3.18	3.02	2.92	2.84	2.77	2.72	2.67	2.63	2.60	2.55	2.51	2.46	2.42	2.38	2.34	2.32	2.28	2.26	2.24	2.22	2.21
	9.07	6.70	5.74	5.20	4.86	4.62	4.44	6.30	4.19	4.10	4.02	3.96	3.85	3.78	3.67	3.59	3.51	3.42	3.37	3.30	3.27	3.21	3.18	3.16

Degrees of Freedom: Denominator	Degrees of Freedom: Numerator																							
	1	2	3	4	5	6	7	8	9	10	11	12	14	16	20	24	30	40	50	75	100	200	500	∞
14	4.60	3.74	3.34	3.11	2.96	2.85	2.77	2.70	2.65	2.60	2.56	2.53	2.48	2.44	2.39	2.35	2.31	2.27	2.24	2.21	2.19	2.16	2.14	2.13
	8.86	**6.51**	**5.56**	**5.03**	**4.69**	**4.46**	**4.28**	**4.14**	**4.03**	**3.94**	**3.86**	**3.80**	**3.70**	**3.62**	**3.51**	**3.43**	**3.34**	**3.26**	**3.21**	**3.14**	**3.11**	**3.06**	**3.02**	**3.00**
15	4.54	3.68	3.29	3.06	2.90	2.79	2.70	2.64	2.59	2.55	2.51	2.48	2.43	2.39	2.33	2.29	2.25	2.21	2.18	2.15	2.12	2.10	2.08	2.07
	8.68	**6.36**	**5.42**	**4.89**	**4.56**	**4.32**	**4.14**	**4.00**	**3.89**	**3.80**	**3.73**	**3.67**	**3.56**	**3.48**	**3.36**	**3.29**	**3.20**	**3.12**	**3.07**	**3.00**	**2.97**	**2.92**	**2.89**	**2.87**
16	4.49	3.63	3.24	3.01	2.85	2.74	2.66	2.59	2.54	2.49	2.45	2.42	2.37	2.33	2.28	2.24	2.20	2.16	2.13	2.09	2.07	2.04	2.02	2.01
	8.53	**6.23**	**5.29**	**4.77**	**4.44**	**4.20**	**4.03**	**3.89**	**3.78**	**3.69**	**3.61**	**3.55**	**3.45**	**3.37**	**3.25**	**3.18**	**3.10**	**3.01**	**2.96**	**2.89**	**2.86**	**2.80**	**2.77**	**2.75**
17	4.45	3.59	3.20	2.96	2.81	2.70	2.62	2.55	2.50	2.45	2.41	2.38	2.33	2.29	2.23	2.19	2.15	2.11	2.08	2.04	2.02	1.99	1.97	1.96
	8.40	**6.11**	**5.18**	**4.67**	**4.34**	**4.10**	**3.93**	**3.79**	**3.68**	**3.59**	**3.52**	**3.45**	**3.35**	**3.27**	**3.16**	**3.08**	**3.00**	**2.92**	**2.86**	**2.79**	**2.76**	**2.70**	**2.67**	**2.65**
18	4.41	3.55	3.16	2.93	2.77	2.66	2.58	2.51	2.46	2.41	2.37	2.34	2.29	2.25	2.19	2.15	2.11	2.07	2.04	2.00	2.98	1.95	1.93	1.92
	8.28	**6.01**	**5.09**	**4.58**	**4.25**	**4.01**	**3.85**	**3.71**	**3.60**	**3.51**	**3.44**	**3.37**	**3.27**	**3.19**	**3.07**	**3.00**	**2.91**	**2.83**	**2.78**	**2.71**	**2.68**	**2.62**	**2.59**	**2.57**
19	4.38	3.52	3.13	2.90	2.74	2.63	2.55	2.48	2.43	2.38	2.34	2.31	2.26	2.21	2.15	2.11	2.07	2.02	2.00	1.96	1.94	1.91	1.90	1.88
	8.18	**5.93**	**5.01**	**4.50**	**4.17**	**3.94**	**3.77**	**3.63**	**3.52**	**3.43**	**3.36**	**3.30**	**3.19**	**3.12**	**3.00**	**2.92**	**2.84**	**2.76**	**2.70**	**2.63**	**2.60**	**2.54**	**2.51**	**2.49**
20	4.35	3.49	3.10	2.87	2.71	2.60	2.52	2.45	2.40	2.35	2.31	2.28	2.23	2.18	2.12	2.08	2.04	1.99	1.96	1.92	1.90	1.87	1.85	1.84
	8.10	**5.85**	**4.94**	**4.43**	**4.10**	**3.87**	**3.71**	**3.56**	**3.45**	**3.37**	**3.30**	**3.23**	**3.13**	**3.05**	**2.94**	**2.86**	**2.77**	**2.69**	**2.63**	**2.56**	**2.53**	**2.47**	**2.44**	**2.42**
21	4.32	3.47	3.07	2.84	2.68	2.57	2.49	2.42	2.37	2.32	2.28	2.25	2.20	2.15	2.09	2.05	2.00	1.96	1.93	1.89	1.87	1.84	1.82	1.81
	8.02	**5.78**	**4.87**	**4.37**	**4.04**	**3.81**	**3.65**	**3.51**	**3.40**	**3.31**	**3.24**	**3.17**	**3.07**	**2.99**	**2.88**	**2.80**	**2.72**	**2.63**	**2.58**	**2.51**	**2.47**	**2.42**	**2.38**	**2.36**
22	4.30	3.44	3.05	2.82	2.66	2.55	2.47	2.40	2.35	2.30	2.26	2.23	2.18	2.13	2.07	2.03	1.98	1.93	1.91	1.87	1.84	1.81	1.80	1.78
	7.94	**5.72**	**4.82**	**4.31**	**3.99**	**3.76**	**3.59**	**3.45**	**3.35**	**3.26**	**3.18**	**3.12**	**3.02**	**2.94**	**2.83**	**2.75**	**2.67**	**2.58**	**2.53**	**2.46**	**2.42**	**2.37**	**2.33**	**2.31**
23	4.28	3.42	3.03	2.80	2.04	2.53	2.45	2.38	2.32	2.28	2.24	2.20	2.14	2.10	2.04	2.00	1.96	1.91	1.88	1.84	1.82	1.79	1.77	1.76
	7.88	**5.66**	**4.76**	**4.26**	**3.94**	**3.71**	**3.54**	**3.41**	**3.30**	**3.21**	**3.14**	**3.07**	**2.97**	**2.89**	**2.78**	**2.70**	**2.62**	**2.53**	**2.48**	**2.41**	**2.37**	**2.32**	**2.28**	**2.26**
24	4.26	3.40	3.01	2.78	2.62	2.51	2.43	2.36	2.30	2.26	2.22	2.18	2.13	2.09	2.02	1.98	1.94	1.89	1.86	1.82	1.80	1.76	1.74	1.73
	7.82	**5.61**	**4.72**	**4.22**	**3.90**	**3.67**	**3.50**	**3.36**	**3.25**	**3.17**	**3.09**	**3.03**	**2.93**	**2.85**	**2.74**	**2.66**	**2.58**	**2.49**	**2.44**	**2.36**	**2.33**	**2.27**	**2.23**	**2.21**
25	4.24	3.38	2.99	2.76	2.60	2.49	2.41	2.34	2.28	2.24	2.20	2.16	2.11	2.06	2.00	1.96	1.92	1.87	1.84	1.80	1.77	1.74	1.72	1.71
	7.77	**5.57**	**4.68**	**4.18**	**3.86**	**3.63**	**3.46**	**3.32**	**3.21**	**3.13**	**3.05**	**2.99**	**2.89**	**2.81**	**2.70**	**2.62**	**2.54**	**2.45**	**2.40**	**2.32**	**2.29**	**2.23**	**2.19**	**2.17**
26	4.22	3.37	2.98	2.74	2.59	2.47	2.39	2.32	2.27	2.22	2.18	2.15	2.10	2.05	1.99	1.95	1.90	1.85	1.82	1.78	1.76	1.72	1.70	1.69
	7.72	**5.53**	**4.64**	**4.14**	**3.82**	**3.59**	**3.42**	**3.29**	**3.17**	**3.09**	**3.02**	**2.96**	**2.86**	**2.77**	**2.66**	**2.58**	**2.50**	**2.41**	**2.36**	**2.28**	**2.25**	**2.19**	**2.15**	**2.13**
27	4.21	3.35	2.96	2.73	2.57	2.46	2.37	2.30	2.25	2.20	2.16	2.13	2.08	2.03	1.97	1.93	1.88	1.84	1.80	1.76	1.74	1.71	1.68	1.67
	7.68	**5.49**	**4.60**	**4.11**	**3.79**	**3.56**	**3.39**	**3.26**	**3.14**	**3.06**	**2.98**	**2.93**	**2.83**	**2.74**	**2.63**	**2.55**	**2.47**	**2.38**	**2.33**	**2.25**	**2.21**	**2.16**	**2.12**	**2.10**
28	4.20	3.34	2.95	2.71	2.56	2.44	2.36	2.29	2.24	2.19	2.15	2.12	2.06	2.02	1.96	1.91	1.87	1.81	1.78	1.75	1.72	1.69	1.67	1.65
	7.64	**5.45**	**4.57**	**4.07**	**3.76**	**3.53**	**3.36**	**3.23**	**3.11**	**3.03**	**2.95**	**2.90**	**2.80**	**2.71**	**2.60**	**2.52**	**2.44**	**2.35**	**2.30**	**2.22**	**2.18**	**2.13**	**2.09**	**2.06**
29	4.18	3.33	2.93	2.70	2.54	2.43	2.35	2.28	2.22	2.18	2.14	2.10	2.05	2.00	1.94	1.90	1.85	1.80	1.77	1.73	1.71	1.68	1.65	1.64
	7.60	**5.42**	**4.54**	**4.04**	**3.73**	**3.50**	**3.33**	**3.20**	**3.08**	**3.00**	**2.92**	**2.87**	**2.77**	**2.68**	**2.57**	**2.49**	**2.41**	**2.32**	**2.27**	**2.19**	**2.15**	**2.10**	**2.06**	**2.03**
30	4.17	3.32	2.92	2.69	2.53	2.42	2.34	2.27	2.21	2.16	2.12	2.09	2.04	1.99	1.93	1.89	1.84	1.79	1.76	1.72	1.69	1.66	1.64	1.62
	7.56	**5.39**	**4.51**	**4.02**	**3.70**	**3.47**	**3.30**	**3.17**	**3.06**	**2.98**	**2.90**	**2.84**	**2.74**	**2.66**	**2.55**	**2.47**	**2.38**	**2.29**	**2.24**	**2.16**	**2.13**	**2.07**	**2.03**	**2.01**
32	4.15	3.30	2.90	2.67	2.51	2.40	2.32	2.25	2.19	2.14	2.10	2.07	2.02	1.97	1.91	1.86	1.82	1.76	1.74	1.69	1.67	1.64	1.61	1.59
	7.50	**5.34**	**4.46**	**3.97**	**3.66**	**3.42**	**3.25**	**3.12**	**3.01**	**2.94**	**2.86**	**2.80**	**2.70**	**2.62**	**2.51**	**2.42**	**2.34**	**2.25**	**2.20**	**2.12**	**2.08**	**2.02**	**1.98**	**1.96**
34	4.13	3.28	2.88	2.65	2.49	2.38	2.30	2.23	2.17	2.12	2.08	2.05	2.00	1.95	1.89	1.84	1.80	1.74	1.71	1.67	1.64	1.61	1.59	1.57
	7.44	**5.29**	**4.42**	**3.93**	**3.61**	**3.38**	**3.21**	**3.08**	**2.97**	**2.89**	**2.82**	**2.76**	**2.66**	**2.58**	**2.47**	**2.38**	**2.30**	**2.21**	**2.15**	**2.08**	**2.04**	**1.98**	**1.94**	**1.91**
36	4.11	3.26	2.86	2.63	2.48	2.36	2.28	2.21	2.15	2.10	2.06	2.03	1.98	1.93	1.87	1.82	1.78	1.72	1.69	1.65	1.62	1.59	1.56	1.55
	7.39	**5.25**	**4.38**	**3.89**	**3.58**	**3.35**	**3.18**	**3.04**	**2.94**	**2.86**	**2.78**	**2.72**	**2.62**	**2.54**	**2.43**	**2.35**	**2.26**	**2.17**	**2.12**	**2.04**	**2.00**	**1.94**	**1.90**	**1.87**
38	4.10	3.25	2.85	2.62	2.46	2.35	2.26	2.19	2.14	2.09	2.05	2.02	1.96	1.92	1.85	1.80	1.76	1.71	1.67	1.63	1.60	1.57	1.54	1.53
	7.35	**5.21**	**4.34**	**3.86**	**3.54**	**3.32**	**3.15**	**3.02**	**2.91**	**2.82**	**2.75**	**2.69**	**2.59**	**2.51**	**2.40**	**2.32**	**2.22**	**2.14**	**2.08**	**2.00**	**1.97**	**1.90**	**1.86**	**1.84**
40	4.08	3.23	2.84	2.61	2.45	2.34	2.25	2.18	2.12	2.07	2.04	2.00	1.95	1.90	1.84	1.79	1.74	1.69	1.66	1.61	1.59	1.55	1.53	1.51
	7.31	**5.18**	**4.31**	**3.83**	**3.51**	**3.29**	**3.12**	**2.99**	**2.88**	**2.80**	**2.73**	**2.66**	**2.56**	**2.49**	**2.37**	**2.29**	**2.20**	**2.11**	**2.05**	**1.97**	**1.94**	**1.88**	**1.84**	**1.81**

(Continued)

(Continued)

Degrees of Freedom: Denominator	Degrees of Freedom: Numerator																							
	1	*2*	*3*	*4*	*5*	*6*	*7*	*8*	*9*	*10*	*11*	*12*	*14*	*16*	*20*	*24*	*30*	*40*	*50*	*75*	*100*	*200*	*500*	∞
42	4.07	3.22	2.83	2.59	2.44	2.32	2.24	2.17	2.11	2.06	2.02	1.99	1.94	1.89	1.82	1.78	1.73	1.68	1.64	1.60	1.57	1.54	1.51	1.49
	7.27	**5.15**	**4.29**	**3.80**	**3.49**	**3.26**	**3.10**	**2.96**	**2.86**	**2.77**	**2.70**	**2.64**	**2.54**	**2.46**	**2.35**	**2.26**	**2.17**	**2.08**	**2.02**	**1.94**	**1.91**	**1.85**	**1.80**	**1.78**
44	4.06	3.21	2.82	2.58	2.43	2.31	2.23	2.16	2.10	2.05	2.01	1.98	1.92	1.88	1.81	1.76	1.72	1.66	1.63	1.58	1.56	1.52	1.50	1.48
	7.24	**5.12**	**4.26**	**3.78**	**3.46**	**3.24**	**3.07**	**2.94**	**2.84**	**2.75**	**2.68**	**2.62**	**2.52**	**2.44**	**2.32**	**2.24**	**2.15**	**2.06**	**2.00**	**1.92**	**1.88**	**1.82**	**1.78**	**1.75**
46	4.05	3.20	2.81	2.57	2.42	2.30	2.22	2.14	2.09	2.04	2.00	1.97	1.91	1.87	1.80	1.75	1.71	1.65	1.62	1.57	1.54	1.51	1.48	1.46
	7.21	**5.10**	**4.24**	**3.76**	**3.44**	**3.22**	**3.05**	**2.92**	**2.82**	**2.73**	**2.66**	**2.60**	**2.50**	**2.42**	**2.30**	**2.22**	**2.13**	**2.04**	**1.98**	**1.90**	**1.86**	**1.80**	**1.76**	**1.72**
48	4.04	3.19	2.80	2.56	2.41	2.30	2.21	2.14	2.08	2.03	1.99	1.96	1.90	1.86	1.79	1.74	1.70	1.64	1.61	1.56	1.53	1.50	1.47	1.45
	7.19	**5.08**	**4.22**	**3.74**	**3.42**	**3.20**	**3.04**	**2.90**	**2.80**	**2.71**	**2.64**	**2.58**	**2.48**	**2.40**	**2.28**	**2.20**	**2.11**	**2.02**	**1.96**	**1.88**	**1.84**	**1.78**	**1.73**	**1.70**
50	4.03	3.18	2.79	2.56	2.40	2.29	2.20	2.13	2.07	2.02	1.98	1.95	1.90	1.85	1.78	1.74	1.69	1.63	1.60	1.55	1.52	1.48	1.46	1.44
	7.17	**5.06**	**4.20**	**3.72**	**3.41**	**3.18**	**3.02**	**2.88**	**2.78**	**2.70**	**2.62**	**2.56**	**2.46**	**2.39**	**2.26**	**2.18**	**2.10**	**2.00**	**1.94**	**1.86**	**1.82**	**1.76**	**1.71**	**1.68**
55	4.02	3.17	2.78	2.54	2.38	2.27	2.18	2.11	2.05	2.00	1.97	1.93	1.88	1.83	1.76	1.72	1.67	1.61	1.58	1.52	1.50	1.46	1.43	1.41
	7.12	**5.01**	**4.16**	**3.68**	**3.37**	**3.15**	**2.98**	**2.85**	**2.75**	**2.66**	**2.59**	**2.53**	**2.43**	**2.35**	**2.23**	**2.15**	**2.06**	**1.96**	**1.90**	**1.82**	**1.78**	**1.71**	**1.66**	**1.64**
60	4.00	3.15	2.76	2.52	2.37	2.25	2.17	2.10	2.04	1.99	1.95	1.92	1.86	1.81	1.75	1.70	1.65	1.59	1.56	1.50	1.48	1.44	1.41	1.39
	7.08	**4.98**	**4.13**	**3.65**	**3.34**	**3.12**	**2.95**	**2.82**	**2.72**	**2.63**	**2.56**	**2.50**	**2.40**	**2.32**	**2.20**	**2.12**	**2.03**	**1.93**	**1.87**	**1.79**	**1.74**	**1.68**	**1.63**	**1.60**
65	3.99	3.14	2.75	2.51	2.36	2.24	2.15	2.08	2.02	1.98	1.94	1.90	1.85	1.80	1.73	1.68	1.63	1.57	1.54	1.49	1.46	1.42	1.39	1.37
	7.04	**4.95**	**4.10**	**3.62**	**3.31**	**3.09**	**2.93**	**2.79**	**2.70**	**2.61**	**2.54**	**2.47**	**2.37**	**2.30**	**2.18**	**2.09**	**2.00**	**1.90**	**1.84**	**1.76**	**1.71**	**1.64**	**1.60**	**1.56**
70	3.98	3.13	2.74	2.50	2.35	2.23	2.14	2.07	2.01	1.97	1.93	1.89	1.84	1.79	1.72	1.67	1.62	1.56	1.53	1.47	1.45	1.40	1.37	1.35
	7.01	**4.92**	**4.08**	**3.60**	**3.29**	**3.07**	**2.91**	**2.77**	**2.67**	**2.59**	**2.51**	**2.45**	**2.35**	**2.28**	**2.15**	**2.07**	**1.98**	**1.88**	**1.82**	**1.74**	**1.69**	**1.62**	**1.56**	**1.53**
80	3.96	3.11	2.72	2.48	2.33	2.21	2.12	2.05	1.99	1.95	1.91	1.88	1.82	1.77	1.70	1.65	1.60	1.54	1.51	1.45	1.42	1.38	1.35	1.32
	6.96	**4.88**	**4.04**	**3.56**	**3.25**	**3.04**	**2.87**	**2.74**	**2.64**	**2.55**	**2.48**	**2.41**	**2.32**	**2.24**	**2.11**	**2.03**	**1.94**	**1.84**	**1.78**	**1.70**	**1.65**	**1.57**	**1.52**	**1.49**
100	3.94	3.09	2.70	2.46	2.30	2.19	2.10	2.03	1.97	1.92	1.88	1.85	1.79	1.75	1.68	1.63	1.57	1.51	1.48	1.42	1.39	1.34	1.30	1.28
	6.90	**4.82**	**3.98**	**3.51**	**3.20**	**2.99**	**2.82**	**2.69**	**2.59**	**2.51**	**2.43**	**2.36**	**2.26**	**2.19**	**2.06**	**1.98**	**1.89**	**1.79**	**1.73**	**1.64**	**1.59**	**1.51**	**1.46**	**1.43**
125	3.92	3.07	2.68	2.44	2.29	2.17	2.08	2.01	1.95	1.90	1.86	1.83	1.77	1.72	1.65	1.60	1.55	1.49	1.45	1.39	1.36	1.31	1.27	1.25
	6.84	**4.78**	**3.94**	**3.47**	**3.17**	**2.95**	**2.79**	**2.65**	**2.56**	**2.47**	**2.40**	**2.33**	**2.23**	**2.15**	**2.03**	**1.94**	**1.85**	**1.75**	**1.68**	**1.59**	**1.54**	**1.46**	**1.40**	**1.37**
150	3.91	3.06	2.67	2.43	2.27	2.16	2.07	2.00	1.94	1.89	1.85	1.82	1.76	1.71	1.64	1.59	1.54	1.47	1.44	1.37	1.34	1.29	1.25	1.22
	6.81	**4.75**	**3.91**	**3.44**	**3.14**	**2.92**	**2.76**	**2.62**	**2.53**	**2.44**	**2.37**	**2.30**	**2.20**	**2.12**	**2.00**	**1.91**	**1.83**	**1.72**	**1.66**	**1.56**	**1.51**	**1.43**	**1.37**	**1.33**
200	3.89	3.04	2.65	2.41	2.26	2.14	2.05	1.98	1.92	1.87	1.83	1.80	1.74	1.69	1.62	1.57	1.52	1.45	1.42	1.35	1.32	1.26	1.22	1.19
	6.76	**4.71**	**3.88**	**3.41**	**3.11**	**2.90**	**2.73**	**2.60**	**2.50**	**2.41**	**2.34**	**2.28**	**2.17**	**2.09**	**1.97**	**1.88**	**1.79**	**1.69**	**1.62**	**1.53**	**1.48**	**1.39**	**1.33**	**1.28**
400	3.86	3.02	2.62	2.39	2.23	2.12	2.03	1.96	1.90	1.85	1.81	1.78	1.72	1.67	1.60	1.54	1.49	1.42	1.38	1.32	1.28	1.22	1.16	1.13
	6.70	**4.66**	**3.83**	**3.36**	**3.06**	**2.85**	**2.69**	**2.55**	**2.46**	**2.37**	**2.29**	**2.23**	**2.12**	**2.04**	**1.92**	**1.84**	**1.74**	**1.64**	**1.57**	**1.47**	**1.42**	**1.32**	**1.24**	**1.19**
1000	3.85	3.00	2.61	2.38	2.22	2.10	2.02	1.95	1.89	1.84	1.80	1.76	1.70	1.65	1.58	1.53	1.47	1.41	1.36	1.30	1.26	1.19	1.13	1.08
	6.66	**4.62**	**3.80**	**3.34**	**3.04**	**2.82**	**2.66**	**2.53**	**2.43**	**2.34**	**2.26**	**2.20**	**2.09**	**2.01**	**1.89**	**1.81**	**1.71**	**1.61**	**1.54**	**1.44**	**1.38**	**1.28**	**1.19**	**1.11**
∞	3.84	2.99	2.60	2.37	2.21	2.09	2.01	1.94	1.88	1.83	1.79	1.75	1.69	1.64	1.57	1.52	1.46	1.40	1.35	1.28	1.24	1.17	1.11	1.00
	6.64	**4.60**	**3.78**	**3.32**	**3.02**	**2.80**	**2.64**	**2.51**	**2.41**	**2.32**	**2.24**	**2.18**	**2.07**	**1.99**	**1.87**	**1.79**	**1.69**	**1.59**	**1.52**	**1.41**	**1.36**	**1.25**	**1.15**	**1.00**

NOTE: The specific F distribution must be identified by the number of degrees of freedom characterizing the numerator and the denominator of F. The values of F corresponding to 5% of the area in the upper tail are shown in roman type and those corresponding to 1%, in boldface type.

Appendix E

Studentized Range Statistic (for Tukey HSD)

		k = Number of Groups								
df_{with}	α	2	3	4	5	6	7	8	9	10
5	.05	3.64	4.60	5.22	5.67	6.03	6.33	6.58	6.80	6.99
	.01	5.70	6.98	7.80	8.42	8.91	9.32	9.67	9.97	10.24
6	.05	3.46	4.34	4.90	5.30	5.63	5.90	6.12	6.32	6.49
	.01	5.24	6.33	7.03	7.56	7.97	8.32	8.61	8.87	9.10
7	.05	3.34	4.16	4.68	5.06	5.36	5.61	5.82	6.00	6.16
	.01	4.95	5.92	6.54	7.01	7.37	7.68	7.94	8.17	8.37
8	.05	3.26	4.04	4.53	4.89	5.17	5.40	5.60	5.77	5.92
	.01	4.75	5.64	6.20	6.62	6.96	7.24	7.47	7.68	7.86
9	.05	3.20	3.95	4.41	4.76	5.02	5.24	5.43	5.59	5.74
	.01	4.60	5.43	5.96	6.35	6.66	6.91	7.13	7.33	7.49
10	.05	3.15	3.88	4.33	4.65	4.91	5.12	5.30	5.46	5.60
	.01	4.48	5.27	5.77	6.14	6.43	6.67	6.87	7.05	7.21
11	.05	3.11	3.82	4.26	4.57	4.82	5.03	5.20	5.35	5.49
	.01	4.39	5.15	5.62	5.97	6.25	6.48	6.67	6.84	6.99
12	.05	3.08	3.77	4.20	4.51	4.75	4.95	5.12	5.27	5.39
	.01	4.32	5.05	5.50	5.84	6.10	6.32	6.51	6.67	6.81
13	.05	3.06	3.73	4.15	4.45	4.69	4.88	5.05	5.19	5.32
	.01	4.26	4.96	5.40	5.73	5.98	6.19	6.37	6.53	6.67
14	.05	3.03	3.70	4.11	4.41	4.64	4.83	4.99	5.13	5.25
	.01	4.21	4.89	5.32	5.63	5.88	6.08	6.26	6.41	6.54
15	.05	3.01	3.67	4.08	4.37	4.59	4.78	4.94	5.08	5.20
	.01	4.17	4.84	5.25	5.56	5.80	5.99	6.16	6.31	6.44
16	.05	3.00	3.65	4.05	4.33	4.56	4.74	4.90	5.03	5.15
	.01	4.13	4.79	5.19	5.49	5.72	5.92	6.08	6.22	6.35
17	.05	2.98	3.63	4.02	4.30	4.52	4.70	4.86	4.99	5.11
	.01	4.10	4.74	5.14	5.43	5.66	5.85	6.01	6.15	6.27
18	.05	2.97	3.61	4.00	4.28	4.49	4.67	4.82	4.96	5.07
	.01	4.07	4.70	5.09	5.38	5.60	5.79	5.94	6.08	6.20
19	.05	2.96	3.59	3.98	4.25	4.47	4.65	4.79	4.92	5.04
	.01	4.05	4.67	5.05	5.33	5.55	5.73	5.89	6.02	6.14
20	.05	2.95	3.58	3.96	4.23	4.45	4.62	4.77	4.90	5.01
	.01	4.02	4.64	5.02	5.29	5.51	5.69	5.84	5.97	6.09
24	.05	2.92	3.53	3.90	4.17	4.37	4.54	4.68	4.81	4.92
	.01	3.96	4.55	4.91	5.17	5.37	5.54	5.69	5.81	5.92
30	.05	2.89	3.49	3.85	4.10	4.30	4.46	4.60	4.72	4.82
	.01	3.89	4.45	4.80	5.05	5.24	5.40	5.54	5.65	5.76

(Continued)

(Continued)

		k = Number of Groups								
df_{with}	α	2	3	4	5	6	7	8	9	10
40	.05	2.86	3.44	3.79	4.04	4.23	4.39	4.52	4.63	4.73
	.01	3.82	4.37	4.70	4.93	5.11	5.26	5.39	5.50	5.60
60	.05	2.83	3.40	3.74	3.98	4.16	4.31	4.44	4.55	4.65
	.01	3.76	4.28	4.59	4.82	4.99	5.13	5.25	5.36	5.45
120	.05	2.80	3.36	3.68	3.92	4.10	4.24	4.36	4.47	4.56
	.01	3.70	4.20	4.50	4.71	4.87	5.01	5.12	5.21	5.30
∞	.05	2.77	3.31	3.63	3.86	4.03	4.17	4.29	4.39	4.47
	.01	3.64	4.12	4.40	4.60	4.76	4.88	4.99	5.08	5.16

Appendix F

Chi-Square Table

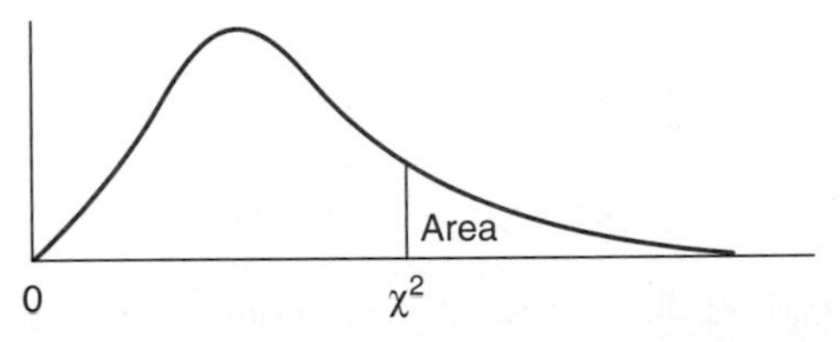

	Area in the Upper Tail				
df	*.10*	*.05*	*.025*	*.01*	*.005*
1	2.71	3.84	5.02	6.63	7.88
2	4.61	5.99	7.38	9.21	10.60
3	6.25	7.81	9.35	11.34	12.84
4	7.78	9.49	11.14	13.28	14.86
5	9.24	11.07	12.83	15.09	16.75
6	10.64	12.59	14.45	16.81	18.55
7	12.02	14.07	16.01	18.48	20.28
8	13.36	15.51	17.53	20.09	21.96
9	14.68	16.92	19.02	21.67	23.59
10	15.99	18.31	20.48	23.21	25.19
11	17.28	19.68	21.92	24.72	26.76
12	18.55	21.03	23.34	26.22	28.30
13	19.81	22.36	24.74	27.69	29.82
14	21.06	23.68	26.12	29.14	31.32
15	22.31	25.00	27.49	30.58	32.80
16	23.54	26.30	28.85	32.00	34.27
17	24.77	27.59	30.19	33.41	35.72
18	25.99	28.87	31.53	34.81	37.15
19	27.20	30.14	32.85	36.19	38.58
20	28.41	31.41	34.17	37.57	40.00
21	29.62	32.67	35.48	38.93	41.40
22	30.81	33.92	36.78	40.29	42.80
23	32.01	35.17	38.08	41.64	44.18
24	33.20	36.42	39.36	42.98	45.56
25	34.38	37.65	40.65	44.31	46.93
26	35.56	38.89	41.92	45.64	48.29
27	36.74	40.11	43.19	46.96	49.64
28	37.92	41.34	44.46	48.28	50.99
29	39.09	42.56	45.72	49.59	52.34
30	40.26	43.77	46.98	50.89	53.67

(Continued)

(Continued)

	Area in the Upper Tail				
df	*.10*	*.05*	*.025*	*.01*	*.005*
40	51.81	55.76	59.34	63.69	66.77
50	63.17	67.50	71.42	76.15	79.49
60	74.40	79.08	83.30	88.38	91.95
70	85.53	90.53	95.02	100.42	104.22
80	96.58	101.88	106.63	112.33	116.32
90	107.56	113.14	118.14	124.12	128.30
100	118.50	124.34	129.56	135.81	140.17
120	140.23	146.57	152.21	158.95	163.64

NOTE: The first column identifies the specific χ^2 distribution according to its number of degrees of freedom. Other columns give the proportion of the area under the entire curve that falls above the tabled value of χ^2.

Appendix G

Correlation Table

	Levels of Significance for a One-Tailed Test			
	.05	*.025*	*.01*	*.005*
	Levels of Significance for a Two-Tailed Test			
df	*.10*	*.05*	*.02*	*.01*
1	.988	.997	.9995	.9999
2	.900	.950	.980	.990
3	.805	.878	.934	.959
4	.729	.811	.882	.917
5	.669	.754	.833	.874
6	.622	.707	.789	.834
7	.582	.666	.750	.798
8	.549	.632	.716	.765
9	.521	.602	.685	.735
10	.497	.576	.658	.708
11	.476	.553	.634	.684
12	.458	.532	.612	.661
13	.441	.514	.592	.641
14	.426	.497	.574	.623
15	.412	.482	.558	.606
16	.400	.468	.542	.590
17	.389	.456	.528	.575
18	.378	.444	.516	.561
19	.369	.433	.503	.549
20	.360	.423	.492	.537
21	.352	.413	.482	.526
22	.344	.404	.472	.515
23	.337	.396	.462	.505
24	.330	.388	.453	.496
25	.323	.381	.445	.487
26	.317	.374	.437	.479
27	.311	.367	.430	.471
28	.306	.361	.423	.463
29	.301	.355	.416	.456
30	.296	.349	.409	.449
32	.287	.339	.397	.436
34	.279	.329	.386	.424
36	.271	.320	.376	.413
38	.264	.312	.367	.403
40	.257	.304	.358	.393
42	.251	.297	.350	.384
44	.246	.291	.342	.376
46	.240	.285	.335	.368
48	.235	.279	.328	.361
50	.231	.273	.322	.354
55	.220	.261	.307	.339
60	.211	.250	.295	.325
65	.203	.240	.284	.313
70	.195	.232	.274	.302
75	.189	.224	.265	.292
80	.183	.217	.256	.283
85	.178	.211	.249	.275
90	.173	.205	.242	.267
95	.168	.200	.236	.260
100	.164	.195	.230	.254
120	.150	.178	.210	.232
150	.134	.159	.189	.208
200	.116	.138	.164	.181
300	.095	.113	.134	.148
400	.082	.098	.116	.128
500	.073	.088	.104	.115
1000	.052	.062	.073	.081

NOTE: Values of the correlation coefficient required for different levels of significance when H_0: $r = 0$.

Appendix H

Odd Solutions to Textbook Exercises

Module 1. Math Review, Vocabulary, and Symbols

1.

Fraction	*Percentage*	*Decimal*
$33^1/_3$	33.33%	.333 . . .
$12^1/_2$	12.5%	.125
$66^7/_{10}$	66.7%	.667

3.

Fraction	*Percentage*	*Decimal*
1/20	5%	.05
95/100	95%	.95
1/4	25%	.25

5.

a. 26.41
b. 62.74
c. 36.85

7.

a. 95.56
b. 0.02
c. 48.95

9.

a. $4 \times 5 + 3 \times 2 = 20 + 6 = 26$
b. $4(5 + 3) \times 2 = (4)(8)(2) = 64$
c. $((4 \times 5) + 3)(2) = ((20) + 3)(2) = (23)(2) = 46$
d. $\sqrt{4(5+3)(2)} = \sqrt{4(8)(2)} = \sqrt{64} = 8$
e. $4^2(5 + 3)(2) = (16)(8)(2) = (16)(16) = 256$

11.

a. $4\sqrt{36} \times (2 \times 3) + 2 = 146$

b. $6^2/12 - 4/2 = 1$

c. $\sqrt{(8-4)} \times (6/2 - 1) = 4$

d. $[(30)(.5) - 6] \times 2 + 4 = 22$

e. $5 + 3 \times 7 - 4 = 22$

13.

a. $16/5 - 0.246 = (16)(1/5) - .246 = 3.2 - .246 = 2.954 = 2.95$

b. $68 + 68/3 = 68 + (68)(1/3) = 68 + 22.667 = 90.667 = 90.67$

c. $2/3 - 1/5 = (2)(1/3) - 1/5 = .667 + .20 = .867 = .87$

15.

a. $6/3 - 0.95 = (6)(1/3) - 0.95 = 2 - 0.95 = 1.05$

b. $12.50 - 8/2 = 12.50 - (8)(1/2) = 12.50 - 4 = 8.50$

c. $25/5 + 6 = (25)(1/5) + 6 = 5 + 6 = 11$

17.

a. $64/4 + b = 30 \rightarrow 64 + 4b = 120 \rightarrow 4b = 56 \rightarrow b = 56/4 \rightarrow b = 14$

b. $.30c = 20 \rightarrow c = 20/.30 \rightarrow c = 66.67$

c. $y = bX + a \rightarrow y - bX = a$ . . . OR . . . $a = y - bX$

19.

a. $b + 24/8 = 3 \rightarrow b = 3 - 24/8 \rightarrow b = 3 - 3 \rightarrow b = 0$

b. $T = 12c - 4 \rightarrow c = (T + 4)/12$

c. $12 = 3a \rightarrow a = 4$

21.

a. $\Sigma(X + Y)^2 = \Sigma(X^2 + Y^2 + 2XY) = \Sigma X^2 + \Sigma Y^2 + 2\Sigma XY$

b. $\Sigma(bXY + 1)^2 = \Sigma(bXY)^2 + N1^2 + 2b\Sigma XY = b^2\Sigma X^2Y^2 + N + 2b\Sigma XY$

c. $\Sigma(XY + 1) = \Sigma XY + N1 = \Sigma XY + N$

Module 2. Measurement Scales

1.

a. Feet of snow = R

b. Brands of carbonated soft drinks = N

c. Class rank at graduation = O

d. GPA = I

e. Speed of a baseball pitch = R

3.

a. Hair length = R
b. Species of tree = N
c. Military rank = O
d. Political party membership = N
e. Yearly income = R

5.

a. Literature genres = N
b. Relative academic position in one's graduating class = O
c. Hair length = R
d. Job salary = R
e. Self-confidence = I

7.

a. Inches of rainfall = C
b. GPA = C
c. Speed of a baseball pitch = C
d. Time to solve a puzzle = C
e. Level of depression = C

9.

a. Number of TVs in the household = D
b. Level of extraversion = C
c. Color saturation level = C
d. How tired you feel right now = C
e. How many awards you won as a child = D

11.

a. IQ = 115.5 and 116.5
b. Height = 66.25 and 66.75
c. Age = 15 and 25
d. Driving speed = 62.5 and 67.5
e. Olympic performance = 9.65 and 9.75

13.

a. GPA = 3.195 and 3.205
b. Test score = 83.35 and 83.45
c. Miles to work = 17.5 and 22.5
d. Feet of snow = 2.5 and 3.5
e. Shoe size = 10.75 and 11.25

Module 3. Frequency and Percentile Tables

1.

Number of Dates	*Frequency*	*Cumulative Frequency*
8	1	40
[7	0	39] may omit
6	1	39
5	2	38
4	2	36
3	6	34
2	7	28
1	6	21
0	15	15

3.

Number of Pets	*Frequency*	*Cumulative Frequency*
7	1	20
6	0	19
5	1	19
4	3	18
3	4	15
2	5	11
1	4	6
0	2	2

5.

Number of Dates	*Frequency*	*Relative Frequency*	*Cumulative Relative Frequency*
8	1	.025	1.00
[7	0	.000	.975 may omit]
6	1	.025	.975
5	2	.050	.950
4	2	.050	.900
3	6	.150	.850
2	7	.175	.700
1	6	.150	.525
0	15	.375	.375

7.

Number of Pets	*Frequency*	*Cumulative Frequency*	*Percentage*	*Cumulative Percentage*
7	1	20	5.0	100.0
6	0	19	0.0	95.0

(Continued)

(Continued)

Number of Pets	*Frequency*	*Cumulative Frequency*	*Percentage*	*Cumulative Percentage*
5	1	19	5.0	95.0
4	3	18	15.0	90.0
3	4	15	20.0	75.0
2	5	11	25.0	55.0
1	4	6	20.0	30.0
0	2	2	10.0	10.0

9.

a.

Feet	*Frequency*
105–109	1
100–104	2
95–99	2
90–94	3
85–89	7
80–84	4
75–79	2
70–74	3
65–69	4
60–64	2
55–59	3
50–54	3

b.

Feet	*Frequency*
100–109	3
90–99	5
80–89	11
70–79	5
60–69	6
50–59	6

c.

Feet	*Frequency*
90–109	8
70–89	16
50–69	12

d. The 10-ft intervals give the clearest picture of the data.

11.

a. b. c.

2-Meal Interval		5-Meal Interval		10-Meal Interval	
No. of Skips	*Frequency*	*No. of Skips*	*Frequency*	*No. of Skips*	*Frequency*
14–15	1	15–19	1	10–19	3
12–13	1	10–14	2	0–9	27
10–11	1	5–9	14		
8–9	3	0–4	13		
6–7	7				
4–5	10				
2–3	4				
0–1	3				

d. The 2-meal interval seems most informative, although the 5-meal interval might also be informative. The 10-meal interval is not informative.

13. Answers will vary. Here is the calculation for a score of 83:

$$PR = \frac{18 + (2/1)(83 - 82.5)}{50} \times 100$$

$$PR = \frac{18 + (2)(.5)}{50} \times 100$$

$$PR = \frac{18 + 1}{50} \times 100$$

$$PR = \frac{19}{50} \times 100$$

$$PR = (.38)(100)$$

$$PR = 38$$

15. Exact score falling at the 25th percentile rank in Table 3.6:

$$\begin{aligned} X_{25} &= LL + (i/f_i)(\text{cum } f_{UL} - \text{cum } f_{LL})(0.5) \\ &= 79.5 + (1/2)(28 - 24)(0.5) \\ &= 79.5 + 1.0 \\ &= 80.5 \end{aligned}$$

Module 4. Graphs and Plots

1.

a.

Stem	*Leaf*
1	7
2	0 1 3 4 5 5 5 6 6 7 7 7 7
3	0 1 2 3 4 4 5 5 5 6 6 7 8 9

OR

Stem	*Leaf*
1	
1	7
2	0 1 3 4
2	5 5 5 6 6 7 7 7 7
3	0 1 2 3 4 4
3	5 5 5 6 6 7 8 9
4	

b.

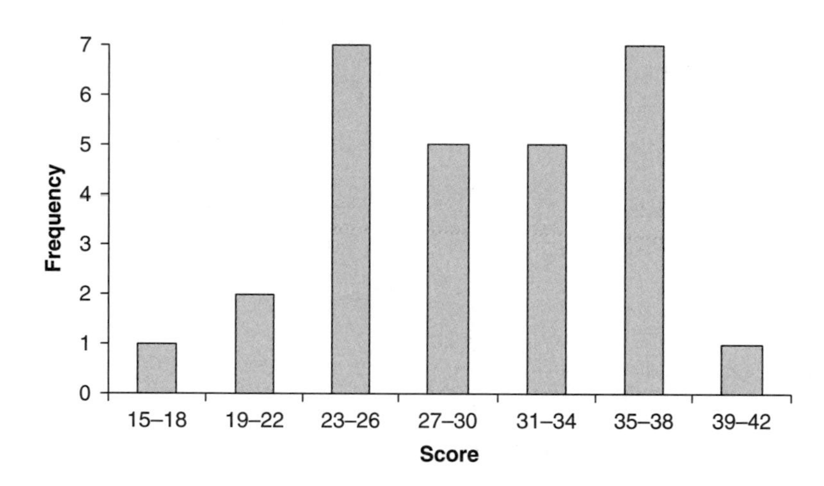

c.

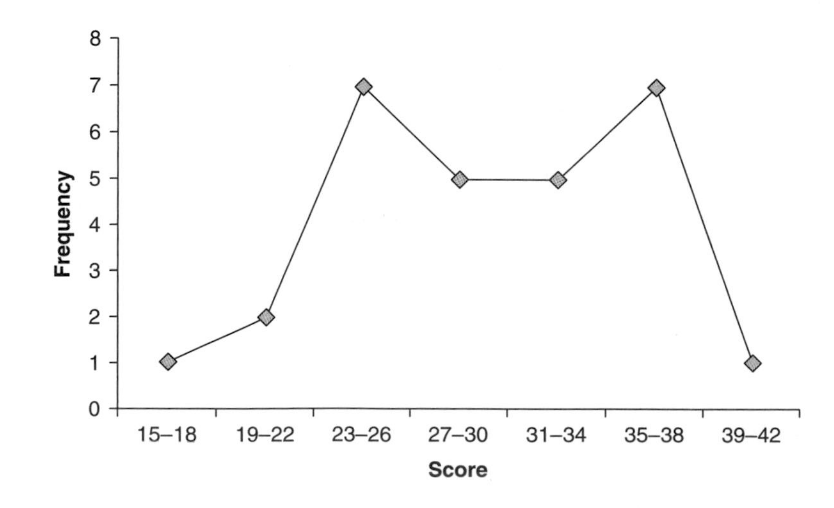

3.

a.

Stem	*Leaf*
2	1
1	2 2 2 3 3 3 3 4 5 5 5 5 5 5 6 7 7 8
0	9

b.

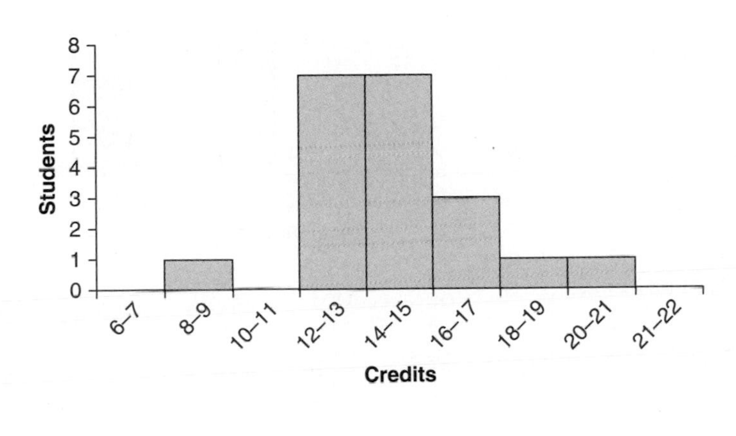

c.

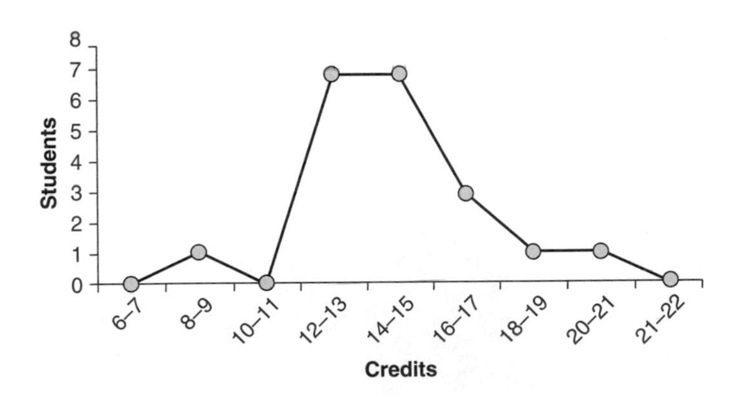

5.

a. SAT scores for graduate students = NS
b. Age at which babies start to walk = NO
c. Self-esteem of Olympic medal winners = NS
d. Age of enlisted military personnel = PS
e. Running speed for sprinters in the Olympics and in the Special Olympics = BI

7. Positively skewed.
9. Positively skewed.
11.
   a.

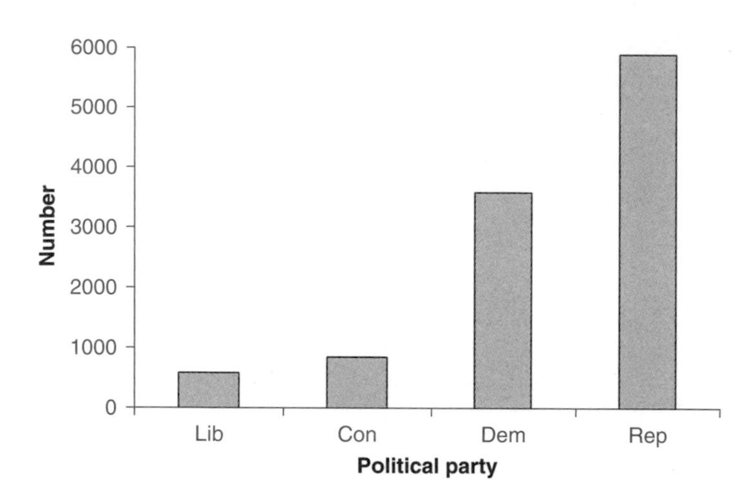

   b. Republican = 53.85%
      Democratic = 32.93%
      Conservative = 7.64%
      Liberal = 5.58%

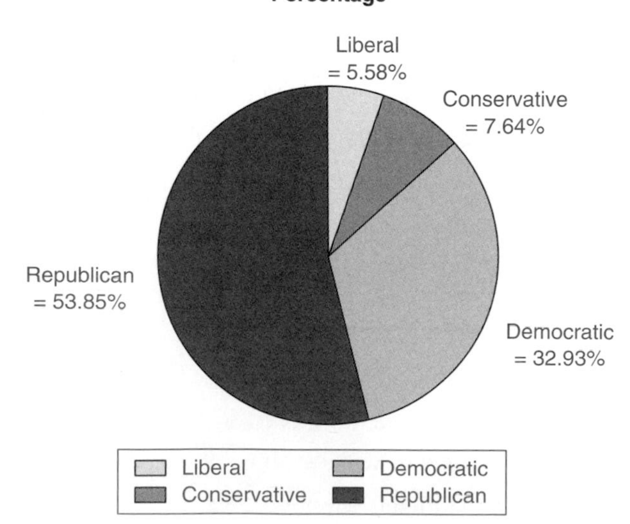

13.

a.

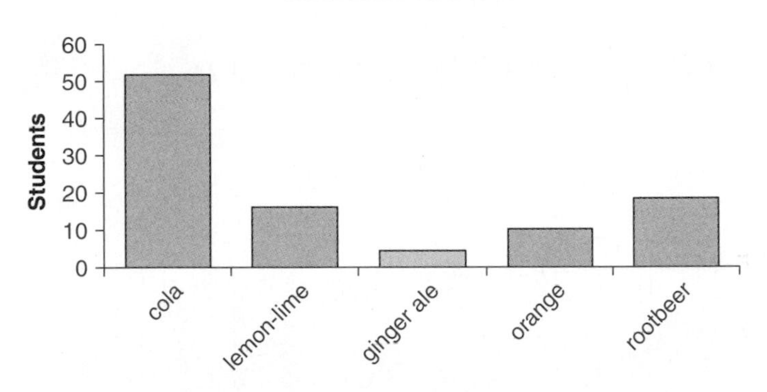

b.

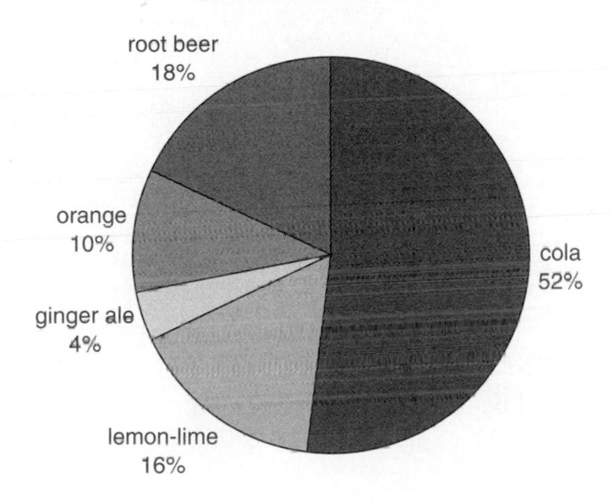

15.

a. Feet of snowfall per storm last year = SL, H, F
b. Brands of carbonated soft drinks drunk = B, P
c. High school class rank of entering college freshmen = SL, H, F
d. GPA of 500 freshmen = SL, H, F
e. Speed of a baseball pitch per team member = SL, H, F

17.

a. Consumer preference for 2-door, 4-door, or hatchback style cars = B, P
b. Number of miles rowed by 30 campers = SL, H, F
c. Number of votes cast per candidate in a televised talent contest = SL, H, F
d. Number of ribbons won in a series of athletic contests = SL, H, F
e. Perfume fragrance preferences of 50 female shoppers = B, P

Module 5. Mode, Median, and Mean

1.

   a. 3

   b. $2.5 + 1\frac{(68 - 27)}{54} = 2.5 + \frac{41}{54} = 2.5 + .76 = 3.26$

3.

   a. 7

   b. $6.5 + 1\frac{(25 - 17)}{10} = 6.5 + .8 = 7.3$

5.

   a. 84

   b. $84.5 + 1\frac{(25 - 23)}{5} = 84.5 + \frac{2}{5} = 84.5 + .4 = 84.9$

   c. $\frac{4173}{50} = 83.46$

7.

   a. 2

   b. 2

   c. 2.5

9.

   a. Negatively skewed

   b. The mean should be lower than the median.

11.

   a. Negatively skewed

   b. Not quite, so the data must be somewhat "lumpy."

13. Median, because the data will probably be negatively skewed due to improvement in speed as days continue.

15. Mean, because the data will probably be normally distributed and further statistics can be calculated from the mean.

Module 6. Range, Variance, and Standard Deviation

1.

   a. 98 – 56 = 42

   b.

X	$X - M$	$(X - M)^2$	X	$X - M$	$(X - M)^2$
92	8.54	72.9316	73	–10.46	109.4116
83	–0.46	0.2116	84	0.54	0.2916

X	X – M	(X – M)²	X	X – M	(X – M)²
98	14.54	211.4116	96	12.54	157.2516
76	–7.46	55.6516	88	4.54	20.6116
84	0.54	0.2916	62	–21.46	460.5316
87	3.54	12.5316	75	–8.46	71.5716
66	–17.46	304.8516	79	–4.46	19.8916
82	–1.46	2.1316	87	3.54	12.5316
93	9.54	91.0116	91	7.54	56.8516
85	1.54	2.3716	90	6.54	42.7716
86	2.54	6.4516	90	6.54	42.7716
87	3.54	12.5316	86	2.54	6.4516
56	–27.46	754.0516	85	1.54	2.3716
94	40.54	111.0916	89	5.54	30.6916
89	5.54	30.6916	76	–7.46	55.6516
73	–10.46	109.4116	69	–14.46	209.0916
80	–3.46	11.9716	83	–0.46	0.2116
82	–1.46	2.1316	84	0.54	0.2916
85	1.54	2.3716	86	2.54	6.4516
88	4.54	20.6116	94	10.54	111.0916
80	–3.46	11.9716	Σ = 4173		Σ = 3534.4200
84	0.54	0.2916			
72	–11.46	131.3316			
88	4.54	20.6116			
79	–4.46	19.8916			
92	8.54	72.9316			
89	5.54	30.6916			
84	0.54	0.2916			
87	3.54	12.5316			
85	1.54	2.3716			

$M = 4{,}173/50 = 83.460$

$s^2 = 3534.42/50 = 70.688$ (using $/N$)

OR

$s^2 = 3534.42/49 = 72.131$ (using $/n - 1$)

c. $s = \sqrt{70.688} = 8.408$ (using$/N$)

OR

$s = \sqrt{72.131} = 8.493$ (using$/n - 1$)

3.

a. 12 – 4 = 8

b.

X	X – M	(X – M)²
4	–2.9	8.41
8	1.1	1.21
7	0.1	0.01
5	–1.9	3.619

(Continued)

(Continued)

12	5.1	26.01
9	2.1	4.41
8	1.1	1.21
5	−1.9	3.61
7	0.1	0.01
4	−2.9	8.41
Σ = 69		Σ = 56.90

$M = 69/10 = 6.90$

$s^2 = 56.90/10 = 5.690$ (using $/N$)

OR

$s^2 = 56.90/9 = 6.322$ (using $/n - 1$)

c. $s = \sqrt{56.90/9} = 2.385$ (using$/N$)

OR

$s = \sqrt{6.322} = 2.514$ (using$/n - 1$)

5.

a. range = 7
b. variance = 2.850 (3.000, using $/n - 1$)
c. standard deviation = 1.688 (1.732, using $/n - 1$)

7.

a. range = 9
b. variance = 3.930 (4.010, using $/n - 1$)
c. standard deviation = 1.982 (2.003, using $/n - 1$)

Module 7. Percent Area and the Normal Curve

1. 13.59 + 34.13 + 34.13 + 13.59 = 95.44%
3. 34.13% and 34.13%. They are the same.
5. 34.13%
7. Expect average SAT score of 500.
9. 0.13 and 0.13. They are the same.

Module 8. *z* Scores

1. $z = \frac{136 - 100}{15} = \frac{36}{15} = 2.40$

a. 99.18%
b. .82%
c. 49.18%

3. $z = \frac{430 - 500}{100} = \frac{-70}{100} = -.70$

a. 24.20%
b. 75.80%
c. 15.80%

5. $z = \frac{68 - 69.5}{2} = \frac{-1.5}{2} = -.75$

a. 22.66%
b. 77.34%
c. 27.34%

7. $z = \frac{82 - 0}{163} = \frac{82}{163} = +.503 = +.50$

a. 69.15%
b. 30.85%
c. 19.15%

9. $z = \frac{236 - 0}{163} = \frac{236}{163} = +1.45$

a. 92.65%
b. 7.35%
c. 42.65%

11. $z = \frac{270 - 266}{12} = +0.33$

a. at the same point in pregnancy as or earlier than Portia = 62.93%
b. at a later point in pregnancy than Portia = 37.07%
c. between their due date and the point at which Portia gave birth = 12.93%

13. Elm Street: $z = \frac{1.64 - 1.45}{.08} = \frac{.19}{.08} = +2.38$

Third Avenue: $z = \frac{1.67 - 1.53}{.11} = \frac{.14}{.11} = +1.27$

Devon Drive: $z = \frac{1.65 - 1.56}{.12} = \frac{.09}{.12} = +.75$

Morton Lane: $z = \frac{1.62 - 1.49}{.13} = \frac{.13}{.13} = +1.00$

The Elm Street station is charging the highest price relative to its cost.

15. Human: $z = \frac{251 - 266}{12} = \frac{-15}{12} = -1.25$

Chicken: $z = \frac{20.5 - 21}{.5} = \frac{-.5}{.5} = -1.00$

Coral snake: $z = \frac{68.5 - 70.0}{1.00} = \frac{-1.5}{1.00} = -1.50$

Pig: $z = \frac{110 - 114}{2.5} = \frac{-4}{2.5} = -1.60$

Horse: $z = \frac{323 - 340}{17.00} = \frac{-17}{17} = -1.00$

The pig was born most prematurely.

Module 9. Score Transformations and Their Effects

1.

X	$X - M$	$(X - M)^2$
75	18.9	357.21
74	17.9	320.41
62	5.9	34.81
61	4.9	24.01
59	2.9	8.41
58	1.9	3.61
54	–2.1	4.41
50	–6.1	37.21
38	–18.1	327.61
30	–26.1	681.21
Σ = 561		Σ = 1798.90

$M = 561/10 = 56.1$

$s^2 = 1798.90/10 = 179.890$

$s = \sqrt{179.890} = 13.41$

$= 13.41$

a. The mean goes up by 2 points. Addition does change the mean.

b. The standard deviation stays the same. Addition does not change the standard deviation.

3. (Calculations not provided)

a. When we multiply every score by a constant and/or subtract a constant from every score, the mean changes by those same constants. In this case, the mean doubles and then goes down 10 points. The original mean was 91.10, so the new mean is [(2)(91.10)] – 10, which is 172.20.

b. When we double every score, the standard deviation is also doubled. Subtracting points from every score does not change the standard deviation. Therefore, the new standard deviation is (2)(8.53), which is 17.06.

5.

X	$X - M$	$(X - M)^2$
50.0	13.5	182.25
38.5	2.0	4.00
38.0	1.5	2.25
37.5	1.0	1.00
36.5	0.0	0.00
36.0	−0.5	0.25
35.5	−1.0	1.00
34.5	−2.0	4.00
33.5	−3.0	9.00
25.0	−11.5	132.25
Σ = 365.0		Σ = 336.00

$M = 365.0/10 = 36.5$

$s^2 = 336.00/10 = 33.60$

$s = \sqrt{33.60} = 5.797$

a. The mean is first 5 points more than it was previously and then divided into half. Both addition and division change the mean.

b. The standard deviation is half of what it was (within rounding error). Subtraction does not change the standard deviation.

7. 500 + 1.8(100) = 500 + 180 = 680

9.

Bip = 500 + 1.4(100) = 500 + 140 = 640
Bop = 500 + −.3(100) = 500 − 30 = 470
Difference = 640 − 470 = 170.

11. 50 + (−0.3)(10) = 50 − 3 = 47

13.

Barry = 50 + 1.5(10) = 50 + 15 = 65
Harry = 50 + −.2(10) = 50 − 2 = 48
Difference = 65 − 48 = 17

Module 10. Probability Definitions and Theorems

1.

a. Yes
b. Yes
c. No
d. Yes
e. No

3.

a. Yes
b. No

c. No
d. No
e. Yes

5.

a. Yes
b. No
c. Yes
d. No
e. Yes

7.

a. Yes
b. Yes
c. No
d. Yes
e. No

9.

a. MT
b. AT
c. MT
d. AT

11.

a. AT
b. AT
c. MT
d. MT

Module 11. The Binomial Distribution

1.

TTT

TTH

THT

THH

HTT

HTH

HHT

HHH

The probability of each 3-toss outcome is (.5)(.5)(.5) = .125.

a. 1 way to get 3 heads. Therefore, the probability is .125.

b. 3 ways to get 2 heads, 3 ways to get 1 head, and 1 way to get 0 heads. Therefore, the probability is 7 times .125 = .875.

c. 1 way to get 3 heads. Therefore, the probability is .125.

3.

BBBB
BBBG
BBGB
BBGG
BGBB
BGBG
BGGB
BGGG
GBBB
GBBG
GBGB
GBGG
GGBB
GGBG
GGGB
GGGG

The probability of each 4-child outcome is (.5)(.5)(.5)(.5) = .0625.

a. 5 ways to get 2 boys. Therefore, the probability is 5 times .0625 = .3125.

b. 4 ways to get 1 boy, and 1 way to get 0 boys. Therefore, the probability is 5 times .0625 = .3125.

c. 6 ways to get 2 boys, 4 ways to get 3 boys, and 1 way to get 4 boys. Therefore, the probability is 11 times .0625 = .6875.

5.

WWW
WWL
WLW
WLL
LWW
LWL
LLW
LLL

The probability of each 3-race outcome is (.5)(.5)(.5) = .125.

a. 3 ways to win 1 race. Therefore, the probability is 3 times .125 = .375.

b. 3 ways to win 2 races. Therefore, the probability is 3 times .125 = .375.

c. 1 way to win 3 races. Therefore, the probability is .125.

d. 1 way to win 0 races. Therefore, the probability is .125.

e. 3 ways to win 1 race, 3 ways to win 2 races, and 1 way to win 3 races. Therefore, the probability is 7 times .125 = .875.

7. The draws are independent, so the probability remains at .50 regardless of what previously occurred.

9.

a. NNN = 1/8 = .125

b. BBN, BNB, NBB = 3/8 = .375

c. BBB, BBN, BNB, BNN, NBB, NBN, NNB = 7/8 = .875

11.

$$(p+q)^3 = p^3 + 3p^{3-1}q^1 + \frac{3\times 2}{1\times 2}p^{3-2}q^2 + (.5)^3$$

$$(.5+.5)^3 = (.5)^3 + 3(.5)^2(.5)^1 + \frac{3\times 2}{1\times 2}(.5)^1(.5)^2 + (.5)^3$$

$$= 1(.5)^3 + 3(.5)^2(.5)^1 + 3(.5)^1(.5)^2 + (.5)^3$$

a. 0 heads = $(.5)^3 = (.5)(.5)(.5) = .125$

b. 0 heads = $(.5)(.5)(.5) = .125$

1 head = $3(.5)^1(.5)^2 = 3(.5)(.5)(.5) = 3(.125) = .375$

0 heads + 1 head = .125 + .375 = .50

c. 2 heads = $3(.5)^2 + (.5)^1 = 3(.5)(.5)(.5) = 3(.125) = .375$

3 heads = $(.5)^3 = (.5)(.5)(.5) = .125$

2 heads + 3 heads = .375 + .125 = .50

13.

$$(p+q)^4 = p^4 + 4p^3q^1 + \frac{4\times 3}{1\times 2}p^2q^2 + \frac{4\times 3\times 2}{1\times 2\times 3}p^1q^3 + q^4$$

$$(.5+.5)^4 = 1(.5)^4 + 4(.5)^3(.5)^1 + \frac{4\times 3}{1\times 2}(.5)^2(.5)^2 + \frac{4\times 3\times 2}{1\times 2\times 3}(.5)^1(.5)^3 + (.5)^4$$

$$= 1(.5)^4 + 4(.5)^3(.5)^1 + 6(.5)^2(.5)^2 + 4(.5)^1(.5)^3 + 1(.5)^4$$

a.

2 races = $6(.5)^2(.5)^2 = 6(.5)(.5)(.5)(.5) = 6(.0625) = .375$

1 race = $4(.5)^1(.5)^3 = 4(.5)(.5)(.5)(.5) = 4(.0625) = .25$

0 races = $1(.5)^4 = (.5)(.5)(.5)(.5) = .0625$

2 races + 1 race + 0 races = .375 + .25 + .0625 = .6875

b. 2 races = $6(.5)^2(.5)^2 = 6(.5)(.5)(.5)(.5) = 6(.0625) = .375$

c. 4 races = $1(.5)^4 = (.5)(.5)(.5)(.5) = .0625$

d. 0 races = $1(.5)^4 = (.5)(.5)(.5)(.5) = .0625$

e.

1 race = $4(.5)^1(.5)^3 = 4(.5)(.5)(.5)(.5) = 4(.0625) = .25$

2 races = $6(.5)^2(.5)^2 = 6(.5)(.5)(.5)(.5) = 6(.0625) = .375$

3 races = $4(.5)^3(.5)^1 = 4(.5)(.5)(.5)(.5) = 4(.0625) = .25$

4 races = $1(.5)^4 = (.5)(.5)(.5)(.5) = .0625$

1 race + 2 races + 3 races + 4 races = .25 + .375 + .25 + .0625 = .9375

15. $(p+q)^4 = p^4 + 4p^3q^1 + \frac{4\times 3}{1\times 2}p^2q^2 + \frac{4\times 3\times 2}{1\times 2\times 3}p^1q^3 + q^4$

a. $1(.5)^4 = .0625$

b. $4(.5)^3(.5)^1 = 4(.125)(.5) = .25$

c. $1(.5)^4 + 4(.5)^3(.5)^1 + [4(3)/1 \times 2](.5)^2(.5)^2 = .0625 + .25 + 6(.25)(.25) = .6875$

17.

a. .016

b. .234 + .094 + .016 = .344

c. .094 + .016 = .110

19.

a. .156 + .031 = .187

b. .156

c. .031

d. .031

e. .312 + .312 + .156 + .031 = .811

21.

a. .0312

b. .3125

c. .3125 + .1562 = .4687

23.

a. .009 + .001 + .000 + .000 = .010

b. .300

c. .000

d. .117 + .039 = .156

25.

a. .0305 + .0021 + .0001 + .0000 + .0000 = .0327

b. .0021

c. .7351

27. .006 + .000 = .006

29. $N = 4$, $p = .25$. At least three correct = .0469 + .0039 = .0508

31. There is no column for $p = .167$ (which is 1/6).

33. .016 + .003 + .000 + .000 + .000 = .019

Module 12. Sampling, Variables, and Hypotheses

1.
   a. Cluster
   b. Convenience
   c. Stratified random
   d. Simple random

3.
   a. There was no sampling because every package was X-rayed.
   b. No inference from sample to population is necessary because every package was X-rayed.

5. Simple random.

7. Convenience

9.

   IV = holding a job or not

   DV = No. of video games played per week

11.

   IV = heavy drinkers or not

   DV = size of hippocampus

13.

   IV = gender

   DV = number of recreational activities signed up for

15.

   IV = asking questions or not

   DV = student retention of material

17. 9 (job), 11 (drinking), 12 (bullied), 13 (gender), and 14 (socioeconomic) are post hoc studies. When the treatments are "found" rather than randomly assigned, there can be confounding variables.

19.
   a. Students who hold a job while attending school will play fewer video games per week than those who don't hold a job while attending school.
   b. Students who hold a job while attending school will differ in the number of video games they play per week from students who don't hold a job while attending school.
   c. There will be no significant difference in the number of video games played per week between students who hold a job while attending school and those who don't hold a job while attending school.

21.

a. Adolescents and young adults who are heavy drinkers will have a smaller hippocampus than those who are not heavy drinkers.

b. The size of the hippocampus will differ between adolescents and young adults who are heavy drinkers and those who are not.

c. There will be no significant difference in the size of the hippocampus between adolescents and young adults who are heavy drinkers and those who are not.

23.

a. Boys sign up for significantly more recreational activities than girls do.

b. Boys and girls sign up for a significantly different number of recreational activities.

c. There is no significant difference in the number of recreational activities that girls and boys sign up for.

25.

a. Student retention of material is higher when the teacher asks questions at the start of class.

b. Student retention of material will differ when the teacher asks questions at the start of class from when she does not ask questions at the start of class.

c. Asking questions at the start of class has no effect on student retention of material.

Module 13. Errors and Significance

1.

a. Top row

b. Right column

c. Type 2

3. Type 1—he really does not have the mono germ, but the test found that he does.

5. Type 2—there really is a difference in the number of fish in the two lakes, but Abe and Gabe thought that there was no difference in the number of fish.

7. Type 2—males really do get more speeding citations, but the officer did not find the difference in this study.

Module 14. The *z* Score as a Hypothesis Test

1.

a. Toward the center of the distribution

b. In one of the tails of the distribution

3.

a. There is no significant difference between Monica's SAT score and the average SAT score.

b. Monica's SAT score is significantly below average.

c. $Z = \frac{440 - 500}{100} = \frac{-60}{100} = -.60$

About 27% of SAT takers score as low as or lower than what Monica scored.

d. SAT scores this low occur too often for Monica's score to be considered significantly below average, so I would not reject the null hypothesis.

5.

a. Null hypothesis: There is no significant difference between Dominic's height and that of the average U.S. adult male.

b. Dominic is significantly taller than the average U.S. adult male.

c. z = (72.0 – 69.5)/2.00 = +1.25. The probability of being Dominic's height (or taller) from a population with the given average height is 10.56%.

d. Dominic is taller than average but more than 10% of the population also is as tall as or taller than Dominic. His height does not seem rare enough to reject the null hypothesis.

Module 15. Standard Error of the Mean

1. It will be normally distributed. According to the central limit theorem, the sampling distribution of the mean will be normally distributed regardless of the shape of the underlying distribution, as long as sample size is sufficient.

3.

a. $\frac{100}{\sqrt{25}} = \frac{100}{5} = 20.00$

b. $\frac{100}{\sqrt{121}} = \frac{100}{11} = 9.09$

c. $\frac{100}{\sqrt{400}} = \frac{100}{20} = 5.00$

5.

a. $\frac{0.4}{\sqrt{9}} = \frac{0.4}{3} = 0.133$

b. $\frac{0.4}{\sqrt{16}} = \frac{0.4}{4} = 0.100$

c. $\frac{0.4}{\sqrt{25}} = \frac{0.4}{5} = 0.080$

7. It will be normally distributed because of the central limit theorem.

9. Answers will vary. However, the Size 3 standard error of the mean will be larger than the Size 7 standard error of the mean. Typical values fall between 1.3 and 1.6 for Size 3 samples and between 0.8 and 1.1 for Size 7 samples.

Module 16. Normal Deviate *Z* Test

1. The instructor needs to know the average resting heart rate for the population and the standard deviation of the resting heart rate for the population.

3.

   a. There is no significant difference between the average weight of this sample of cereal boxes and the company's announced average weight.

   b. $M = 17.53$

   $$\sigma_M = \frac{.50}{\sqrt{32}} = \frac{.50}{5.66} = 0.09$$

   $$Z_{\text{norm dev}} = \frac{17.53 - 18.0}{.09} = \frac{-.47}{.09} = -5.22$$

   c. Much less than 1/10 of 1%.

5. $M = 72$

   $$\sigma_M = \frac{10}{\sqrt{30}} = \frac{10}{5.477} = +1.83$$

   $$Z_{\text{norm dev}} = \frac{72 - 50}{1.83} = \frac{22}{1.83} = +12.02$$

   The probability is less than 1/10 of 1%.

7. $M = 9$ months

   $$\sigma_M = \frac{\sigma}{\sqrt{n}} = \frac{2}{\sqrt{36}} = \frac{2}{6} = 0.333$$

   $$Z_{\text{norm dev}} = \frac{9 - 10}{0.333} = \frac{-1}{0.333} = -3.00$$

   The probability is less than .0013.

Module 17. One-Sample *t* Test

1. Because she knows the population standard deviation, she will not need to estimate it via the sample standard deviation.

3.

   a. The actual service time is significantly longer than the advertised service time.

   b. Directional critical t at .05 $\alpha = 1.761$

   c. $M = 9.267$

   $$\sigma^2_{\text{est}} = 6.209$$

   $$\sigma_{\text{est}} = 2.492$$

$$\sigma_M = \frac{2.492}{\sqrt{15}} = \frac{2.492}{3.873} = 0.643$$

$$t = \frac{9.267 - 8.0}{.643} = \frac{1.267}{.643} = 1.970$$

d. Yes

5.

a. The national standard deviation for contributions

b. Ms. DuGood's local contributors give significantly more per person than national contributors do.

c. Directional critical t at .01 α = 2.326

d. $$\sigma_M = \frac{7.82}{\sqrt{632}} = \frac{7.82}{25.14} = 0.31$$

$$t = \frac{28.19 - 27.32}{.31} = \frac{.87}{.31} = +2.81$$

e. Yes

7.

a. Nondirectional hypothesis = There is a difference in the lifespan of this flock of geese and that of the species.

b. Critical t for α .05 = 2.0452

c. $\Sigma(X - M)^2$ = 4980.80

$$\sigma_{est} = \frac{\sum(X - M)^2}{n - 1} = 13.10541$$

[*Note:* The large *SD* is due to the few very large life spans. The life span distribution is not symmetric, and so a *t*-test result might be misleading.]

$$\sigma_M = \frac{\sigma}{\sqrt{n}} = \frac{13.10541}{\sqrt{30}} = \frac{13.10541}{5.477} = 2.393$$

$$t = \frac{(M - \mu)}{\sigma_M} = \frac{8.8 - 7.2}{2.393} = \frac{1.6}{2.393} = 0.67$$

d. No, the life span of this flock of Canada geese is not significantly different from that of the species.

9.

a. Nondirectional hypothesis: This year's contestants run the race in a significantly different amount of time than previous years' contestants.

b. critical t for α .05 = 2.0086.

c.

$$\sigma_M = \frac{\sigma}{\sqrt{n}} = \frac{1.2}{\sqrt{51}} = \frac{1.2}{7.141} = 0.168$$

$$t = \frac{(M - \mu)}{\sigma_M} = \frac{12.1 - 12.3}{0.168} = \frac{-0.2}{0.168} = -0.19$$

d. No, this year's runners did not complete the race in a significantly different time than previous years' runners.

11.

a. No (2.70 needed)

b. No (2.02 needed)

c. Yes (1.68 needed)

Module 18. Interpreting and Reporting One-Sample *t*: Error, Confidence, and Parameter Estimates

1.

a. 14 *df*

b. 3.5%

c. 96.5%

d. $t(14) = 2.38, p = .035$

3.

a. $df = 31 - 1 = 30$

b. Actual Type 1 error = .75% (that's ¾ of 1%)

c. Confidence = 100 − .75 = 99.25%

d. $t(30) = 2.60, p = .0075$

5. 92%

7.

a. 86

b. 86

c. CI = 86 ± (1.984)(1.50) = 86 ± 2.976 = 83.024 and 88.976

d. Smaller, because we are less confident that the interval includes the actual mean for all 700 freshmen.

e. Larger, because as sample size decreases, the standard error and the tabled critical *t* both increase, and hence the confidence interval increases.

9.

a. Point estimate = 27.8 years

b. 27.8 years will fall in the center of the interval

c. 95% CI = $M \pm (t_{\text{crit at directional } \alpha})(\sigma_M)$ = 27.8 ± (2.093)(0.492) = 27.8 ± 1.030 = 26.77 and 28.83

d. With 30 felons rather than 20, the confidence interval will be narrower because the σ_M will decrease, which in turn happens due to the larger divisor in the σ_M formula.

11. CI = .65 ± (2.08)(.03) = .65 ± .06 = .59 and .71

13.

   a. $28.19

   b. CI = $28.19 ± (2.58)(.31) = 28.19 ± .80 = $27.39 and $28.99

15.

   a. Point estimate = 8.8 years

   b. 95% CI = 8.8 ± (2.045)(2.393) = 8.8 ± 4.894 = 3.904 and 13.694. [*Note:* This wide interval is due to the large size of the standard deviation, which in turn was due to the nonsymmetric life span distribution with several extreme scores. The standard deviation is the numerator of the σ_M formula, which resulted in a large σ_M, and hence a wide confidence interval. I had previously warned in this exercise in Module 17 that the nonsymmetric life span distribution might lead to misleading t-test results. It also leads to an overly large confidence interval.]

17. 99% CI = 12.1 ± (2.6778)(0.168) = 12.1 ± 0.4498 = 11.9502 and 12.5498

Module 19. Standard Error of the Difference Between the Means

1. Small sample sizes lead to more error variance, which in turn increases the size of the $\sigma_{M1} - \sigma_{M2}$.

3. If sample size is small, the $\sigma_{M1} - \sigma_{M2}$ is large, and if the $\sigma_{M1} - \sigma_{M2}$ is large, the t will be small. This is because the $\sigma_{M1} - \sigma_{M2}$ is the divisor in the t test. For this reason, small sample sizes can mask the difference that exists in the numerator.

5. $\sigma_{M_1-M_2} = \sqrt{3.27^2 + 2.54^2} = \sqrt{10.69 + 6.45} = \sqrt{17.14} = 4.14$

7. $\sigma_{M_1-M_2} = \sqrt{2.16^2 + 2.56^2} = \sqrt{4.6656 + 6.5536} = \sqrt{11.2192} = 3.350$

9. $\sigma_{M_1-M_2} = \sqrt{1.14^2 + 1.56^2} = \sqrt{1.2996 + 2.4336} = \sqrt{3.7332} = 1.932$

Module 20. t Test With Independent Samples and Equal Sample Sizes

1. We can reject H_0 at .05 α, but cannot reject H_0 at .01 α.

3.

   a. Directional

   b. Customers will rate scented cleaners as more effective than unscented cleaners.

   c. Unscented: Σ = 72, M = 6.00, σ = 1.27, σ_M = 0.33

      Scented: Σ = 89, = 7.42, σ = 1.31, σ_M = 0.38

      $\sigma_{M_1-M_2}$ = 0.50

      t = −2.84

      Critical t = 1.72. Reject H_0.

5.

a. Directional

b. Male college students study less than female college students.

c. Male: $\Sigma = 111$, $M = 11.10$, $\sigma = 3.24$, $\sigma_M = 1.03$

Female: $\Sigma = 109$, $M = 10.90$, $\sigma = 2.64$, $\sigma_M = 0.84$

$\sigma_{M_1 - M_2} = 1.33$

$t = 0.15$

Critical $t = 1.73$. Do not reject H_0

d. Because sex is not assigned, coexisting extraneous variables may explain the results.

7.

Boys: $N = 10$; $M = 63$; $\Sigma(X - M)^2 = 14{,}846$; $\sigma^2_{est} = 1{,}649.556$; $\sigma_{est} = 40.61472$; $\sigma_M = 12.8435$

Girls: $N = 10$; $M = 47$; $\Sigma(X - M)^2 = 8{,}918$; $\sigma^2_{est} = 990.8889$; $\sigma_{est} = 31{,}47839$; $\sigma_M = 9.95434$

$\sigma_{M_1 - M_2} = \sqrt{12.8435^2 + 9.95434^2} = \sqrt{264.0443} = 16.24944$

$t = (63 - 47)/16.24944 = 0.984649$

No, boys and girls do not differ significantly in the time they stay in a video arcade.

9. Doctor visits

Higher co-pay: $N = 14$; $M = 1.357$; $\Sigma(X - M)^2 = 27.2143$; $\sigma^2_{est} = 2.0934$; $\sigma_{est} = 1.4469$; $\sigma_M = 0.3867$

Lower co-pay: $N = 14$; $M = 2.6429$; $\Sigma(X - M)^2 = 37.2143$; $\sigma^2_{est} = 2.8626$; $\sigma_{est} = 1.6919$; $\sigma_M = 0.4522$

$\sigma_{M_1 - M_2} = \sqrt{0.3867^2 + 0.4522^2} = 0.5950$

$t = (1.357 - 2.6429)/0.5950 = -2.16$

Yes, the lower co-pay results in significantly more doctor visits. [*Note:* The calculated t barely met the tabled critical t. Had this test instead been conducted at the .01 α level, we could not have rejected the null hypothesis. This is a good example of the arbitrariness of dichotomous decision making, as mentioned at the end of Module 17 and discussed at length in Modules 18 and 23.]

Module 21. *t* Test With Unequal Sample Sizes

1. The researcher should use an unequal sample size t test.

3. The letter should include these points:

   a. The amount by which students who were taught by phonics outscored students who were taught by whole language was very small and did not reach statistical significance.

b. There may be systematic differences between students in Eastside versus Westside schools that will confound interpretation of the proposed study.

5. Isolated: $M = 5.111$, $SS = 27.776$

   With peers: $M = 3.562$, $SS = 57.94$

   $t = 2.77$

   Critical $t = 2.04$

   Children complete significantly fewer tasks while in the presence of other children than while alone.

7. Boys: $N = 10$; $M = 63$; $\Sigma(X - M)^2 = 14{,}846$

   Girls: $N = 14$; $M = 46$; $\Sigma(X - M)^2 = 10{,}228$

   $\sigma_{M_1 - M_2} = \sqrt{195.3818} = 13.97780$

   $t = (63 - 46)/13.9780 = 1.22$

   No, boys and girls do not differ significantly in the time they stay in a video arcade.

9. Higher co-pay: $N = 14$; $M = 1.357$; $\Sigma(X - M)^2 = 27.2143$

   Lower co-pay: $N = 18$; $M = 2.555$; $\Sigma(X - M)^2 = 46.4444$

   $\sigma_{M_1 - M_2} = \sqrt{0.31178} = 0.5584$

   $t = (1.357 - 2.146)/0.5584 = -2.14$

   Yes, the lower co-pay results in significantly more doctor visits. [*Note:* The calculated t barely met the tabled critical t. Had this test instead been conducted at the .01 α level, we could not have rejected the null hypothesis. This is a good example of the arbitrariness of dichotomous decision making, as mentioned at the end of Module 17 and discussed at length in Modules 18 and 23.]

Module 22. *t* Test With Related Samples

1.
   a. Independent samples
   b. Repeated measures

3.
   a. Independent
   b. Related—matched on socioeconomic status

5. Video game playing time: Cannot be conducted as a repeated measures study because the IV is gender, which cannot change. Participants are able to experience only one of the conditions.

7. Mean of $D = 4.50$, $\Sigma D = 90$, $\Sigma D^2 = 824$

   $t = 4.285$

   Critical $t = 2.54$

   The children who viewed the violent films showed significantly more aggression.

9. Mean of $D = 6.071$, $\Sigma D = 85$, $\Sigma D^2 = 1985$

   $t = -2.349$

   Critical $t = 1.77$

   a. Students attained significantly higher bowling scores after the course than before the course.

   b. Compare the students' scores before and after the course with those of students at the same two times but who do not take the course.

11. Mean of $D = -1.2857$; $\Sigma D = -18$; $\Sigma D^2 = 88$

$$t = \frac{\text{Mean of } D}{\sqrt{\frac{\Sigma D^2 - \frac{(\Sigma D)^2}{n}}{n(n-1)}}}$$

$$t = \frac{-1.2857}{\sqrt{\frac{88 - \frac{(-18)^2}{14}}{14(14-1)}}}$$

$= -2.154$

No, the lower co-pay did not result in significantly more doctor visits. [*Note:* The calculated t barely missed being statistically significant. Had the table included a column for a .06 α level and we conducted the test at that level, we would have been able to reject the null hypothesis. This is a good example of the arbitrariness of dichotomous decision making, as mentioned at the end of Module 17 and discussed at length in Modules 18 and 23.]

Module 23. Interpreting and Reporting Two-Sample *t*: Error, Confidence, and Parameter Estimates

1.
   a. 26 *df*
   b. 6%
   c. 94%
   d. $t(26) = 1.986$, $p = .06$

3.
   a. 28 *df*
   b. 3.5%
   c. 96.5%
   d. $t(28) = 2.26$, $p = .035$

5.
   a. 14 *df*
   b. 3.75%
   c. 96.25%
   d. $t(14) = 1.95$, $p = .0375$

7. 93%
9. 97.5%
11.
    a. point estimate = .80 min
    b. .80 min will fall in the center of the interval
    c. CI = .80 ± (1.73)(.47) = .80 ± .81 = −.01 and +1.61
    d. Larger, because we are more confident that the interval includes the actual difference in mingle time between the next group of transfers from nearby and distant colleges.
    e. Smaller, because as sample size increases, the standard error and the tabled critical t both decrease, and hence the confidence interval decreases.

13.
    a. point estimate = 16 min
    b. 16 min will fall in the center of the interval
    c. 99% confidence = 16 ± (2.8784)(16.2944) = 16 ± 46.9018 = −30.9018 (note the sign flip; this means the girls are spending longer than boys) and 62.9018 (with boys spending longer than girls)

15.
    a. point estimate = −1.2858 visits
    b. −1.2858 will fall in the center of the interval
    c. 95% CI = −1.2858 ± (2.0484)(0.59498) = −1.2858 ± 1.2188 = −2.4046 and −0.067

17.
    a. point estimate = 1.72 points
    b. 1.72 points will fall in the center of the interval
    c. CI = 1.72 ± (2.07)(.50) = 1.72 ± 1.04 = .68 and 2.76

19. CI = .19 ± (2.09)(.048) = .19 ± .100 = .09 and .29

Module 24. ANOVA Logic: Sums of Squares, Partitioning, and Mean Squares

1.

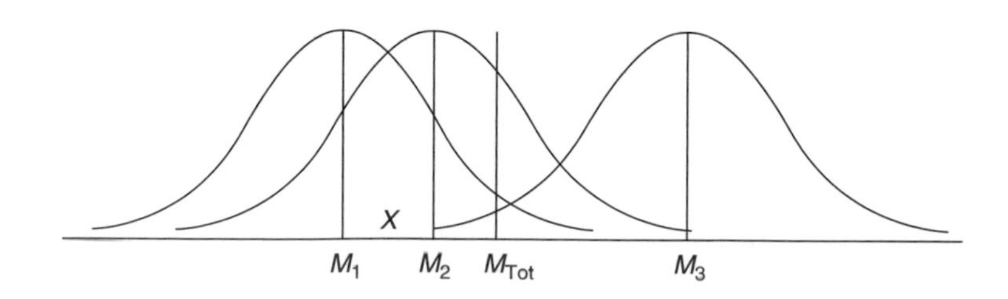

    a. $X - M_{tot}$
    b. $X - M_1$
    c. $M_1 - M_{tot}$

3. $348.76 + 872.45 = 1221.21$
5. $492.56 - 306.55 = 186.01$
7. There must be a calculation error because squared values (variances) are always positive.
9. $800/380 = 2.105$
11. $2.00 = 500/MS_{with}$. Thus, $MS_{with} = 250$
13. $3.00 = MS_{bet}/200$. Thus, $MS_{bet} = 600$
15.
   a. $1,600/380 = 4.21$
   b. $800/760 = 1.05$
   c.

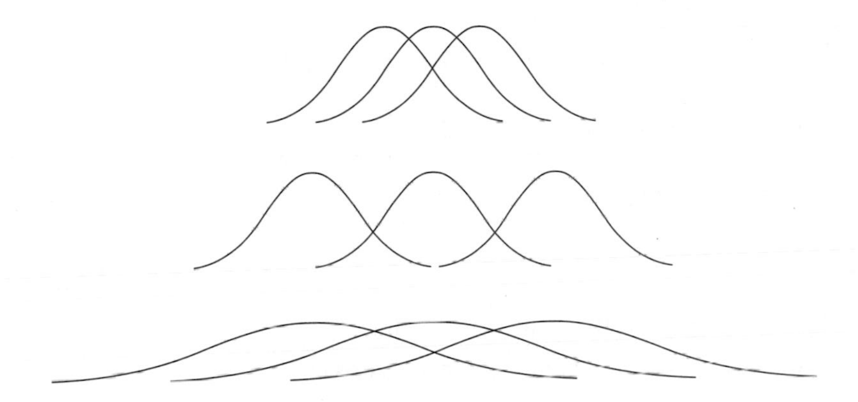

17. The numerator increases while the denominator stays the same. This will increase the value of F.

Module 25. One-Way ANOVA: Independent Samples and Equal Sample Sizes

1.
   a.
      IV = type of diet regime
      DV = number of pounds lost
   b.
      H_0: There will be no significant difference in the amount of weight lost between subjects in the three different diet regimens.
      H_A: There will be a significant difference in the amount of weight lost between subjects in the three different diet regimens.
   c.
      Diet and exercise: $\Sigma X = 99$, $\Sigma X^2 = 989$
      Diet, exercise, and personal trainer: $\Sigma X = 186$, $\Sigma X^2 = 3,400$
      Diet, exercise, and support group: $\Sigma X = 132$, $\Sigma X^2 = 1,626$
      $\Sigma\Sigma X = 417$

$\Sigma\Sigma X^2 = 6{,}015$

$SS_{bet} = 321.5$

$SS_{with} = 863.25$

$MS_{bet} = 160.75$

$MS_{with} = 26.16$

$F = 6.145$

Critical $F(2, 33) = 2.89$ at .05 α, 4.44 at .01 α

Reject H_0 at .01 α

d.

Source	*SS*	*df*	*MS*	*F*	*Sig.*
Between	321.5	2	160.75	6.145**	.01
Within	863.25	33	26.16		
Total	1184.75	35			

e. $F(2, 33) = 6.145, p < .01$

3.

a.

IV = position of difficult (or easy) items

DV = test score (number of questions correct)

b.

H_0: Position of the difficult (or easy) items makes no significant difference in test score.

H_A: Position of the difficult (or easy) items makes a significant difference in test scores.

c.

Difficult first: $\Sigma X = 135$, $\Sigma X^2 = 1{,}945$

Easiest first: $\Sigma X = 143$, $\Sigma X^2 = 2{,}137$

Random order: $\Sigma X = 139$, $\Sigma X^2 = 2{,}031$

$\Sigma\Sigma X = 417$

$\Sigma\Sigma X^2 = 6{,}113$

$SS_{bet} = 3.2$

$SS_{with} = 313.5$

$MS_{bet} = 1.6$

$MS_{with} = 11.61$

$F = 0.138$

Critical $F(2, 27) = 3.35$ at .05 α, 5.49 at .01 α

Do not reject H_0 at either .01 α or .05 α

d.

Source	*SS*	*df*	*MS*	*F*	*Sig.*
Between	3.2	2	1.6	0.138	n.s.
Within	313.5	27	11.61		
Total	316.7	29			

e. $F(2, 27) = .138$, p = n.s.

5.

a. There is no significant difference in weight gain due to where freshmen eat.

b.

	Sum of Squares	*df*	*Mean Square*	*F*	*Sig.*
Between Eating Locations	87.2	2	43.6	4.557	<.05
Within Eating Locations	258.3	27	9.567		
Total	345.5	29			

c. $F(2, 27) = 4.557$, $p < .05$

7.

a. There is no significant difference in time to heat a room due to energy source.

b.

	Sum of Squares	*df*	*Mean Square*	*F*	*Sig.*
Between Energy Sources	36	2	18	0.049	>.10
Within Energy Sources	7,715.625	21	367.4107		
Total	7,751.625	23			

c. $F(2, 21) = 0.049$, $p > .10$

9.

a.

Source	*SS*	*df*	*MS*	*F*
Between	85.37	2	42.685	14.17
Within	171.69	57	3.012	
Total	257.06	59		

b. 3.16

c. >99% confident, $p < .01$

11.

	Sum of Squares	*df*	*Mean Square*	*F*	*Sig.*
Between Groups	246.33	3	82.110	6.020	<.01
Within Groups	927.47	68	13.639		
Total	1173.80	71			

a. 4 groups

b. 18 subjects in each group

c. critical F at .01 = 4.09

Module 26. Tukey HSD Test

1.

Diet and exercise: M = 8.25

Diet, exercise, and personal trainer: M = 15.50

Diet, exercise, and support group: M = 11.00

$$\text{HSD} = 3.48\sqrt{\frac{26.16}{12}} = 3.48\sqrt{2.18} = (3.48)(1.48) = 5.14$$

	D and E	*D, E, and PT*	*D, E, and SG*
D and E	—		
D, E, and PT	6.86	—	
D, E, and SG	2.75	4.50	—

The statistical significance is between diet and exercise versus diet, exercise, and personal trainer.

3.

Difficult first: M = 13.5

Easiest first: M = 14.3

Random order: M = 13.9

$$\text{HSD} = 3.51\sqrt{\frac{11.61}{10}} = 3.51\sqrt{1.161} = (3.51)(1.08) = 3.79$$

	DF	*EF*	*RO*
DF	—		
EF	.80	—	
RO	.40	.40	—

There is no statistical significance between any of the pairs. This is as it should be, because the overall F was not statistically significant.

5.

Home: M = 3.5

Campus dining: M = 6.9

Prepare own: M = 3.1

$$\text{HSD} = 3.51\sqrt{\frac{9.567}{10}} = 3.51\sqrt{0.9567} = (3.51)(0.978) = 3.43$$

The significant F is due to the difference between *campus dining* and *preparing own.* [*Note:* The difference between campus dining and home misses statistical significance by a few hundredths of a pound. With a different rounding stringency, we might find this pair to meet the critical value.]

7. No post hoc test because there was no statistical significance in the overall F.

9. $\text{HSD} = 3.51\sqrt{\frac{22.22}{10}} = 3.51\sqrt{2.222} = (3.51)(1.49) = 5.23$

	G1	*G2*	*G3*
G1	—		
G2	10	—	
G3	20	10	—

The statistical significance lies between all pairs: (1) G1 and G2, (2) G1 and G3, and (3) G2 and G3.

Module 27. Scheffé Test

1.

D and E versus D, E, and PT:

$SS = 315.375$

$MS = 157.6875$

$F = 6.03$

D and E versus D, E, and SG:

$SS = 45.375$

$MS = 22.6875$

$F = 0.87$

D, E, and PT versus D, E, and SG:

$SS = 121.5$

$MS = 60.75$

$F = 2.32$

Critical $F(2, 33) = 2.89$

The statistical significance is between diet and exercise versus diet, exercise, and personal trainer.

3.

Difficult versus easy:

$SS = 3.2$

$MS = 1.6$

$F = 0.14$

Difficult versus random:

$SS = 0.8$

$MS = 0.4$

$F = 0.03$

Easy versus random:

$SS = 0.8$

$MS = 0.4$

$F = 0.03$

Critical $F(2, 27) = 2.96$

There is no statistical significance between any of the pairs. This is as it should be because the overall F was not statistically significant.

Module 28. Main Effects and Interaction Effects

1. Answer will vary.
3. Three
5. Two
7. Four
9. One
11.
    a.
        IV1 = number of hours per week
        IV2 = type of show
        DV = aggression level
    b. Two possible main effects and one possible interaction effect
    c. Two-way ANOVA
    d. 4×3
    e.

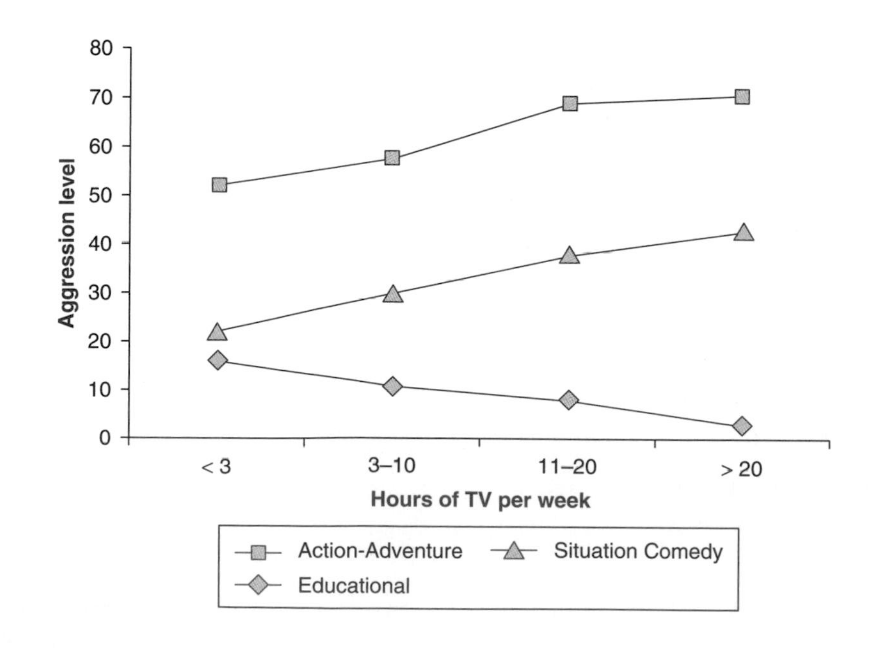

f.

Educational	Action-Adventure	Situation Comedy	
16	52	22	30
11	58	30	33
8	69	38	38.33
3	71	43	39
9.5	62.5	33.25	

g. There appears to be a main effect for type of show. The line heights differ markedly, as do the column means.

h. There appears to be an interaction between type of show and number of hours watched. One of the lines is not parallel, and the column means increase and decrease inconsistently by type of show.

13.

a. IVs = parental marital status and highest parental education; DV = age at which male child commits first crime

b. Two possible main effects and one possible interaction effect

c. Two-way ANOVA or a 2 × 4 ANOVA

d. Tree diagram of design

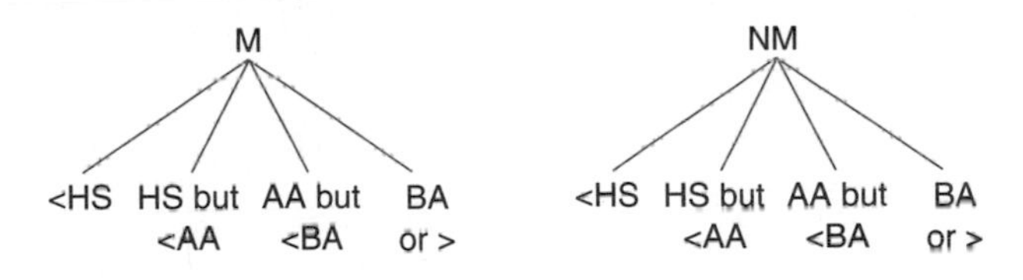

e. Graph of group means

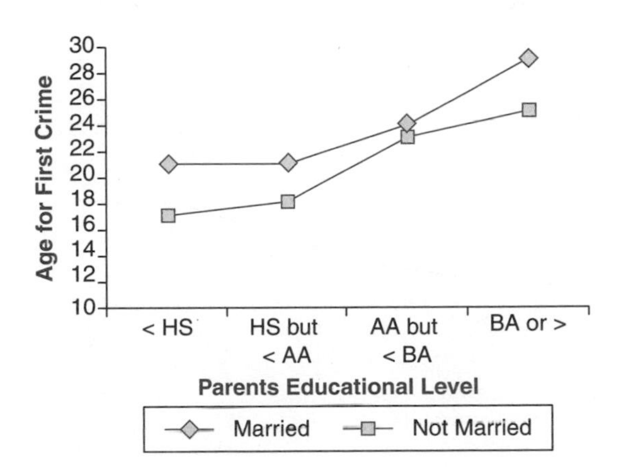

f. Table of group means

	< HS	*HS but < AA*	*AA but < BA*	*BA or >*	*Row Totals*
Married	21	21	24	29	23.75
Not married	17	18	23	25	20.75
Column Totals	19	19.5	23.5	27	

g. Row totals suggest a main effect for parental marital status. Column totals suggest a main effect for parental educational level. Graph shows the same average mean differences for each IV.

h. Means within the body of the chart follow a consistent pattern across levels of the two IVs, so there does not appear to be an interaction effect. Graph shows reasonably parallel lines, so no indication of an interaction.

15.1.

a. No
b. Yes
c. No

15.2.

a. No
b. No
c. Yes

15.3.

a. Yes
b. No
c. No

15.4.

a. Yes
b. No
c. Yes

17.1.

a. No
b. No
c. Yes

17.2.

a. No
b. Yes
c. Yes

17.3.

a. Yes
b. No
c. No

17.4.

a. No
b. Yes
c. No

19.1.

a. No main effect for IV1
b. No main effect for IV2
c. No interaction effect

19.2.

a. No main effect for IV1

b. Yes, main effect for IV2

c. Yes, interaction effect

19.3.

a. Yes, main effect for IV1

b. Yes, main effect for IV2

c. Yes, interaction effect

19.4.

a. Yes, main effect for IV1

b. Slight main effect for IV2, but probably not statistically significant

c. Slight interaction effect, but probably not statistically significant

Module 29. Factorial ANOVA

1.

a.

DV = time to complete a task

IV1 = gender

IV2 = noise level

b. 2 × 2 ANOVA

c. Two possible main effects and one possible interaction effect

d.

$X_{M,\ VN} = 3.52$

$X_{M,\ NN} = 2.79$

$X_{F,\ VN} = 3.39$

$X_{F,\ NN} = 2.52$

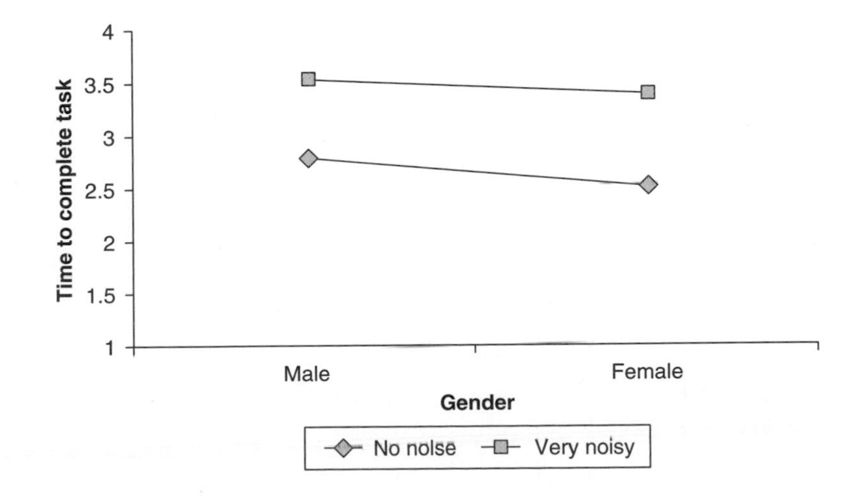

e. I expect a main effect for noise level, but not for gender. I expect no interaction effect.

f., g.

Source	*SS*	*df*	*MS*	*F*	*Sig*
Gender	0.238	1	0.238	3.217	
Noise	3.848	1	3.848	52.016	**
Gender × noise	0.027	1	0.027	0.370	
Within	1.480	20	0.074		
Total	5.593	23			

h. There is a main effect for noise level.

i. There is no interaction effect.

j.

$F_{gender}(1, 20) = 0.238, p > .05$ n.s.

$F_{noise}(1, 20) = 3.848, p < .01$

$F_{gender \times noise}(1, 20) = .027, p > .05$ n.s.

3.

a.

DV = milliseconds

IV1 = age

IV2 = number of stimuli on memory card

b. 3 × 3 ANOVA

c. There are two possible main effects and one possible interaction effect.

d.

$X_{9,1} = 484.2$

$X_{9,3} = 635.4$

$X_{9,5} = 848.2$

$X_{13,1} = 388.2$

$X_{13,3} = 577.0$

$X_{13,5} = 690.0$

$X_{17,1} = 264.4$

$X_{17,3} = 411.0$

$X_{17,5} = 538.0$

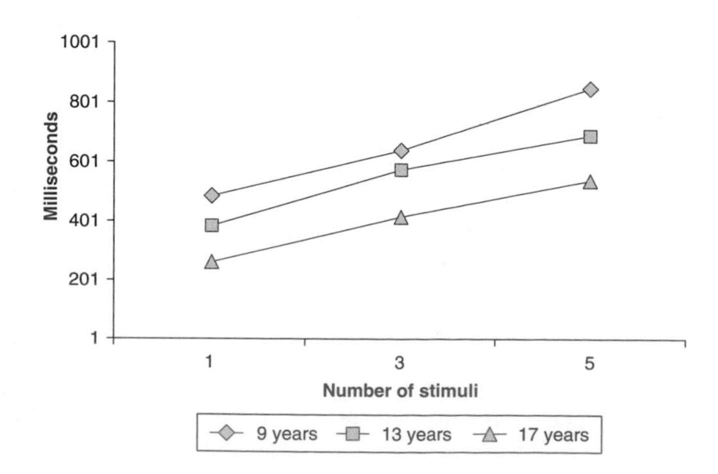

e. I expect a main effect for age and a main effect for number of stimuli. I do not expect an interaction effect.

f., g.

Source	*SS*	*df*	*MS*	*F*	*Sig*
Stimuli	735,710.978	2	367,855.489	182.02	<.01
Age	478,902.978	2	239,451.489	118.48	<.01
Stimuli × age	18,649.822	4	4,662.456	2.31	>.05
Within	72,754.800	36	2,020.967		
Total	1,306,018.578	44			

h. There is a main effect for number of stimuli and a main effect for age.

i. There is no interaction effect.

j.

$F_{stimuli}(2, 36) = 182.02, p < .01$

$F_{age}(2, 36) = 118.48, p < .01$

$F_{stimuli \times age}(4, 36) = 2.31, p > .05$ n.s.

5.

a. ANOVA summary table

Source	*Sums of Squares*	*df*	*Mean Square*	*F*	*Sig.*
Gender	3.267	1	3.267	0.325	.571
Eating locations	136.633	2	68.317	6.796	.002
Gender × eating locations	3.033	2	1.517	0.151	.860
Within	542.800	54	10.052		
Total	685.733	59			

b.

$F_{gender}(1, 54) = 0.325, p = .571$

$F_{location}(2, 54) = 6.796, p = .002$

$F_{gender \times location}(2, 54) = 0.151, p = .860$

7.

a. ANOVA summary table

Source	*Sums of Squares*	*df*	*Mean Square*	*F*	*Sig.*
Construction efficiency	2,961.021	1	2.961.021	7.446	.009
Fuel source	6.292	2	3.146	0.008	.992
Construction Efficiency × fuel source	36.292	2	18.246	0.046	.995
Within	16,702.375	42	397.696		
Total	19,705.979	47			

b.

$F_{const\ effic}(1, 42) = 7.446, p = .009$ [*Note:* The one-way ANOVA in Module 25 was not significant, without this second IV.]

$F_{fuel\ source}(2, 42) = .008, p = .992$

$F_{const\ eff \times fuel\ source}(2, 42) = .046, p = .995$

Module 30. One-Variable Chi-Square: Goodness of Fit

1.

a. There is a significant difference between the proportion of students changing majors at this college and nationwide.

b. 1 *df*

c.

Changed Major		*Didn't Change Major*	
Obs	Exp	Obs	Exp
268	233	379	414

d. $\frac{(268-233)^2}{233}+\frac{(379-414)^2}{414}=8.22$

e. Critical χ^2 = 6.63. The research hypothesis is supported. Reject the null hypothesis.

3.

a. There is a significant difference in the percentage of textbooks sold at five grade levels between LaTawna's business and City University's campus bookstore.

b. 4 *df*

c.

100 Level		*200 Level*		*300 Level*		*400 Level*		*500 Level*	
Obs	Exp	Obs	Exp	Obs	Exp	Obs	Exp	Obs	Exp
387	326	147	159	180	88	135	168	33	141

d. $$\frac{(387-326)^2}{326}+\frac{(147-159)^2}{159}+\frac{(180-88)^2}{88}$$
$$+\frac{(135-168)^2}{168}+\frac{(33-141)^2}{141}=197.70$$

e. Critical χ^2 = 13.28. The research hypothesis is supported. Reject the null hypothesis.

5.

a.

Before 25		*25–30*		*30–35*		*After 35 or Never*	
Obs	Exp	Obs	Exp	Obs	Exp	Obs	Exp
28	44	40	32	19	15	13	9

b. $\frac{(28-44)^2}{44}+\frac{(40-32)^2}{32}+\frac{(19-15)^2}{15}+\frac{(13-9)^2}{9}=10.67$

c. Critical χ^2 = 7.81 at .05 α and 11.34 at .01 α. The research hypothesis is supported at the .05 level but not at the .01 level. Confidence is about = 98%.

7. Expected:

Hotel = (800)(.57) = 456

Rented Condo = (800)(.26) = 208

Campground = (800)(.10) = 80

Relatives = (800)(.07) = 56

a.

Hotel		*Rented Condo*		*Campground*		*Relatives*	
Obs	Exp	Obs	Exp	Obs	Exp	Obs	Exp
504	456	190	208	82	80	24	56

b. $\chi^2 = 3[(O–E)^2/E] = (504–456)^2/456 + (190–208)^2/208 + (82–80)^2/80 + (24–56)^2/56$
$= (48)^2/456 + (-18)^2/208 + (2)^2/80 + (-32)^2/56$
$= 2304/456 + 324/208 + 4/80 + 1024/56$
$= 5.053 + 1.558 + 0.05 + 18.286$
$= 24.947$

c. Reject at <.005 α; confidence is more than 99.5%

9. Expected:
   Hamburgers = (189)(.43) = 81.27
   Hot dogs = (189)(.37) = 69.93
   Sausage rolls = (189)(.12) = 22.68
   Veggie burgers = (189) (.08) = 15.12

a. Table

Hamburgers		*Hot Dogs*		*Sausage Rolls*		*Veggie Burgers*	
Obs	Exp	Obs	Exp	Obs	Exp	Obs	Exp
72	81.27	71	69.93	27	22.68	19	15.12

b. $\chi^2 = \Sigma[(Obs - Exp)^2/Exp] = (72 - 81.27)^2/81.27 + (71 - 69.93)^2/69.93 + (27 - 22.68)^2/22.68 + (19 - 15.12)^2/15.12$
$= (-9.27)^2/81.27 + (1.07)^2/69.93 + (4.32)^2/22.68 + (3.88)^2/15.12$
$= 85.933/81.27 + 1.145/69.93 + 18.662/22.68 + 15.054/15.12$
$= 1.057 + 0.016 + 0.823 + 0.996$
$= 2.892$

c. $p > .10$, so cannot reject null hypothesis

Module 31. Two-Variable Chi-Square: Test of Independence

1. Gender and preferred car color

3. Type of hobby and whether friends or classmates

5. What action taken prior to drifting off to sleep and introversion/extraversion

7.
   a. Class year and vegetable preference
   b. There is no significant relationship between class year and vegetable preference.

c. $(4 - 1)(2 - 1) = (3)(1) = 3\ df$

d.

	Carrots	*Beans*	*Corn*	*Spinach*	
Under class	40 (48)	32 (26)	43 (37)	8 (12)	123
Upper class	38 (30)	10 (16)	17 (23)	12 (8)	77
	78	42	60	20	200

e.

$$\chi^2 = \frac{(40-48)^2}{48} + \frac{(32-26)^2}{26} + \frac{(43-37)^2}{37} + \frac{(8-12)^2}{12} + \frac{(38-30)^2}{30} + \frac{(10-16)^2}{16} + \frac{(17-23)^2}{23} + \frac{(12-8)^2}{8} = 12.43$$

f. Critical $\chi^2 = 11.34$. Yes, reject the null hypothesis.

9.

a. the sex of the dorm's residents and the type of vending machine snack

b. There is no significant relationship between the sex of the dorm's residents and the type of vending machine snack they prefer.

c. $(2 - 1)(2 - 1) = (1)(1) = 1\ df$

d.

	Chips	*Candy*	*Row Total*
Male Dorm	222 (218.7)	94 (97.3)	316
Female Dorm	268 (271.3)	124 (120.7)	392
Column Total	490	218	708

e.

$$\chi^2 = \frac{(222-218.7)^2}{218.7} + \frac{(94-97.3)^2}{97.3} + \frac{(268-271.3)^2}{271.3} + \frac{(124-120.7)^2}{120.7} = 0.292$$

f. Critical $\chi^2 = 3.84$ at $\alpha = .05$ and 6.63 at $\alpha = .01$. Retain the null hypothesis. There is no relationship between sex (as measured by dorm residency) and the snacks preferred by that sex.

11.

	Hotel Room	*Rented Condo*	*Campground*	*Relative's House*	*Row Total*
Below 30 years	58 (74.7)	93 (91.9)	60 (43.8)	39 (39.7)	250
30–50 years	87(104.6)	131(128.6)	62 (61.2)	70 (55.6)	350
Above 50 years	94 (59.8)	70 (73.5)	18 (35.0)	18 (31.8)	200
Column Total	239	294	140	127	800

$df = (R - 1)(C - 1) = 2 \times 3 = 6$

$\chi^2 = \Sigma[(\text{Obs} - \text{Exp})^2/\text{Exp}] = 50.556$. Yes, there is a significant relationship between type of accommodation and age.

13.

	Hamburger	*Hot Dog*	*Sausage Roll*	*Veggie Burger*	*Row Total*
Midweek	70 (72.3)	63 (65.7)	32 (24.4)	17 (19.6)	182
Weekend	96 (93.7)	88 (85.3)	24 (31.6)	28 (25.4)	236
Column Total	166	151	56	45	418

$df = (R - 1)(C - 1) = 1 \times 3 = 3$

$\chi^2 = \Sigma[(\text{Obs} - \text{Exp})^2/\text{Exp}] = 5.153$. No, there is not a significant relationship between vendor food type and time of week.

Module 32. Measures of Effect Size

1.

a. $d = \frac{476.40 - 514.60}{760.75} = -\frac{-38.2}{760.75} = -0.05$

$$r = \sqrt{\frac{(.107)^2}{(.107)^2 + 18}} = \sqrt{\frac{.011}{.011 + 18}} = \sqrt{\frac{.011}{18.011}} = \sqrt{.0006} = .000$$

b. There is no practical difference. Effect size is very small.

3.

a. $d = \frac{11.1 - 10.9}{2.81} = \frac{-.2}{2.81} = 0.07$

$$r = \sqrt{\frac{(.15)^2}{(.15)^2 + 18}} = \sqrt{\frac{.0225}{.0225 + 18}} = \sqrt{\frac{.0225}{18.0225}} = \sqrt{.0013} = .04$$

b. There is no practical difference. Effect size is very small.

5.

a. $\eta = \sqrt{\frac{3.20}{316.70}} = \sqrt{.010} = .10$

b. There is no practical difference. Effect size is small.

7.

a. $\sqrt{\frac{197.70}{(882)(5 - 1)}} = \sqrt{\frac{197.70}{3528}} = \sqrt{.056} = .24$

b. There is no practical difference. Effect size is small, although almost medium.

9.

a. $\sqrt{\dfrac{10.67}{(100)(4-1)}} = \sqrt{\dfrac{10.67}{300}} = \sqrt{.036} = .19$

b. There is no practical difference. Effect size is small.

11.

a. $\phi = \sqrt{\dfrac{36.61}{103}} = \sqrt{.355} = .60$

b. Yes, there is a practical difference. Effect size is large.

Module 33. Power and the Factors Affecting It

1.

a. 20%

b. 88%

3.

a. Decrease power

b. The denominator of the *t* test will be larger.

5.

a. Decrease power

b. The denominator of the *t* test will be larger.

7.

a. Increase power

b —

9. 33 subjects

11. $d = 0.50$

13. 80%

15. 0.287

17. 120 subjects

19. 80%

Module 34. Relationship Strength and Direction

1.

a. Yes

b. No

c. Yes

d. No

3.

a. Yes
b. Yes
c. No
d. No

5.

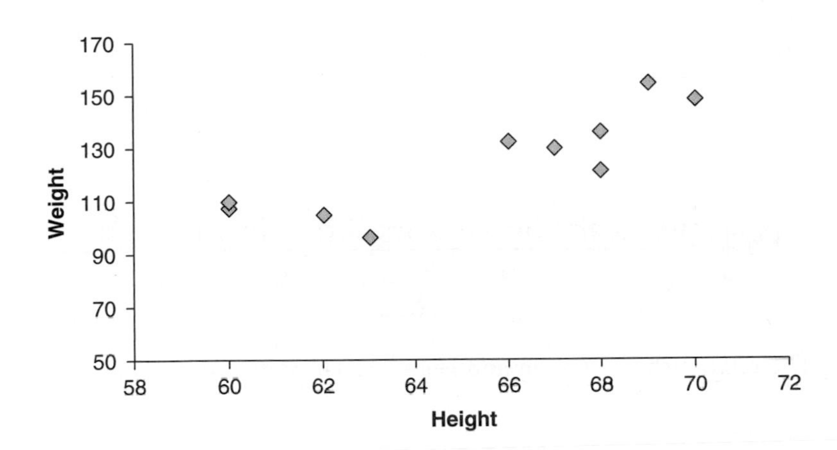

7.

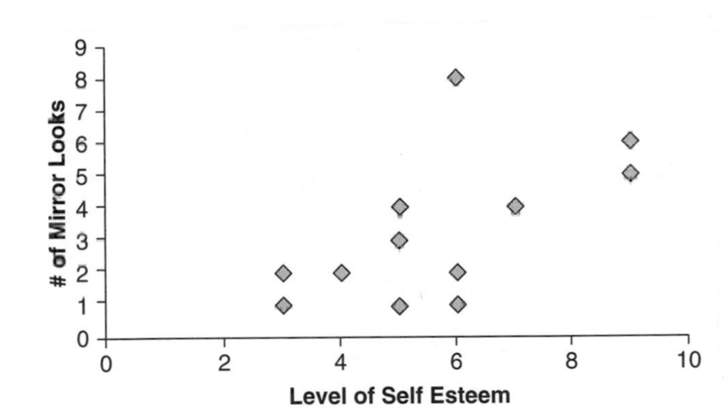

9.

a. +
b. +
c. 0
d. –
e. +

11.

a. 0
b. +
c. –
d. 0
e. +

13.

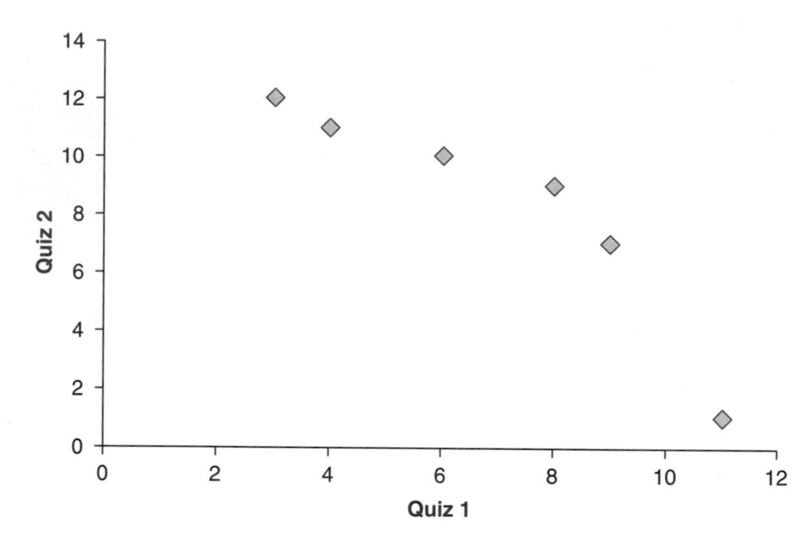

The relationship is strong and negative. There are no outliers.

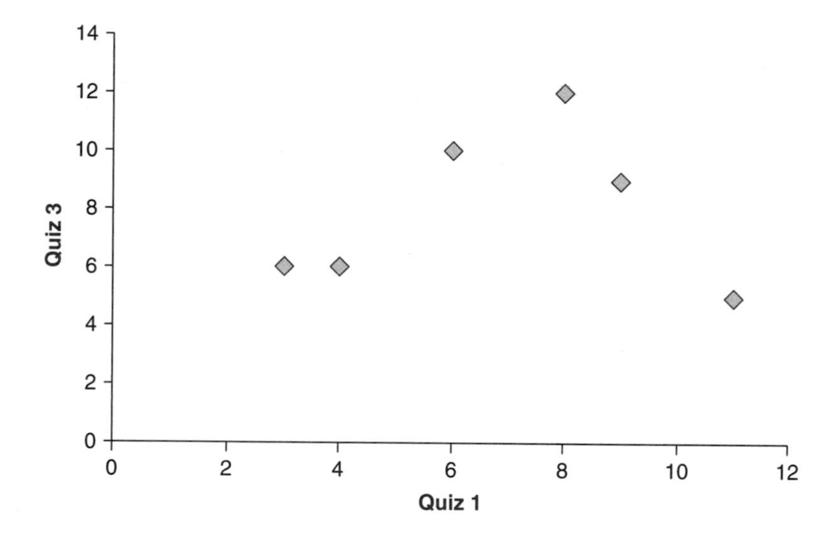

The relationship is either (1) curvilinear or (2) moderate and positive with the removal of Ann as an outlier.

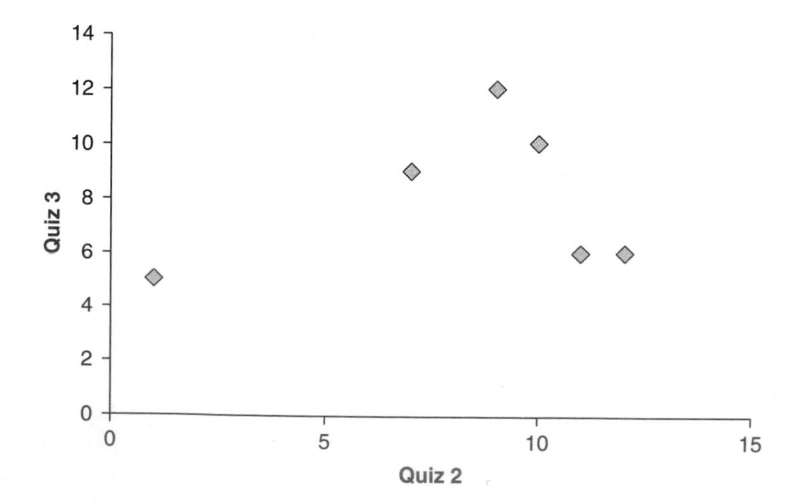

The relationship is either (1) moderate and positive or (2) moderate and negative with the removal of Ann as an outlier.

15. Answers will vary.

Module 35. Pearson *r*

1.

a. No—data are nominal across multiple categories.

b. Yes

c. Yes

d. No—data consist of means rather than raw scores, there are four groups rather than one group, and there is only one measured variable rather than two.

e. No—data are ranks.

3.

a.

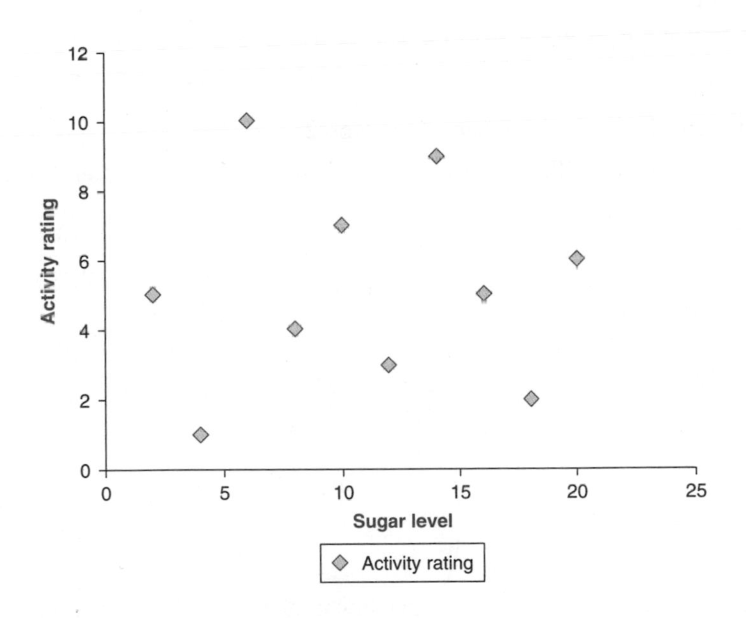

b. I expect a correlation of about zero—little or no relationship.

c. There is no significant relationship between the amount of sugar children eat and their activity level.

d. $\Sigma X = 110$ $\Sigma Y = 52$ $\Sigma XY = 574$

$\Sigma X^2 = 1{,}540$ $\Sigma Y^2 = 346$ $r = +.01$

e. Critical r for a one-tailed hypothesis = .549 at .05 α and .716 at .01 α. Do not reject the null hypothesis.

f. $r(8) = .01, p > .10$ (n.s.)

5.

X = High school GPA; Y = college GPA

a. $\Sigma X = 55.10$ $\Sigma Y = 51.90$ $\Sigma XY = 191.13$

$\Sigma X^2 = 203.07$ $\Sigma Y^2 = 180.37$ $r = +.66$

b. Critical r for a one-tailed hypothesis = .441 at .05 α and .592 at .01 α. Reject the null hypothesis with >99% confidence.

7.

X = Days of rainfall; Y = hotel occupancy rate

$\Sigma X = 39$	$\Sigma Y = 950$	$\Sigma XY = 2{,}930$
$\Sigma X^2 = 163$	$\Sigma Y^2 = 76{,}030$	$r_{XY} = -.91$

9.

X = Mirror looks; Y = self-esteem

$\Sigma X = 50$	$\Sigma Y = 87$	$\Sigma XY = 321$
$\Sigma X^2 = 222$	$\Sigma Y^2 = 551$	$r_{XY} = +.612$

11. Answers will vary.

Module 36. Correlation Pitfalls

1. The score range was too restricted, or the sample of subjects was too homogeneous, or the measuring instrument was unreliable, or mere error variance (chance) occurred.

3. The sample size is too large → Type 1

   The sample is too homogeneous → Type 2

   The researcher is willing to reject the null hypothesis with even a very small effect size → Type 1

5.

a. $(-0.90)^2 = .81$

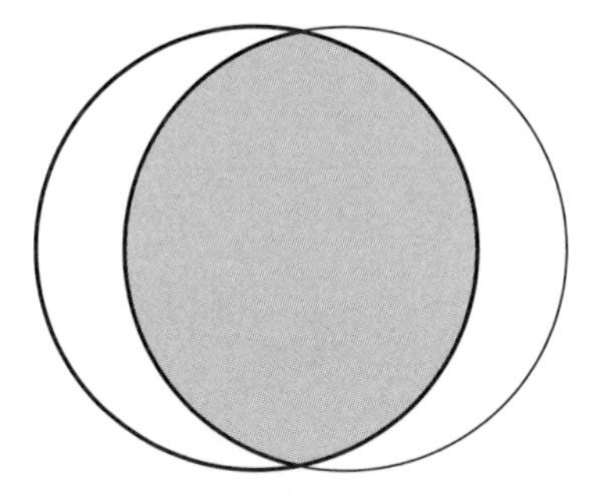

b. $(+0.25)^2 = .0625$

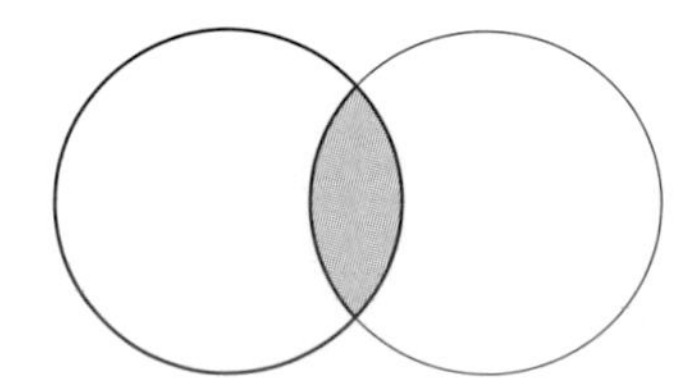

c. $(-1.00)^2 = 1.00$

7. The heterogeneity of the sample will increase the calculated correlation coefficient.

9.

   a. It makes them feel better, increases their physical conditioning, gets the endorphins flowing, improves the mood, and so on.

   b. Most older people are retired and so have more available time to volunteer.

   c. Wealth buys better health care, which leads to living longer. Wealth also allows people to work less, and hence have more leisure time for volunteering.

11. "Twenty percent of the variation in activity level is related to meat in the diet." Not a valid conclusion. Correlation is not common variance. The common variance is only $(.20)^2 = .04$.

Module 37. Linear Prediction

1.

   a.

$$\begin{aligned} Y' &= .72\left(\frac{.57}{180}\right)X - .72\left(\frac{.57}{180}\right)1050 + 2.14 \\ &= .72(.00316)X - .72(.00316)(1050) + 2.14 \\ &= .00228X - .00228(1050) + 2.14 \\ &= .00228X - 2.389 + 2.14 \\ &= .00228X - .249 \end{aligned}$$

   b. $$\begin{aligned} Y' &= .00228(700) - .249 \\ &= 1.596 - .249 \\ &= 1.347 \end{aligned}$$

   c. $$\begin{aligned} Y' &= .00228(1300) - .249 \\ &= 2.964 - .249 \\ &= 2.715 \end{aligned}$$

3.

a.

$$
\begin{aligned}
Y' &= .69\left(\frac{1.853}{1.912}\right)X - .69\left(\frac{1.853}{1.912}\right)(8.1) + 7.1 \\
&= .69(.96914)X - .69(.96914)(8.1) + 7.1 \\
&= .66871X - .66871(8.1) + 7.1 \\
&= .66871X - 5.417 + 7.1 \\
&= .66871X + 1.683
\end{aligned}
$$

b. $Y' = .66871(7) + 1.683$
$= 4.6810 + 1.683$
$= 6.364$

5.

a.

$$
\begin{aligned}
Y' &= .66\left(\frac{.238}{.219}\right)X - .66\left(\frac{.238}{.219}\right)(3.673) + 3.460 \\
&= .66(1.0868)X - .66(1.0868)(3.673) + 3.460 \\
&= .71729X - .71729(3.673) + 3.460 \\
&= .71729X - 2.635 + 3.460 \\
&= .71729X + .825
\end{aligned}
$$

b. $Y' = .71729(3.8) + .825$
$= 2.726 + .825$
$= 3.539$

7.

a.

$$
\begin{aligned}
Y' &= -.913\left(\frac{8.643}{1.815}\right)X - (-.913)\left(\frac{8.643}{1.815}\right)(3.250) + 79.167 \\
&= -.913(4.76198)X - (-.913)(4.76198)(3.250) + 79.167 \\
&= -4.34769X - (-14.12999) + 79.167 \\
&= -4.34769X + 14.12999 + 79.167 \\
&= -4.34769X + 93.29699
\end{aligned}
$$

b. $Y' = -4.34769(3) + 93.29699$
$= -13.04307 + 93.29699$
$= 80.25\%$ hotel occupancy

9.

a.

$$
\begin{aligned}
Y' &= -.612\left(\frac{1.988}{1.821}\right)X - .612\left(\frac{1.988}{1.821}\right)(5.80) + 3.333 \\
&= .612(1.09171)X - .612(1.09171)(5.80) + 3.33 \\
&= .66813X - 3.87513 + 3.33 \\
&= .66813X - .54513
\end{aligned}
$$

b. $Y' = .66813(4) - .54513$
$= 2.67252 - .54513$
$= 2.13$ mirror looks

Module 38. Standard Error of Prediction

1. CI = 78.5 ± (2)2.4 = 78.5 ± 4.8 = 73.7 and 83.3

3.

a.

$$\begin{aligned} s_{YX} &= 1.853\sqrt{1-(.69)^2} \\ &= 1.853\sqrt{1-.476} \\ &= 1.853\sqrt{.524} \\ &= (1.853)(.724) \\ &= 1.34 \end{aligned}$$

b. 6.364 ± (1)(1.34)
= 6.364 ± 1.34
= 5.024 and 7.704

5.

a.

$$\begin{aligned} s_{YX} &= .238\sqrt{1-(-.66)^2} \\ &= .238\sqrt{1-.436} \\ &= .238\sqrt{.564} \\ &= (.238)(.751) \\ &= 0.179 \end{aligned}$$

b. 3.539 ± (3)(.179)
= 3.539 ± .537
= 3.002 and 4.076

7.

a.

$$\begin{aligned} s_{YX} &= 8.643\sqrt{1-(-.913)^2} \\ &= 8.643\sqrt{1-.8336} \\ &= 8.643\sqrt{.1664} \\ &= 8.643(.4079) \\ &= 3.525 \end{aligned}$$

b. 95% CI = 80.25 ± (2)(3.525) = 80.25 ± 7.05 = 73.20% and 87.30% occupancy

9.

a.

$$\begin{aligned} s_{YX} &= 1.988\sqrt{1-(.612)^2} \\ &= 1.988\sqrt{1-.3745} \\ &= 1.988\sqrt{.6255} \\ &= 1.988(.79088) \\ &= 1.57 \end{aligned}$$

b. 68% CI = 2.21 ± (1)(1.57) = 2.21 ± 1.57 = 0.64 and 3.78 mirror looks

11. It will decrease the size of the standard error of prediction.
13. The decreased correlation coefficient will increase the width of the confidence interval.

Module 39. Introduction to Multiple Regression

1. 1 – .74 = .26
3. This is not possible. R^2 + residual must add to 1.00
5. The R^2 should go down. One source of criterion prediction has been lost.
7. The R^2 should stay the same. The order of the predictors will change their individual prediction values, but their sum will remain the same.
9.
   a. .388 or 38.8%
   b. The regression *SS* divided by the total *SS*
   c. 1 – .388 = .612 or 61.2%
   d. Poor choice of predictor variables
11.
   a. The teacher's years of teaching experience
   b. The larger the class size, the lower the student achievement.
   c. No, the unstandardized weights cannot be compared directly because their units differ.
   d. $Y' = (-.456)(7) + (.232)(12) + (.129)(24) + 78.241 = 80.929$
13.
   a. 125.74/269.01 = .4674
   b. $\sqrt{.4674} = .6837$
   c. 2
   d. $N = 12$. There were 6 subjects per predictor, which is too few for the results to be stable.
   e. No. $F = 3.949$, with an α of 5.5%.

Module 40. Selecting the Appropriate Analysis

1.
   Pearson *r*:
   a. One group
   b. Interval scale
   c.

Subject	*Var1*	*Var2*
S_1		
S_2		
S_3		
S_4		
S_5		
(etc.)		

t test:

a. Two groups

b. Interval scale

c.

Group 1	*Group 2*
M	*M*

F test:

a. Three or more groups

b. Interval scale

c.

Group 1	*Group 2*	*Group 3*
M	*M*	*M*

χ^2 test:

a. One group

b. Nominal scale

c.

	1A	*1B*	*1C*
2A			
2B			

References

Cohen, J. (1988). *Statistical power analysis for the behavioral sciences* (2nd ed.). Mahwah, NJ: Lawrence Erlbaum.

De Belis, M. D., Clark, D. B., Beers, S. R, Soloff, P., Boring, A. M., Hall, J., et al. (2000). Hippocampal volume in adolescent-onset alcohol use disorders. *American Journal of Psychiatry, 157,* 737–744.

Olweus, D. (1993). *Bullying at school.* Cambridge, MA: Blackwell.

Rosnow, R. L., & Rosenthal, R. (1989). Statistical procedures and the justification of knowledge in psychological science. *American Psychologist, 44,* 1276–1284.

Stevens, S. S. (1946). On the theory of scales of measurement in statistics. *Science, 103,* 677–680.

Index